STUDENT'S SOLUTIONS MANUAL

BRIAN BEAUDRIE
Northern Arizona University

BARBARA BOSCHMANS
Northern Arizona University

to accompany

A PROBLEM SOLVING APPROACH TO MATHEMATICS FOR ELEMENTARY SCHOOL TEACHERS

THIRTEENTH EDITION

Rick Billstein
University of Montana

Barbara Boschmans
Northern Arizona University

Shlomo Libeskind
University of Oregon

Johnny W. Lott
University of Montana

Pearson

Reproduced by Pearson from electronic files supplied by the author.

Pearson

ISBN-13: 978-0-13-518420-2
ISBN-10: 0-13-518420-7

ScoutAutomatedPrintCode

Contents

CHAPTER 1

AN INTRODUCTION TO PROBLEM SOLVING

Assessment 1-1A: Mathematics and Problem Solving

1. (a) List the numbers:

$$\begin{array}{r} 1 + 2 + \cdots + 98 + 99 \\ 99 + 98 + \cdots + 2 + 1 \\ \hline 100 + 100 + \cdots + 100 + 100 \end{array}$$

There are 99 sums of 100. Thus the total can be found by computing $\frac{99 \cdot 100}{2} = \mathbf{4950}$.

(Another way of looking at this problem is to realize there are $\frac{99}{2} = 49.5$ pairs of sums, each of 100; thus $49.5 \cdot 100 = 4950$.)

(b) The number of terms in any sequence of numbers may be found by subtracting the first term from the last, dividing the result by the common difference between terms, and then adding 1 (because both ends must be accounted for). Thus $\frac{1001-1}{2} + 1 = 501$ terms.

List the numbers:

$$\begin{array}{r} 1 + 3 + \cdots + 999 + 1001 \\ 1001 + 999 + \cdots + 3 + 1 \\ \hline 1002 + 1002 + \cdots + 1002 + 1002 \end{array}$$

There are 501 sums of 1002. Thus the total can be found by computing $\frac{501 \cdot 1002}{2} = \mathbf{251,001}$.

(c) The number of terms in any sequence of numbers may be found by subtracting the first term from the last, dividing the result by the common difference between terms, and then adding 1 (because both ends must be accounted for). Thus $\frac{300-3}{3} + 1 = 100$ terms.

List the numbers:

$$\begin{array}{r} 3 + 6 + \cdots + 297 + 300 \\ 300 + 297 + \cdots + 6 + 3 \\ \hline 303 + 303 + \cdots + 303 + 303 \end{array}$$

There are 100 sums of 303. Thus the total can be found by computing $\frac{100 \cdot 303}{2} = \mathbf{15,150}$.

(d) The number of terms in any sequence of numbers may be found by subtracting the first term from the last, dividing the result by the common difference between terms, and then adding 1 (because both ends must be accounted for). Thus $\frac{400-4}{4} + 1 = 100$ terms.

List the numbers:

$$\begin{array}{r} 4 + 8 + \cdots + 396 + 400 \\ 400 + 396 + \cdots + 8 + 4 \\ \hline 404 + 404 + \cdots + 404 + 404 \end{array}$$

There are 100 sums of 404. Thus the total can be found by computing $\frac{100 \cdot 404}{2} = \mathbf{20,200}$.

2. (a)

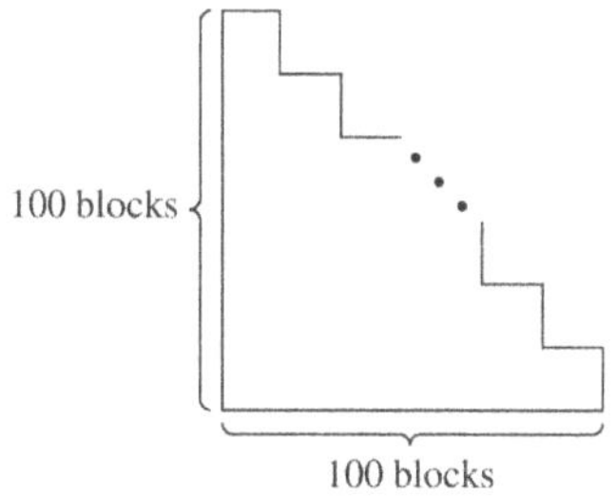

(b)

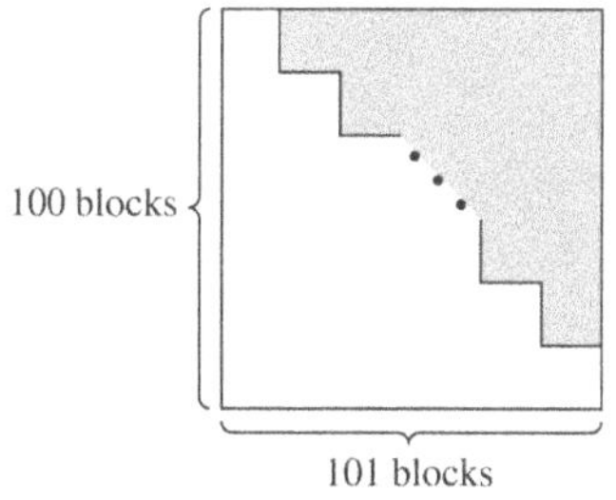

When the stack in (a) and a stack of the same size is placed differently next to the original stack in (a), a rectangle containing 100 (101) blocks is created. Since each block is represented twice, the desired sum is $100(101)/2 = \mathbf{5050}$.

While the above represents a specific example, the same thinking can be used for any natural number n to arrive at a formula $n(n+1)/2$.

3. There are $\frac{147-36}{1} + 1 = 112$ terms.

List the numbers:

$$\begin{array}{ccccccccc} 36 & + & 37 & + & \cdots & + & 146 & + & 147 \\ 147 & + & 146 & + & \cdots & + & 37 & + & 36 \\ \hline 183 & + & 183 & + & \cdots & + & 183 & + & 183 \end{array}$$

There are 112 sums of 183. Thus the total can be found by computing $\frac{112 \cdot 183}{2} = \mathbf{10,248}$.

4. (a) Make a table as follows; there are 9 rows so there are **9 different ways**.

6-cookie packages	2-cookie packages	single-cookie packages
1	2	0
1	1	2
1	0	4
0	5	0
0	4	2
0	3	4
0	2	6
0	1	8
0	0	10

(b) Make a table as follows; there are 12 rows so there are **12 different ways**.

6-cookie packages	2-cookie packages	single-cookie packages
2	0	0
1	3	0
1	2	2
1	1	4
1	0	6
0	6	0
0	5	2
0	4	4
0	3	6
0	2	8
0	1	10
0	0	12

5. If each layer of boxes has 7 more than the previous layer we can add powers of 7:

$7^0 = 1$ (red box)

$7^1 = 7$ (blue boxes)

$7^2 = 49$ (black boxes)

$7^3 = 343$ (yellow boxes)

$7^4 = 2401$ (gold boxes)

$1 + 7 + 49 + 343 + 2401 =$ **2801 boxes** altogether.

6. Using strategies from Poyla's problem solving list identify subgoals (solve simpler problems) and make diagrams to solve the original problem.

1 triangle; name this the "unit" triangle.

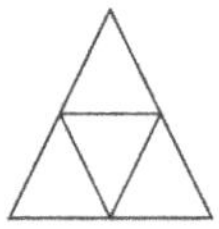

This triangle is made of 4 unit triangles. Counting the large triangle there are 5 triangles

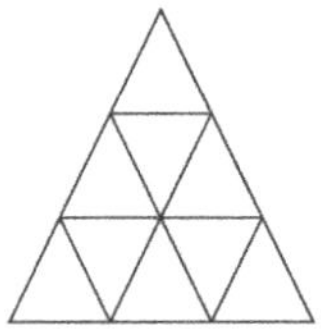

Unit triangles	4 unit triangles	9 unit triangles
9	3	1

13 total triangles

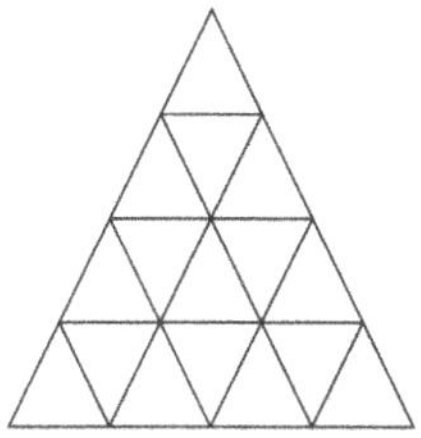

Unit triangles	4 unit triangles	9 unit triangles	16 unit triangles
16	7	3	1

There are 27 triangles in the original figure.

7. Observe that $E = (1+1)+(3+1)+\cdots+(97+1) = O+49$. Thus, E is 49 more than O.

   Alternative strategy:

$$O + E = 1 + 2 + 3 + 4 + 5 + 6 + \cdots + 97 + 98$$
$$= \frac{98(99)}{2} = 49(99)$$
$$E = 2(1 + 2 + 3 + 4 + \cdots + 49) = 2\left(\frac{49(50)}{2}\right) = 49(50)$$
$$O = O + E - E = 49(99) - 49(50) = 49(49).$$

   So O is 49 less than E.

8. Bubba is last; Cory must be between Alababa and Dandy; Dandy is faster than Cory. Listing from fastest to slowest, the finishing order is then **Dandy, Cory, Alababa, and Bubba**.

9. Make a table.

\$20 bills	\$10 bills	\$5 bills
2	1	0
2	0	2
1	3	0
1	2	2
1	1	4
1	0	6
0	5	0
0	4	2
0	3	4
0	2	6
0	1	8
0	0	10

   There are twelve rows so there are **twelve different ways**.

10. The diagonal from the left, top corner to the right, bottom corner sums to $17 + 22 + 27 = 66$.

    The first row sums to $17 + a + 7 = 24 + a$. So $a = 66 - 24 = \mathbf{42}$. The last column sums to $7 + b + 27 = 34 + b$. So $b = 66 - 34 = \mathbf{32}$. The first column sums to $17 + 12 + c = 29 + c$. So $c = 66 - 29 = \mathbf{37}$. The second column sums to $42 + 22 + d = 64 + d$. So $d = 66 - 64 = \mathbf{2}$.

11. Debbie and Amy began reading on the same day, since 72 pages for Debbie ÷ 9 pages per day = 8 days. Thus Amy is on 6 pages per day × 8 days = **page 48.**

12. The last three digits must sum to 20, so the second to last digit must be $20 - (7 + 4) = 9$. Since the sum of the 11th, 12th, and 13th digits is also 20, the 11th digit is $20 - (7 + 9) = 4$.

A		7								4	7	9	4

    We can continue in this fashion until we find that A is 9, or we can observe the repeating pattern from back to front, 4, 9, 7, 4, 9, 7, … and discover that **A is 9**.

13. Choose the box labeled Oranges and Apples (Box B). Retrieve a fruit from Box B. Since Box B is mislabeled, Box B should be labeled as having the fruit you retrieved. For example, if you retrieved an apple, then Box B should be labeled Apples. Since Box A is mislabeled, the Oranges and Apples label should be placed on Box A. These leave only one possibility for Box C; it should be labeled Oranges. If an orange was retrieved from Box B, then Box C would be labeled Oranges and Apples and Box A should be labeled Apples.

14. The electrician made \$1315 for 4 days at \$50 per hour. She spent \$15 per day on gasoline so 4 • \$15 = \$60 on gasoline. The total is then \$1315 + \$60 = \$ 1375. At \$50 per hour, she worked $\frac{1375}{50} =$ **27.5 hours.**

15. Working backward: Top – 6 rungs – 7 rungs + 5 rungs – 3 rungs = top – 11 rungs, which is located at the middle. From the middle rung travel up 11 to the top or down 11 to the bottom. Along with the starting rung, then, there are 11 + 11 + 1 = **23 rungs.**

16. There are several different ways to solve this. One is to use a variable. So, let *a* be equal to the number of apple pies that are baked. This means the number of cherry pies that are baked is

represented by 4 – a. So, using 9 slices for an apple pie and 7 slices for a cherry pie, we get:

$9a+7(4-a)=34$

$9a+28-7a=34$

$2a=6$

$a=3$

So the number of **apple pies is 3**. Therefore, the number of **cherry pies is 1**.

Since the number of pies is small, another solution strategy is to make a table of all possible cases:

Apple Pies	Cherry Pies	Apple slices	Cherry slices	Total slices
0	4	0	28	28
1	3	9	21	30
2	2	18	14	32
3	**1**	**27**	7	**34**
4	0	36	0	36

From the table, we can see there are 34 slices when there are three apple pies and one cherry pie.

17. **Al made $50**. Examine it step by step. First, Al spent $100 on the CD player. At this point, he is down $100. Once he sold it for $125, he is now $25 ahead. When he later bought it back for $150, he was down $125; but after selling it again for $175, he is now ahead $50.

18. Since the bat is $49 more than the ball, and the total spent is $50, we can use the guess and check method to solve. The table below represents possible guesses:

Cost of bat	Cost of ball	sum
$49.00	$0.00	$49.00
$49.25	$0.25	$49.50
$49.50	**$0.50**	**$50.00**

Another method would be to use a variable. Let the cost of the ball (in dollars) be the variable b. The cost of a bat in dollars, therefore, would be b + 49. Together the two costs must add up to 50, so:

$b+(b+49)=50$

$2b+49=50$

$2b=1$

$b=½$. In terms of money, b = $0.50, **50 cents**. Since the ball costs 50 cents, **the bat must cost $49.50**, and the sum of the price of bat and ball does equal $50.

Assessment 1-2A: Explorations with Patterns

1. **(a)** Each figure in the sequence adds one box each to the top and bottom rows. The next would be:

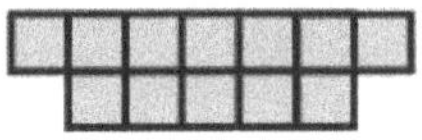

(b) Each figure in the sequence adds one upright and one inverted triangle. The next would be:

(c) Each figure in the sequence adds one box to the base and one row to the overall triangle. The next would be:

2. **(a)** Terms that continue a pattern are **17, 21, 25**, , … . This is an **arithmetic** sequence because each successive term is obtained from the previous term by addition of 4.

(b) Terms that continue a pattern are **220, 270, 320**, … . This is **arithmetic** because each successive term is obtained from the previous term by addition of 50.

(c) Terms that continue a pattern are **27, 81, 243**, … . This is **geometric** because each successive term is obtained from the previous term by multiplying by 3.

(d) Terms that continue a pattern are $\mathbf{10^9, 10^{11}, 10^{13}}$, … . This is **geometric** because each successive term is obtained from the previous term by multiplying by 10^2.

(e) Terms that continue a pattern are $\mathbf{193+10\times 2^{30}, 193+11\times 2^{30}, 193+12\times 2^{30}}$, … . This is **arithmetic** because each successive term is obtained from the previous term by addition of 2^{30}.

3. In these problems, let a_n represent the nth term in a sequence, a_1 represent the first term, d represent the common difference between terms in an arithmetic sequence, and r represent the common ratio between terms in a geometric sequence. In an arithmetic sequence,

$a_n = a_1 + (n-1)d$; in a geometric sequence $a_n = a_1 r^{n-1}$. Thus:

(a) Arithmetic sequence: $a_1 = 1$ and $d = 4$:

(*i*) $a_{100} = 1 + (100-1)\cdot 4$
$= 1 + 99 \cdot 4 = \mathbf{397}$.

(*ii*) $a_n = 1 + (n-1)\cdot 4$
$= 1 + 4n - 4 = \mathbf{4n - 3}$.

(b) Arithmetic sequence:
$a_1 = 70$ and $d = 50$:

(*i*) $a_{100} = 70 + (100-1)\cdot 50$
$= 70 + 99 \cdot 50 = \mathbf{5020}$.

(*ii*) $a_n = 70 + (n-1)\cdot 50$
$= 70 + 50n - 50$ or $\mathbf{50n + 20}$.

(c) Geometric sequence: $a_1 = 1$ and $r = 3$:

(*i*) $a_{100} = 1 \cdot 3^{100-1} = \mathbf{3^{99}}$.

(*ii*) $a_n = 1 \cdot 3^{n-1} = \mathbf{3^{n-1}}$.

(d) Geometric sequence:
$a_1 = 10$ and $r = 10^2$:

(*i*) $a_{100} = 10 \cdot (10^2)^{(100-1)} = 10 \cdot (10^2)^{99}$
$= 10 \cdot 10^{198} = \mathbf{10^{199}}$.

(*ii*) $a_n = 10 \cdot (10^2)^{(n-1)}$
$= 10 \cdot 10^{(2n-2)} = \mathbf{10^{2n-1}}$.

(e) Arithmetic sequence:
$a_1 = 193 + 7 \cdot 2^{30}$ and $d = 2^{30}$:

(*i*)
$a_{100} = 193 + 7 \cdot 2^{30} + (100-1)\cdot 2^{30}$
$= 193 + 7 \cdot 2^{30} + 99 \cdot 2^{30}$
$= \mathbf{193 + 106 \times 2^{30}}$.

(*ii*) $a_n = 193 + 7 \cdot 2^{30} + (n-1)\cdot 2^{30}$
$= \mathbf{193 + (n + 6) \times 2^{30}}$.

4. **2, 7, 12,** … . Each term is the 5th number on a clock face (clockwise) from, the preceding term.

5. **(a)** Make a table.

Number of term	Term
1	$1 \cdot 1 \cdot 1 = 1$
2	$2 \cdot 2 \cdot 2 = 8$
3	$3 \cdot 3 \cdot 3 = 27$
4	$4 \cdot 4 \cdot 4 = 64$
5	$5 \cdot 5 \cdot 5 = 125$
6	$6 \cdot 6 \cdot 6 = 216$
7	$7 \cdot 7 \cdot 7 = 343$
8	$8 \cdot 8 \cdot 8 = 512$
9	$9 \cdot 9 \cdot 9 = 729$
10	$10 \cdot 10 \cdot 10 = 1000$
11	$11 \cdot 11 \cdot 11 = 1331$

The 11th term **1331** is the least 4-digit number greater than 1000.

(b) The 9th term **729** is the greatest 3-digit number in this pattern.

(c) $10^4 = 10{,}000$; The greatest number less than 10^4 is $21 \cdot 21 \cdot 21 = \mathbf{9261}$.

(d) The cell A14 corresponds to the 14th term, which is $14 \cdot 14 \cdot 14 = \mathbf{2744}$.

6. **(a)** The number of matchstick squares in each windmill form an arithmetic sequence with $a_1 = 5$ and $d = 4$. The number of matchstick squares required to build the 10th windmill is thus $5 + (10-1)\cdot 4 = 5 + 9 \cdot 4 = $ **41 squares**.

(b) The *n*th windmill would require $5 + (n-1)\cdot 4 = 5 + 4n - 4 = \mathbf{4n + 1}$ **squares**.

(c) There are 16 matchsticks in the original windmill. Each additional windmill adds 12 matchsticks.

This is an arithmetic sequence with $a_1 = 16$ and $d = 12$, so $a_n = 16 + (n-1)\cdot 12 = \mathbf{12n + 4}$ **matchsticks**.

7. **(a)** Each cube adds four squares to the preceding figure; or 6, 10, 14, … . This is an arithmetic sequence with $a_1 = 6$ and $d = 4$. Thus $a_{15} = 6 + (15-1)\cdot 4 = $ **62 squares** to be painted in the 10th figure.

(b) This is an arithmetic sequence with $a_1 = 6$ and $d = 6$. The *n*th term is thus: $a_n = 6 + (n-1)\cdot 4 = \mathbf{4n + 2}$.

8. Since the first year begins with 700 students, after the first year there would be 760, after the second there would be 820, … , and after the twelfth year the number of students would be the 13th term in the sequence.

This then is an arithmetic sequence with $a_1 = 700$ and $d = 60$, so the 13th term (current enrollment + twelve more years) is:

$700 + (13 - 1) \cdot 60 =$ **1420 students.**

9. Using the general expression for the nth term of an arithmetic sequence with $a_1 = 24{,}000$ and $a_9 = 31{,}680$ yields:

$$31680 = 24000 + (9 - 1)d$$
$$31680 = 24000 + 8d \Rightarrow d = 960,$$

the amount by which Juan's income increased each year.

To find the year in which his income was \$45120:

$$45120 = 24000 + (n - 1) \cdot 960$$
$$45120 = 23040 + 960n$$
$$\Rightarrow n = 23.$$

Juan's income was \$45,120 in his **23rd year**.

10. The number that fits into the last triangle is **8**. The numbers inside the triangle are found by multiplying the number at the top of the triangle by the number at the bottom left of the triangle; then subtracting from that the number at the bottom right of the triangle. So, $2 \times 5 - 2 =$ **8**.

11. (a) To build an up-down-up staircase with 3 steps up and 3 steps down, each current step will have a block placed on it; plus a block will be added to each end. In total, there will be five more blocks, for a total of 9 blocks used. To build an up-down-up staircase with 4 steps up and 4 steps down, each step from the previous iteration will have a block placed on it (5 blocks) plus a block at each end (2 blocks), adding seven total blocks, bringing the total to 9 + 7 = 16 blocks. The table below illustrates the pattern:

Steps up/down	Blocks added	Total blocks
1	0	1
2	3	4
3	5	9
4	7	16

Therefore, to build an up-down-up staircase with 5 steps up and 5 steps down, 9 blocks need to be added to the previous 16 blocks to arrive at a total of **25 blocks**.

(b) Based on the table and answer above, the total blocks are always the square of the number of steps up and down. So, if the number of steps up and down are n, then the total number of blocks will be $\boldsymbol{n^2}$.

12. (a) Using the general expression for the nth term of an arithmetic sequence with $a_1 = 51$, $a_n = 251$, and $d = 1$ yields:

$$251 = 51 + (n - 1) \cdot 1 \Rightarrow$$
$$251 = 50 + n \Rightarrow n = 201.$$

There are **201 terms** in the sequence.

(b) Using the general expression for the $\boldsymbol{n}$th term of a geometric sequence with $a_1 = 1$, $a_n = 2^{60}$, and $r = 2$ yields:

$$2^{60} = 1(2)^{n-1} = 2^{n-1}$$
$$\Rightarrow n - 1 = 60 \Rightarrow n = 61.$$

There are **61 terms** in the sequence.

(c) Using the general expression for the $\boldsymbol{n}$th term of an arithmetic sequence with $a_1 = 10$, $a_n = 2000$, and $d = 10$ yields:

$$2000 = 10 + (n - 1) \cdot 10 \Rightarrow$$
$$2000 = 10n \Rightarrow n = 200.$$

There are **200 terms** in the sequence.

(d) Using the general expression for the nth term of a geometric sequence with $a_1 = 1$, $a_n = 1024$, and $r = 2$ yields:

$$1024 = 1(2)^{n-1} \Rightarrow$$
$$2^{10} = 2^{n-1} \Rightarrow n - 1 = 10$$
$$\Rightarrow n = 11.$$

There are **11 terms** in the sequence.

13. (a)

First term:	$(1)^2 + 2 =$ **3**;
Second term:	$(2)^2 + 2 =$ **6**;
Third term:	$(3)^2 + 2 =$ **11**;
Fourth term:	$(4)^2 + 2 =$ **18**; and
Fifth term:	$(5)^2 + 2 =$ **27**.

(b) First term: $5(1) + 1 = \mathbf{6}$;
Second term: $5(2) + 1 = \mathbf{11}$;
Third term: $5(3) + 1 = \mathbf{16}$;
Fourth term: $5(4) + 1 = \mathbf{21}$; and
Fifth term: $5(5) + 1 = \mathbf{26}$.

(c) First term: $10^{(1)} - 1 = \mathbf{9}$;
Second term: $10^{(2)} - 1 = \mathbf{99}$;
Third term: $10^{(3)} - 1 = \mathbf{999}$;
Fourth term: $10^{(4)} - 1 = \mathbf{9999}$; and
Fifth term: $10^{(5)} - 1 = \mathbf{99999}$.

(d) First term: $3(1) - 2 = \mathbf{1}$;
Second term: $3(2) - 2 = \mathbf{4}$;
Third term: $3(3) - 2 = \mathbf{7}$;
Fourth term: $3(4) - 2 = \mathbf{10}$; and
Fifth term: $3(5) - 2 = \mathbf{13}$.

14. Answers may vary; examples are:

(a) If $n = 5$, then $\frac{5+5}{5} = 2 \neq 5 + 1 = 6$.

(b) If $n = 2$, then $(2 + 4)^2 = 6^2 = 36$ does not equal $2^2 + 4^2 = 20$.

15. (a) There are 1, 5, 11, 19, 29 tiles in the five figures. Each figure adds 2n tiles to the preceding figure, thus a_6 the 6th term has $29 + 12 =$ **41 tiles.**

(b) $n^2 = 1, 4, 9, 16, 25, \ldots$. Adding $(n - 1)$ to n^2 yields $1, 5, 11, 19, 29, \ldots$, which is the proper sequence. Thus the *nth* term has $\mathbf{n^2 + n - 1}$.

(c) If $n^2 + (n - 1) = 1259$;

Then $n^2 + n - 1260 = 0$. This implies $(n - 35)(n + 36) = 0$, so $n = 35$.

There are 1259 tiles in the **35th figure**.

16. The nth term of the arithmetic sequence is $200 + n(200)$. The sequence can also be generated by adding 200 to the previous term. The nth term of the geometric sequence is 2^n. The sequence can also be generated by multiplying the previous term by 2. Make a table.

Number of the term	Arithmetic term	Geometric term
7	1600	128
8	1800	256
9	2000	512
10	2200	1024
11	2400	2048
12	2600	4096

With the **12th term**, the geometric sequence is greater.

17. (a) Start with one piece of paper. Cutting it into five pieces gives us 5. Taking each of the pieces and cutting it into five pieces again gives $5 \cdot 5 = 25$ pieces. Continuing this process gives a geometric sequence: 1, 5, 25, 125, … . After the 5th cut there are $\mathbf{5^5 = 3125}$ pieces of paper.

(b) The number of pieces after the nth cut would be $\mathbf{5^n}$.

18. (a) For an arithmetic sequence there is a common difference between the terms. Between 39 and 69 there are three differences so we can find the common difference by subtracting 39 from 69 and dividing the answer by three: $69 - 39 = 30$ and $30 \div 3 = 10$. The common difference is 10 and we can find the missing terms: 39 – 10 = **29** and 39 + 10 = **49** and 49 + 10 = **59**.

(b) For an arithmetic sequence there is a common difference between the terms. Between 200 and 800 there are three differences so we can find the common difference by subtracting 200 from 800 and dividing the answer by three: $800 - 200 = 600$ and $600 \div 3 = 200$. The common difference is 200 and we can find the missing terms: 200 – 200 = **0** and 200 + 200 = **400** and 400 + 200 = **600**.

(c) For a geometric sequence there is a common ration between the terms. Between 5^4 and 5^{10} there are three common ratios used so we can find the common ratio by dividing 5^{10} by 5^4 and then taking the cube root:

$5^{10} \div 5^4 = 5^6$ and $(5^6)^{\left(\frac{1}{3}\right)} = 5^2$. The common ratio is 52 and we can find the missing terms:

$5^4 \div 5^2 = \mathbf{5^2}, 5^4 \cdot 5^2 = \mathbf{5^6}, 5^6 \cdot 5^2 = \mathbf{5^8}$.

19. (a) Let's call the missing terms a, b, c, d, e and f, then the sequence becomes:

$a, b, 1, 1, c, d, e, f$

$b + 1 = 1 \rightarrow b = 0$

$a + b = 1 \rightarrow a + 0 = 1 \rightarrow a = 1$

$1 + 1 = c \rightarrow c = 2$

$1 + c = d \rightarrow 1 + 2 = d \rightarrow d = 3$

$c + d = e \rightarrow 2 + 3 = e \rightarrow e = 5$

$d + e = f \rightarrow 3 + 5 = f \rightarrow = 8.$

The missing terms are **1, 0, 2, 3, 5**, and **8**.

(b) Let's call the missing terms a, b, c, and d, then the sequence becomes:

$a, b, c, 10, 13, d, 36, 59$

$c + 10 = 13 \rightarrow c = 3$

$b + c = 10 \rightarrow b + 3 = 10 \rightarrow b = 7$

$a + b = c \rightarrow a + 7 = 3 \rightarrow = {}^{-}4$

$10 + 13 = d \rightarrow d = 23$

The missing terms are **-4, 7, 3**, and **23**.

(c) If a Fibonacci-type sequence is a sequence in which the first two terms are arbitrary and in which every term starting from the third is the sum of the previous two terms, then we can add 0 and 2 to get the third term and continue the pattern:

$0 + 2 = 2$

$2 + 2 = 4$

$2 + 4 = 6$

$4 + 6 = 10$

$6 + 10 = 16$

$10 + 16 = 26$

The missing terms are **2, 4, 6, 10, 16**, and **26**.

20. (a)

Year 1 $80 + .05(80) = 84$

Year 2 $84 + .05(84) = 88.2$

Year 3 $88.2 + .05(88.2) = 92.61$

Year 4 $92.61 + .05(92.61) = 97.2405$

Year 5 $97.2405 + .05(97.2405) = 102.102525$

$\approx$ **$102.10**.

(b) This is a geometric sequence with $a_1 = 80$ and $\boldsymbol{r} = 1.05$, so the price after $\boldsymbol{n}$ years is $\mathbf{80 \cdot 1.05^n}$.

Mathematical Connections 1-2: Review Problems

15. Order the teams from 1 to 10, and consider a simpler problem of counting how many games are played if each team plays each other once. The first team plays nine teams. The second team also plays nine teams, but one of these games has already been counted. The third team also plays 9 teams, but two of these games were counted in the previous two summands. Continuing in this manner, the total is 10 + 9 + 8 + ... + 3 + 2 + 1 = 9(10) / 2 = 45 games. Double this amount to obtain **90 games must be played** for each team to play each other twice.

17. If the problem is interpreted to stated that at least one 12-person tent is used, **then there are 10 ways**. This can be seen by the table below, which illustrates the ways 2-,3-,5-, and 6-person tents can be combined accommodate 14 people.

6-Person	5-Person	3-Person	2-Person
2	0	0	1
1	1	1	0
1	0	2	1
1	0	0	4
0	2	0	2
0	1	3	0
0	1	1	3
0	0	4	1
0	0	2	4
0	0	0	7

Chapter 1 Review

1. Make a plan. Every 7 days (every week) the day will change from Sunday to Sunday. 365 days per year $\div$ 7 days per week $\approx$ 52 weeks per year $+\frac{1}{7}$ weeks per year. Thus the day of the week will change from Sunday to Sunday 52 times and then change from Sunday to Monday. **July 4 will be a Monday**.

2. $\$5.90 \div 2 = \$2.95$ more on one of the items. That is $\$20 + \$2.95 =$ **$22.95 for the more expensive item** and $\$20 - \$2.95 = \$17.05$ for the less expensive item. Check that both items add up to $40: $\$22.95 + \$17.05 = \$40$.

3. **(a)** **15, 21, 28. Neither**. The successive differences of terms increases by one; e.g., 10 + 5, 15 + 6, … .
   (b) **32, 27, 22. Arithmetic** Subtract 5 from each term to obtain the subsequent term.
   (c) **400, 200, 100. Geometric** Each term is half the previous term.
   (d) **21, 34, 55. Neither** Each term is the sum of the previous two terms—this is the Fibonacci sequence.
   (e) **17, 20, 23. Arithmetic** Add 3 to each term to obtain the subsequent term.
   (f) **256, 1024, 4096. Geometric** Multiply each term by 4 to obtain the subsequent term.
   (g) **16, 20, 24. Arithmetic** Add 4 to each term to obtain the subsequent term.
   (h) **125, 216, 343. Neither** Each term is the 3rd power of the counting numbers = 13, 23, 33, … .

4. **(a)** The successive differences are 3. Each term is 3 more than the previous term. This suggests that it is an arithmetic sequence of the form $3n + ?$. Since the first term is 5, $3(1) + ? = 5$. The nth term would be $\mathbf{3n + 2}$.
   (b) Each term given is 3 times the previous term. This suggests that the sequence is geometric. $\boldsymbol{n^{th}}$ term will be $\mathbf{3^n}$.
   (c) The only number that changes in successive terms is the exponent. For the first term, the exponent is 2; for the second term, the exponent is 3; for the third term, the exponent is 4; and so on. So, for the nth term, the exponent will be $n+1$. Therefore, the nth term will be $\mathbf{2^{n+1} - 1}$.

5. **(a)** $3(1) - 2 = \mathbf{1}$;
   $3(2) - 2 = \mathbf{4}$;
   $3(3) - 2 = \mathbf{7}$;
   $3(4) - 2 = \mathbf{10}$; and
   $3(5) - 2 = \mathbf{13}$.
   (b) $1^2 + 1 = \mathbf{2}$;
   $2^2 + 2 = \mathbf{6}$;
   $3^2 + 3 = \mathbf{12}$;
   $4^2 + 4 = \mathbf{20}$; and
   $5^2 + 5 = \mathbf{30}$.
   (c) $4(1) - 1 = \mathbf{3}$;
   $4(2) - 1 = \mathbf{7}$;
   $4(3) - 1 = \mathbf{11}$;
   $4(4) - 1 = \mathbf{15}$; and
   $4(5) - 1 = \mathbf{19}$.

6. **(a)** $a_1 = 2, d = 2, a_n = 200$.
   So $200 = 2 + (n-1) \cdot 2 \Rightarrow n = 100$.
   Sum is $\frac{100(2+200)}{2} = \mathbf{10{,}100}$.
   (b) $a_1 = 51, \ d = 1, a_n = 151$.
   So $151 = 51 + (n-1) \cdot 1 \Rightarrow n = 101$.
   Sum is $\frac{101 \cdot (51+151)}{2} = \mathbf{10{,}201}$.

7. **(a)** Answers will vary; for example 5 and 3 are odd numbers, but 5 + 3 = 8, which is not odd.
   (b) 15 is odd; and it does not end in a 1 or a 3.
   (c) The sum of any two even numbers is always even. An even number is one divisible by 2, so any even number can be represented by $2 + 2 + 2 + \cdots$.
   Regardless of how many twos are added, the result is always a multiple of 2, or an even number.

8. All rows, columns, and diagonals must add to 34; i.e., the sum of the digits in row 1. Complete rows or columns with one number missing, then two, etc. to work through the square:

16	3	2	13
5	10	**11**	**8**
9	**6**	7	12
4	**15**	14	**1**

9. The ten middle tables will hold two each and the two end tables will hold three each, totaling **26 people.**

10. (a) $\square + 2^{60} = 2^{61}$

$\square = 2^{61} - 2^{60}$

$\square = 2(2^{60}) - 2^{60}$

$\square = 2^{60}(2-1)$

$\square = \mathbf{2^{60}}$

(b) $\square^2 = 625$

$\sqrt{\square^2} = \sqrt{625}$

$\square = \mathbf{25}$

11. $100 \div 5 = 20$ plus $1 = $ **21 posts.** 1 must be added because both end posts must be counted.

12. 1 mile $=$ 5280 feet.

$5280 \div 6$ feet $=$ 880 turns per mile.

880×50000 miles $=$ **44,000,000 turns.**

13. There are 9 students between 7 and 17 (8 through 16). There must be 9 between them in both directions, since they are direct opposites. $9 + 9 + 2 =$ **20 students.**

14. Let l be a large box, m be a medium box, and s be a small box:

$3l + (3l \times 2m \text{ each}) + [(3 \times 2)m \times 5s \text{ each}]$

$3l + 6m + 30s =$ **39 total boxes.**

15. Extend the pattern of doubling the number of ants each day. This is a geometric sequence with $a_1 = 1500$, $a_n = 100{,}000$, and $r = 2$.

$100{,}000 = 1500 \cdot 2^{n-1} \Rightarrow$

$66\frac{2}{3} = 2^{n-1}.$

Since $2^{7-1} < 66\frac{2}{3}$ and $2^{8-1} > 66\frac{2}{3}$, the ant farm will fill sometime **between the 7th and 8th day.**

16. The best strategy would be one of guessing and checking:

(i) Ten 3's + two 5's = 40... close but too low.

(ii) Nine 3's + three 5's = 42... still too low.

(iii) Eight 3's + Four 5's = 44.

They must have answered **four 5-point questions.**

17. Let $\ell =$ length of the longest piece,

$m =$ length of the middle-sized piece, and

$s =$ lenth of the shortest piece.

Then $\ell = 3m$ and $s = m - 10$.

So $\ell + m + s = 90 \Rightarrow$

$3m + m + (m - 10) = 90 \Rightarrow$

$5m = 100.$

Thus $m = \mathbf{20\,cm}$;

$\ell = 3m = \mathbf{60\,cm}$; and

$s = m - 10 = \mathbf{10\ cm}.$

18. Make a diagram that demonstrates all the ways four-digit numbers can be formed from left (thousands place) to right (ones place). **12 four-digit numbers** can be formed.

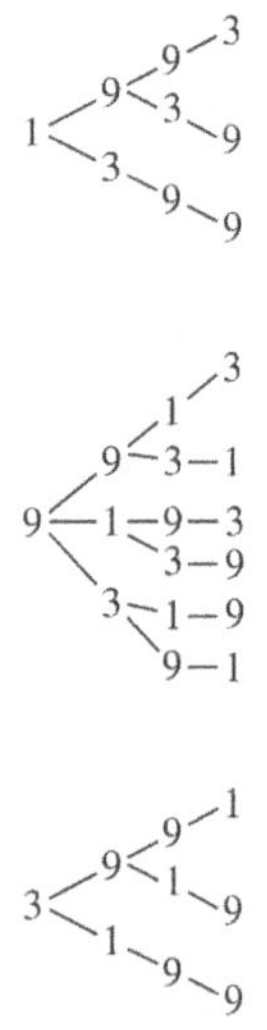

19. Answer may vary. Fill the 4-cup container with water and pour the water into the 7-cup container. Fill the 4-cup container again and pour water into the 7-cup container until it is full. Four minus three (1) cups of water will remain in the 4-cup container. Empty the 7-cup container and pour the contents of the 4-cup container into the 7- cup container. The 7-cup container now holds 1 cup of water. Refill the 4-cup container and pour it into the 7-cup container. The 7-cup container now contains exactly 5 cups of water.

20. A possible pattern is to increase each rectangle by one row of dots and one column of dots to obtain the next term in the sequence. Make a table.

Number of the term	Row of dots	Column of dots	Term (row × column)
1	1	2	2
2	2	3	6
3	3	4	12
4	4	5	20
5	5	6	30
6	6	7	42
7	7	8	56
⋮			
100	100	101	10100
⋮			
n	n	$n+1$	$n(n+1)$

We also observe that the number of the term corresponds to the number of rows in the arrays and that the number of columns in the array is the number of the term plus one. Thus, the next three terms are **30, 42, and 56. The 100th term is 10,100 and the nth term is $n(n+1)$.**

21. A possible pattern is that each successive figure is constructed by adjoining another pentagon to the previous figure

(a)

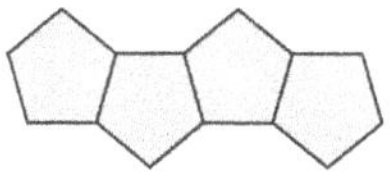

(b) Observe that the perimeter of the first figure is 5 and that when a new pentagon is adjoined 4 new sides are added and one side (where the new pentagon is adjoined) is lost. Make a table.

Number of terms	1-unit sides (perimeter)
1	**5**
2	$5-1+4=$ **8**
3	$8-1+4=$ **11**
4	$11-1+4=$ **14**

(c) and **(d)** Looking at the terms in the sequence and noting that the difference of terms is three, we suspect that the sequence is arithmetic and conjecture that the nth term is $3n+2$. However, we need to be sure. Looking at the 4th term (part a) we observe that the pentagons on the end Contribute 4 sides to the perimeter and the "middle" pentagons contribute 3 sides. Thus, in the nth figure there will be 2 "end" pentagons that contribute 4 1-unit sides and $n-2$ "middle" pentagons that contribute $3(n-2)$ 1 unit sides. The total will be $3(n-2)+2(4)=$ **3n + 2 units**. Thus the 100th term is $3(100)+2=$ **302**.

22. **(a)** The circled terms will constitute an arithmetic sequence because the common difference will be twice the difference in the original series.

(b) The new sequence will be a geometric sequence because the ratio will be the square of the ratio of the original series.

23. When $n=1$, then $n^2-n=1^2-1=0$, so the first term, $a_1=0$. When $n=2$, then $n^2-n=2^2-2$ or 2. Thus $a_1+a_2=2$; hence $a_2=2$. For $n=3$, $n^2-n=3^2-3=6$. Substituting for a_1 and a_2, we get $a_3=6-2=4$. For $n=4$, $4^2-4=12$. Substituting for a_1, a_2, and a_3, we get $0+2+4+a_4=12$. Hence $\boldsymbol{a_4=6}$.

24. **(a)** Let's call the missing terms a, and b then the sequence becomes:

$13, a, b, 27$

$13+a=b \rightarrow a=b-13$

$a+b=27 \rightarrow b-13+b=27$

$\rightarrow 2b=40$

$\rightarrow b=20$

$a=b-13 \rightarrow a=7$

So **7 and 20** are the missing terms.

(b) Let's call the missing terms a, and b then the sequence becomes:

$137, a, b, 163$

$137+a=b \rightarrow a=b-137$

$a+b=163 \rightarrow b-137+b=163$

$\rightarrow 2b=300$

$\rightarrow b=150$

$a=b-137 \rightarrow a=13$

So **13 and 150** are the missing terms.

(c) Let's call the missing terms *x*, and *y*, then the sequence becomes:

$$b, x, y, a$$

$$b + x = y \rightarrow x = y - b$$

$$x + y = a \rightarrow y - b + y = a$$

$$\rightarrow 2y = a + b$$

$$\rightarrow y = \frac{a+b}{2}$$

$$x = \frac{a+b}{2} - b \rightarrow x = \frac{a+b}{2} - \frac{2b}{2}$$

$$\rightarrow x = \frac{a-b}{2}$$

So the missing terms are $\boldsymbol{\frac{a-b}{2}}$ **and** $\boldsymbol{\frac{a+b}{2}}$.

25. The first line tells us that since three cylinders =15, each cylinder must be worth 5. Using that information, we can use the second line to figure out that a circle must be worth 4. That information can be used on the third line to determine that the cup-like shape must be equal to 1. So, putting it all together in the fourth line, we have 4 + 5 + 1 = **10.**

26. The pattern involves taking two squares that are diagonal, summing their numbers, and putting that sum in the square that is below one of the diagonal squares and to the left of the other diagonal square. For example, the top square is 4; the square diagonal to it (to the right) is 2; their sum is 6, which is the number that appears in the square below the 4 and to the left of the 2. You can also see it where 9 + 10 = 19, and 1 + 6 =7. So using this pattern, the question mark must be **equal to 3.**

27. Since 1, 2, and 3 are already on the triangle, the numbers 4-9 will be used to fill in the six question marks. On the left side, 1+2 = 3, so the two question marks on that side must sum to 14. Similarly on the bottom, the two question marks must sum to 13, and on the right side, the two question marks must sum to 12. So on the right side with the available numbers, only two possibilities, 5+7, or 4+8, sum to 12.

If the two numbers on the right side are 5 and 7, the bottom two numbers (which must sum to 13) are 4 and 9, and the two numbers on the left side must be 6 and 8. If the two numbers on the right side are 4 and 8, the two bottom numbers must be 6 and 7, leaving 5 and 9 to be the numbers on the right. Below are the two solutions presented on the triangle:

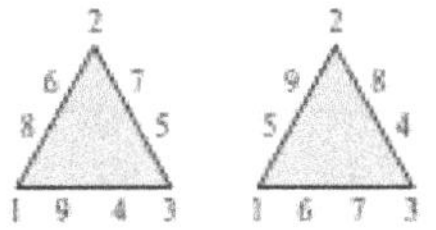

CHAPTER 2

INTRODUCTION TO LOGIC AND SETS

Assessment 2-1A: Reasoning and Logic: An Introduction

1. (a) **False statement.** A statement is a sentence that is either true or false, but not both.
   (b) **False statement.** Los Angeles is a city, not a state.
   (c) **Not a statement.** Questions are not statements.
   (d) **True statement.**
2. (a) **There exists at least one** natural number n such that $n + 8 = 11$.
   (b) **There exists at least one** natural number n such that $n^2 = 4$.
   (c) For **all** natural numbers n, $n + 3 = 3 + n$.
   (d) For **all** natural numbers n, $5n + 4n = 9n$.
3. (a) For **all** natural numbers n, $n + 8 = 11$.
   (b) For **all** natural numbers n, $n^2 = 4$.
   (c) There is **no** natural number x such that $x + 3 = 3 + x$.
   (d) There is **no** natural number x such that $5x + 4x = 9x$.
4. (a) The book **does not have** 500 pages.
   (b) $3 \cdot 5 \neq 15$.
   (c) **Some** dogs **do not have** four legs.
   (d) **No** rectangles are squares.
   (e) **All** rectangles are squares.
   (f) **Some** dogs have fleas.
5. (a) If $n = 4$, or $n = 5$, then $n < 6$ *and* $n > 3$, so the statement is **true**, since it can be shown to work for some natural numbers n.
   (b) All natural numbers are greater than zero; so, since the condition is that $n > 0$ *or* $n < 5$, the statement is **true**.
6. (a)

p	$\sim p$	$\sim(\sim p)$
T	**F**	**T**
F	**T**	**F**

   (b)

p	$\sim p$	$p \vee \sim p$	$p \wedge \sim p$
T	**F**	**T**	**F**
F	**T**	**T**	**F**

   (c) **Yes**. The truth table entries are the same.
   (d) **No**. The truth table entries are not the same.
7. (a)

p	q	$p \rightarrow q$	$\sim p$	$\sim p \vee q$	
T	**T**	**T**	**F**	**T**	
T	**F**	**F**	**F**	**F**	
F	**T**	**T**	**T**	**T**	
F	**F**	**T**	**T**	**T**	

   (b) $p \rightarrow q \equiv \sim p \vee q$. Answers will vary. Here are two possible examples: 1. Let p be "the Bobcats win" and q be "the Bobcats make the playoffs". Then Column 3 would read "If the Bobcats win, then the Bobcats make the playoffs". Column 5 would read "The Bobcats lose or the Bobcats make the playoffs."2. Let p be "it is summer vacation" and q be "I am at home". Then Column 3 would read "If it is summer vacation, then I am at home". Column 5 would read "It is not summer vacation or I am at home."
   (c) In this problem, p is the statement "2 + 3 = 5" and q is the statement "4 + 6 = 10". That would make the statement in this problem be in the form of $\sim p \vee q$. So, it will be logically equivalent to a statement in the form of $p \rightarrow q$ or "if 2 + 3 = 5, then 4 + 6 = 10."
8. (a) **$q \wedge r$**. Both q and r are true.
   (b) **$r \vee \sim q$**. r is true or q is not true.
   (c) **$\sim(q \wedge r)$**. q and r are not both true.
   (d) **$\sim q$**. q is not true.
9. (a) **False**. The statement is a conjunction. In order for a conjunction to be true, both p and q must be true; otherwise, the conjunction is false. The two parts of the conjunction could be stated as such: p is the statement $2 + 3 = 5$ and q is the

statement $4+7=10$. In this situation, p is true, but q is false.

(b) **False**. The United States Supreme Court has nine justices when every seat is filled.

(c) **True**. The only triangles that have three sides of the same length are equilateral triangles. In every case, an equilateral triangle will have two sides the same length as well.

(d) **False**. Isosceles triangles have two sides equal in length, but the third side does not have to be equal to the other two.

10. (a) By DeMorgan's Laws, the negation of $p \wedge q$ is $\sim p \vee \sim q$. Therefore, the answer is $\mathbf{2+3 \neq 5}$ **or** $\mathbf{4+7 \neq 10}$.

(b) **With every seat filled, the Supreme Court of the United States does not have 12 justices.**

(c) By problem #7 and DeMorgan's Laws, the negation of $p \rightarrow q$ is $p \wedge \sim q$. So, the statement would read: **the triangle has three sides of the same length and the triangle does not have two sides of the same length.**

(d) **The triangle has two sides of the same length and the triangle does not have three sides of the same length.**

In both (c) and (d) above, the negation of a conditional statement $p \rightarrow q$ is $p \wedge \sim q$.

11. (a)

p	q	$\sim p$	$\sim q$	$\sim p \vee \sim q$	$\sim(p \vee q)$
T	T	F	F	F	F
T	F	F	T	T	F
F	T	T	F	T	F
F	F	T	T	T	T

Since the truth values for $\sim p \vee \sim q$ are not the same as for $\sim(p \vee q)$, the statements are **not logically equivalent**.

(b)

p	q	$\sim p$	$\sim q$	$p \wedge q$	$\sim(p \wedge q)$	$\sim p \wedge \sim q$
T	T	F	F	T	F	F
T	F	F	T	F	T	F
F	T	T	F	F	T	F
F	F	T	T	F	T	T

Since the truth values for $\sim(p \wedge q)$ are not the same as for $\sim p \wedge \sim q$, the statements are **not logically equivalent.**

12. Alex is a male teacher who does not teach math and is 30 years old or younger.

13. If $p=$ "it is raining" and $q=$ "the grass is wet":

(a) $p \rightarrow q$. **(b)** $\sim p \rightarrow q$.

(c) $p \rightarrow \sim q$.

(d) $p \rightarrow q$. The hypothesis is "it is raining;" the conclusion is "the grass is wet."

(e) $\sim q \rightarrow \sim p$. **(f)** $q \leftrightarrow p$.

14. (a) **Converse**: If a triangle has no two sides of the same length, then the triangle is scalene. **Inverse**: If a triangle is not scalene, then the triangle has (at least) two sides of the same length. **Contrapositive**: If a triangle does not have two sides of the same length, then the triangle is not scalene.

(b) **Converse:** If an angle is a right angle, then it is not acute. **Inverse**: If an angle is acute, then it is not a right angle. **Contrapositive:** If an angle is not a right angle, then it is acute. Note that the original statement and the contrapositive are not true, while the converse and inverse are true.

(c) **Converse**: If Maria is not a citizen of Cuba, then she is a U.S. citizen. **Inverse**: If Maria is not a U.S. citizen, then she is a citizen of Cuba. **Contrapositive**: If Maria is a citizen of Cuba, then she is not a U.S. citizen. Note that the original statement and the contrapositive are true, while the converse and inverse are not true

(d) **Converse**: If a number is not a natural number, then it is a whole number. **Inverse**: If a number is not a whole number, then it is a natural number. **Contrapositive**: If a number is a natural number, then it is not a whole number.

15. The statements **are negations** of each other.

p	q	$\sim q$	$p \wedge \sim q$	$\sim(p \wedge \sim q)$	$p \rightarrow q$	$\sim(p \rightarrow q)$
T	T	F	F	T	T	F
T	F	T	T	F	F	T
F	T	F	F	T	T	F
F	F	T	F	T	T	F

16. The **contrapositive** is logically equivalent: "**If a number is not a multiple of 4 then it is not a multiple of 8.**"

17. **(a)** **Valid.** This is valid by the transitivity property. "All squares are quadrilaterals" is $p \rightarrow q$; "all quadrilaterals are polygons" is $q \rightarrow r$; and "all squares are polygons" is $p \rightarrow r$.

(b) **Invalid.** We do not know what will happen to students who are not freshman. There is no statement "sophomores, juniors, and seniors do not take mathematics."

18. **(a)** Since all students in Integrated Mathematics I make A's, and some of those students are in Beta Club, then **some Beta Club students make A's**.

(b) Let $p =$ I study for the final, $q =$ I pass the final, $r =$ I pass the course, $s =$ I look for a teaching job. Then $p \rightarrow q$, if I study for the final, then I will pass the final. $q \rightarrow r$, if I pass the final, then I will pass the course. $r \rightarrow s$, if I pass the course, I will look for a teaching job. So $p \rightarrow s$, **if I study for the final, then I will look for a teaching job**.

(c) The first statement could be rephrased as "If a triangle is equilateral, then it is isosceles." Let $p =$ equilateral triangle and $q =$ isosceles triangle. So the first statement is $p \rightarrow q$,. The second statement is simply p; then the conclusion should be q or **there exist triangles that are isosceles**.

19. **(a)** If a figure is a square, then it is a rectangle.

(b) If a number is an integer, then it is a rational number.

(c) If a polygon has exactly three sides, then it is a triangle.

20. **(a)** If $\sim p \vee \sim q \equiv \sim(p \wedge q)$, then

$$\mathbf{3 \cdot 2 \neq 6} \textbf{\textit{ or }} \mathbf{1 + 1 = 3}.$$

(b) If $\sim p \wedge \sim q \equiv \sim(p \vee q)$, then **you cannot pay me now and you cannot pay me later**.

Assessment 2-2A: Describing Sets

1. **(a)** Either a list or set-builder notation may be used: $\{\mathbf{a,s,e,m,n,t}\}$ or $\{x \mid x$ **is a letter in the word *assessment***$\}$.

(b) $\{\mathbf{21,22,23,24,\ldots}\}$ or $\{x \mid x$ **is a natural number and** $x > \mathbf{20}\}$ or $\{x \mid x \in N$ **and** $x > \mathbf{20}\}$.

2. **(a)** $\boldsymbol{P = \{p,q,r,s\}}$.

(b) $\{\mathbf{1,2}\} \subset \{\mathbf{1,2,3}\}$. The symbol $\subset$ refers to a proper subset.

(c) $\{\mathbf{0,1}\} \not\subseteq \{\mathbf{1,2,3}\}$. The symbol $\subseteq$ refers to a subset.

3. **(a)** **Yes**. $\{1,2,3,4,5\} \sim \{m,n,o,p,q\}$ because both sets have the same number of elements and thus exhibit a one-to-one correspondence.

(b) **Yes**. $\{a,b,c,d,e,f,\ldots,m\} \sim \{1,2,3,\ldots,13\}$ because both sets have the same number of elements.

(c) **No**. $\{x \mid x$ is a letter in the word *mathematics*$\} \nsim \{1,2,3,4,\ldots,11\}$; there are only eight unduplicated letters in the word *mathematics*.

4. **(a)** The first element of the first set can be paired with any of the six in the second set, leaving five possible pairings for the second element, four for the third, three for the fourth, two for the fifth, and one for the sixth. Thus there are $6 \cdot 5 \cdot 4 \cdot 3 \cdot 2 \cdot 1 = \mathbf{720}$ one-to-one correspondences.

(b) There are $n \cdot (n-1) \cdot (n-2) \cdot \ldots \cdot 3 \cdot 2 \cdot 1 = \mathbf{n!}$ possible one-to-one correspondences. The first element of the first set can be paired with any of the n elements of the second set; for each of those n ways to make the first pairing, there are $n-1$ ways the second element of the first set can be paired with any element of the second set; which means there are $n-2$ ways the third element of the first set can be paired with any element of the third set; and so on. The Fundamental Counting Principle states that the choices can be multiplied to find the total number of correspondences.

5. (a) If x must correspond to 5, then y may correspond to any of the four remaining elements of $\{1, 2, 3, 4, 5\}$, z may correspond to any of the three remaining, etc. Then $1 \cdot 4 \cdot 3 \cdot 2 \cdot 1 = 24$ **one-to-one correspondences**.

(b) There would be $1 \cdot 1 \cdot 3 \cdot 2 \cdot 1 = 6$ **one-to-one correspondences**.

(c) The set $\{x, y, z\}$ could correspond to the set $\{1,3,5\}$ in $3 \cdot 2 \cdot 1 = 6$ ways. The set $\{u, v\}$ could correspond with the set $\{2,4\}$ in $2 \cdot 1 = 2$ ways. There would then be $6 \cdot 2 = 12$ **one-to-one correspondences**.

6. There are three pairs of equal sets:

(i) $A = C$. The order of the elements does not matter.

(ii) $E = H$. They are both the null set.

(iii) $I = L$. Both represent the numbers $1,3,5,7,\ldots$.

7. (a) Assume an arithmetic sequence with $a_1 = 201$, $a_n = 1100$, and $d = 1$. Thus $1100 = 201 + (n - 1) \cdot 1$; solving, $n = 900$. The cardinal number of the set is therefore **900**.

(b) Assume an arithmetic sequence with $a_1 = 1$, $a_n = 101$, and $d = 2$. Thus $101 = 1 + (n - 1) \cdot 2$; solving, $n = 51$. The cardinal number of the set is therefore **51**.

(c) Assume a geometric sequence with $a_1 = 1$, $a_n = 1024$, and $r = 2$. Thus $1024 = 1 \cdot 2^{n-1} \Rightarrow 2^{10} = 2^{n-1} \Rightarrow n - 1 = 10 \Rightarrow n = 11$. The cardinal number of the set is therefore **11**.

(d) If $k = 1,2,3,\ldots,100$, the cardinal number of the set $\{x | x = k^3, k = 1,2,3,\ldots,100\} = \mathbf{100}$, since there are 100 elements in the set.

8. $\overline{A}$ represents all elements in U that are not in A, or **the set of all college students with at least one grade that is not an A**.

9. (a) A proper subset must have at least one less element than the set, so the maximum $\boldsymbol{n(B) = 7}$.

(b) Since $B \subset C$, and $n(B) = 8$ then C could have **any number of elements in it, so long as it was greater than eight**.

10. (a) If $C \subseteq D$ and $D \subseteq C$, then the sets arc equal; so $\boldsymbol{n(D) = 5}$.

(b) Answers vary. For example, **the sets are equal** and/or **the sets are equivalent**.

11. (a) A has 5 elements, thus $2^5 = 32$ **subsets**.

(b) Since A is a subset of A and A is the only subset of A that is not proper, A has $2^5 - 1 = 31$ **proper subsets**.

(c) Let $B = \{b,c,d\}$. Since $B \subset A$, the subsets of B are all subsets of A that do not contain a and e. There are $2^3 = 8$ of these subsets. If we join (union) a and e to each of these subsets there are still **8 subsets**.

Alternative. Start with $\{a,e\}$. For each element b, c, and d there are two options: include the element or do not include the element. So there are $2 \cdot 2 \cdot 2 = 8$ ways to create subsets of A that include a and e.

12. If there are n elements in a set, 2^n subsets can be formed. This includes the set itself. So if there are 127 proper subsets, then there are 128 subsets. Since $2^7 = 128$, the set has **7 elements**.

13. In roster format, $A = \{3,6,9,12,\ldots\}, B = \{6,12,18,24,\ldots\}$, and $C = \{12,24,36,\ldots\}$. Thus, $\boldsymbol{B \subset A, C \subset A}$, **and** $\boldsymbol{C \subset B}$.

Alternatively: $12n = 6(2n) = 3(4n)$. Since $2n$ and $4n$ are natural number $\boldsymbol{C \subset A, C \subset B}$, and $\boldsymbol{B \subset A}$.

14. (a) $\notin$. There are no elements in the empty set.

(b) $\in$. $1024 = 2^{10}$ and $10 \in N$.

(c) $\in$. $3(1001) - 1 = 3002$ and $1001 \in N$.

(d) $\notin$. For example, $x = 3$ is not an element because for $3 = 2^n$, $n \notin N$.

15. (a) $\not\subseteq$. 0 is not a set so cannot be a subset of the empty set, which has only one subset, $\varnothing$.

(b) $\not\subseteq$. 1024 is an element, not a subset.

(c) $\not\subseteq$. 3002 is an element, not a subset.

(d) $\not\subseteq$. x is an element, not a subset.

16. **(a)** **Yes**. Any set is a subset of itself, so if $A = B$ then $A \subseteq B$.

(b) **No**. A could equal B; then A would be a subset but not a proper subset of B.

(c) **Yes**. Any proper subset is also a subset.

(d) **No**. Consider $\mathbf{A = \{1,2\}}$ **and** $\mathbf{B = \{1,2,3\}}$.

17. **(a)** Let $A = \{1,2,3,\ldots,100\}$ and $B = \{1,2,3\}$. Then $n(A) = 100$ and $n(B) = 3$. Since $B \subset A$, $\mathbf{n(B) = 3 < 100 = n(A)}$.

(b) $n(\varnothing) = 0$. Let $A = \{1,2,3\} \Rightarrow n(A) = 3$. $\varnothing \subset A$, which implies that there is at least one more element in A than in $\varnothing$. Thus $\mathbf{0 < 3}$.

18. There are seven senators to choose from and 3 will be chosen. Consider the ways to form subsets with only three members. If we pick the first member, there are 7 senators to choose from. To pick the second member, there are only 6 to choose from, since 1 member has already been chosen. For the third seat, there are 5 to choose from. This yields $7 \cdot 6 \cdot 5$. However, this calculation counts {Able, Brooke, Cox} as a different committee that {Brooke, Able, Cox}. In fact, for any 3 names, there are $3 \cdot 2 \cdot 1$ ways to arrange the names. Thus, the number of unique committees is $\frac{7 \cdot 6 \cdot 5}{3 \cdot 2 \cdot 1} = 7 \cdot 5 = \mathbf{35}$.

19. Answers vary. For example, the set of all odd natural numbers and the set of all even natural numbers are two infinite sets that are equivalent but not equal. Another possibility is the set of all natural numbers and the set of all whole numbers.

20. Each even natural number $2n$ can be paired with each odd natural number $2n - 1$ in a one-to-one correspondence.

21.

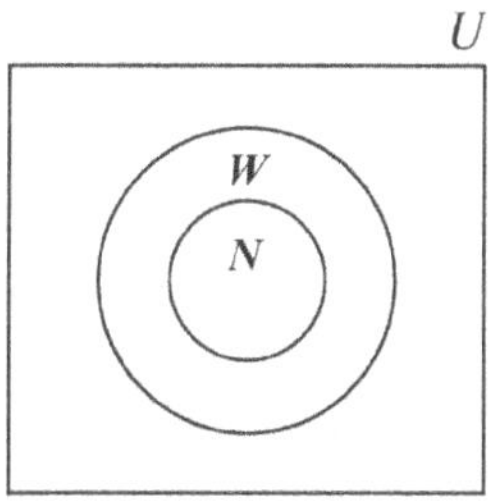

22.

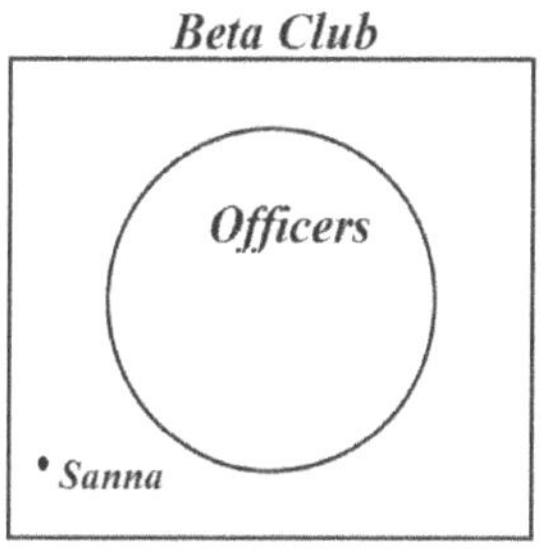

23. **All members of the Beta Club are officers.**

Mathematical Connections 2-2: Review Problems

25. **(a)** **False**. In order for a conjunction to be true, both parts of the conjunction must be true. Since p is false, $p \wedge q$ is false.

(b) **False**. Since q is true, $\sim q$ is false.

(c) **True**. Since p is false, $\sim p$ is true. This makes both parts of the conjunction true, so therefore the conjunction $\sim p \wedge q$ is true.

(d) **True**. In part (a) above, we found out that $p \wedge q$ was false. Therefore, $\sim(p \wedge q)$ must be true.

(e) **False**. In order for a conjunction to be true, both parts of the conjunction must be true. Since q is true, $\sim q$ is false; so $\sim q \wedge \sim p$ is false.

27.

p	q	$\sim q$	$p \vee \sim q$
T	T	**F**	**T**
T	F	**T**	**T**
F	T	**F**	**F**
F	F	**T**	**T**

Assessment 2-3A: Other Set Operations and Their Properties

1. $\mathbf{A=\{1,3,5,\ldots\}}$; $\mathbf{B=\{2,4,6,\ldots\}}$; $\mathbf{C=\{1,3,5,\ldots\}}$

(a) Either $\mathbf{A}$ or $\mathbf{C}$. Every element in $A \cup C$ is either in set A or set C.

(b) $\mathbf{N}$. Every natural number is in either A or B.

(c) $\varnothing$. There are no natural numbers in both A and B.

2. For these problems, as an example, let
$U = \{1,2,3,4,5,6,7\}$,
$A = \{1,5,6\}, B = \{1,4,5,6,7\}$, and
$C = \{1,2,3,4\}$

(a) **Yes.** $A - A$ means the set of all elements that are in A that are also not in A; or more formally,
$A - A = \{x | x \in A \text{ and } x \notin A\} = \varnothing$.

(b) **Yes.** $B - A = \{4,7\}$; $\overline{A} = \{2,3,4,7\}$
which means $B \cap \overline{A} = \{4,7\}$ so the sets are equal.

(c) **No.** $B - A = \{4,7\}$;
$B \cap A = \{1,5,6\}$
so the sets are not equal.

(d) **Yes.** $A \cup A = \{1,5,6\}$; $A \cup \emptyset = \{1,5,6\}$, so the sets are equal. In general, combining a set to itself via set union will not change the elements in the set; combining a set via set union to the empty set will not change the elements in the set.

3. **(a)** **True.** Let $A = \{1,2\}$. $A \cup \varnothing = \{1,2\}$. In general, combining any set with the empty set via set union contributes no extra elements to the set than what is already in the set.

(b) **False.** Let $A = \{1,2\}$ and $B = \{2,3\}$.
$A - B = \{1\}; B - A = \{3\}$.

(c) **False.** Let $U = \{1,2\}$; $A = \{1\}$; $B = \{2\}$.
$\overline{A \cap B} = \{1,2\}; \overline{A} \cap \overline{B} = \varnothing$.

(d) **False.** Let $A = \{1,2\}$; $B = \{2,3,4\}$.
$(A \cup B) - A = \{1,2,3,4\} - \{1,2\} =$
$\{3,4\} \neq B$.

(e) **False.** Let $A = \{1,2\}$; $B = \{2,3\}$.
$(A - B) \cup A = \{1\} \cup \{1,2\} = \{1,2\}$;
$(A - B) \cup (B - A) = \{1\} \cup \{3\} = \{1,3\}$.

4. **(a)** If $B \subseteq A$, all elements of B must also be elements of A, but there may be elements of A that are not elements of B, so $\boldsymbol{A \cap B = B}$.

(b) If $B \subseteq A$, all elements of B must also be elements of A, but there may be elements of A that are not elements of B, so $\boldsymbol{A \cup B = A}$.

5. **(a)** $(A \cap B) \cup (A \cap C)$

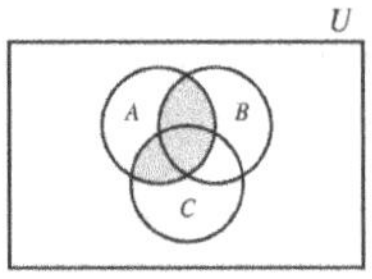

(b) $(A \cup B) \cap \overline{C}$

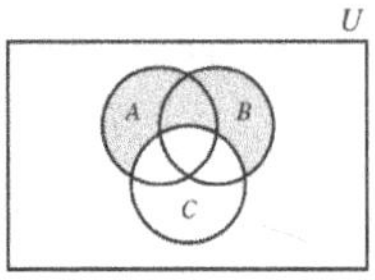

(c) $(A \cap B) \cup C$

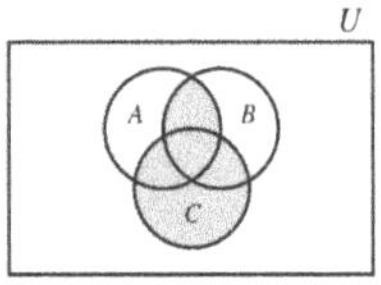

6. **(a)** $S \cup \overline{S} = \{x | x \in S \text{ or } x \in \overline{S}\} = \boldsymbol{U}$.

(b) If U is the universe the complement of U can have no elements, thus $\overline{\boldsymbol{U}} = \varnothing$.

(c) There are no elements common to S and $\overline{S}$, so $\boldsymbol{S \cap \overline{S}} = \varnothing$.

(d) Since there are no elements in the empty set there are none common to it and S, so $\varnothing \cap \boldsymbol{S} = \varnothing$.

7. **(a)** If $A \cap B = \varnothing$ then A and B are disjoint sets and any element in A is not in B, so $A - B = \{x | x \in A \text{ and } x \notin B\} = \boldsymbol{A}$.

(b) Since B is the empty set, there are no elements to remove from A, so $A - B = \boldsymbol{A}$.

(c) If $B = U$ there are no elements in A which are not in B, so $A - B = \varnothing$.

8. **Yes.** By definition, $A - B$ is the set of all elements in A that are not in B. If $A - B = \varnothing$ this means there are no elements in A that are not in B,
thus $A \subseteq B$.

More formally, suppose $A \not\subseteq B$. Then there must be an element in A that is not in B, which implies that it is in $A - B$. This implies that $A - B$ is not empty, which is a contradiction. Thus $A \subseteq B$.

9. Answers may vary; the possible answers are:

(a) $\boldsymbol{B \cap \overline{A}}$ or $\boldsymbol{B - A}$; i.e., $\{x | x \in B$ but $x \notin A\}$.

(b) $\overline{\boldsymbol{A \cup B}}$ or $\overline{\boldsymbol{A}} \cap \overline{\boldsymbol{B}}$; i.e., $\{x | x \notin A \text{ or } B\}$.

(c) $(\boldsymbol{A \cap B}) \cap \overline{\boldsymbol{C}}$ or $(\boldsymbol{A \cap B}) - \boldsymbol{C}$; i.e., $\{x | x \in A \text{ and } B \text{ but } x \notin C\}$.

10. (a) $\overline{A}$ is the set of all elements in U that are not in A. $\overline{A} \cap B$ is the set of all elements common to $\overline{A}$ and B:

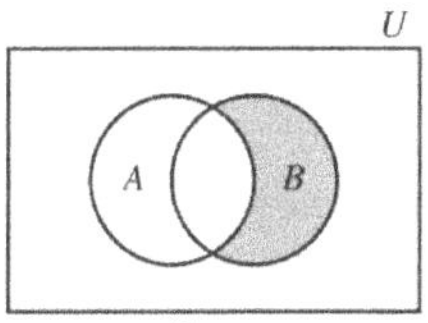

$\overline{A} \cap B$

(b) Answers may vary.
$\overline{A} \cap B = B - (A \cap B)$. or
$\overline{A} \cap B = B - A$. are two common responses.

11. (a) False:
$A \cup (B \cap C) \neq (A \cup B) \cap C$

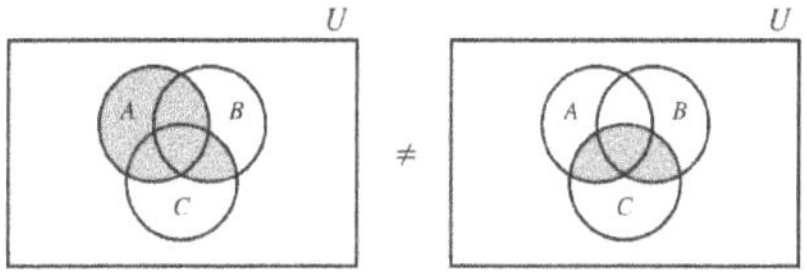

$A \cup (B \cap C)$ $(A \cup B) \cap C$

(b) False:
$A - (B - C) \neq (A - B) - C$

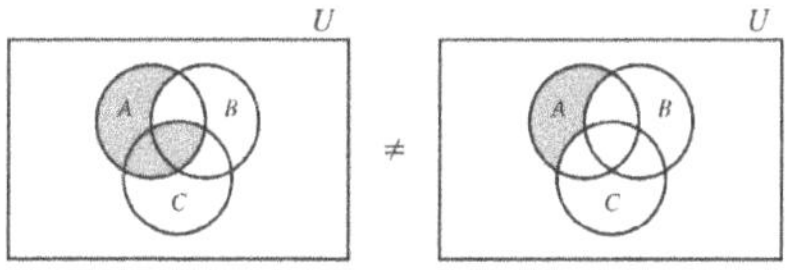

$A - (B - C)$ $(A - B) - C$

12. (a) Yes. Answers will vary. Note that in general, $\boldsymbol{A \cap B \subseteq A \cup B}$ because all elements of $A \cap B$ are included in $A \cup B$. Example: Let $A = \{1,2,3,4,5\}$ and $B = \{4,5,6,7,8\}$. Then $A \cap B = \{4,5\}$ and $A \cup B = \{1,2,3,4,5,6,7,8\}$.

(b) Yes. Answers will vary. Example: Let $A = \{1,2,3,4,5\}$ and $B = \{4,5,6,7,8\}$, where the universal set is $\{1,2,3,4,5,6,7,8\}$ Then $A - B = \{1,2,3\}$ which means $\overline{A - B} = \{4,5,6,7,8\}$, which does contain two elements of A, namely 4 and 5.

13. (a) (*i*) Greatest $n(A \cup B) = n(A) + n(B) = \mathbf{5}$ **if** A **and** B **are disjoint**.

(*ii*) Greatest $n(A \cap B) = n(B) = \mathbf{2}$ if $B \subseteq A$.

(*iii*) Greatest $n(B - A) = n(B) = \mathbf{2}$ if A and B are disjoint.

(*iv*) Greatest $n(A - B) = n(A) = \mathbf{3}$ if A and B are disjoint.

(b) (*i*) Greatest $n(A \cup B) = \boldsymbol{n + m}$ if A and B are disjoint.

(*ii*) Greatest $n(A \cap B) = \boldsymbol{m}$, if $B \subseteq A$, or $\boldsymbol{n}$, if $A \subseteq B$.

(*iii*) Greatest $n(B - A) = \boldsymbol{m}$ if A and B are disjoint.

(*iv*) Greatest $n(A - B) = \boldsymbol{n}$ if A and B are disjoint.

14. (a) (*i*) Greatest $n(A \cup B \cup C) = n(A) + n(B) + n(C) = 4 + 5 + 6 = \mathbf{15}$, if A, B, and C are disjoint.

(*ii*) Least $n(A \cup B \cup C) = n(C) = \mathbf{6}$, if $A \subset B \subset C$.

(b) (*i*) Greatest $n(A \cap B \cap C) = n(A) = \mathbf{4}$, if $A \subset B \subset C$.

(*ii*) Least $n(A \cap B \cap C) = \mathbf{0}$, if A, B, and C are disjoint.

15. (a) There were **18 non-Democrats** in the room. There were a total 13 Democrats in the room out of a total of 31 people; Seven men are non-Democrats, and 11 women are non-Democrats.

(b) There were 7+5=**12 men** in the room.

(c) There were **8 Democratic women** in the room. Of the 13 total Democrats, 5 are men.

(d) There were a total of 8+5+7+11 = **31 total people** in the room.

16. An example Venn Diagram is as follows. Note that the labels of the regions can vary.

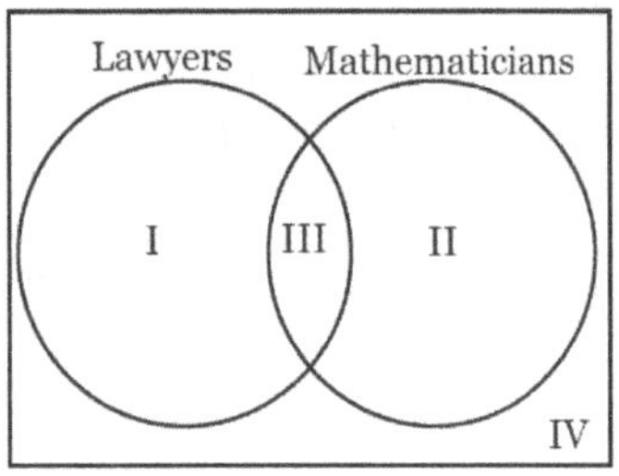

In the Venn Diagram above:

Section I represents people who are **lawyers but not mathematicians;**

Section II represents people who are **mathematicians but not lawyers;**

Section III represents people who are **both lawyers and mathematicians;**

Section IV represents people who are **neither lawyers or mathematicians**.

17. Constructing a Venn diagram will help in visualization:

(a) $B \cap S$ is the set of college **basketball players more than 200 cm tall.**

(b) $\overline{S}$ is the set of humans who are **not college students** or who are **college students less than or equal to 200 cm tall.**

(c) $B \cup S$ is the set of humans who are **college basketball players** or who are **college students taller than 200 cm**.

(d) $\overline{B \cup S}$ is the set of all humans who are **not college basketball players and who are not college students taller than 200 cm**.

(e) $\overline{B} \cap S$ is the set of all **college students taller than 200 cm who are not basketball players**.

(f) $B \cap \overline{S}$ is the set of all **college basketball players less than or equal to 200 cm tall.**

18. Use a three-set Venn diagram, labeling the sets B (for basketball), V (volleyball), and S (soccer):

(i) Enter 2 in the region representing $B \cap V \cap S$ (i.e., there were two who played all three sports);

(ii) Enter 1 in the region representing $(B \cap V) - S$ (i.e., there was one who played basketball and volleyball but not soccer);

(iii) Enter 1 in the region representing $(B \cap S) - V$ (i.e., there was one who played basketball and soccer but not volleyball);

(iv) Enter 2 in the region representing $(V \cap S) - B$ (i.e., there were two who played volleyball and soccer but not basketball);

(v) Enter $7 - (1 + 1 + 2) = 3$ in the region representing $B - (V \cup S)$ (i.e., of the seven who played basketball, one also played volleyball, one also played soccer, and two also played both volleyball and soccer—leaving three who played basketball only);

(vi) Enter $9 - (1 + 2 + 2) = 4$ in the region representing $V - (B \cup S)$ (i.e., of the nine who played volleyball, one also played basketball, two also played soccer, and two also played both basketball and soccer—leaving four who played volleyball only);

(vii) Enter $10 - (1 + 2 + 2) = 5$ in the region representing $S - (B \cup V)$ (i.e., of the ten who played soccer, one also played basketball, two also played volleyball, and two also played both basketball and volleyball—leaving five who played soccer only).

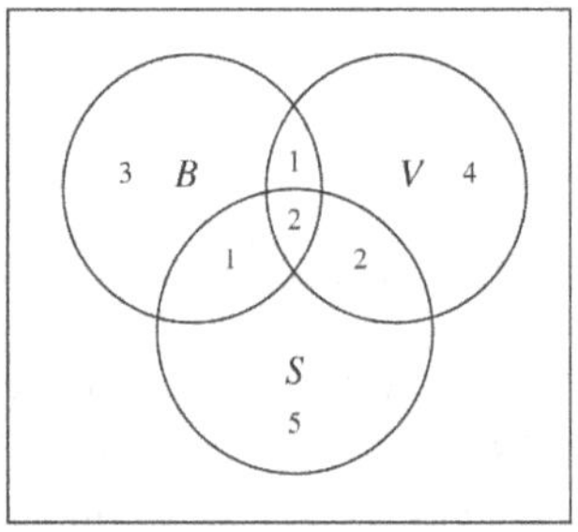

There are then $3 + 4 + 5 + 1 + 1 + 2 + 2 = \mathbf{18}$ who played one or more sports.

19. In the Venn diagram below:

(i) There were 5 members who took both biology and mathematics;

(ii) Of the 18 who took mathematics 5 also took biology, leaving 13 who took mathematics only;

(iii) 8 took neither course, so of the total of 40 members there were $40 - (5 + 13 + 8) = \mathbf{14}$ who took biology but not mathematics.

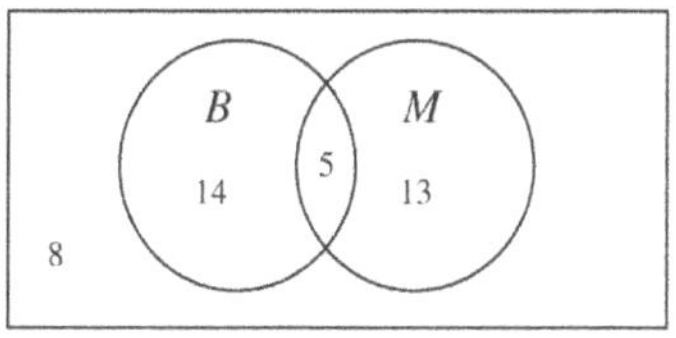

20. **(a)** If all bikes needing new tires also need gear repairs, i.e., if $\{TIRES\} \subset \{GEARS\}$, then $n[\{TIRES\} \cap \{GEARS\}] = \mathbf{20\ bikes}$.

(b) Adding the separate repairs gives $20 + 30 = 50$ bikes, which is the total number of bikes; so it is possible that **zero bikes** needed both repairs, if every bike there needed exactly one repair.

(c) If the maximum number of bikes needed both repairs then all 20 receiving tires would also receive gear work. That would leave 10 additional bikes needing gear work only, leaving **20 bikes** that needed no service.

21. Generate the following Venn diagram in this order:

(i) The 4 who had A, B, and Rh antigens;

(ii) The $5 - 4 = 1$ who had A and B antigens, but who were Rh negative;

(iii) The $31 - 4 = 27$ who had A antigens and were Rh positive;

(iv) The $10 - 4 = 6$ who had B antigens and were Rh positive;

(v) The $50 - 27 - 4 - 1 = 18$ who had A antigens only;

(vi) The $18 - 6 - 4 - 1 = 7$ who had B antigens only, and;

(vii) The $82 - 27 - 4 - 6 = 45$ who were O positive.

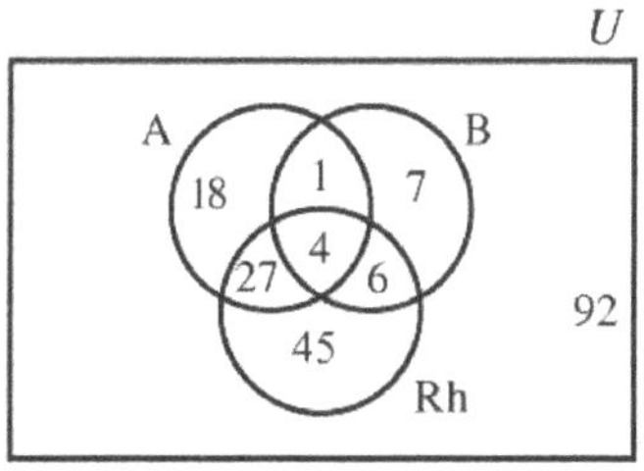

$n(A \cup B \cup \text{Rh}) = 18 + 1 + 7 + 27 + 4 + 6 + 45 = 108$. Thus the set of people who are O-negative is $200 - 108 = \mathbf{92}$.

22. Let M be the set of students taking mathematics, C be the set of students taking chemistry, and P be the set of students taking physics. Generate the following Venn diagram in this order:

(i) The 8 who had all three subjects;

(ii) The $10 - 8 = 2$ students who took mathematics and physics, but not chemistry;

(iii) The $15 - 8 = 7$ students who took chemistry and physics, but not mathematics;

(iv) The $20 - 8 = 12$ students who took mathematics and chemistry, but not physics;

(v) The $45 - 12 - 8 - 2 = 23$ students who took mathematics only;

(vi) The $40 - 12 - 8 - 7 = 13$ students who took chemistry only;

(vii) The $47 - 2 - 8 - 7 = 30$ students who took physics only

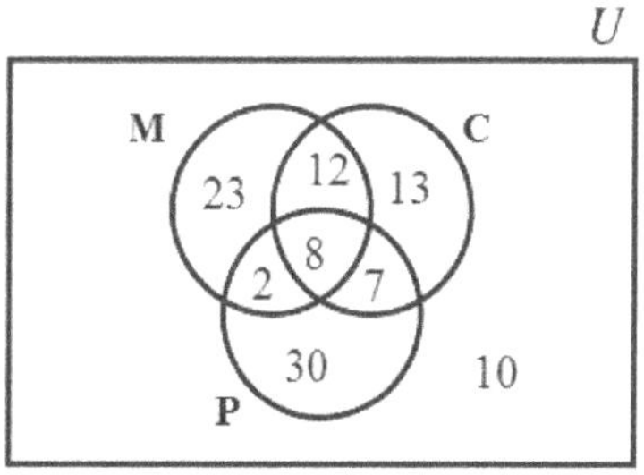

Combined with the 10 students who didn't take any of the three courses, this Venn diagram yields a total number of students as 105, which contradicts John's reported total of 100. So, given that there must be errors somewhere in John's data gathering, **John probably shouldn't be hired for the job.**

23. The following Venn diagram helps in isolating the choices:

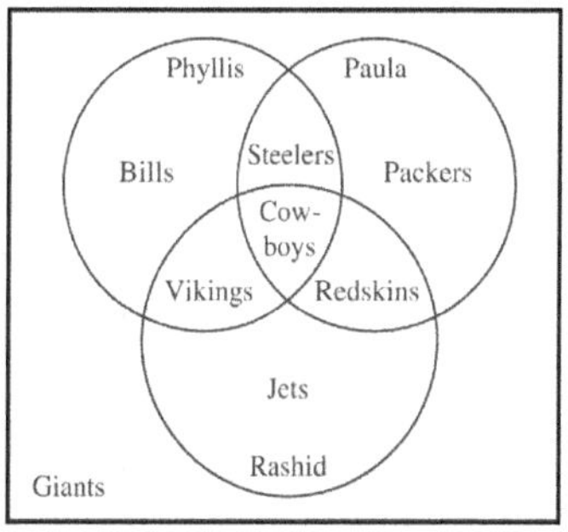

All picked the Cowboys to win their game, so their opponent cannot be among any of the other

choices; the only team not picked was the Giants.

Phyllis and Paula both picked the Steelers, so their opponent cannot be among their other choices. This leaves the Jets.

Phyllis and Rashid both picked the Vikings which leaves the Packers as the only possible opponent.

Paula and Rashid both picked the Redskins which leaves the Bills as the only possible opponent.

Thus we have **Cowboys vs Giants, Vikings vs Packers, Redskins vs Bills, and Jets vs Steelers.**

24. **(a)**

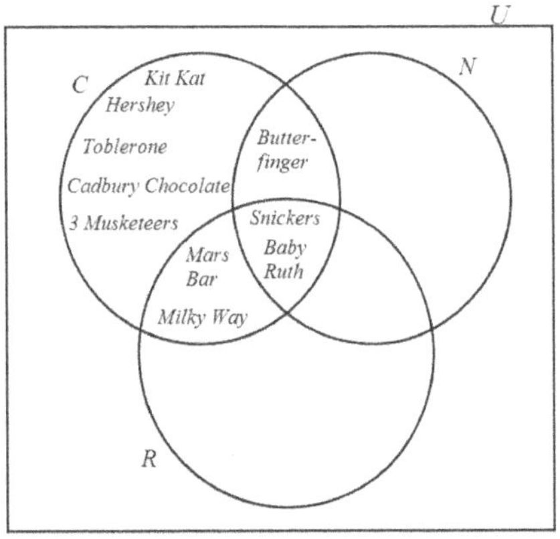

(b) **(*i*)** **none**. Since all candy bars have chocolate, there aren't any in the set of bars without chocolate.

(*ii*) **none**. Same reason as (*i*).

(*iii*) **Chocolate** is the most popular ingredient; **nuts/peanut butter** is the least popular ingredient.

(*iv*) Answers will vary.

25. **57.** Of the 324 first class passengers, you know there were $146 + 4 = 150$ total women and children, and 117 men who were lost. To find the number of men who survived, find $324 - 150 - 117 = 57$

26. **Abby, Harry, and Dick are in one family; Tom, Jane, and Mary are in the other family.** First, note that Abby and Harry have the same characteristics, blue eyes and blond hair. Also note that Mary is their opposite, with brown eyes and brown hair; therefore, Mary is certainly in a different family than Abby and Harry. Now note that Jane and Tom both have blue eyes and brown hair, while Dick has brown eyes and blond hair. Dick cannot be in the same family as Jane and Tom. So, since each family has 3 children, Dick is in the family with Abby and Harry (where they all have blond hair), while Tom, Jane, and Mary are in the other family (where they all have brown hair).

27. If a Cartesian product is the set of all ordered pairs such that the first element of each pair is an element of the first set and the second element of each pair is an element of the second set:

(a) $A \times B = \{(x,a),(x,b),(x,c),(y,a),(y,b),(y,c)\}$.

(b) $B \times A = \{(a,x),(a,y),(b,x),(b,y),(c,x),(c,y)\}$.

(c) **No**. For sets to be equal, all the elements of one must be elements of the other. But (a,x) for example is not a member of $A \times B$. For ordered pairs to be equal, their first coordinates must be equal and their second coordinates must be equal.

28. **(a)** The first element of each ordered pair is a, so $C = \{a\}$. The second elements in the ordered pairs are, respectively, b, c, d, and e, so $D = \{b, c, d, e\}$.

(b) The first element in the first three ordered pairs is 1; in the second three is 2, so $C = \{1,2\}$. The second element in the ordered pairs is, respectively, 1, 2, and 3, so $D = \{1,2,3\}$.

(c) The numbers 0 and 1 appear in each ordered pair, so $C = D = \{0,1\}$. (The order of the numbers in these sets is irrelevant.)

Mathematical Connections 2-3: Review Problems

15. **(a)** Therefore, Mary will change the lunch menu. (Modus Ponens, or law of detachment)

(b) Therefore, Samuel stays after school.

(c) Therefore, the lake is not frozen. (Modus Tollens)

17. **(a)** **6**; c, o, m, n, r, and e are the elements in the set. c and m are duplicated once, and o is duplicated twice.

(b) **6**; c, o, m, i, t, e are the elements in the set. m, t, and e are duplicated once.

19. Answers may vary. Some possibilities:

(a) The set of 18 holes on a golf course, and the set of 18 flagsticks on a golf course.

(b) The set of letters in the English alphabet (26) with the set of letters in the Greek alphabet (24).

Chapter 2 Review

1. Answers will vary. Some statements: i) Arizona is a state; ii) The current president of the United States is Barack Obama; iii) $2+12=24$; $2 \times 12 = 24$; iv) France is not a country. Some non-statements: i) Las Vegas is a fun city; ii) Teaching is a tough job; iii) How old are you?; iv) Zane Grey is the best writer ever.

2. In statement (i), every student in the class earned an A, B, or C grade on the final exam; in statement (ii), there was at least one student who earned an A, B, or C grade on the final exam. In statement (i), no student could have received a grade of D or F on the final exam; while in statement (ii), it is possible that a student (or several students) did receive a grade of D or F on the final exam.

3. **(a)** **Yes**. Even though it is false, it is still a statement, by the definition of a statement.

(b) **No**. This sentence is neither true or false, since it depends upon the value of the variable n. Since its truth value cannot be definitely ascertained, it is not a statement.

(c) **Yes**. It can definitely be determined whether or not this statement is true or false.

4. **(a)** **Some women smoke**. The original statement means that there exists no women who smoke. So to negate it, one would have to write a statement that implied there exists at least one woman who smokes.

(b) **3 + 5 = 8**. The original statement is false; to negate it, write a statement that is true.

(c) **Bach wrote some music that was not classical**. The reasoning behind this answer is similar to the answer in part (a). The original statement implies that Bach only wrote classical music. To negate it, a statement must be written to imply that there exists at least one piece of Bach's music that was not classical.

5. The original statement is in the form $p \to q$.
Converse $(q \to p)$: **If someone will read a tweet, then the whole world is tweeting.**
Inverse $(\sim p \to \sim q)$: **If the whole world is not tweeting, then no one will read a tweet.**
Contrapositive $(\sim q \to \sim p)$: **If no one will read a tweet, then the whole world is not tweeting.**

6.

p	q	$\sim p$	$\sim q$	$\sim p \to q$	$\sim q \to p$
T	**T**	**F**	**F**	**T**	**T**
T	**F**	**F**	**T**	**T**	**T**
F	**T**	**T**	**F**	**T**	**T**
F	**F**	**T**	**T**	**F**	**F**

7. **(a)**

p	q	$\sim q$	$p \vee \sim q$	$p \vee q$	$(p \vee \sim q) \wedge (p \vee q)$
T	T	F	T	T	T
T	F	T	T	T	T
F	T	F	F	T	F
F	F	T	T	F	F

(b)

p	q	$\sim q$	$p \vee \sim q$	$(p \vee \sim q) \wedge \sim q$	$[(p \vee \sim q) \wedge \sim q] \to p$
T	T	F	T	F	T
T	F	T	T	T	T
F	T	F	F	F	T
F	F	T	T	T	F

8. **(a)** $\therefore$ **Alfinia loves Mom and apple pie**. This is true by direct reasoning. The first statement could be read as "If a person is a Eurasian, then that person loves Mom and apple pie". We can then make p be the statement "a person is a Eurasian" and q the statement "A person loves Mom and apple pie." So, the first statement is denoted in logical notation as $p \to q$; the second statement, Alfinia is a Eurasian, is denoted as p. So, it makes sense that the conclusion should be q.

(b) $\therefore$ **The Washington Monument will eventually crack**. This is true by direct reasoning, similar to (a). The first statement could be written in $p \to q$ form as "if an object is made of marble and granite, then it will eventually crack."

(c) $\therefore$ **Therefore, Josef passed the math for elementary teachers course**. The first sentence consists of two statements "Josef passed the math for elementary teachers course" is one statement; "Josef dropped out of school" is the other statement. The "or" between the two statements implies that one or the other is true. The second sentence "Josef did not drop out of school" would imply that the second part of that first sentence isn't true…so the first part must be.

9. This argument is valid by modus tollens. Let p be the statement "Bob passes the course" and q be the statement "Bob scored at least a 75 on the final exam". Then the first sentence is $p \to q$; the second sentence is $\sim q$; therefore the conclusion is $\sim p$.

10 (a) **If students do not have at least a B average, then they are not eligible for scholarships.**

(b) **If you attended the Oprah Winfrey show, then you received a gift.**

(c) **If a player suited up for the basketball game, then the player played.**

11. A set with n elements has 2^n subsets. $2^4 = 16$. This includes A. There are $2^4 - 1 =$ **15 proper subsets.**

12. There are 4 elements in the set, thus there are $2^4 =$ **16 subsets**: $\{\}, \{m\}, \{a\}, \{t\}, \{h\}, \{m,a\}, \{m,t\}, \{m,h\}, \{a,t\}, \{a,h\}, \{t,h\}, \{m,a,t\}, \{m,a,h\}, \{m,t,h\}, \{a,t,h\}, \{m,a,t,h\}$.

13. (a) $A \cup B = \{r,a,v,e\} \cup \{a,r,e\} = \{r,a,v,e\} = \boldsymbol{A}$.

(b) $C \cap D = \{l,i,n,e\} \cap \{s,a,l,e\} = \boldsymbol{\{l,e\}}$.

(c) $\overline{D} = \overline{\{s,a,l,e\}} = \boldsymbol{\{u,n,i,v,r\}}$.

(d) $A \cap \overline{D} = \{r,a,v,e\} \cap \{u,n,i,v,r\} = \boldsymbol{\{r,v\}}$.

(e) $\overline{B \cup C} = \overline{\{a,r,e\} \cup \{l,i,n,e\}} = \overline{\{a,e,i,l,n,r\}} = \boldsymbol{\{s,u,v\}}$.

(f) $B \cup C = \{a,e,i,l,n,r\} \Rightarrow (B \cup C) \cap D = \{a,e,i,l,n,r\} \cap \{s,a,l,e\} = \boldsymbol{\{a,l,e\}}$.

(g) $\overline{A} \cup B = \{u,n,i,s,l\} \cup \{a,r,e\} = \{u,n,i,e,r,s,a,l\}$.

$C \cap \overline{D} = \{l,i,n,e\} \cap \{u,n,i,v,r\} = \{i,n\}$.

$\Rightarrow (\overline{A} \cup B) \cap (C \cap \overline{D}) = \boldsymbol{\{i,n\}}$.

(h) $(C \cap D) \cap A = (\{l,i,n,e\} \cap \{s,a,l,e\}) \cap \{r,a,v,e\} = \{l,e\} \cap \{r,a,v,e\} = \boldsymbol{\{e\}}$.

(i) $n(B - A) = \mathbf{0}$. There are no elements in B that are not also in A.

(j) $n(\overline{C}) = n\{u,v,r,s,a\} = \mathbf{5}$.

(k) $n(C \times D) = 4 \cdot 4 = \mathbf{16}$. Each of the four elements in C can be paired with each of the four in D.

14. (a) $A \cap (B \cup C)$ includes the elements in A that are common to the union of B and C:

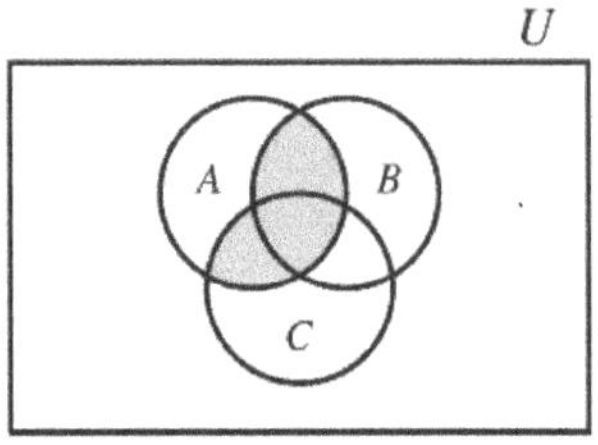

(b) $\overline{(A \cup B)} \cap C$ is the same as $C - (A \cup B)$:

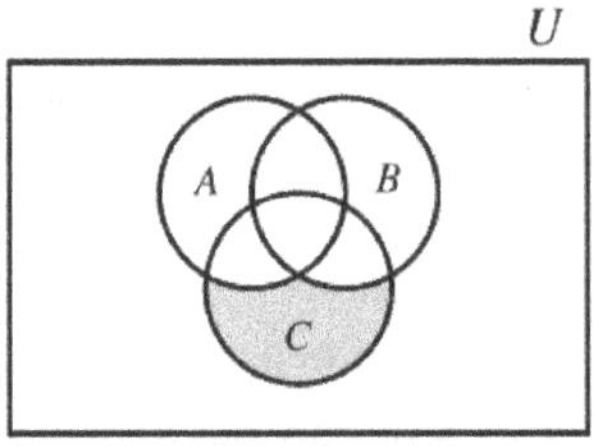

15. Using a Venn Diagram helps to solve the problem. The two circles of the Venn Diagram represent hardcopy subscriptions and online subscriptions.

Since 52 people have both, that number goes in the region that represents the intersection of the two circles. Since 180 total have hardcopy subscriptions and 52 of those have online too, 128 people have only hard copy subscriptions. Finally, noting that 22 have neither, the Venn Diagram thus far would look like this:

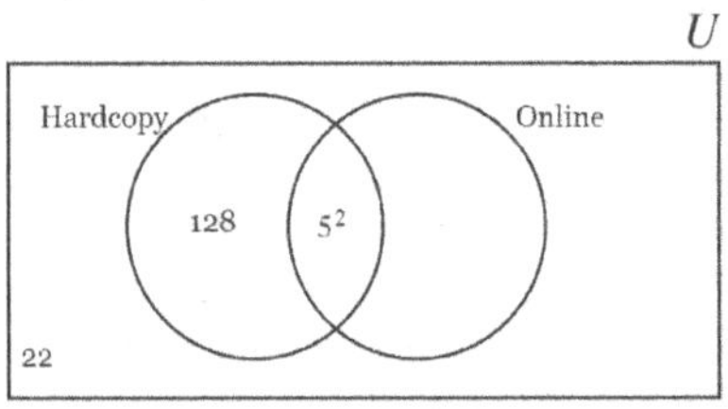

Since there are 300 total people in the poll, the total that have only online subscriptions equals the answer to 300 – (128+52+22) = **98 people.**

16. Since all 5 letters are distinct, consider seven "slots" in which to put the letters. There are 5 letters which could go in the first slot, then 4 left which could go in the second slot, and so on. So, the number of possible arrangements is then $5 \cdot 4 \cdot 3 \cdot 2 \cdot 1 = \mathbf{120}$

17. **(a)** Answers may vary. **One possible** correspondence is $\boldsymbol{t \leftrightarrow e}$, $\boldsymbol{h \leftrightarrow n}$, **and** $\boldsymbol{e \leftrightarrow d}$.

(b) There are three possible one-to-one correspondences between D and the e in E, two possible between D and the n in E, and then only one possible remaining for the d
in E. Thus $3 \cdot 2 \cdot 1 = \mathbf{6}$ one-to-one correspondences are possible.

18. $A \cup (B - C) \quad \neq \quad (A \cup B) - C$

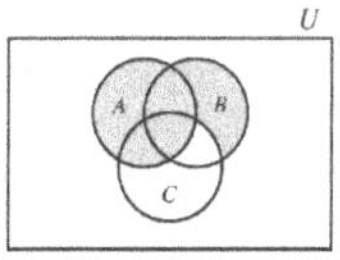

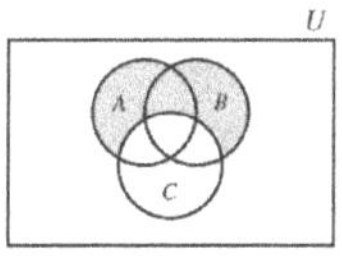

19. **(a)** **False**. Let $U = \{1,2,3,4\}$, $A = \{1,2\}$, and $B = \{1,2,3\}$ Then $\overline{A} = \{3,4\}$ and $\overline{B} = \{4\}$. So $\overline{A} \not\subseteq \overline{B}$.

(b) **True**.

$\overline{A}$

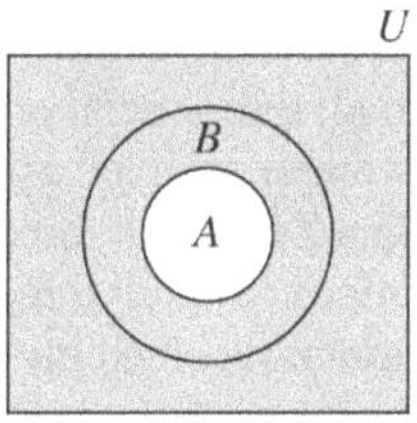

$\overline{B}$

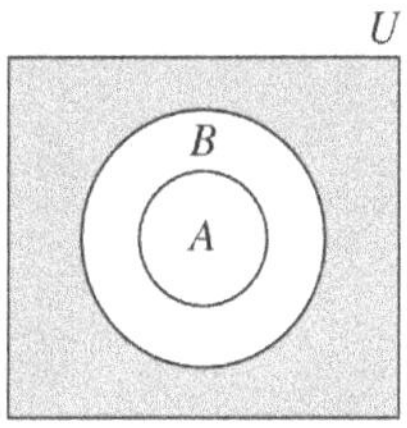

(c) **True**. Since $A \subseteq B$, every element in A is in B, So combining A and B does not include new elements to B.

(d) **True**. Since $A \subseteq B$ and $A \cap B = \{x | x \in A \text{ and } x \in B\}$, the intersection, the elements common to A and B, is A.

(e) **True**. Since $\overline{B} \subseteq \overline{A}$, this property was established in d.

(f) **True,** This is because $\overline{B} \subseteq \overline{A}$, as established in b.

20. **(a)** **False.** For example, let $U = \{1,2,3,4,5\}$; $A = \{1,2,3,4\}$; $B = \{3,4,5\}$ and $C = \{3,4\}$. Then
$A - B = \{1,2\} = A - C$; but $B \neq C$

(b) **False.** For example, let $U = \{1,2,3,4\}$; $A = \varnothing$; $B = \{1,2\}$ and $C = \{3,4\}$. Then
$A \times B = \varnothing$; $A \times C = \varnothing$; but $B \neq C$.

21. **(a)** $\boldsymbol{A}$. By the distributive property of set intersection over set union $(A \cap B) \cup (A \cap \overline{B})$ can be written as $A \cap (B \cup \overline{B})$. Since $(B \cup \overline{B}) = U$, we have $A \cap U$, which equals A.

(b) $\boldsymbol{A \cup \overline{B}}$. By the distribution property of set union over set intersection, $(A \cap B) \cup \overline{B}$ can be rewritten as $(A \cup \overline{B}) \cap (B \cup \overline{B})$. Since $(B \cup \overline{B}) = U$, we have $(A \cup \overline{B}) \cap U$, which equals $A \cup \overline{B}$.

22. **(a)** **False**. The sets could be disjoint.

(b) **False**. The empty set is not a proper subset of itself.

(c) **False**. $A \sim B$ only requires the same number of elements—not necessarily the same elements.

(d) **False**. The set is in one-to-one correspondence with the set of natural numbers, so it increases without limit.

(e) **False**. For example, the set $\{5,10,15,...\}$ is a proper subset of the natural numbers and is equivalent since there is a one-to-one correspondence.

(f) **False**. Let $B = \{1,2,3\}$ and $A =$ the set of natural numbers.

(g) **True**. If $A \cap B \neq \varnothing$, then the sets are not disjoint.

(h) **False**. The sets may be disjoint but not empty.

23. **(a)** **True**. Venn diagrams show that $A - B$, $B - A$, and $A \cap B$ are all disjoint sets, so $n(A - B) + n(B - A) + n(A \cap B) = n(A \cup B)$.

(b) **True**. Venn diagrams show that $A - B$ and B are disjoint sets, so $n(A - B) + n(B) = n(A \cup B)$. Likewise, Venn diagrams show that $(B - A)$ and A are disjoint, so $n(B - A) + n(A) = n(A \cup B)$.

24. **(a)** **17**, if $P = Q$.

(b) **34**, if P and Q are disjoint.

(c) **0**, if P and Q are disjoint.

(d) **17**, if $P = Q$.

25. $A \times B \times C$ is the set of ordered triples (a, b, c), where $a \in A$, $b \in B$, and $c \in C$. There are 3 possibilities for a, the first entry, 4 possibilities for the second, and 2 for the third. So $n(A \times B \times C) =$ **24**.

26. $n(\text{Crew}) + n(\text{Swimming}) + n(\text{Soccer}) = 57$. The 2 lettering in all three sports are counted three times, so subtract 2 twice, giving 53. $n(\text{students}) = 46$, so $53 - 46 = 7$ were counted twice; i.e., **7 lettered in exactly two sports.**

27. Use the following Venn diagram place values in appropriate areas starting with the fact that 3 students liked all three subjects; 7 liked history and mathematics, but of those 7, three also liked English so 4 is placed in the H ∩ M region; etc.

(a) A total of **36 students** were in the survey.

(b) **6 students** liked only mathematics.

(c) **5 students** liked English and mathematics but not history.

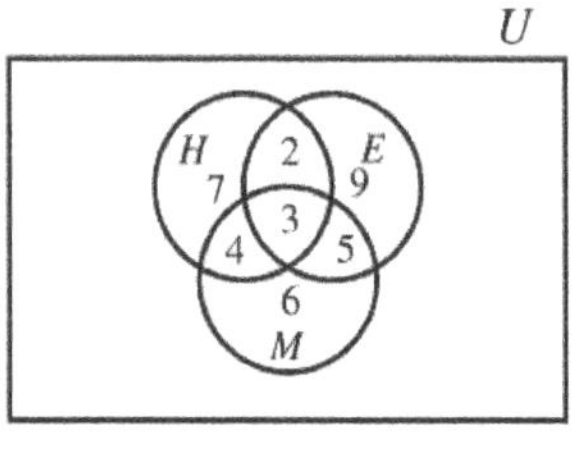

28. Answers may vary. Possibilities are:

(a) The shaded areas show $B \cup (A \cap C)$

(b) The shaded areas show $B - C$ or $B \cap \bar{C}$

29. 2 slacks × 3 blouses × 2 sweaters = **12** outfits.

30. **(a)** Let $A = \{1, 2, 3, ..., 13\}$ and $B = \{1, 2, 3\}$. $B \subset A$, so B has fewer elements than A. **Then $n(B) < n(A)$ and thus 3 < 13.**

(b) Let A $= \{1, 2, 3, ..., 12\}$ and $B = \{1, 2, 3, ..., 9\}$. $B \subset A$, so A has more elements than B. **Then $n(B) > n(A)$** and **thus 12 > 9.**

31. Given that 37 total women were in the set of data, and 5 received no invitations, that means 32 received at least one invitation. The sum total of the invitations sent from all three social organizations is $20 + 21 + 12 = 53$. The difference, $53 - 32 = 21$, is the number of "duplicate" invitations; (the number of invitations after each of the 32 received their first invitation; those invitations that were people's second and third invitation). Given that 10 received invitations from Alpha Pi and Beta Zeta; 5 received invitations from Beta Zeta and Gamma Iota and 9 from Alpha Pi and Gamma Iota, $10 + 9 + 5 = 24$. This means $24 - 21 =$ **3 women** must have received invitations from all three social organizations. Since 10 total women received invitations from Alpha Pi and Beta Zeta, with three women receiving invitations from all three, then $10 - 3 =$ **7 women** received invitations from Alpha Pi and Beta Zeta, but not Gamma Iota.

CHAPTER 3

NUMERATION SYSTEMS AND WHOLE NUMBER OPERATIONS

Assessment 3-1A: Numeration Systems

1. **(a)** $\overline{\overline{\text{M}}}$**CDXXIV**. The double bar over the M represents $1000 \cdot 1000 \cdot 1000$ while a single bar over the M would represent only $1000 \cdot 1000$.
   (b) **46,032**. The 4 in 46,032 represents 40,000 while the 4 in 4632 represents only 4000.
   (c) < ▼▼. < in the first number represents 10, while in the second number it represents $10 \cdot 60$ because of the space.
   (d) [symbol] ∩ I. [symbol] has a place value of 1000 while [symbol] has a place value of only 100.
   (e) [symbol]. [symbol] represents three groups of 20 plus zero 1's, or 60, while [symbol] represents three 5's and three 1's, or 18.
2. **(a)** MCMXLIX represents 1949; thus one more is 1950, or **MCML**; one less is 1948, or **MCMXLVIII**.
   (b) < <   < ∇ is $20 \cdot 60 + 11 = 1211$; thus one more is 1212, or < <   < ∇∇ one less is 1210, or < <   <.
   (c) [symbol] is $1000 + 100 + 100 = 1200$; thus one more is 1201, or [symbol]; one less is 1199, or [symbol].
   (d) [symbol] is $7 \cdot 20 + 13 \cdot 1 = 153$; thus one more is 154, or [symbol]; one less is 152, or [symbol].
3. MCMXXII is $1000 + 900 + 20 + 2$, or the year **1922**.
4. **(a)** CXXI, or $100 + 20 + 1$.
   (b) XLII, or $50 - 10 + 2$.
   (c) XCI, or $100 - 10 + 1$.
   (d) MMXIV, or $2000 + 10 + 5 - 1$.
5. Using the place value table in (1):
   (a) The hundreds place value is 5.
   (b) The ten thousands place value is 3.
6. **(a)** **Hundreds**. From the decimal point moving to the left: units → tens → hundreds.
   (b) **Tens**. From the decimal point moving to the left: units → tens.
7. **(a)** $3{,}000{,}000 + 4000 + 5 =$ **3,004,005**.
   (b) $20{,}000 + 1 =$ **20,001**.
8. Using the following place value table,

Billions	Hundred Millions	Ten Millions	Millions	Hundred Thousands	Ten Thousands	Thousands	Hundreds	Tens	Units

   we can write the numerals in words.
   (a) Fifty-six million, two hundred eighty-three thousand, nine hundred fourteen.
   (b) Five billion, three hundred sixty-five million, two hundred ninety-five thousand, two hundred thirty-four.
9. There are 15 digits in the standard form of $625 \bullet 10^{12}$: 625,000,000,000,000.
10. Either **811** or **910**. Each number satisfies the conditions that the hundreds digit must be 8 or 9, the tens digit must be odd, and the sum of the digits must be 10.
11. **(a)** $(3 \cdot 25) + (2 \cdot 5) + (1 \cdot 1) =$ **86**.
    (b) $(1 \cdot 8) + (0 \cdot 4) + (1 \cdot 2) + (1 \cdot 1) =$ **11**.
12. The top row of blocks contains one set of 64 blocks plus four sets of 16 each, or $2 \cdot 4^3$ blocks; the second row of blocks contains one set of 16 blocks, or $1 \cdot 4^2$, plus four single blocks, or $1 \cdot 4^1$ blocks, plus two single blocks, or $2 \cdot 4^0 = \mathbf{2112}_{four}$.

13. **(a)** Place values represent powers of 4; e.g.,

$1 \cdot 4^0 = 1_{four}$

$1 \cdot 4^1 + 0 \cdot 4^0 = 10_{four}$

$1 \cdot 4^2 + 0 \cdot 4^1 + 0 \cdot 4^0 = 100_{four}$, etc.

Thus the first 15 counting numbers are:
$(1, 2, 3, 10, 11, 12, 13, 20, 21, 22, 23, 30, 31, 32, 33)_{four}$.

(b) Place values represent powers of 7; e.g.,

$1 \cdot 7^0 = 1_{seven}$

$1 \cdot 7^1 + 0 \cdot 7^0 = 10_{seven}$

$1 \cdot 7^2 + 0 \cdot 7^1 + 0 \cdot 7^0 = 100_{seven}$, etc.

Thus the first 15 counting numbers are:
$(1, 2, 3, 4, 5, 6, 10, 11, 12, 13, 14, 15, 16, 20, 21)_{seven}$.

14. **20** digits. One digit is needed for each of the units from 1 to 19, and one digit for 0.

15. $2032_{four} = (2 \cdot 10^3 + 0 \cdot 10^2 + 3 \cdot 10^1 + 2 \cdot 10^0)_{four}$. 2032_{four} may also be expanded as $(2 \cdot 4^3 + 0 \cdot 4^2 + 3 \cdot 4^1 + 2 \cdot 4^0)_{ten} = \mathbf{142}$.

16. **(a)** $\mathbf{111}_{two}$. 1 is the largest units digit in base two.

(b) 666_{seven}. is the largest units digit in base seven.

17. **(a)** $EE0_{twelve} = 11 \cdot 12^2 + 11 \cdot 12^1 + 0 \Rightarrow$

(i)
$(EE0 - 1)_{twelve} =$
$11 \cdot 12^2 + 10 \cdot 12^1 + 11 = \boldsymbol{ETE}_{twelve}$.

(ii) $(EE0 + 1)_{twelve} =$
$11 \cdot 12^2 + 11 \cdot 12^1 + 1 = \boldsymbol{EE1}_{twelve}$.

(b) ***(i)***
$(100000 - 1)_{two} = 1 \cdot 2^5 + 0 \cdot 2^4 +$
$0 \cdot 2^3 + 0 \cdot 2^2 + 0 \cdot 2^1 + 0 - 1 =$
$1 \cdot 2^4 + 1 \cdot 2^3 + 1 \cdot 2^2 + 1 \cdot 2^1 + 1 =$
$\mathbf{11111}_{two}$ [analogous to
$(100{,}000 - 1)_{ten} = 99{,}999_{ten}$].

(ii) $(100000 + 1)_{two} = 1 \cdot 2^5 + 0 \cdot 2^4 +$
$0 \cdot 2^3 + 0 \cdot 2^3 + 0 \cdot 2^1 + 0 + 1 =$
$\mathbf{100001}_{two}$.

(c) ***(i)*** $(555 - 1)_{six} = 5 \cdot 6^2 + 5 \cdot 6^1 +$
$5 - 1 = \mathbf{554}_{six}$.

(ii) $(555 + 1)_{six} = 5 \cdot 6^2 + 5 \cdot 6^1 +$
$5 + 1 = 1 \cdot 6^3 + 0 \cdot 6^2 +$
$0 \cdot 6^1 + 0 = 1000_{six}$ [analogous to
$(999 + 1)_{ten} = 1000_{ten}$].

18. **(a)** There is **no numeral 4 in base four**. The numerals are 0, 1, 2, and 3.

(b) There are **no numerals 6 or 7 in base five.** The numerals are 0, 1, 2, 3, and 4.

19. 231_{five} is represented by two flats $(2 \cdot 5^2)$, three longs $(3 \cdot 5^1)$, and one unit.

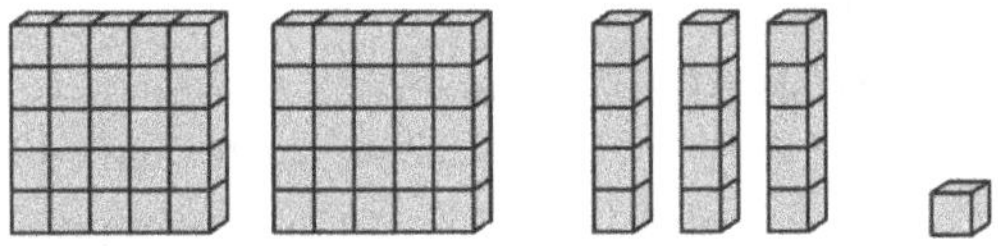

20. **(a)** Since both numerals have the same groups of 64s, we only need to compare lesser place values.

$100_{four} = (1 \cdot 4^2 + 0 \cdot 4 + 0 \cdot 1)_{ten} = 16$

$030_{four} = (0 \cdot 4^2 + 3 \cdot 4 + 0 \cdot 1)_{ten} = 12$.

So, 3030_{four} **is the lesser.**

(b) Again, we only need to compare TE_{twelve} to ET_{twelve}.

TE_{twelve} has ten groups of 12 and 11 groups of 1.

ET_{twelve} has 11 groups of 12 and 10 groups of 1.

Thus, $\boldsymbol{EOTE}_{twelve}$ **is lesser.**

21. **(a)** 10 flats = 1 **block.** $1 \cdot 10^3 = \mathbf{1000}$.

(b) 20_{twelve} flats = 12 flats + 8 flats = **1 block + 8 flats.** When written in base twelve notation, it looks like
$1 \cdot 12^3 = 8 \cdot 10^2 = \mathbf{1800}_{twelve}$

22. **(a)**

```
125 | 456 | 3
    -375
 25 |  81 | 3
     -75
  5 |   6 | 1
      -5
       1
```

$456 = \mathbf{3311}_{five}$

How many groups of 125 in 456?
How many groups of 25 in 81?

How many groups of 5 in 6?

(b)

$$\begin{array}{r|r|r} 1728 & 1782 & 1 \\ & -1728 & \\ \hline 12 & 54 & 4 \\ & -48 & \\ \hline & 6 & \end{array}$$

$1782 = \mathbf{1046_{twelve}}$

How many groups of $1728 = 12^3$ in 1782?
How many groups of 12 in 54?

(c)

$$\begin{array}{r|r|r} 32 & 32 & 1 \\ & -32 & \\ \hline & 0 & \end{array}$$

$32 = \mathbf{100000_{two}}$

How many groups of $2^5 = 32$ in 32?

23. **(a)** $432_{five} = 4 \cdot 5^2 + 3 \cdot 5^1 + 2 = 100 + 15 + 2 = \mathbf{117}$.

(b) $101101_{two} = 1 \cdot 2^5 + 1 \cdot 2^3 + 1 \cdot 2^2 + 1 = 32 + 8 + 4 + 1 = \mathbf{45}$.

(c) $92E_{twelve} = 9 \cdot 12^2 + 2 \cdot 12 + 11 = 1296 + 24 + 11 = \mathbf{1331}$.

24. Using the place value in base six,

6^3	6^2	6^1	1

we have $a \bullet 6^3 + b \bullet 6^2 + c \bullet 6^1 + d \bullet 1$.

25. To find out the base, add the units digits: $2_b + 6_b$. In base 10 that equals 8 but in base b it equals a numbers with units digit 1. Take out a group of b and have one left over. The only choice is **base seven** and then $2_7 + 6_7 = 11_7$

and
$$\begin{aligned} 12_7 + 26_7 &= (10_7 + 20_7) + (2_7 + 6_7) \\ &= (30_7) + (11_7) \\ &= 41_7 \end{aligned}$$

26. To give the fewest number of prizes, the dollar amount of each must be maximized. Thus

$900 =

1 prize or $625 with $275 left over;
2 prizes of $125 with $25 left over; and
1 prize of $25 with nothing left over.
I.e., $900 = 12100_{five}$.

27. **(a)** $b = \mathbf{6}$. There are 6 groups of 7 in 44.

(b) $b = \mathbf{1}$. Subtracting 5 groups of 144 from 734 leaves 14; there is 1 group of 12 in 14.

28. **(a)** $3 \cdot 5^4 + 3 \cdot 5^2 = 3 \cdot 5^4 + 0 \cdot 5^3 + 3 \cdot 5^2 + 0 \cdot 5^1 + 0 \cdot 5^0 = \mathbf{30300_{five}}$.

(b) $2 \cdot 12^5 + 8 \cdot 12^3 + 12 = 2 \cdot 12^5 + 0 \cdot 12^4 + 8 \cdot 12^3 + 0 \cdot 12^2 + 1 \cdot 12^1 + 0 \cdot 12^0 = \mathbf{208010_{twelve}}$.

29. **(a)**

$$\begin{array}{r|ll} 2 & 37 & \\ 2 & 18 & 1 \\ 2 & 9 & 0 \\ 2 & 4 & 1 \\ 2 & 2 & 0 \\ & 1 & 0 \end{array} \qquad \begin{array}{r|ll} 5 & 37 & \\ 5 & 7 & 2 \\ & 1 & 2 \end{array} \qquad \begin{array}{r|ll} 7 & 37 & \\ & 5 & 2 \end{array}$$

$\mathbf{100101_{two}}$ $\mathbf{122_{five}}$ $\mathbf{25_{seven}}$

(b) $1101_{two} = 1 \bullet 2^3 + 1 \bullet 2^2 + 0 \bullet 2^1 + 1 \bullet 2^0 = 8 + 4 + 0 + 1 = \mathbf{13}$ ***(base ten)***

$$\begin{array}{r|ll} 5 & 13 & \\ & 2 & 3 \end{array} \qquad \begin{array}{r|ll} 7 & 13 & \\ & 1 & 6 \end{array}$$

$\mathbf{23_{five}}$ $\mathbf{16_{seven}}$

(c) $423_{five} = 4 \bullet 5^2 + 2 \bullet 5^1 + 3 \bullet 5^0 = 100 + 10 + 3 = \mathbf{113}$ ***(base ten)***

$$\begin{array}{r|ll} 2 & 113 & \\ 2 & 56 & 1 \\ 2 & 28 & 0 \\ 2 & 14 & 0 \\ 2 & 7 & 0 \\ 2 & 3 & 1 \\ & 1 & 1 \end{array} \qquad \begin{array}{r|ll} 7 & 113 & \\ 7 & 16 & 1 \\ & 2 & 2 \end{array}$$

$\mathbf{1110001_{two}}$ $\mathbf{221_{seven}}$

(d) $26_{seven} = 2 \bullet 7^1 + 6 \bullet 7^0 = 14 + 6 = \mathbf{20}$ ***(base ten)***

2|20
2|10 0
2|5 0
2|2 1
1 0

10100_{two}

5|20
4 0

40_{five}

30. Trade two groups of 4 units for two longs. Now there is one flat, eight longs, and one unit. Now trade two groups of 4 longs for two flats. 301_{four}.

Assessment 3-2A: Addition of Whole Numbers

1. (a) **True**.
   $n(A) = 3; n(B) = 2; n(A) + n(B) = 5$.
   $A \cup B = \{a, b, c, d, e\}; n(A \cup B) = 5$.
   $\mathbf{n(A) + n(B) = n(A \cup B)}$ because the sets are disjoint.

   (b) **False**.
   $n(A) = 3; n(B) = 2; n(A) + n(B) = 5$.
   $A \cup B = \{a, b, c\}; n(A \cup B) = 3$.
   $\mathbf{n(A) + n(B) \neq n(A \cup B)}$ because the sets are not disjoint.

   (c) **True**.
   $n(A) = 3; n(B) = 0; n(A) + n(B) = 3$.
   $A \cup B = \{a, b, c\}; n(A \cup B) = 3$.
   $\mathbf{n(A) + n(B) = n(A \cup B)}$ because the sets are disjoint.

2. The sets are not disjoint because $n(A) + n(B) \neq n(A \cup B)$. $n(A \cap B)$ must therefore equal $n(A) + n(B) - n(A \cup B) = 3 + 5 - 6 = 2$. In the following Venn diagram:

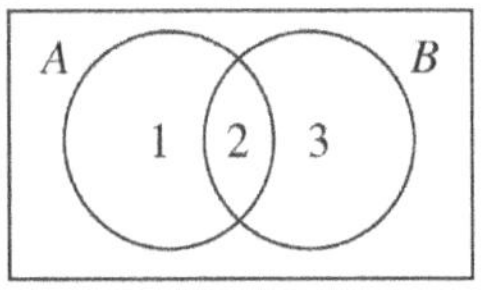

   $n(A) = 3, n(B) = 5, n(A \cup B) = 6$, and $\mathbf{n(A \cap B) = 2}$.

3. If A and B are not disjoint:

   Let $A = \{1, 2\}, B = \{2, 3\}$, so $A \cup B = \{1, 2, 3\}$.
   Then $n(A) = 2, n(B) = 2, n(A \cup B) = 3$.
   But $n(A) + n(B) = 4 \neq 3 = n(A \cup B)$.

4. (a) (*i*) $\mathbf{n(B) = 3}$ if the sets are disjoint.
   (*ii*) $\mathbf{n(B) = 4}$ if $n(A \cap B) = 1$.
   (*iii*) $\mathbf{n(B) = 5}$ if $n(A \cap B) = 2$.
   (*iv*) $\mathbf{n(B) = 6}$ if $n(A \cap B) = 3$.

   (b) If $n(A \cap B) = \emptyset$ the sets are disjoint and $\mathbf{n(B) = 3}$.

5.

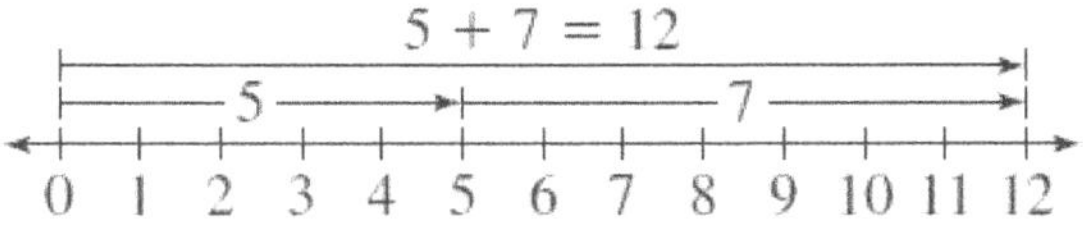

6. The student is adding **13 and 4**. Start with thirteen and add one each time until seventeen then notice that four numbers were added.

7. $7 + 8 = 7 + 7 + 1 = 14 + 1 = 15$

8. (a)

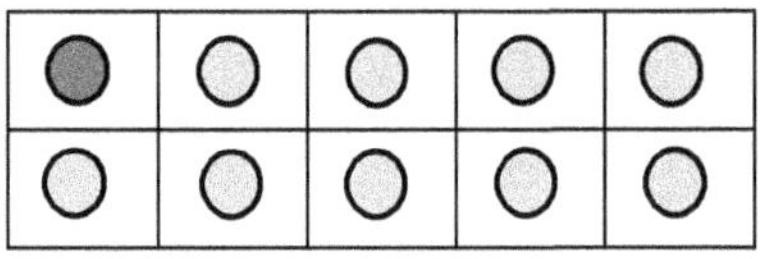

   $1 + 9 = 10$

   (b)

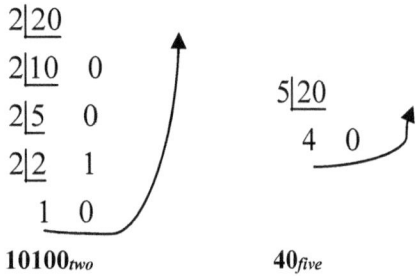

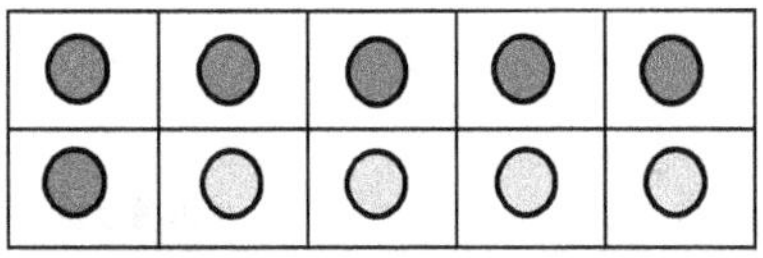

   $6 + 4 = 10$

9. Answers vary. A possible answer is $2 + 2 = \mathbf{4}$, $4 + 2 = \mathbf{6}$, $4 + 4 = \mathbf{8}$, $2 + 3 = \mathbf{5}$, $5 + 2 = \mathbf{7}$, and $7 + 2 = \mathbf{9}$.

10. (a) **Closed**. $0 + 0 = 0$, and $0 \in \{0\}$.

    (b) **Closed**. Assuming the arithmetic sequence $0, 3, 6, \ldots$, any element of T added to any other element in T is whole and divisible by 3.

    (c) **Closed**. N is the set of natural numbers and any natural number added to any other natural number is an element of N.

(d) **Not Closed**. $3 + 7 \notin \{3,5,7\}$.

(e) **Closed**. $\{W\}$ is the set of whole numbers and any whole number greater than 10 added to any other whole number greater than 10 is an element of W.

(f) **Not Closed.** $1 + 1 \notin \{0,1\}$.

11. (a) 1, commutative property of addition.

(b) 7, commutative property of addition.

(c) 0, additive identity.

(d) 7, associative and commutative properties of addition.

12. (a) **Commutative property of addition**; i,e., if a and b are any whole numbers, then $a + b = b + a$.

(b) **Associative property of addition**; i.e., if $a, b,$ and c are any whole numbers, then $(a + b) + c = a + (b + c)$.

(c) **Commutative property of addition**; i.e., $(6 + 3) = (3 + 6)$.

(d) **Identity property of whole numbers**; i.e., for any whole number $a, a + 0 = a$.

(e) **Commutative property of addition.**

(f) **Associative property of addition.**

(g) **Closure property of addition.**

13. (a) By the associative property of addition, $x + (y + z) = (x + y) + z$. By the commutative property of addition, $(x + y) + z = z + (x + y)$.

(b) By the commutative property of addition, $x + (y + z) = (y + z) + x$.

By the associative property of addition, $(y + z) + x = y + (z + x)$.

By the commutative property of addition, $y + (z + x) = y + (x + z)$.

14. (a) $3 + (4 + 7) = (3 + x) + 7$

$(3 + 4) + 7 = (3 + x) + 7$

$3 + 4 = 3 + x$

$\mathbf{4} = x$.

(b) $8 + 0 = x \Rightarrow \boldsymbol{x} = \mathbf{8}$.

(c) $5 + 8 = 8 + x$

$5 + 8 = x + 8$

$\mathbf{5} = x$.

(d) $x + 8 = 12 + 5$

$x + 8 = 17$

$x + 8 - 8 = 17 - 8$

$\boldsymbol{x} = \mathbf{9}$.

(e) $x + 8 = 5 + (x + 3)$

$x + 8 = x + 8$

$\boldsymbol{x} = x$.

The solution is all whole numbers.

(f) $x - 2 = 9$

$\boldsymbol{x} = \mathbf{11}$.

(g) $x - 3 = \boldsymbol{x} + 1$

$-3 = 1$

There are no solutions in set of whole numbers.

(h) $0 + x = x + 0$

$\boldsymbol{x} = \boldsymbol{x}$.

All whole numbers are solutions.

15. Step 1 → Expanded form;

Step 2 → Commutative and associative properties of addition;

Step 3 → Distributive property of multiplication over addition;

Step 4 → Closure property of addition

Step 5 → Expanded form condensed.

16. Base ten blocks:

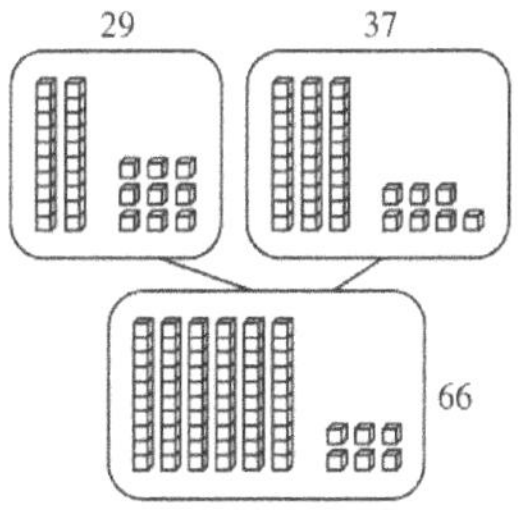

17. (a) (*i*)

$$\begin{array}{r} 6\ 8\ 7 \\ +\ 5\ 4\ 9 \\ \hline 1\ 6 \\ 1\ 2\ \ \\ 1\ 1\ \ \ \ \\ \hline 1\ 2\ 3\ 6 \end{array}$$

(ii)
```
      3  5  9
   +  6  7  3
   _____
         1  2
      1  2
      9
   _____
   1  0  3  2
```

(b) The algorithm works because the placement of partial sums under their addends accounts for place value.

18. **(a)**
```
   4  3  5  8
 + 3  8  6  4
 _____
```
$$\frac{0}{7} \quad \frac{1}{1} \quad \frac{1}{1} \quad \frac{1}{2}$$
```
   8  2  2  2
```

(b)
```
   4  9  2  3
 + 9  8  9  7
 _____
```
$$\frac{1}{3} \quad \frac{1}{7} \quad \frac{1}{1} \quad \frac{1}{0}$$
```
 1 4  8  2  0
```

19.

```
   2 │ 5 │ 6
 + 1 │ 5 │ 4
 ─────────────
   3 │ 10│ 10
   3 │ 11│ 0
   4 │ 1 │ 0
```

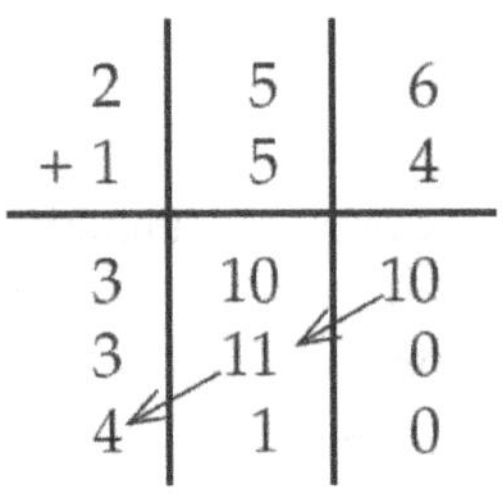

20.

$$1102 \xrightarrow{-2} 1100 \xrightarrow{-100} 1000$$

$$1809 \xrightarrow{+2} 1811 \xrightarrow{+100} \begin{array}{r} 1911 \\ 2911 \end{array}$$

21. Remember that in base five the number 11, for example, means $1 \cdot 5 + 1$, corresponding to 6 in base ten.

(a)
```
       1
     4  3  five
   + 2  3  five
   _____
   1  2  1  five
```

Alternative approach: An example of a different algorithm for checking our answers would be to write in expanded form.

$$43_{five} + 23_{five} = (4 \cdot 5 + 3 \cdot 5^0)$$
$$+ (2 \cdot 5^1 + 3 \cdot 5^0)$$
$$= 6 \cdot 5^1 + 6 \cdot 5^0$$
$$= (5 + 1)5^1 + (5 + 1)5^0$$
$$= 5^2 + 5 + 5 + 1$$
$$= 1 \cdot 5^2 + 2 \cdot 5 + 1 \cdot 5^0$$
$$= (121)_{five}$$

(b)
```
   1  1
   4  3  2   five
 +       2  3  five
 _____
 1  0  1  0   five
```

(c)
```
      1
   1  1  0   two
 +       1  1  two
 _____
 1  0  0  1   two
```

22. There is **no numeral 5** in base five;
$$2_{five} + 3_{five} =$$
$$10_{five}; 22_{five} + 33_{five} = 110_{five}.$$

23. **(a)** $567 + 38$:
   (i) $567 + 30 = 597$ (add 567 to the tens value of 38).
   (ii) $597 + 8 = \mathbf{605}$ (add 597 to the units value of 38).

   (b) $418 + 215$:
   (i) $418 + 200 = 618$ (add 418 to the hundreds value of 215).
   (ii) $618 + 10 = 628$ (add 618 to the tens value of 215).
   (iii) $628 + 5 = 633$ (add 628 to the units value of 215).

24. **(a)** $87 + 33 \Rightarrow (87 + 3) + (33 - 3) = 90 + 30 = \mathbf{120}$.
   (b) $58 + 39 \Rightarrow (58 + 2) + (39 - 2) = 60 + 37 = \mathbf{97}$.

25. **(a)** 5280:   The number is between 5200 and 5300;

   The midpoint is 5250;

   The number is greater than the midpoint;

   So it rounds up to **5300**.

(b) $\underline{1}15,234$: The number is between 100,000 and 200,000;

The midpoint is 150,000;

The number is less than the midpoint;

So it rounds down to **100,000**.

(c) $1\underline{1}5,234$: The number is between 110,000 and 120,000;

The midpoint is 115,000;

The number is greater than the midpoint;

So it rounds up to **120,000**.

(d) $23\underline{2}5$: The number is between 2320 and 2330;

The midpoint is 2325;

When the number is at the midpoint it is conventional to round up (to **2330**).

26. Answers vary. For example:

(a) $878 + 2340 \approx 900 + 2300 = 3200$

(b) $2215 + 3023 + 5967 + 975 \approx 2000 + 3000 + 6000 + 1000 =$ **12,000**.

27. (a) $2215 + 3023 + 5987 + 975$:

(*i*) $2 + 3 + 5 + 0 = 10$ (add front-end digits);

(*ii*) 10,000 (place value);

(*iii*) $215 + 23 + 987 + 975 \approx 200 + 0 + 1000 + 1000 = 2200$ (adjust);

(*iv*) $10,000 + 2200 =$ **12,200** (adjusted estimate).

(b) $234 + 478 + 987 + 319 + 469$:

(*i*) $2 + 4 + 9 + 3 + 4 = 22$ (add front-end digits)

(*ii*) 2200 (place value);

(*iii*) $34 + 78 + 87 + 19 + 69 \approx 30 + 80 + 90 + 20 + 70 = 290$ (adjust);

(*iv*) $2200 + 290 =$ **2490** (adjusted estimate).

28. (a) (*i*) **No**. The numbers are not clustered.

(*ii*) **Yes**. The numbers are clustered around 500.

(b) Estimate may vary:

(*i*) Front-end: $0 + 1 + 0 + 2 = 3$; place value 3000; adjust by $500 + 500 + 100 + 400 = 1500$; estimate $3000 + 1500 =$ **4500**.

Grouping to nice numbers: $474 + 1467 \approx 2000$; $64 + 2445 \approx 2500$; sum $\approx 2000 + 2500 =$ **4500**.

Rounding: Sum $\approx 500 + 1500 + 100 + 2400 =$ **4500**.

(*ii*) Front-end: $4 + 4 + 5 + 5 + 5 = 23$; place value 2300; adjust by $80 + 80 + 30 + 0 + 30 = 220$; estimate $2300 + 220 =$ **2520**.

Grouping to nice numbers: $483 + 475 \approx 1000$; $530 + 528 \approx 1000$; $503 \approx 500$; sum $\approx 1000 + 1000 + 500 =$ **2500**.

Rounding: sum $\approx 500 + 500 + 500 + 500 + 500 =$ **2500**.

29. (a) The range is $100 + 600 = 700$ to $200 + 700 = 900$. Then $700 < (145 + 678) < 900$.

(b) The range is $200 + 0 = 200$ to $300 + 100 = 400$. Then $200 < (278 + 36) < 400$.

30. Answers vary. For example: $3300 - 100 - 300 - 400 - 500 =$ **2000**. The estimate is **high** because the amounts were rounded to the hundreds place. \$8 was taken away from the check amounts while \$13 was added to the \$3287.

33. (a) False. Sweden would be estimated to be about 44,000 square miles larger than Finland.

(b) False. Twice Norway's size would be about 250,000 square miles.

(c) False. France would be estimated to be about 85,000 square miles larger than Norway.

(d) True. About 195,000 square miles to about 174,000 square miles.

32. In a magic square each row, column, and diagonal must have the same sum. $8+5+2=15$ so each row, column, and diagonal must add to 15.

8	**3**	**4**
1	**5**	**9**
6	**7**	**2**

or

8	**1**	**6**
3	**5**	**7**
4	**9**	**2**

33. **(a)** **9**. A number greater than 9 would have two digits.

(b) **8**. If A were greater, C would have two digits.

(c) **3**. A and B must be 1 and 2 (in any order); no lesser single digit numbers are available.

(d) **6 or 8**. A and B must be 2 and 4 or 2 and 6; greater even numbers would make C have two digits.

(e) **5**. If $A+B=C$ and C is 5 more than A, then $A+5=C$.

(f) **4 or 8**. B could be 1 and A must then be 3; or B could be 2 and then A must be 6.

(g) **9**. B must be 2 and A must be 7.

34. Since the way the domino is positioned doesn't matter, i.e., 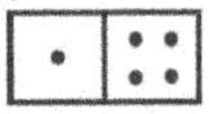 is the same domino as 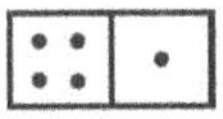, each number put on the left side gets paired with each of the 9 choices for the right

Number printed on left	Number of choices for right
0	9
1	8
2	7
3	6
4	5
5	4
6	3
7	2
8	1

Sum the right column and we have $\frac{10(9)}{2}=45$ dominos.

35. **(a)** In the units column, $1+\underline{1}=2$.

In the tens column, $\underline{8}+2=10$ (regroup).

In the hundreds column, $1+\underline{9}+4=14$.

$$\begin{array}{rrrr} & \underline{9} & \underline{8} & 1 \\ + & 4 & 2 & \underline{1} \\ \hline \underline{1} & 4 & 0 & 2 \end{array}$$

(b) In the units column, $5+6+8=19$ (regroup).

In the tens column, $1+2+\underline{9}+4=16$ (regroup).

In the hundreds column, $1+0+1+1=\underline{3}$.

In the thousands column, $\underline{2}+1+3=6$.

$$\begin{array}{rrrrr} & \underline{2} & 0 & 2 & 5 \\ & 1 & 1 & \underline{9} & 6 \\ + & 3 & 1 & 4 & 8 \\ \hline & 6 & \underline{3} & 6 & \underline{9} \end{array}$$

36. **(a)** Answers vary. For the unique greatest sum, the larger numbers must be in the hundreds column:

$$\begin{array}{rrrr} & \boxed{7} & \boxed{6} & \boxed{2} \\ + & \boxed{8} & \boxed{5} & \boxed{3} \\ \hline 1 & 6 & 1 & 5 \end{array}$$

(b) Answers vary. For the unique least sum the smaller numbers must be in the hundreds column:

$$\begin{array}{rrrr} & \boxed{2} & \boxed{6} & \boxed{7} \\ + & \boxed{3} & \boxed{5} & \boxed{8} \\ \hline & 6 & 2 & 5 \end{array}$$

37. The ones digit must be a sum of 9, so the options are 3-6, 4-5, or 8-1. There is no way that a regrouping will take place, so the tens column must sum to 5 or 15. But 15 is not an option with the given digits, so it must sum to 5; the only options are 4-1 and 2-3. Since 15 was not an option, there is no regrouping into the hundreds column, meaning the sum must be 10. The only option for the hundreds column is 4-6. Placing these values in the correct boxes eliminates the options for the ones and tens columns, so the ones column must be 8-1 and the tens column must be 2-3. This leaves only 3-5 for the thousands column.

$$\begin{array}{rrrrr} & \boxed{3} & \boxed{4} & \boxed{2} & \boxed{8} \\ + & \boxed{5} & \boxed{6} & \boxed{3} & \boxed{1} \\ \hline & 9 & 0 & 5 & 9 \end{array}$$

38. **(a)** The sum of each row, column, and diagonal is **34**. For example, the sum of the entries in row 1 is $1 + 15 + 14 + 4 = \mathbf{34}$.

(b) $6 + 7 + 10 + 11 = \mathbf{34}$.

(c) $1 + 4 + 16 + 13 = \mathbf{34}$.

(d) **Yes**. Adding five to each number in the square will increase the sum of any four numbers in the square by 20.

(e) **Yes**. Subtracting 1 from each number in the square will decrease the sum of any four numbers in the square by 4.

39. **(a)** If 1 is placed in the middle then we can pair the numbers 2 through 9 to sum to eleven so that the sum in each of the four directions is twelve.

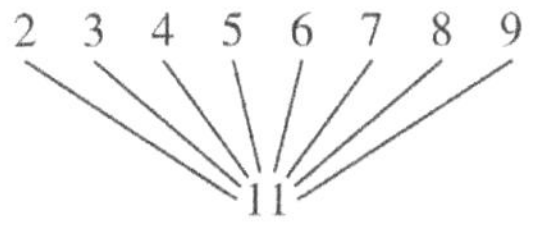

Similar arrangements can be made for 5 and 9 in the middle.

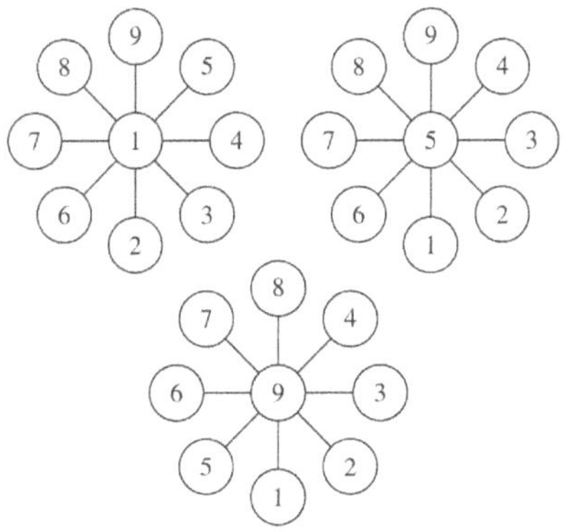

(b) The three **{1,5,9}** are the only numbers that can be placed in the middle. To observe that 2, for example, can't be placed in the middle, pair 9 with each of the remaining numbers and observe that there is no way to pair the other numbers to form a common sum.

40. **(a)** Using the scratch algorithm, whenever a sum is 10 or more, scratch a line through the last digit added and write the number of units next to the scratched digit; count the number of scratches in each column and add to the column to the left.

		1	1	
		4	3	2
		$\not{9}_4$	$\not{7}_1$	6
+	^{1}1	4	1	$\not{8}_6$
	2	8	2	6

(b) A base five addition table might be helpful:

+	0	1	2	3	4
0	0	1	2	3	4
1	1	2	3	4	10
2	2	3	4	10	11
3	3	4	10	11	12
4	4	10	11	12	13

	3	3		
		$\not{3}_1$	2	*five*
		1	$\not{3}_0$	*five*
		2	2	*five*
		$\not{4}_3$	$\not{3}_0$	*five*
		$\not{2}_0$	3	*five*
+		1	$\not{2}_0$	*five*
	3	1	0	*five*

Mathematical Connections 3-2: Review Problems

23. **1410.** $M = 1000, CD = 500 - 100$, and $X = 10$.

25. $12^4 + 12^2 + 13 = 1 \cdot 12^4 + 0 \cdot 12^3 + 1 \cdot 12^2 + 1 \cdot 12 + 1 = \mathbf{10111_{twelve}}$.

Assessment 3-3A: Subtraction of Whole Numbers

1. **(a)** Take away:

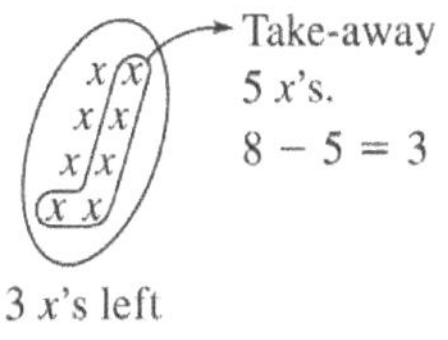

(b) Missing addend:

$\square + 5 = 8 \Rightarrow \square = 3.$

(c) Comparison:

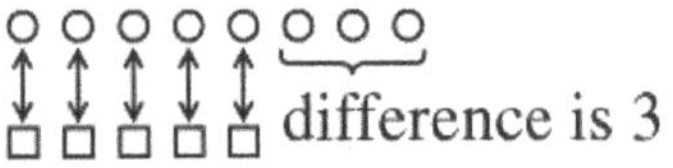

(d) Number line:

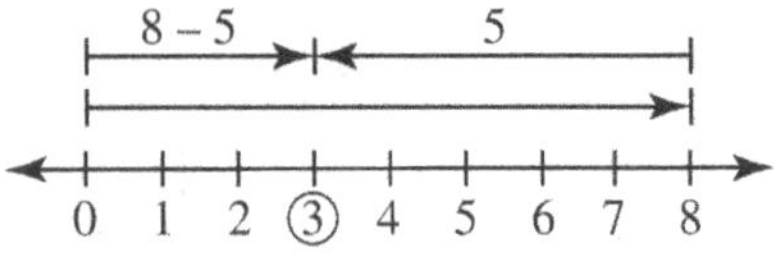

2. (a) **Answers vary.** For example: "Kaden has 12 cookies and Tristan takes away 7 cookies. How many cookies does Kaden have left?"

(b) **Answers vary.** For example: "Jasmine has 7 yards of fabric and makes a blanket for her sister Ebony with 3 yards of **fabric. How much fabric is left over?"**

3. (a) The other three members of the fact family are:

$5+6=11$

$11-6=5$

$11-5=6$

(b)

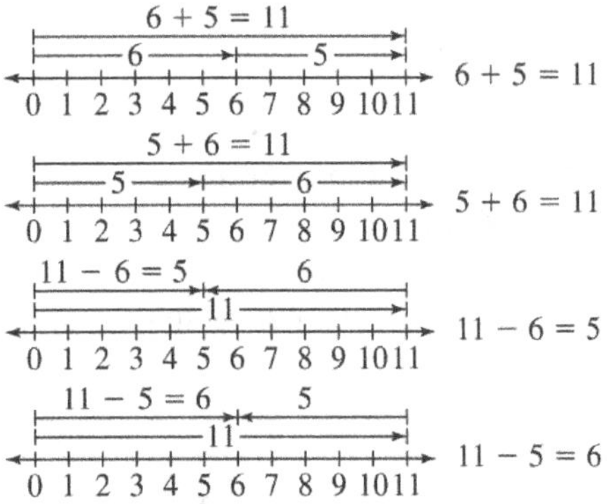

4. $a \geq b$; if $b > a$ then $a - b$ would not be a whole number.

5. (a) Add 7 to each side: $\mathbf{9 = 7 + x}$.

(b) Add 6 to each side: $\mathbf{x = 3 + 6}$

(c) Add x to each side: $\mathbf{9 = x + 2}$.

6. (a) Solve for x:

i. $15-9=15-10+x$

$6=5+x$

$1=x$

ii. $27-9=27-10+x$

$18=17+x$

$1=x$

iii. $32-9=32-10+x$

$23=22+x$

$1=x$

(b) A strategy for subtracting 9 from a whole number would be subtracting 10 then adding 1.

7. Kelsey has $\boldsymbol{a - (b + c)}$ more marbles than Gena and Noah combined.

8. Answers vary. Some possibilities are:

(a) $8 + 5 = 13$ and 13 was written down with no regrouping. $2 + 7 = 9$ and 9 was simply placed in front of the 13.

(b) $8 + 5 = 13$, but instead of writing 3 and regrouping with the 1, the 1 was written and the 3 was regrouped.

(c) Only the difference in the units $(9 - 5 = 4)$, tens $(5 - 0 = 5)$, and the hundreds $(3 - 2 = 1)$ was recorded, without taking into account the signs of the numbers.

(d) Three hundreds was regrouped as 2 hundreds and 10 tens, but 10 tens was not regrouped as $9 \cdot 10 + 15$ in order to obtain $15 - 9 = 6$ in the ones place.

9. (a) Order the players as follows:

$\boxed{\text{Kent}} < \boxed{\text{Mischa}} < \boxed{\text{Sally}} < \boxed{\text{Vera}}$

Kent is shortest; Vera is tallest.

(b) Answers vary. As long as the player's heights increase in the order:

Kent $\rightarrow$ Mischa $\rightarrow$ Sally $\rightarrow$ Vera.

10. By dinner time Tom had consumed $90 + 120 + 119 + 185 + 110 + 570 = 1194$ calories. Subtracting 1194 from 1500 gives $1500 - 1194 = 306$. Tom may have fish or salad, but **not both**. (He may have tea with either.)

11. (a)

$$\begin{array}{r} 93 \\ -\ 37 \\ \hline \end{array} \Rightarrow \begin{array}{r} 93+3 \\ -\ 37+3 \\ \hline \end{array} \Rightarrow \begin{array}{r} 96 \\ -\ 40 \\ \hline 56 \end{array}$$

(b)

$$\begin{array}{r} 321 \\ -\ 38 \\ \hline \end{array} \Rightarrow \begin{array}{r} 321+2 \\ -\ 38+2 \\ \hline \end{array} \Rightarrow \begin{array}{r} 323 \\ -\ 40 \\ \hline \end{array}$$

$$\Rightarrow \begin{array}{r} 323+60 \\ -\ 40+60 \\ \hline \end{array} \Rightarrow \begin{array}{r} 383 \\ -\ 100 \\ \hline 283 \end{array}$$

12. (a) $28+2=30$; $30+20=50$; $50+3=53$. Then $2+20+3=\mathbf{25}$.

(b) $47 + 3 = 50;\ 50 + 10 = 60;\ 60 + 3 = 63$. Then $3 + 10 + 3 = \mathbf{16}$.

13.

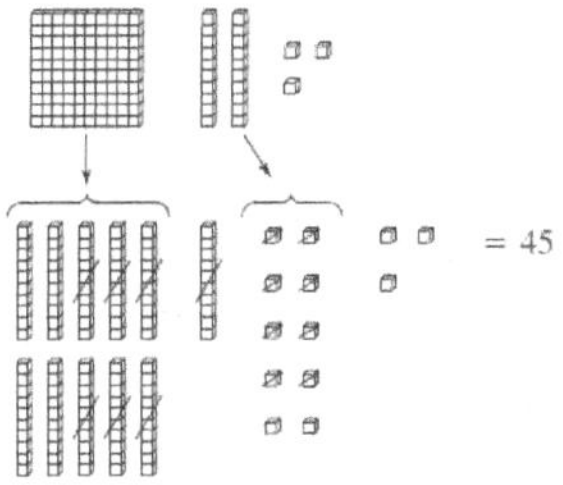

14.

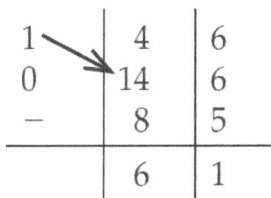

1 → 0	4 → 14	6
−	8	5
	6	1

15.

$$\begin{array}{r} \boxed{8}\boxed{7}\boxed{6} \\ -\ \boxed{2}\boxed{3}\boxed{5} \\ \hline 6\ 4\ 1 \end{array}$$

16. For an example of how to use the table, move down the rows in the + column to 3 and then across that row to the column headed by 6. This will give the sum of 3 + 6, or 11_{eight}.

+	0	1	2	3	4	5	6	7
0	0	1	2	3	4	5	6	7
1	1	2	3	4	5	6	7	10
2	2	3	4	5	6	7	10	11
3	3	4	5	6	7	10	11	12
4	4	5	6	7	10	11	12	13
5	5	6	7	10	11	12	13	14
6	6	7	10	11	12	13	14	15
7	7	10	11	12	13	14	15	16

(a)

$$\begin{array}{rrrl} 4 & 16 & 13 & \\ \not{5} & \not{7} & \not{3} & _{eight} \\ - & 7 & 7 & _{eight} \\ \hline 4 & 7 & 4 & _{eight}. \end{array}$$

We used the table to complete the subtraction. For example $13_{eight} - 7_{eight}$ can be found by observing that $4_{eight} + 7_{eight} = 13_{eight}$ in the table.

(b)

$$\begin{array}{rrrl} 6 & 15 & 15 & \\ \not{7} & \not{6} & \not{5} & _{eight} \\ - & 7 & 6 & _{eight} \\ \hline 6 & 6 & 7 & _{eight} \end{array}$$

17. (a) Subtract 2132_{five} from 3423_{five}.

$$\begin{array}{rrrrl} 3 & 4 & 2 & 3 & _{five} \\ -\ 2 & 1 & 3 & 2 & _{five} \\ \hline \mathbf{1} & \mathbf{2} & \mathbf{4} & \mathbf{1} & _{\mathbf{five}} \end{array}$$

(b) Subtract 11011_{two} from 100000_{two}.

$$\begin{array}{rrrrrrl} 1 & 0 & 0 & 0 & 0 & 0 & _{two} \\ - & 1 & 1 & 0 & 1 & 1 & _{two} \\ \hline & & & \mathbf{1} & \mathbf{0} & \mathbf{1} & _{\mathbf{two}} \end{array}$$

(c) Subtract 1 from TEE_{twelve}:

$$\begin{array}{rrrl} T & E & E & _{twelve} \\ - & & 1 & _{twelve} \\ \hline \boldsymbol{T} & \boldsymbol{E} & \boldsymbol{T} & _{\mathbf{twelve}} \end{array}$$

(d) Subtract 1000_{five} from 10000_{five}.

$$\begin{array}{rrrrrl} 1 & 0 & 0 & 0 & 0 & _{five} \\ - & 1 & 0 & 0 & 0 & _{five} \\ \hline & \mathbf{4} & \mathbf{0} & \mathbf{0} & \mathbf{0} & _{\mathbf{five}} \end{array}$$

18.

$$\begin{array}{rlrlrl} 4 & & 95 \\ & & 35 & & 98 & \\ \not{5} & \text{hours} & \not{3}\not{6} & \text{minutes} & \not{3}\not{8} & \text{seconds} \\ -\ 3 & \text{hours} & 56 & \text{minutes} & 58 & \text{seconds} \\ \hline 1 & \text{hour} & 39 & \text{minutes} & 40 & \text{seconds} \end{array}$$

19. The information in (*a*) and (*b*) complete the first column.

Teams	1
Hawks	14
Elks	18 = 14 + 4

The information in (*c*) and (*d*) complete the second column.

Teams	1	2
Hawks	14	22 = 23 − 1
Elks	18	23 = 18 + 5

The information in (g) and (h) complete the third column.

Teams	1	2	3
Hawks	14	22	$36 = 2(18)$
Elks	18	23	$36 = 14 + 22$

The information in (f) tells us that $14 + 22 + 36 + (4^{th}$ quarter score$) = 120$.

So the 4th quarter score for the Hawks is 48. The information in (e) tells us that the Elks scored $48 + 6$ points in the 4th quarter.

	Quarter				
Teams	1	2	3	4	Final
Hawks	14	22	36	48	120
Elks	18	23	36	54	131

20. Answers vary. Possibilities include:

(a) $180 + 97 - 23 + 20 - 140 + 26$

$= 180 + 97 + 20 + 26 - 23 - 140$

$= (180 + 90 + 20 + 20 + 7 + 6)$

$- 23 - 140$

$= (310 + 13) - 23 - 140$

$= 323 - 23 - 140 = 300 - 140 =$ **160.**

(b) $87 - 42 + 70 - 38 + 43$

$= 87 + 70 + 43 - 42 - 38$

$= (80 + 70 + 40 + 7 + 3)$

$- 40 - 30 - 2 - 8$

$= (190 + 10) - 70 - 10$

$= 200 - 80 =$ **120.**

21. (a) Answers vary. Possibilities include:

Compatible numbers

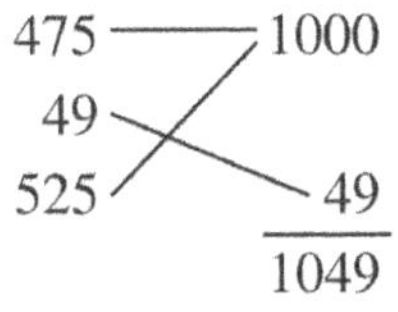

(b) Answers vary; possibilities include:

Breaking up and bridging.

$375 - 76 = 375 - 75 - 1$

$= 300 - 1 = 299.$

(c) Answers vary; possibilities include:

Compatible numbers

$230 + 60 + 70 + 44 + 6 =$

$(230 + 70) + (60 + 40) + (4 + 6) =$

$300 + 100 + 10 =$

410

22. $997 - 32$:

(*i*) $997 - 30 = 967$ (subtract the tens value of 32 from 997).

(*ii*) $967 - 2 =$ **965** (subtract the units value of 32 from 967).

Mathematical Connections 3-3: Review Problems

31. (a) 12113_{four}

$= 1 \cdot 4^4 + 2 \cdot 4^3 + 1 \cdot 4^2 + 1 \cdot 4^1 + 3 \cdot 4^0$

$= 256 + 128 + 16 + 4 + 3$

$= 407$

(b) 52031_{six}

$= 5 \cdot 6^4 + 2 \cdot 6^3 + 0 \cdot 6^2 + 3 \cdot 6^1 + 1 \cdot 6^0$

$= 6480 + 432 + 0 + 18 + 1$

$= 9931$

(c) 10000_{five}

$= 1 \cdot 5^4 + 0 \cdot 5^3 + 0 \cdot 5^2 + 0 \cdot 5^1 + 0 \cdot 5^0$

$= 625 + 0 + 0 + 0 + 0$

$= 625$

(d) 56_{seven}

$= 5 \cdot 7^1 + 6 \cdot 7^0$

$= 35 + 6$

$= 41$

33. For example: $(2 + 3) + 4 = 2 + (3 + 4)$.

Assessment 3-4A: Multiplication of Whole Numbers

1. (a)

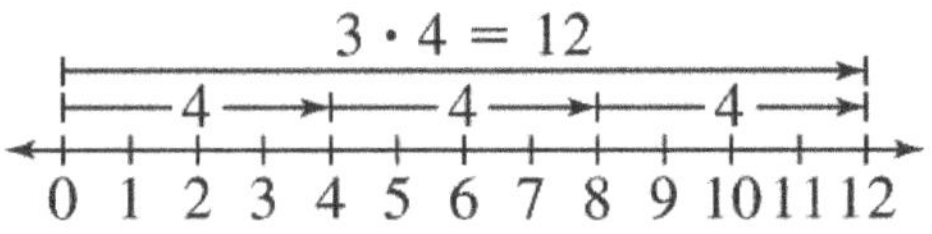

(b)

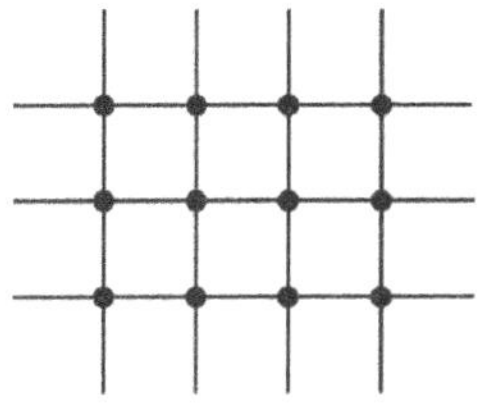

(c)

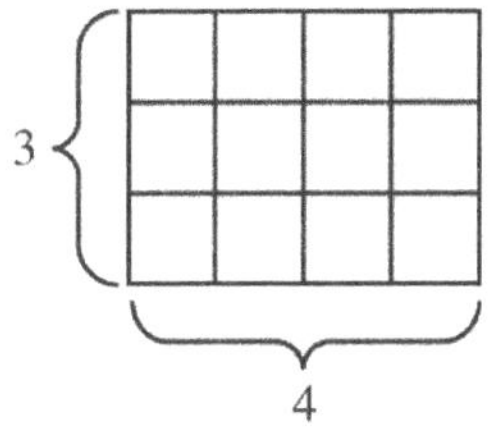

(d) $A = \{a, b, c\}$

$B = \{w, x, y, z\}$

$A \times B = \{(a,w),(a,x),(a,y),(a,z),$
$(b,w),(b,x),(b,y),(b,z),$
$(c,w),(c,x),(c,y),(c,z)\}$

$n(A) = 3, n(B) = 4$

$n(A \times B) = 3 \cdot 4 = 12.$

2. **(a)** $A = \{a, b\}$

$B = \{x, y, z\}$

$A \times B = \{(a,x),(a,y),(a,z),$
$(b,x),(b,y),(b,z)\}$

(b) $n(A) = 2, n(B) = 3$

$n(A \times B) = 2 \cdot 3 = 6$

(c) $2 \cdot 3 = 6$ **.**

3. **(a)** Illustrated is 4 groups of 2 *x*s: $\mathbf{4 \cdot 2} = 8$.

(b) Illustrated is a 4 by 2 array or a 2 by 4 array: $\mathbf{2 \cdot 4 = 8}$ **or** $\mathbf{4 \cdot 2 = 8}$**.**

4. **(a)** Use the repeated addition model: $3 + 3 + 3 + 3 + 3 = 15$; there are five threes, so $3 \cdot \boxed{5} = 15$.

(b) $18 - 6 = 6 - 6 + 3 \cdot \square$ (note the order of operations specifying that $6 + 3$ is not permitted) $\Rightarrow 12 = 3 \cdot \square$. Now use the repeated addition model: $3 + 3 + 3 + 3 = 12$; there are four threes, so $18 = 6 + 3 \cdot \boxed{4}$.

(c) The distributive property of multiplication over addition, where n is **any whole number,** specifies that $a(b + c) = ab + ac$. Thus $\boxed{n} \cdot (5 + 6) = \boxed{n} \cdot \mathbf{5} + \boxed{n} \cdot \mathbf{6}$.

5. Use the Cartesian-product model, where s is shirts, p is pants, and v is vests. Then $n(s) = 6$, $n(p) = 4$, and $n(v) = 3$. The Fundamental Counting Principle indicates that the number of ordered triplets in $s \times p \times v = n(s) \cdot n(p) \cdot n(v) = 6 \cdot 4 \cdot 3 =$ **72 possible outfits.**

6. Answers may vary; one possibility is:

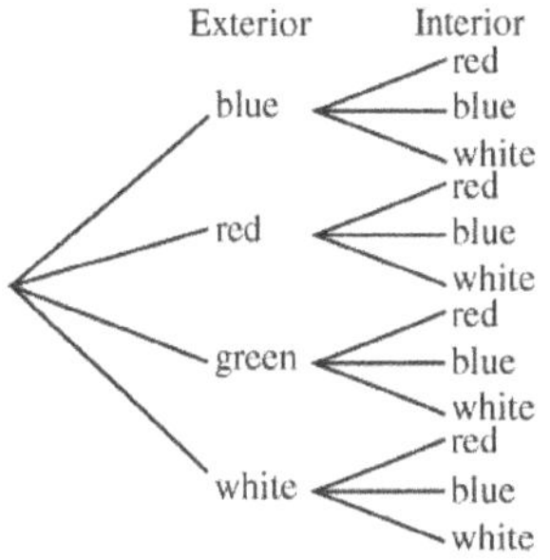

Resulting in $4 \cdot 3 =$ **12 color schemes.**

7. **(a)** **Closed.** $0 \cdot 0 = 0, 0 \cdot 1 = 0, 1 \cdot 0 = 0$, and $1 \cdot 1 = 1$ are all products contained in $\{0, 1\}$.

(b) **Closed.** The product of any two even numbers is also an even number.

(c) **Closed.** Since the set can be written as $\{3n + 1 \mid n \in W\}$, for any whole numbers m and $n, (3m + 1)(3n + 1) = 3(3mn + m + n) + 1$ is in the set since the whole numbers are closed under multiplication and addition.

8. **(a)** **No.** $2 + 3 = 5$.

(b) **Yes.** There will be no numbers in the set that will multiply to give a product of 5.

9. **(a)** Applying the distributive property gives $(a + b)(c + d) = a(c + d) + b(c + d) = \mathbf{ac + ad + bc + bd}$.

(b) $\square(\Delta + O) = \square \cdot \Delta + \square \cdot O$.

(c) $a(b + c) - ac = \mathbf{ab + ac - ac} = ab$.

10. **(a)** $xy + y^2 = \mathbf{y(x + y)}$, factoring y from each term.

(b) $xy + x = \mathbf{x(y + 1)}$, factoring x from each term.

(c) $a^2b + ab^2 = \mathbf{ab(a + b)}$, factoring ab from each term.

11. (a) Associative property of multiplication of whole numbers; i.e., $(a \cdot b) \cdot c = a \cdot (b \cdot c)$.

(b) Commutative property of multiplication of whole numbers; i.e., $a \cdot b = b \cdot a$.

(c) Commutative property of multiplication.

(d) Identity property of multiplication of whole numbers; i.e., $1 \cdot a = a$.

(e) Zero multiplication property of whole numbers; i.e., $a \cdot 0 = 0$.

(f) Distributive property of multiplication over addition for whole numbers; i.e., $(a + b)(c + d) = (a + b) \cdot c + (a + b) \cdot d$.

12. (a) Distributive property of multiplication over addition for whole numbers; i.e., $a(b + c) = ab + ac$.

(b) (*i*) $12 \cdot 32 = 12(30 + 2) = 12 \cdot 30 + 12 \cdot 2 = 360 + 24 = \mathbf{384}$, or

(*ii*) $32 \cdot 12 = 32(10 + 2) = 32 \cdot 10 + 32 \cdot 2 = 320 + 64 = \mathbf{384}$.

13. (a) $9(10 - 2) = 9 \cdot 10 - 9 \cdot 2 = 90 - 18 = \mathbf{72}$.

(b) $20(8 - 3) = 20 \cdot 8 - 20 \cdot 3 = 160 - 60 = \mathbf{100}$.

14. (a)

$(a + b)^2 = (a + b)(a + b) = a(a + b) + b(a + b) = a^2 + ab + ba + b^2 = a^2 + 2ab + b^2$.

(b)

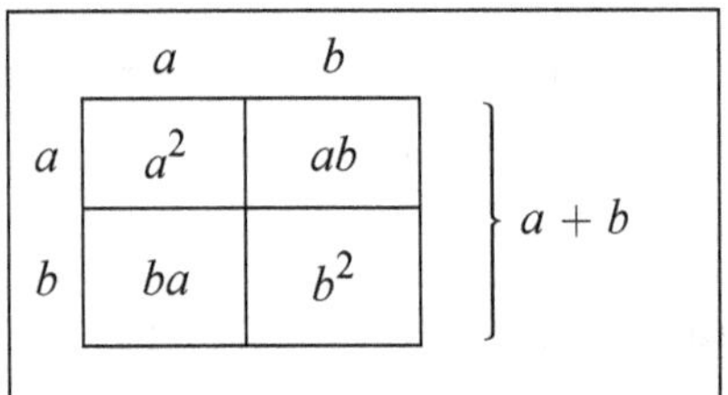

The area of a square with sides $a + b$ can be expressed as $(a + b) \cdot (a + b)$, and also as the sum of areas of the four regions: two squares, $a \cdot a$ and $b \cdot b$, and two rectangles $a \cdot b$ and $b \cdot a$. Thus:

$$(a + b)^2 = a \cdot a + a \cdot b + b \cdot a + b \cdot b = a^2 + 2ab + b^2.$$

15. The question is really to show that the area of the large square minus the area of the small square is the same as the area of the four rectangles.

The area of the large square is $(a + b)^2$. The area of the small square is $(a - b)^2$. The difference between the two areas is the four rectangles, each with area ab; the total area of the four is $4ab$.

Therefore $(a + b)^2 - (a - b)^2 = 4ab$.

16. (a) $51^2 = (50 + 1)(50 + 1)$
$= 50^2 + 2 \cdot 50 \cdot 1 + 1^2$
$= 2500 + 100 + 1 = 2601$

(b) $102^2 = (100 + 2)(100 + 2)$
$= 100^2 + 2 \cdot 100 \cdot 2 + +2^2$
$= 10{,}000 + 400 + 4 = 10{,}404$

17. To factor is to reverse the process of the distributive property of multiplication over addition or subtraction. The result is to find the factors which, when multiplied together, will yield a given product.

(a) $xy - y^2 = y(x - y)$; i.e., if y and $x - y$ were to be multiplied, using the distributive property of multiplication over subtraction, the product would be $xy - y^2$.

(b) $47 \cdot 101 - 47 = \mathbf{47(101 - 1)}$. i.e., 47 is a factor of both $47 \cdot 101$ and $47 \cdot 1$.

(c) $ab^2 - ba^2 = \mathbf{ab(b - a)}$.

18. (a) There are $7 + 12 = 19$ factors of 5. Thus $5^7 \cdot 5^{12} = 5^{7+12} = \mathbf{5^{19}}$.

(b) $6^{10} \cdot 6^2 \cdot 6^3 = 6^{10+2+3} = \mathbf{6^{15}}$.

(c) $10^{296} \cdot 10^{17} = 10^{296+17} = \mathbf{10^{313}}$.

(d) $2^7 \cdot 10^5 \cdot 5^7 = 2^7 \cdot 5^7 \cdot 10^5$
$= (2 \cdot 5)^7 \cdot 10^5$
$= 10^7 \cdot 10^5$
$= 10^{7+5} = \mathbf{10^{12}}$.

19. (a) $\mathbf{2^{100}}$ is greater.
$2^{80} + 2^{80} = 2^{80}(1 + 1) = 2^{80} \cdot 2 = 2^{81} < 2^{100}$.

(b) $\mathbf{2^{102}}$ is greatest.

$2^{102} = 2^2 \cdot 2^{100} > 3 \cdot 2^{100} >$

$2^{101} = 2 \cdot 2^{100}.$

20. We can use the repeated addition model of multiplication and make three groups of 42:

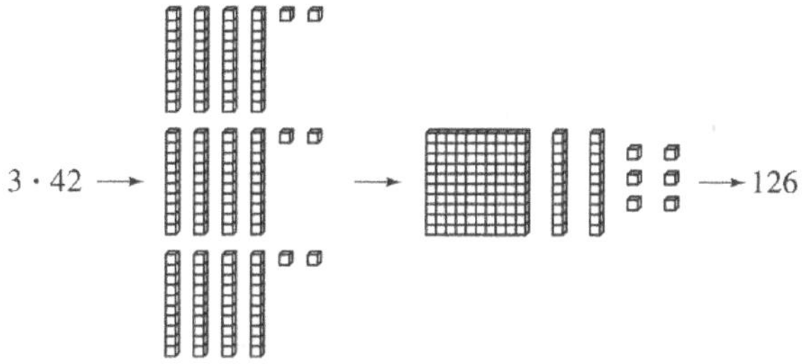

21. (a) The following partial products are obtained through the distributive property of multiplication over addition. The parenthetical computations below illustrate the corresponding area notions from the model:

$$\begin{array}{rrrl} & 2 & 2 & \\ \times & 1 & 3 & \\ \hline & & 6 & (3 \times 2) \\ & 6 & 0 & (3 \times 20) \\ & 2 & 0 & (10 \times 2) \\ 2 & 0 & 0 & (10 \times 20) \\ \hline 2 & 8 & 6 & \end{array}$$

(b) 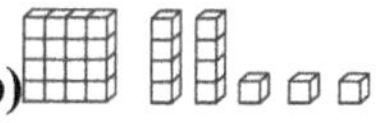

$$\begin{array}{rrrl} & 1 & 5 & \\ \times & 2 & 1 & \\ \hline & & 5 & (1 \times 5) \\ & 1 & 0 & (1 \times 10) \\ 1 & 0 & 0 & (20 \times 5) \\ 2 & 0 & 0 & (20 \times 10) \\ \hline 3 & 1 & 5 & \end{array}$$

Which is illustrated by:

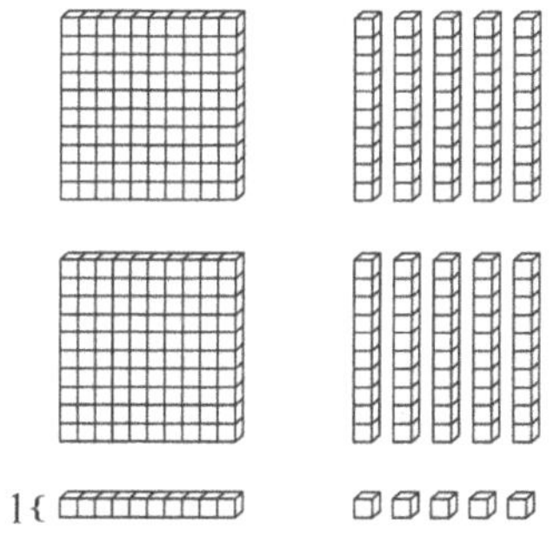

(c)

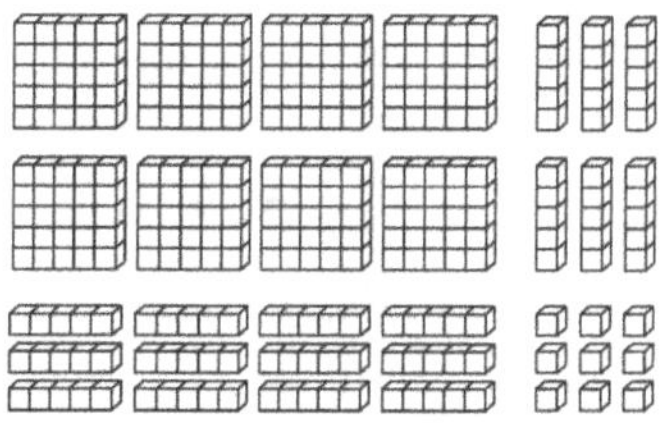

To find $43_{five} \cdot 23_{five}$ count the number of flats, longs, and ones. There are $4 \cdot 2$ flats, $2 \cdot 3 + 3 \cdot 4$ longs, and $3 \cdot 3$ units. 5 units = 1 long, 5 longs = 1 flat, and 5 flats = 1 block, thus $43_{five} \cdot 23_{five} =$ 2 blocks, 1 flat, 4 longs, and 4 units, or $\mathbf{2144_{five}}$.

22. (a) $728 \times 94 = 68{,}432$:

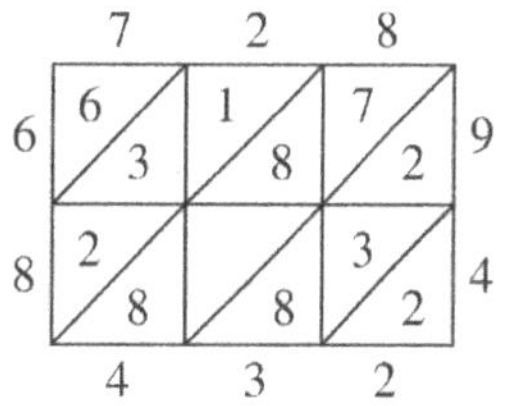

(b) $306 \times 24 = 7344$:

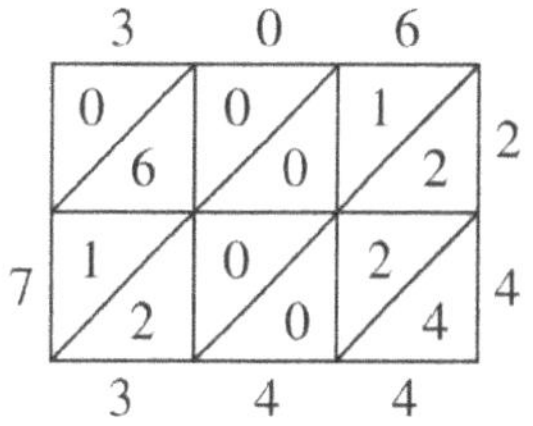

23. $323_{five} \cdot 42_{five} = \mathbf{30221_{five}}$:

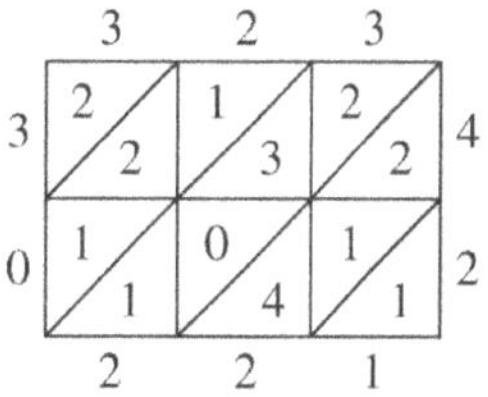

24. $32_a = 23_b \Rightarrow 3a + 2 = 2b + 3 \Rightarrow b = \frac{3a-1}{2}$. The smallest value of $a (a > 1)$ for b to be whole is $a = 3 \Rightarrow a = 3$ and $b = 4$. But 32_{three} is not possible because there is no numeral 3 in base three. The smallest possible solution is thus $a = 5$ and $b = 7$, or $32_{five} = 23_{seven}$.

25. **(a)** In base two, two is 10_{two}. To illustrate the property consider a three digit number in base two:

$abc_{two} \cdot 10_{two} = (a \cdot 10^2_{two} + b \cdot 10_{two} + c \cdot 1_{two}) \cdot 10_{two} = a \cdot 10^3_{two} + b \cdot 10^2_{two} + c \cdot 10_{two} + 0 \cdot 1_{two} = abc0_{two}$. So multiplying abc_{two} by 10_{two} "annexed" the numeral abc_{two} with a 0 in the "ones" place.

(b) In base two, 4 can be expressed as $10_{two} \cdot 10_{two}$. Thus, given what we learned in (b), multiplying by 4 in base two "annexes" a base two numeral by 00 by annexing the original number by 0 twice.

(c)

$$\begin{aligned} 110_{two} \cdot 11_{two} &= 110_{two}(10_{two} + 1_{two}) \\ &= 110_{two} \cdot 10_{two} + 110_{two} \cdot 1_{two} \\ &= 1100_{two} + 110_{two} \\ &= \mathbf{10010}_{two}. \end{aligned}$$

26. **(a)** 5 was multiplied by 6 to obtain 30. The 3 was regrouped, then 3 was multiplied by 2 to obtain 6. The regrouping was added to obtain 9 which was recorded in the tens place.

(b) When 1 was brought down the quotient of 0 was not recorded.

27. $56 \cdot 10$

$= (5 \cdot 10 + 6) \cdot 10$	expanded form.
$= (5 \cdot 10) \cdot 10 + 6 \cdot 10$	distributive property.
$= 5(10 \cdot 10) + 6 \cdot 10$	associative property.
$= 5 \cdot 10^2 + 6 \cdot 10$	definition of a^n.
$= 5 \cdot 10^2 + 6 \cdot 10 + 0 \cdot 1$	additive identity.
$= \mathbf{560}$	place value.

28. **(a)** The greatest product requires the largest multiplicands which can be formed using the four numbers: $8 \times 763 > 7 \times 863$ because $8 \times 700 = 7 \times 800$ but $8 \times 63 > 7 \times 63$. Thus:

$$\begin{array}{r} \boxed{7}\ \boxed{6}\ \boxed{3} \\ \times \quad\quad \boxed{8} \\ \hline 6\ 1\ 0\ 4 \end{array}$$

(b) The least product requires the smallest multiplicands which can be formed using the four numbers; $3 \times 678 < 6 \times 378$ because $3 \times 600 = 6 \times 300$ but $3 \times 78 < 6 \times 78$. Thus:

$$\begin{array}{r} \boxed{6}\ \boxed{7}\ \boxed{8} \\ \times \quad\quad \boxed{3} \\ \hline 2\ 0\ 3\ 4 \end{array}$$

29. **(a)** 3 hrs skiing $\times$ 444 calories/hr $=$ **1332** calories.

(b) Veronica:
2 hrs $\times$ 462 calories/hr $=$ 924 calories.

Carolyn: 3 hrs $\times$ 198 calories/hr $=$ 594 calories.

Thus Veronica burned $924 - 594 =$ **330** more.

(c) Lyle: 3 hrs $\times$ 708 calories/hr $=$ 2124 calories.

Maurice: 5 hrs $\times$ 444 calories/hr $=$ 2220 calories.

Thus Maurice burned $2220 - 2124 =$ **96** more.

30. Let p be the number of pennies in box 3. Then $3p$ is the number of pennies in box 1 and $2(3p) = 6p$ is the number of pennies in box 2. Thus $3p + 6p + p = 4520 \Rightarrow 10p = 4520 \Rightarrow p = 452$.

So box 1 $= 3(452) =$ **1356** pennies;
box 2 $= 6(452) =$ **2712** pennies;
box 3 $=$ **452** pennies.

31. **(a)** Start with multiplication of 4_6 by 3. The _ must be 2, since 1 must be regrouped from $3 \cdot 6$ and only $3 \cdot 2 + 1 = 7$. Then $3 \cdot 4 = 12$ for 2 in the __ of the first partial product. Similar reasoning gives:

$$\begin{array}{r} 4\ \underline{2}\ 6 \\ \times\ 7\ 8\ 3 \\ \hline 1\ \underline{2}\ 7\ 8 \\ 3\ 4\ 0\ 8\ \ \\ \underline{2}\ 9\ 8\ 2\ \ \ \ \\ \hline 3\ 3\ 3\ 5\ \underline{5}\ 8 \end{array}$$

(b) The __ in the multiplier must be 4, since only $4 \cdot 7$ gives 8 in the units place of the second partial product. Similar reasoning gives:

$$\begin{array}{rrrrrr} & & & 3 & 2 & 7 \\ & & \times & 9 & \underline{4} & 1 \\ \hline & & & 3 & 2 & 7 \\ & 1 & \underline{3} & 0 & 8 & \\ \underline{2} & 9 & \underline{4} & 3 & & \\ \hline 3 & 0 & \underline{7} & \underline{7} & 0 & 7 \end{array}$$

32. 8 hours $\times$ 62 mph $= (8 \cdot 60) + (8 \cdot 2) = 480 + 16 =$ **496 miles** (i.e., the front-end multiplying method).

33. Estimates may vary. The range would be from $30 \cdot 20 = 600$ to $40 \cdot 30 = 1200$. Rounding, $38 \approx 35$ or 40 and $23 \approx 20$ or 25, thus about $35 \cdot 20 = 700$ seats, or $40 \cdot 25 = 1000$ seats.

700 would be low (rounded down) and 1000 would be high (rounded up).

34. Answers vary. For example:

(a) **Different**. One factor is the same in each and the other is 4 times larger.

(b) **Same**. 22 was divided by 2 to obtain 11 while 32 was multiplied by 2 to obtain 64. The result is to multiply the original computation by $\frac{2}{2} = 1$ which does not change it.

(c) **Same**. 13 was multiplied by 3 and 33 was divided by 3, thus the original computation was multiplied by 1.

35. Answers vary. One strategy to find $(n5)^2$ would be to write $n(n+1)$ and append 25 (because 5^2 is always 25); e.g., $65^2 = (6 \cdot 7 = 42)$ and append $25 = 4225$ and $75^2 = (7 \cdot 8 = 56)$ and append $25 = 5625$.

36. Answers vary. If the estimated product is 42,000 and 42 = 6 • 7 so 42,000 = 600 • 70, then an estimate of $42{,}000 \approx 612 \bullet 73$.

Mathematical Connections 3-4: Review Problems

21. **No**. For example, $5 - 2 \neq 2 - 5$.

23. $\mathbf{10000}_{three} = 3^4 = 81$. Any power of 3 in base ten can be written as a 1 followed by an appropriate number of zeros in base three.

Assessment 3-5A: Division of Whole Numbers

1. **(a)** $40 \div 8 = 5 \Rightarrow \mathbf{40 = 8 \cdot 5}$.

(b) $326 \div 2 = x \Rightarrow \mathbf{326 = 2 \cdot} x$.

2. **(a)** $18 \div 3 = \boxed{6}$ (since $3 \cdot 6 = 18$).

(b) $\boxed{0} \div 76 = 0$ (since $0 \cdot 76 = 0$).

(c) $28 \div \boxed{4} = 7$ (since $7 \cdot 4 = 28$).

3. If $108 / a = b$ then $108 = a \bullet b$ and $108 / b = a$.

4. The complete fact family:

$72 / 8 = 9$

$72 / 9 = 8$

$8 \bullet 9 = 72$

$9 \bullet 8 = 72$

5. **(a)** Repeated subtraction

$$\begin{array}{r|l} 8 & 6\ 2\ 3 \\ & \underline{5\ 6\ 0} \quad 70 \text{ eights} \\ & \quad\ 6\ 3 \\ & \quad\ \underline{5\ 6} \quad \underline{7} \text{ eights} \\ & \qquad 7 \quad 77 \text{ reminder } 7 \end{array}$$

Standard algorithm

$$\begin{array}{r|l} & \quad\ 7\ 7 \quad \text{remainder } 7 \\ 8 & 6\ 2\ 3 \\ & \underline{5\ 6\ 0} \\ & \quad\ 6\ 3 \\ & \quad\ \underline{5\ 6} \\ & \qquad 7 \end{array}$$

(b) Repeated subtraction

$$\begin{array}{r|l} 3\ 6 & 2\ 9\ 8 \\ & \underline{2\ 8\ 8} \quad \underline{8} \ \ 36\text{'s} \\ & \quad 1\ 0 \quad 8 \ \ \text{remainder } 10 \end{array}$$

Standard algorithm

$$\begin{array}{r|l} & \qquad 8 \quad \text{remainder } 10 \\ 3\ 6 & 2\ 9\ 8 \\ & \underline{2\ 8\ 8} \\ & \quad 1\ 0 \end{array}$$

(c) Repeated subtraction

$$\begin{array}{r|rrrr l} 3\ 9\ 1 & 4 & 0 & 0 & 1 & \\ \hline & 3 & 9 & 1 & 0 & 10 \text{ 391's} \\ \hline & & & 9 & 1 & 10 \text{ remainder } 91 \end{array}$$

Standard algorithm

$$\begin{array}{r|rrrr l} & & 1 & 0 & & \text{reminder } 91 \\ 3\ 9\ 1 & 4 & 0 & 0 & 1 & \\ & 3 & 9 & 1 & & \\ \hline & & & 9 & 1 & \end{array}$$

6. Answers vary, but for example:

(a) There is no associative property in division; e.g., $(8 \div 4) \div 2 \neq 8 \div (4 \div 2)$.

(b) There is no distributive property in division; e.g.,
$8 \div (2 + 2) \neq (8 \div 2) + (8 \div 2)$.

7. Suppose there are two bags of marbles containing a marbles in one and b marbles in the other. It is desired to divide the marbles equally among c boys. Then the number of marbles that each boy receives can be found in two different ways:

Put all the marbles in one bag. There will be $a + b$ marbles and each boy will receive $(a + b) \div c$ marbles.

Divide the marbles in the first bag first and then the second. Each boy would receive $(a \div c) + (b \div c)$ marbles.

Let $a \div c = x$ and $b \div c = y$.
Then $a = cx$ and $b = cy$.
So $a + b = cx + cy = c(x + y)$.
By the definition of division:

$$(x + y) = (a + b) \div c.$$

Substituting for x and y:

$$(a \div c) + (b \div c) = (a + b) \div c.$$

8. Yes. The natural numbers must be $14, 24, 34, \ldots, 94$ to leave a remainder of 4 when divided by 10. Then $\mathbf{64} \div 47 = 1$ with a remainder of 17.

9. (a) The divisor will be the least digit value and the dividend will be arranged so that the greatest is placed for the greatest place value. $\mathbf{3\overline{)754}}$

(b) The thinking in (a) is reversed. $7\overline{)\mathbf{345}}$

10. Reverse the operation: $300 \div 10 = 30$ and $30 \div 10 = \mathbf{3}$.

11.

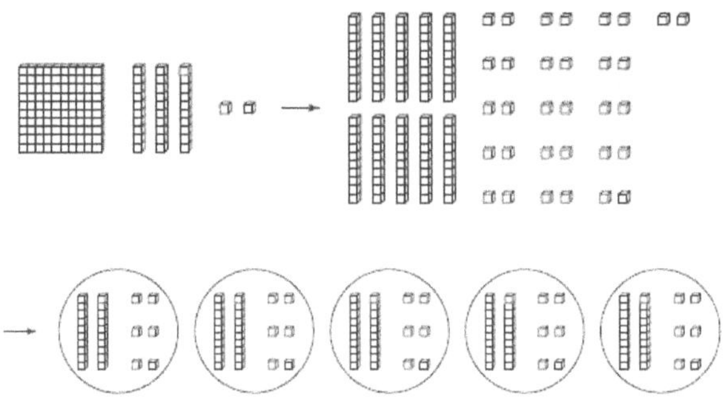

26 R2

12.

$$\begin{array}{r l} 15\overline{)2\,5\,8\,9} & \\ -1\,5\,0\,0 & 15 \times 100 \\ \hline 1\,0\,8\,9 & \\ -9\,0\,0 & 15 \times 60 \\ \hline 1\,8\,9 & \\ -1\,5\,0 & 15 \times 10 \\ \hline 3\,9 & \\ -3\,0 & 15 \times 2 \\ \hline 9 & 172 \text{ R } 9 \end{array}$$

13. In each case it may be helpful to generate a multiplication table in the appropriate base; e.g. in base 5:

×	0	1	2	3	4
0	0	0	0	0	0
1	0	1	2	3	4
2	0	2	4	11	13
3	0	3	11	14	22
4	0	4	13	22	31

(a)

$$\begin{array}{r} {}^{1} \\ 3\ 2_{five} \\ \times\ \ 4_{five} \\ \hline 2\ 3\ 3_{five} \end{array}$$

(b)

$$\begin{array}{r} 4_{five} \quad \text{remainder } 1_{five} \\ 4_{five}\overline{)3\ 2}_{five} \\ 3\ 1 \\ \hline 1 \end{array}$$

(c)

$$\begin{array}{r} 43_{six} \\ \times\ 23_{six} \\ \hline 213 \\ 130 \\ \hline 1513_{six} \end{array}$$

(d)

$$\begin{array}{r} 31_{five} \\ 3_{five}\,\overline{)\,143_{five}} \\ \underline{14} \\ 03 \\ \underline{03} \\ 0 \end{array}$$

(e)

$$\begin{array}{r} 110_{two} \\ 11_{two}\,\overline{)\,10010_{two}} \\ \underline{11} \\ 110 \\ 110 \\ \hline 0 \end{array}$$

(f)

$$\begin{array}{r} 10110_{two} \\ \times\ 101_{two} \\ \hline 10110 \\ 10110 \\ \hline 1101110_{two} \end{array}$$

14. Working the long division backwards, in step 1:
13
$\underline{\times 2}$. So 2 • 3 = 6 in base ten, but there's a 1 in
31
the units place. 6 is 11 in base five, so the long division is done in base five.

15. There were 10 teams with 12 on each team. So, there were $10(12) = 120$ people. Divide them into teams of 8 people and there are $120 \div 8 =$ **15 teams**.

16. These numbers will be multiples of 4 plus 1, $\{4n + 1 | n \in W\}$ or $\{1, 5, 9, 13, \ldots\}$.

17. Assuming the price is \$30 per \$1000 per year, there are $\$50{,}000 \div \$1{,}000 = 50$ installments of \$30. Each quarter, the cost is $50 \cdot \$30 \div 4 =$ **\$375**.

18. **(a)** Finding a number that leaves a remainder of 3 upon division by 4 is equivalent to thinking of a number multiplied by four and then add 3. For example 15 is a number that leaves a remainder of 3 upon division by 4:

$15 \div 4 = 3$ with a remainder of 3 because

$15 = 4 \cdot 3 + 3 \Rightarrow 15 = 12 + 3 \Rightarrow 15 = 15.$

In general, $4n + 3$ is all whole numbers that leave a remainder of 3 when divided by 4. In set builder notation $\{4n + 3 | n \in W\}$.

(b) $4n + 3$

$4(0) + 3 = 3$

$4(1) + 3 = 7$

$4(2) + 3 = 11$

$4(3) + 3 = 15$

$4(4) + 3 = 19$

etc.

3, 7, 11, 15, 19, ...

(c) Arithmetic because each subsequent term is the result of adding 4 to the previous term.

19. **(a)** **(5 + 6) · 3** = 33. Without parentheses, the result would be 23.

(b) **No parentheses** are needed; the order of operations specifies that addition and subtraction are performed in order from left to right.

(c) **No parentheses** are needed; the order of operations specifies that division is performed before addition or subtraction.

(d) **(9 + 6) ÷ 3** = 5. Without parentheses, the result would be 11.

20. **(a)** **Subtract 18 from 45**. Order of operations specifies operations within parentheses first, then multiplication before addition or subtraction.

(b) **Divide 54 by 9**. Order of operations specifies operations within parentheses first.

(c) **Add 11 to 48**. Operations within parentheses first.

(d) **Add 8 to 61**. Multiplication or division before addition.

21. The process always results in the original numbers.

Step 1: "Think of a number." Name this number x.

Step 2: $5x$

Step 3: $5x + 5$

Step 4: $\frac{5x+5}{5} = x + 1$

Step 5: $x + 1 - 1 = x$.

22. Note that in parts (b) through (d), a larger divisor produces a lower quotient when the dividend stays the same and a larger dividend produces a higher quotient when the divisor stays the same.

(a) **High**. $299 \cdot 300 < 300 \cdot 300$.

(b) **Low**. $6001 \div 299 > 6000 \div 300$.

(c) **Low**. $6000 \div 299 > 6000 \div 300$.

(d) **Low**.
$10 \cdot 99 = 990 < 999 \Rightarrow 999 \div 99 > 10$.

Mathematical Connections 3-5: Review Problems

21. Answers vary. For example: $3 + 0 = 3 = 0 + 3$.

23. $59{,}260$ miles $-$ $52{,}281$ miles $=$ **6979 miles** traveled.

Chapter 3 Review

1. **(a)** Tens.

(b) Thousands.

(c) Hundreds.

2. **(a)** $\overline{\text{CD}}\text{XLIV} = 1000 \cdot \text{CD} + \text{XLIV} = 1000 \cdot 400 + 40 + 4 = \mathbf{400{,}044}$.

(b) $432_{five} = 4 \cdot 5^2 + 3 \cdot 5 + 2 = 100 + 15 + 2 = \mathbf{117}$.

(c) $\text{ET0}_{twelve} = 11 \cdot 12^2 + 10 \cdot 12 + 0 = 1584 + 120 = \mathbf{1704}$.

(d) $1011_{two} = 1 \cdot 2^3 + 1 \cdot 2 + 1 = 8 + 2 + 1 = \mathbf{11}$.

(e) $4136_{seven} = 4 \cdot 7^3 + 1 \cdot 7^2 + 3 \cdot 7 + 6 = 1372 + 49 + 21 + 6 = \mathbf{1448}$.

3. **(a)** $3^{4+7+6} = \mathbf{3^{17}}$.

(b) $2^{10+11} = \mathbf{2^{21}}$.

4. $1000_{three} + 200_{three} + 100_{three} + 20_{three} = 1000_{three} + 1000_{three} + 20_{three} = \mathbf{2020_{three}}$.

5. $51_{ten} = (1 \cdot 10^3)_{three} + (2 \cdot 10^2)_{three} + (2 \cdot 10)_{three} + (0 \cdot 1)_{three} \Rightarrow$ **<u>1</u> block, <u>2</u> flats, <u>2</u> longs, <u>0</u> units**. So the fewest blocks needed is **5**.

6. **(a)** 123_{four}:

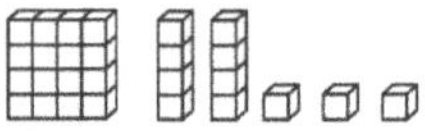

(b) 24_{five}:

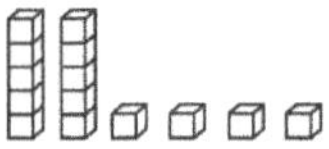

7. **(a)** $40{,}000{,}000{,}000 = 4 \cdot 10^{10}$, so the place value of 4 is $\mathbf{10^{10}}$.

(b) A number in base five having ten digits is $m \cdot 5^9 + n \cdot 5^8 + \cdots$. The place value of the second digit is therefore $\mathbf{5^8}$.

(c) 30 zeros and a 1 in base two represents a number $n = 1 \cdot 2^{31} + 0 \cdot 2^{30} + \cdots + 1$. The place value of the lead digit is therefore $\mathbf{2^{31}}$.

8. Answers vary. One example would be selling pencils by units, dozens, and gross; i.e., base twelve.

9. **(a)** There is 1 group of $125(5^3)$ in 128, with remainder 3.

There are 0 groups of 25 in 3, with remainder 0.

There are 0 groups of 5 in 3, with remainder 0.

There are 3 groups of 1 in 3, with remainder 0.

Thus $128 = \mathbf{1003_{five}}$.

(b) There is 1 group of $128(2^7)$ in 128, with remainder 0.

There are 0 groups of 64 in 0, with remainder 0.

There are 0 groups of 32 in 0, with remainder 0.

There are 0 groups of 16 in 0, with remainder 0.

There are 0 groups of 8 in 0, with remainder 0.

There are 0 groups of 4 in 0, with remainder 0.

There are 0 groups of 2 in 0, with remainder 0.

There are 0 groups of 1 in 0, with remainder 0.

Thus $128 = \mathbf{10000000_{two}}$.

(c) There are $10(T)$ groups of 12 in 128, with remainder 8. There are 8 groups of 1 in 8, with remainder 0.

Thus $128 = \boldsymbol{T8_{twelve}}$.

10. Place value is determined by powers of each base.

(a) $2^{10} + 2^3 = \mathbf{10000001000_{two}}$.

(b) $11 \cdot 12^5 + 10 \cdot 12^3 + 20 = 11 \cdot 12^5 + 10 \cdot 12^3 + 1 \cdot 12 + 8 = \boldsymbol{E0T018_{twelve}}$.

11. $1 \cdot b^2 + 2b + 3 = 83 \Rightarrow b^2 + 2b - 80 = 0 \Rightarrow (b - 8)(b + 10) = 0 \Rightarrow b = 8$ or $b = -10$. Since the base must be positive, $\boldsymbol{b = 8}$.

12. **(a)** **Distributive** property of multiplication over addition.

(b) **Commutative** property of addition.

(c) **Identity** property of multiplication for whole numbers.

(d) **Distributive** property of multiplication over addition.

(e) **Commutative** property of multiplication.

(f) **Associative** property of multiplication.

13. **(a)** $13 = 3 + 10$. Since 10 is a natural number, $3 < 13$.

(b) $12 = 3 + 9$. Since 3 is a natural number, $12 > 9$.

14. **(a)** $4 \cdot \boxed{10 \le \textit{whole number} \le 15} - 37 < 27$.

(b) $398 = \boxed{10} \cdot 37 + 28$.

(c) $\boxed{n} \cdot (3 + 4) = \boxed{n} \cdot 3 + \boxed{n} \cdot 4$, where $n \in W$ is any whole number.

(d) $42 - \boxed{\textit{Any whole number} \le 26} \ge 16$.

15. **(a)** $3a + 7a + 5a = (3 + 7 + 5)a = \mathbf{15}\boldsymbol{a}$.

(b) $3x^2 + 7x^2 - 5x^2 = (3 + 7 - 5)x^2 = \mathbf{5}\boldsymbol{x^2}$.

(c) $x(a + b + y) = \boldsymbol{xa + xb + xy}$.

(d) $(x+5)3+(x+5)y = \boldsymbol{3x+15+xy+5y}$

or

$(x + 5)3 + (x + 5)y = \boldsymbol{(x + 5)(3 + y)}$.

(e) $3x^2 + x = x(3x + 1)$.

(f) $2x^5 + x^3 = x^3(2x^2 + 1)$.

16 60 people × 8 ounces = 480 ounces required.
480 ÷ 12 ounces per can = **40** twelve-ounce cans.

17. 2 slacks × 3 blouses × 2 sweaters = **12** outfits.

18. Work backward from 93 using inverse operations:

Subtract 89, giving 4;

Add 20, giving 24;

Divide by 12, giving 2; then

Multiply by 13, giving **26** as the original number.

19. \$80 per person × 80 people = \$6400. The **\$6000** package is less expensive.

20. 30 hours per week × \$5 per hour + 8 hours overtime × \$8 per overtime hour = **\$214**.

21. Let q be the amount from the first question. Then winnings are $q + 2q + 4q + \cdots$ which is geometric sequence with $a_1 = q, r = 2$, and $n = 5$. $6400 = q(2)^{5-1} \Rightarrow 16q = 6400 \Rightarrow q = \mathbf{\$400}$.

22. **(a)** Let n be the original number. Then

$$\frac{2[2(n + 17) - 4] + 20}{4} - 20 = \frac{4(n + 17) - 8 + 20}{4} - 20 = \frac{4n + 80}{4} - 20 = n + 20 - 20 = \boldsymbol{n}.$$

(b) Answers may vary. For example, if n is the original number:

$4(n + 18) - 7 = 4n + 65$.

Then two more steps might be:

$4n + 65 - 65$ (subtract 65);

$\frac{4n}{4}$ (divide by 4).

(c) Answers may vary; use the techniques of parts (a) and (b).

23. Scratch:

$$\begin{array}{cccc} & 1 & & \\ & 3 & 1 & 6 \\ & \not{7}_1 & 1 & 2 \\ + & & \not{9}_1 & 1 \\ \hline 1 & 1 & 1 & 9 \end{array}$$

Traditional:

$$\begin{array}{r} {}^{1} \\ 316 \\ 712 \\ +\ \ 91 \\ \hline 1119 \end{array}$$

24. Traditional:

$$\begin{array}{r} 613 \\ \times\ \ 98 \\ \hline 4904 \\ 5517 \\ \hline 60074 \end{array}$$

Lattice:

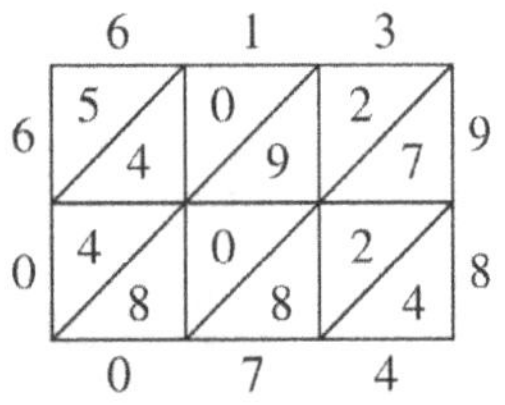

25. **(a)** Repeated subtraction:

$$\begin{array}{r|l} 912 & 4803 \\ & \underline{4560} \quad \text{5-912's} \\ & 243 \quad \underline{}\ \text{5-912's} \Rightarrow 5 \text{ r } 243 \end{array}$$

Traditional:

$$\begin{array}{r|l} & 5 \Rightarrow 5 \text{ r } 243 \\ 912 & 4803 \\ & \underline{4560} \\ & 243 \end{array}$$

(b) Repeated subtraction:

$$\begin{array}{r|l} 11 & 1011 \\ & \underline{990} \quad \text{90-11's} \\ & 21 \\ & \underline{11} \quad 1-11 \\ & 10 \quad \text{91-11's} \Rightarrow 91 \text{ r } 10 \end{array}$$

Traditional:

$$\begin{array}{r|l} & 91 \Rightarrow 91 \text{ r } 10 \\ 11 & 1011 \\ & \underline{990} \\ & 21 \\ & \underline{11} \\ & 10 \end{array}$$

(c) Repeated subtraction:

$$\begin{array}{r|l} 23_{five} & 3312_{five} \\ & \underline{2300} \quad (100\text{-}23\text{'s})_{five} \\ & 1012 \\ & \underline{1010} \quad \underline{(20\text{-}23\text{'s})_{five}} \\ & 2_{five} \quad (120\text{-}23\text{'s})_{five} \end{array}$$

$\Rightarrow 120_{five}$ remainder 2_{five}

Traditional:

$$\begin{array}{r|l} & 120_{five} \\ 23_{five} & 3312_{five} \\ & \underline{23} \\ & 101 \\ & \underline{101} \\ & 02_{five} \end{array}$$

$\Rightarrow 120_{five}$ remainder 2_{five}

(d) Repeated subtraction:

$$\begin{array}{r|l} 11_{two} & 1011_{two} \\ & \underline{110} \quad (10\text{-}11\text{'s})_{two} \\ & 101 \\ & \underline{11} \quad \underline{(1\text{-}11)_{two}} \\ & 10_{two} \quad (11\text{-}11\text{'s})_{two} \end{array}$$

$\Rightarrow 11_{two}$ remainder 10_{two}

Traditional:

$$\begin{array}{r|l} & 11_{two} \\ 11_{two} & 1011_{two} \\ & \underline{11} \\ & 101 \\ & \underline{11} \\ & 10\ two \end{array}$$

$\Rightarrow 11_{two}$ remainder 10_{two}

26. **(a)** If $4803 \div 912 = 5$ remainder 243
Then $912 \cdot 5 + 243 = 4803$.

(b) If $1011 \div 11 = 91$ remainder 10 Then $11 \cdot 91 + 10 = 1011$.

(c) If
$(3312 \div 23)_{five} = (120 \text{ remainder } 2)_{five}$
Then $(23 \cdot 120)_{five} + 2_{five} = 3312_{five}$.

(d) If $(1011 \div 11)_{two} = (11 \text{ remainder } 10)_{two}$

Then $(11 \cdot 11)_{two} + 10_{two} = 1011_{two}$.

27. **(a)** $19 \cdot 5 \cdot 194 \cdot 2 = (19 \cdot 194) \cdot 10 = \mathbf{36{,}860}$.

(b) $379 \cdot 4 \cdot 193 \cdot 25 = (379 \cdot 193) \cdot 100 = \mathbf{7{,}314{,}700}$.

(c) $8 \cdot 481 \cdot 73 \cdot 125 = (481 \cdot 73) \cdot 1000 = \mathbf{35{,}113{,}000}$.

(d) $374 \cdot 200 \cdot 893 \cdot 50 = (374 \cdot 893) \cdot 10{,}000 = \mathbf{3{,}339{,}820{,}000}$.

28. $(\$320 \times 6 \text{ mos}) + (\$410 \times 6 \text{ mos}) = \mathbf{\$4380}$.

29. 15,600 cans per hour $\div$ 24 cans per case $=$ 650 cases per hour. Then 650×4 hours $=$ **2600 cases**.

30. There are 8 groups of 3 in 24 (or two dozen) apples. $8 \times 69¢$ per group $= 552¢$ on sale; $32¢$ each $\times$ 24 apples $= 768¢$ regular price. $768¢ - 552¢ =$ **216¢ saved** (or \$2.16).

31. Let b be the number of bicycles and t be the number of tricycles. Then

$2b + 3t = 126$ wheels $\quad (i)$

$2b + 2t = 108$ pedals $\quad (ii)$

Subtracting (ii) from $(i) \Rightarrow t =$ **18 tricycles**.

Substituting $t = 18$ into (i) $\Rightarrow b =$ **36 bicycles**.

32. **(a)**

$$\begin{array}{rrrl} & {\scriptstyle 1} & {\scriptstyle 1} & \\ 1 & 2 & 3 & _{five} \\ + & 3 & 4 & _{five} \\ \hline \mathbf{2} & \mathbf{1} & \mathbf{2} & _{five} \end{array}$$

(b)

$$\begin{array}{rrrrl} {\scriptstyle 10} & {\scriptstyle 0} & {\scriptstyle 10} & & \\ \not{1} & 0 & \not{1} & 0 & _{two} \\ - & 1 & 0 & 1 & _{two} \\ \hline & \mathbf{1} & \mathbf{0} & \mathbf{1} & _{two} \end{array}$$

(c)

$$\begin{array}{rrrrl} & & 2 & 3 & _{five} \\ & \times & 3 & 4 & _{five} \\ \hline & 2 & 0 & 2 & \\ 1 & 2 & 4 & & \\ \hline \mathbf{1} & \mathbf{4} & \mathbf{4} & \mathbf{2} & _{five} \end{array}$$

(d)

$$\begin{array}{rrrrrrl} & & 1 & 0 & 0 & 1 & _{two} \\ & & \times & 1 & 0 & 1 & _{two} \\ \hline & & 1 & 0 & 0 & 1 & \\ & 0 & 0 & 0 & 0 & & \\ 1 & 0 & 0 & 1 & & & \\ \hline \mathbf{1} & \mathbf{0} & \mathbf{1} & \mathbf{1} & \mathbf{0} & \mathbf{1} & _{two} \end{array}$$

33. $44_{five} \cdot 34_{five} = (4 \cdot 10_{five} + 4) \cdot 34_{five}$

$= 4 \cdot 34_{five} \cdot 10_{five} + 4 \cdot 34_{five}$

$= 3010_{five} + 301_{five}$

$= \mathbf{3311_{five}}$

34.

$$\begin{array}{rll} 4_{five}\overline{)\,434_{five}} & & \\ -\,400_{five} & & (100 \cdot 4)_{five} \\ \hline 34_{five} & & \\ -\,31_{five} & & (4 \cdot 4)_{five} \\ \hline 3_{five} & & \end{array}$$

Thus, $\mathbf{434_{five} = 104_{five} \cdot 4_{five} + 3_{five}}$

35. Answers vary. For example:

(a) $(26 + 24) + (37 - 7) = 50 + 30 = \mathbf{80}$.

(b) $(7 \cdot 9) \cdot (4 \cdot 25) = 63 \cdot 100 = \mathbf{6300}$.

36. Methods vary. For example:

(a) $63 \cdot 7 = (7 \cdot 60) + (7 \cdot 3) = 420 + 21 = \mathbf{441}$ (front-end multiplying).

(b) $85 - 49 = (85 + 1) - (49 + 1) = 86 - 50 = \mathbf{36}$ (trading off).

(c) $(18 \cdot 5) \cdot 2 = 18 \cdot (5 \cdot 2) = 18 \cdot 10 = \mathbf{180}$ (using compatible numbers).

(d) $2436 \div 6 = (2400 \div 6) + (36 \div 6) = 400 + 6 = \mathbf{406}$ (breaking up the dividend).

37. Answers may vary; for example:

(a) Front-end: $5 + 3 + 2 + 4 + 9 = 23$; place value 2300; adjustments $40 + 100 + 60 + 0 + 100 = 300$; adjusted sum $= 2300 + 300 = 2600$.

(b) Rounding: $500 + 400 + 300 + 400 + 1000 = 2600$.

In this case, both estimates give the same result (the actual sum is 2602, so both are reasonable).

38. The addends cluster around 2400, so one would estimate the sum to be $4 \cdot 2400 = \mathbf{9600}$.

39. **(a)** $999 \cdot 47 + 47 = 47(999 + 1) = 47 \cdot 1000 = \mathbf{47{,}000}$.

(b) $43 \cdot 59 + 41 \cdot 43 = 43(59 + 41) = 43 \cdot 100 = \mathbf{4300}$.

(c) $1003 \cdot 79 - 3 \cdot 79 = 79(1003 - 3) = 79 \cdot 1000 = \mathbf{79{,}000}$.

(d) $1001 \cdot 113 - 113 = 113(1001 - 1) = 113 \cdot 1000 = \mathbf{113{,}000}$.

(e) $101 \cdot 35 = (100 + 1) \cdot 35 = 35 \cdot 100 + 35 \cdot 1 = 3500 + 35 = \mathbf{3535}$.

(f) $98 \cdot 35 = (100 - 2) \cdot 35 = 35 \cdot 100 - 35 \cdot 2 = 3500 - 70 = \mathbf{3430}$.

40. **(a)**

$$\begin{array}{rccccccc} & 3x^3 & + & 4x^2 & + & 7x & + & 8 \\ + & & & 5x^2 & + & 2x & + & 1 \\ \hline & 3x^3 & + & 9x^2 & + & 9x & + & 9 \end{array}$$

(b) Answers vary. For example:

$$\begin{array}{rccccccc} & 3 \cdot 10^3 & + & 5 \cdot 10^2 & + & 7 \cdot 10 & + & 8 \\ - & & & 4 \cdot 10^2 & + & 2 \cdot 10 & + & 1 \\ \hline & 3 \cdot 10^3 & + & 1 \cdot 10^2 & + & 5 \cdot 10 & + & 7 \end{array}$$

Is equivalent (when $x = 10$) to:

$$\begin{array}{rccccccc} & 3x^3 & + & 5x^2 & + & 7x & + & 8 \\ - & & & 4x^2 & + & 2x & + & 1 \\ \hline & 3x^3 & + & x^2 & + & 5x & + & 7 \end{array}$$

(c) Answers vary. For example:

$25 \cdot 10^2 = (2 \cdot 10 + 5) \cdot 10^2 = 2 \cdot 10^3 + 5 \cdot 10^2 + 0 \cdot 10 + 0 = 2500$.

Is equivalent to:

$(2x + 5)x^2 = 2x^3 + 5x^2$.

41. Suppose $a, b \in B$. Then $a = 5j$ and $b = 5k$, where $j, k \in W$. Therefore $a + b = 5j + 5k = 5(j + k)$, where $(j + k) \in W$. Therefore $a + b \in B$ and B is closed under addition.

42. After 0 seconds 1 person knew (the principal).

After 30 seconds the principal had called 1 person (2 members notified).

After 60 seconds the principal and one member called 2 other members (4 members notified).

After 90 seconds the one member and the 2 new members called another 3 members (7 members notified).

Continuing this pattern:

After 120 seconds 12 members notified.

After 150 seconds 20 members notified.

After 180 seconds 33 members notified.

After 210 seconds 54 members notified.

After 240 seconds 88 members notified.

It took 240 seconds = **4 minutes** to notify all 85 members.

43. She started with 124 cookies and there were 7 left so the children took 124 – 7= 117. A number between 10 and 30 that evenly divides into 117 is 13. $117 \div 13 = 9$. There were **13 children** who each took 9 cookies.

44. **(a)** $\mathbf{6 \times (4+1) = 30}$

(b) $((48 \div 6) + 5) \times 3 = 39$

(c) $(12 \div (3 \times 2)) - 1 = 1$

(d) $50 / ((10 + 10) / 2) = 5$

45. $20 - 3 \times (2 + 4) = 2$ **Correct Solution**

$(20 - 3) \times 2 + 4 = 38$

$20 - (3 \times 2) + 4 = 18$

$20 - ((3 \times 2) + 4) = 10$

46. Answers vary. For example:

$888 + 88 + 8 + 8 + 8 = 1000$

CHAPTER 4

NUMBER THEORY

Assessment 4-1A: Divisibility

1. **Odd. In chapter 3, it was learned that the product of an even number multiplied by an even number is always an even number. Adding one to any even number creates an odd number**. For example: 2 times 2 plus 1 is 5 which is odd.

2. (a) **False; there is no value $c \in W$ such that $24c = 5$.**

   (b) **True**; $30 \div 10 = 3$.

   (c) **False; there is no value $c \in W$ such that $8c = 324$.**

   (d) **False; there is no value $c \in W$ such that $0c = 24$.**

3. Use these tests for each number:

   (*i*) $2|n$ if the units digit is divisible by 2.

   (*ii*) $3|n$ if the sum of the digits is divisible by 3.

   (*iii*) $4|n$ if the last two digits are divisible by 4.

   (*iv*) $5|n$ if the units digit is 0 or 5.

   (*v*) $6|n$ if $2|n$ and $3|n$.

   (*vi*) $8|n$ if the last three digits are divisible by 8.

   (*vii*) $9|n$ if the sum of the digits is divisible by 9.

   (*viii*) $10|n$ if the units digit is 0.

   (*ix*) $11|n$ if the sum of the digits in places that are even powers of 10 minus the sum of the digits in the places that are odd powers of 10 is divisible by 11.

		2	3	4	5	6	8	9	10	11
(a)	4,201,012	Y	N	Y	N	N	N	N	N	N
(b)	1573	N	N	N	N	N	N	N	N	Y
(c)	15,810	Y	Y	N	Y	Y	N	N	Y	N

4. (a) **Yes**. The question is really, "does $9|1379$?" Using the divisibility test for 9, $9 \nmid (1 + 3 + 7 + 9)$ so $9 \nmid 1379$ and there will be a remainder; i.e., a group of less than 9 players.

   (b) **No**. The question is really, "does $11|1354$?" Using the divisibility test for 11, $11 \nmid [(1 + 5) - (3 + 4)]$ so $11 \nmid 1354$ and there will be a remainder; i.e. trees left over.

   (c) **Yes**. The question is really, "does $8|1216$?" Using the divisibility test for 8, $8|216$ so there will be no remainder.

5. (a) **Any digit from 0 to 9**. The units digit of $1,427,4\square2$ is divisible by 2.

   (b) **1, 4, or 7**. $3|(1+4+2+7+4+\square+2) \Rightarrow 3|(20+\square) \Rightarrow \square = 1, 4, \text{or } 7$.

   (c) **1, 3, 5, 7, or 9**. $4|\square 2 \Rightarrow \square = 1, 3, 5, 7, \text{or } 9$.

   (d) 7. $9|(1 + 4 + 2 + 7 + 4 + \square + 2) \Rightarrow 9|(20 + \square) \Rightarrow \square = 7$.

6. (a) $3|74\underline{7}$. $3|(7 + 4 + 7 = 18)$. The _ could be filled by 1, 4, or 7, but 7 is the greatest.

   (b) $9|83\underline{7}45$. $9|(8 + 3 + 7 + 4 + 5 = 27)$. Only 7 makes this number divisible by 9.

   (c) $11|6\underline{6}55$. $11|(6 + 5) - (6 + 5)$. 6_55 in this case must result in the sum of the digits in places that are even powers of 10 being equal to the sum of the digits in places that are odd powers of 10.

   (d) $5|136\underline{5}$. The _ could be filled in with 0 or 5, but 5 is the greatest.

   (e) $6|2\underline{8}44$. $2|$ 2_44 and $3|(2+\_+4+4) \Rightarrow 3|(10+\_)$ so _ could be 2,5,8, but 8 is the greatest.

   (f) $8|38\underline{6}4$. The _ could be filled in with 2 or 6, but 6 is the greatest.

7. (a) $7|280$ **because $280 = 7 \times 40$ (definition of division**).

(b) $19|(3800 + 19)|$ **because 19|3800 and 19 |19**. [Theorem 4-2(a)].

(c) $15|(2^4 \cdot 3^5 \cdot 5)$ **because** $\mathbf{2^4 \times 3^5 \times 5 =}$ $\mathbf{(3 \times 5) \times 2^4 \times 3^4}$ (definition of division).

(d) $19 \nmid (3800 + 37)$ **because** $\mathbf{19|3800}$ **but** $\mathbf{19 \nmid 37}$ [Theorem 4-2(b)].

8. **(a)** **True**. Break into parts of compatible number: $390026 = 390000 + 26 = 13(30000) + 13(2) = 13(30002)$.

(b) **True**. By Theorem 4-2(b), $13 \mid 260000$ and $13 \nmid 33$. Thus $13 \nmid (260000 + 33)$.

(c) **False.** Because $17 \mid 34{,}000$ and $17 \nmid 15$; 17 does not divide the sum.

(d) **True**. Because $17 \mid 34{,}000$ and $17 \mid 51$; 17 divides the sum.

(e) **False.** Because $19 \mid 19{,}000$ and $19 \nmid 31$; 19 does not divide the sum.

(f) **True.** $93^{11} = 93 \cdot 93^{10} = 31 \cdot 3 \cdot 93^{10} = 31(3 \cdot 93^{10})$. Now $(3 \cdot 93^{10})$ is a whole number; so by Theorem 4.1, since $31 \mid 31$, then $31 \mid 31(3 \times 93^{10})$.

9. **(a)** **True** by Theorem 4-1. For any whole numbers a and d, if $d|a$ and n is any whole number, then $d|na$.. So if $7 \mid 21$, then $7 \mid (21 \cdot 21)$ or $7 \mid 21^2$.

(b) **False**. Let $a = 4, b = 2$, and $c = 3$. 2 divides 4 $(b|a)$, but $(2 + 3)$ does not divide $(4 + 3)$. So $(b + c) \nmid (a + c)$

(c) **True**. Because $b|a$, there is a whole number c such that $a = bc$. Then, $a^3 = b^3c^3 = (bc^3)b^2$, therefore $b^2|a^3$.

(d) **True**. Because $b|a$, there exist $c \in W$ such that $a = bc$. Thus, $a + b = bc + b = b(c + 1)$. Since $c + 1 \in W$, b divides $a + b$, written $b|(a + b)$.

10. **(a)** A number divisible by 2 and 3 but not by 5 has to end on 2, 4, 6, or 8 and the sum of its digits has to be divisible by 3. **Answers vary; for example 6.**

(b) A number divisible by 2 and 4 but not by 8 has to end on 0, 2, 4, 6, or 8 and the last two digits are divisible by 4. However, the last three digits cannot be divisible by 8. **Answers vary; for example 12.**

(c) A number divisible by 5 and 10 has to end in 0 but if it ends in 0, then it is divisible by 2. Such a number **does not exist.**

11. **19¢.** $209 = 11 \cdot 19$ (both prime numbers). For whole cent pricing 1¢, 11¢, 19¢, and 209¢ are possibilities, but the problem statement is for pencils (plural) and the pencils must cost more than 12¢.

12. (a) **True**. For example $3 \mid (1 \cdot 2 \cdot 3)$. In general for three consecutive numbers n, $n + 1$, $n + 2$ every other number is even, at least one (or two) of the three numbers n, $n + 1$, $n + 2$ will be even. Since every third number is divisible by three, exactly one of n, $n + 1$, $n + 2$ will have a factor of three. So in $n \cdot (n+1) \cdot (n+2)$ there will be at least one factor of two and exactly one factor of three.

(b) **False.** For example $4 \nmid (5 \cdot 6)$. The product of two consecutive numbers is divisible by four only if one of the consecutive numbers is also divisible by 4.

13. There are **11 ways**: 99911, 99731, 99551, 99533, 97751, 97733, 97553, 95555, 77771, 77753, 77555.

14. **(a)** **True**. Let $n = 3(a \cdot 10^k + b \cdot 10^{k-1} + \cdots + x \cdot 10^0)$, where $3a$, $3b$, …$3x$ are all single digits. Then every digit is divisible by 3, and the number itself is divisible by 3.

(b) **False**. Twelve is divisible by 3 but its digits {1, 2} are not divisible by 3.

(c) **True**. This is Theorem 4.8, the divisibility test for division by 3.

15. **(a)**

$\begin{array}{r} 2414271 \\ 3\overline{)7242815} \\ -6000000 \\ \hline 1242815 \\ -1200000 \\ \hline 42815 \\ -30000 \\ \hline 12815 \\ -12000 \\ \hline 815 \\ -600 \\ \hline 215 \\ -210 \\ \hline 5 \\ -3 \\ \hline \mathbf{2} \end{array}$	$7 + 2 + 4 + 2 + 8 + 1 + 5 = 29$ $\begin{array}{r} 9 \\ 3\overline{)\ 29} \\ -27 \\ \hline \mathbf{2} \end{array}$

(b) $7242815 = 7 \cdot 1{,}000{,}000 + 2 \cdot 100{,}000 + 4 \cdot 10{,}000 + 2 \cdot 1{,}000 + 8 \cdot 100 + 1 \cdot 10 + 5$

$= 7(999{,}999 + 1) + 2(99{,}999 + 1) + 4(9{,}999 + 1) + 2(999 + 1) + 8(99 + 1) + 1(9 + 1) + 5$

$= 7 \cdot 999{,}999 + 2 \cdot 99{,}999 + 4 \cdot 9{,}999 + 2 \cdot 999 + 8 \cdot 99 + 1 \cdot 9 + 7 + 2 + 4 + 2 + 8 + 1 + 5$

Because a 3 can be factored out of the first six terms, which is equivalent to dividing by 3, applying the division algorithm to the sum of the digits at the end will yield the remainders.

(c) **Yes.** The key idea in (b) was that each place value greater than $1(10^{\circ})$ can be written as the digit times 9…9 plus 1, for example $d \cdot (9{,}999 + 1) = d \cdot 10^4$. Thus, any number can be rewritten as 9 times some whole number plus the sum of the digits.

16. **1, 3, 5, 9, and 15 divide *n*.** 1 divides every number. Because $n = 45 \cdot d$, $d \in W$, then $n = (3 \cdot 15)\, d$ which implies n is divisible by 3 and 15. And, $n = (5 \cdot 9)d$ which implies n is divisible by 5 and 9.

17. **16|*n* if 16|(last four digits of *n*).** This continues the pattern of divisibility by 2, 4, and 8.

18. The term "casting out nines" implies that any 9 or sum of digits equaling 9 in n may be "cast out." The remaining digit is the remainder when n is divided by 9. Then the remainder when n is divided by 9 is the same as the remainder when the sum of the digits of n is divided by 9 ("casting out" as needed).

(a) *(i)* $12{,}343 + 4546 + 56 = 16{,}945.$

(ii) $12343 = 9 \cdot 1371 + 4$

$4546 = 9 \cdot 505 + 1$

$56 = 9 \cdot 6 + 2$

and $4 + 1 + 2 = 7.$

(iii) $16945 = 9 \cdot 1882 + 7.$

(b) *(i)* $987 + 456 + 8765 = 10{,}208.$

(ii) $987 = 9 \cdot 109 + 6$

$456 = 9 \cdot 50 + 6$

$8765 = 9 \cdot 973 + 8$

and $6 + 6 + 8 = 20$; $20 = 9 \cdot 2 + 2.$

(iii) $10208 = 9 \cdot 1134 + 2.$

(c) $1003 - 46 = 957.$ 1003 has a remainder of 4 when divided by 9; 46 has a remainder of 1 when divided by 9; $4 - 1 = 3$ has a remainder of 3 when divided by 9. $9 + 5 + 7 = 21$ with a remainder of 3 when divided by 9.

(d) $345 \cdot 56 = 19{,}320.$ 345 has a remainder of 3 when divided by 9; 56 has a remainder of 2 when divided by 9; $3 \cdot 2 = 6$ has a remainder of 6 when divided by 9. $1 + 9 + 3 + 2 + 0 = 15$ has a remainder of 6 when divided by 9.

(e) Answers may vary. The division may not have a whole number quotient, in which case the test fails.

19. **(a)** **No**. The palindrome 12121 is not divisible by 11.

(b) **Yes.** Answers may vary. Let a, b, and c be digits. All six digit palindromes can be written in the form *abccba*. The sum of the digits in places with even power is $b + c + a$. The sum of the digits in places with odd powers is $a + c + b$.

This difference is zero, which is divisible by 11.

Alternative: *abccba*

$= a \cdot 10^5 + b \cdot 10^4 + c \cdot 10^3 + c \cdot 10^2 + b \cdot 10 + a \cdot 1$

$= (a \cdot 10^5 + a) + (b \cdot 10^4 + b \cdot 10) + (c \cdot 10^3 + c \cdot 10^2)$

$= a \cdot 100{,}001 + b \cdot 10{,}010 + c \cdot 1{,}100$

$= 11(a \cdot 9091 + b \cdot 910 + c \cdot 100).$

20. Any five-digit number may be written as $a \cdot 10^4 + b \cdot 10^3 + c \cdot 10^2 + d \cdot 10 + e = a(9999 + 1) + b(999 + 1) + c(99 + 1) + d(9 + 1) + e$
$= (9999a + 999b + 99c + 9d) + (a + b + c + d + e).$

The first group is divisible by 9 [Theorem 4-2(a)]. Then the five-digit number is divisible by 9 if and only if the second group is divisible by 9 [Theorem 4-2(b)]; i.e., if the sum of the digits is divisible by 9.

21. **a = 2.** If you sum the digits of *aa*248, you arrive at $2a + 14$. Let $2a + 14$ be divisible by 9; i.e., $9 \mid (2a + 14)$. This also means $9 \mid (2(a + 7))$. For this to be true, $a + 7$ must be divisible by 9; therefore, $a = 2$.

Assessment 4-2A: Prime and Composite Numbers

1. (a)

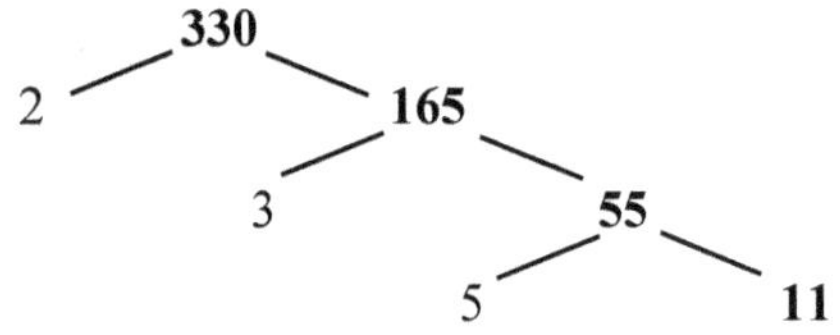

(b) **$2 \cdot 3 \cdot 5 \cdot 11 = 330$**

2. (a)

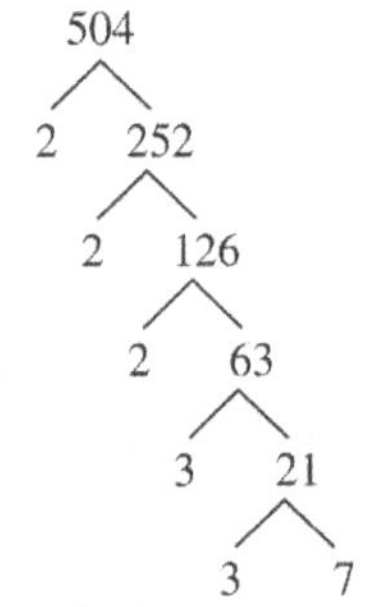

$504 = 2^3 \cdot 3^2 \cdot 7$

(b)

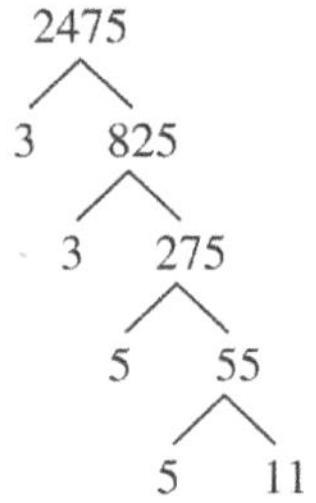

$2475 = 3^2 \cdot 5^2 \cdot 11$

(c)

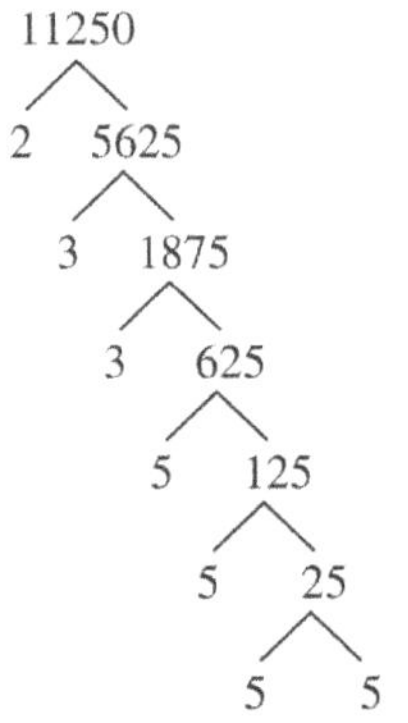

$11250 = 2 \cdot 3^2 \cdot 5^4$

3. (a) $1 \cdot 2 \cdot 3 \cdot 2^2 \cdot 5 \cdot (2 \cdot 3) \cdot 7 \cdot 2^3 \cdot 3^2 = \mathbf{2^7 \cdot 3^4 \cdot 5 \cdot 7}.$

(b) $(2^2 \cdot 5^2) \cdot (13) \cdot (7^2)^{10} = \mathbf{2^2 \cdot 5^2 \cdot 7^{20} \cdot 13}.$

(c) 251 is prime, so its factorization is **251**.

(d) $100^{10} = (2^2 \cdot 5^2)^{10} = \mathbf{2^{20} \times 5^{20}}.$

4. **23.** $23^2 < 769$, but $29^2 > 769$ (23 is prime and 29 is the next largest prime).

5. (a) **Prime.** Fails divisibility test for primes up to 7; no need to test for primes >7 because $11^2 > 103$.

(b) **Not prime.** $7 \cdot 17 = 119$.

(c) **Prime.** Fails divisibility test for primes up to 5; no need to test for primes > 5 because $7^2 > 31$.

(d) **Not prime.** $101 \cdot 3 = 303$.

(e) **Prime.** Fails divisibility test for primes up to 19; no need to test for primes >19 because $23^2 > 463$.

(f) **Prime.** Fails divisibility test for primes up to 7; no need to test for primes >7 because $11^2 > 97$.

(g) **Prime.** $2 \cdot 3 \cdot 5 \cdot 7 + 1 = 211$, which fails divisibility test for primes up to 13; ; no need to test for primes >13 because $17^2 > 211$.

(h) **Not prime.** $2 \cdot 3 \cdot 5 \cdot 7 + 11 = 221 = 13 \cdot 17$.

6. The three smallest primes are 2, 3, and 5, so the least number divisible by three different primes is $2 \cdot 3 \cdot 5 = \mathbf{30}$.

7. The only types of numbers that would have exactly 5 factors would be of the form k^4, when k is prime. Starting with $k = 2$ since 2 is the least prime, $2^4 = 16$ does have 5 factors: 1, 2, 4, 8, and 16. But 16 is a two digit number. Trying $k = 3$, $3^4 = 81$ also has 5 factors (1, 3, 9, 27, 81) but 81 is also a two-digit number. However, $\mathbf{5^4 = 625}$ has five factors 1, 5, 25, 125, and 625 and is a three-digit number.

8. **(a)** The Fundamental Theorem of Arithmetic tells us that each composite number, n, can be written as a product of primes in one and only one way. Since $2|n$ and $3|n$, we know that 2 and 3 must both occur in n's prime decomposition. Therefore, $6|n$.

(b) The Fundamental Theorem of Arithmetic tells us that each composite number, n, can be written as a product of primes in one and only one way. Since $4|n$ and $25|n$, we know that 4 and 5^2 must both occur in n's prime decomposition. Therefore $100|n$.

(c) **Yes.** If $a \mid n$ and $b \mid n$ then there exist natural numbers k and l such that $n = ak$ and $n = bl$. Thus, $n^2 = ak \cdot bl = ab \cdot kl$.

9. **(a)** $36^{10} \cdot 49^{20} \cdot 6^{15}$

$= (2^2 \cdot 3^2)^{10} \cdot (7^2)^{20} \cdot (2 \cdot 3)^{15}$

$= 2^{20} \cdot 3^{20} \cdot 7^{40} \cdot 2^{15} \cdot 3^{15}$

$= \mathbf{2^{35} \cdot 3^{35} \cdot 7^{40}}$.

(b) $100^{60} \cdot 300^{40} = (2^2 \cdot 5^2)^{60} \cdot (2^2 \cdot 3 \cdot 5^2)^{40}$

$= 2^{120} \cdot 5^{120} \cdot 2^{80} \cdot 3^{40} \cdot 5^{80}$

$= \mathbf{2^{200} \cdot 3^{40} \cdot 5^{200}}$.

(c) $(2 \cdot 3^4 \cdot 5^{110} \cdot 7) + (2^2 \cdot 3^4 \cdot 5^{110})$

$= 2 \cdot 3^4 \cdot 5^{110} \cdot (7 + 2)$

$= 2 \cdot 3^4 \cdot 5^{110} \cdot 3^2$

$= \mathbf{2 \cdot 3^6 \cdot 5^{110}}$.

(d) $2 \cdot 3 \cdot 5 \cdot 7 \cdot 11 + 1 = \mathbf{2311}$, which is prime.

10. **(*i*)** $2^3 \cdot 3^2 \cdot 25^3$ is not a prime factorization because 25 is not prime.

(*ii*) $25^3 = (5^2)^3 = 5^6$, so **prime factorization** is $2^3 \cdot 3^2 \cdot 5^6$.

11. **No.** $8^z = (2^3)^z = 2^{3z}$. In other words, 8^z will always have a unique prime factorization that contains only 2*s*. On the other hand $3^x \cdot 5^y$ will always have a unique prime factorization that contains only 3*s* and 5*s*.

12. $32\,n = 2^6 \cdot 3^5 \cdot 5^4 \cdot 7^3 \cdot 11^7$

$= 2^5 \cdot 2 \cdot 3^5 \cdot 5^4 \cdot 7^3 \cdot 11^7$

$= 32 \cdot 2 \cdot 3 \cdot 5 \cdot 7 \cdot 11^6 \cdot 3^4 \cdot 5^3 \cdot 7^2 \cdot 11$

The Fundamental Theorem of Arithmetic tells us that

$n = 2 \cdot 3 \cdot 5 \cdot 7 \cdot 11^6 \cdot (3^4 \cdot 5^3 \cdot 7^2 \cdot 11)$.

Since $3^4 \cdot 5^3 \cdot 7^2 \cdot 11$ is a whole number, $2 \cdot 3 \cdot 5 \cdot 7 \cdot 11^6$ divides n.

13. **Yes.** $7^5 \cdot 11^3 = 7(7^4 \cdot 11^3)$.

14. **(a)** **1 × 48; 2 × 24; 3 × 16; or 4 × 12**, all pairs of divisors of 48 other than 6 by 8.

(b) **One.** 47 is prime, so the only possibility would be 1×47.

15. **(a)** **1, 2, 3, 5, 6, 10, 15, or 30** rows. The prime factorization of 30 is $2 \cdot 3 \cdot 5$, so there are $(1 + 1) \cdot (1 + 1) \cdot (1 + 1) = 8$ divisors.

(b) **1, 2, 4, 7, 14, or 28** rows. The prime factorization 28 is $2^2 \cdot 7$, so there are $(2 + 1) \cdot (1 + 1) = 6$ divisors.

(c) **1 or 23** rows. 23 is prime, so there are only two divisors.

(d) **1, 2, 3, 4, 5, 6, 8, 10, 12, 15, 20, 24, 30, 40, 60, or 120** rows. The prime factorization of 120 is $2^3 \cdot 3 \cdot 5$, so there are $(3 + 1) \cdot (1 + 1) \cdot (1 + 1) = 16$ divisors.

16. There are 70 days in 10 weeks. Briah swims on day 1 and then every other day (1, 3, 5, 7, 9, 11, 13, 15, 17, 19, 21, 23, 25, 27, 29, 31, 33, 35, 37, 39, 41, 43, 45, 47, 49, 51, 53, 55, 57, 59, 61, 63, 65, 67, 69). Emma swims on day 1 and then every three days (1, 4, 7, 10, 13, 16, 19, 22, 25, 28, 31, 34, 37, 40, 43, 46, 49, 52, 55, 58, 61, 64, 67, 70). **They swim together 12 times** {1, 7, 13, 19, 25, 31, 37, 43, 49, 55, 61, 67}.

    Another way to think of this: they swim together on the first day. Afterwards, they should swim together every $2 \times 3 = 6$th day. So, starting at 1 and adding 6 each time would give you the same 12 days as before {1, 7, 13, 19, etc.}

17. **91 eggs.** Let n be the number of eggs in the basket. Then $n-1$ is a multiple of 3 and of 5. Hence, $n-1 = (3 \cdot 5)k$ for some whole number k. Thus, $n = 15k+1$. We know that $n \le 100$ and that $7 \mid n$. We substitute $k = 1, 2, 3, \ldots$ to obtain the following values for n that are less than or equal to 100: 16, 31, 46, 61, 76, 91. Among these values, only 91 is divisible by 7. Hence, $n =$ **91**.

18. The least number of coins the pirates stole is a number that is a multiple of 15 that has a remainder of 3 when divided by 17 and a remainder of 9 when divided by 16. On a spreadsheet we can create a list of multiples of 15 (15, 30, 45, 60, 75, …) and a list of multiples of 16 plus 9 (25, 41, 57, 73, 89, …) and a list of multiples of 17 plus 3 (20, 37, 54, 71, 88, …) to find out that the least number of coins is **105**.

19. Consider all of the two-digit primes: 11, 13, 17, 19, 23, 29, 31, 37, 41, 43, 47, 53, 59, 61, 67, 71, 73, 79, 83, 89, 97. The two-digit primes with the tens digit greater than the units digit are 31, 41, 43, 53, 61, 71, 73, 83, and 97. In this list, there is only one where the sum of the two digits is also a two-digit prime: 83.

    Now consider the three-digit number. The digits are all different: 1, 3, 5, 7, 9. The sum of the three digits is palindromic. Only $1+3+7=11$ works. The sum of the first and third digit is one-half the sum of the first and second. The only order possible is 173 because $1+7=\frac{(1+3)}{2}$. So the **license plate number is 83-173.**

20. There is no analytic method; one must work through the list of primes. The twin primes less than 200 are: **3 and 5; 5 and 7; 11 and 13; 17 and 19; 29 and 31; 41 and 43; 59 and 61; 71 and 73; 101 and 103; 107 and 109; 137 and 139; 149 and 151; 179 and 181; 191 and 193; 197 and 199.**

21. **(a)** **(i)** $1+2+3+4+6=16$, so **12 is not perfect.**

    (ii) $1+2+4+7+14=28$, so **28 is perfect.**

    (iii) $1+5+7=13$, so **35 is not perfect.**

    (b) **Answers vary.** The next three perfect numbers after 28 are 496, 8128, and 2,096,128.

Mathematical Connections 4-2: Review Problems

23. Use the standard divisibility tests for all except 7; for 7 just perform the division:

    (a) Divisors are **2, 3, and 6.**

    (b) Divisors are **2, 3, 5, 6, 9, and 10.**

Assessment 4-3A: Greatest Common Divisor and Least Common Multiple

1. *(i)* The 2 rods can be used to build both the 6 rod and the 8 rod $\Rightarrow GCD(6,8) = \mathbf{2}$.

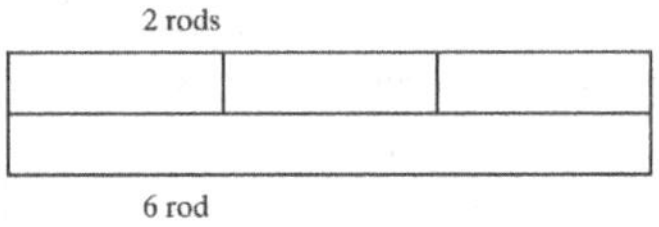

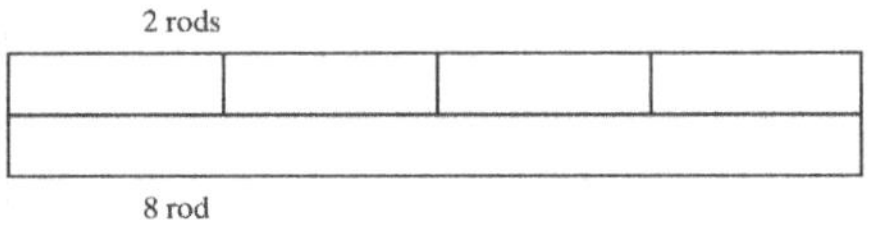

   (ii) Four 6 rods (length 24) are the same length as three 8 rods (length 24) $\Rightarrow LCM(6,8) = \mathbf{24}$.

2. **(a)** *(i)* $D_{18} = \{1,2,3,6,9,18\}$;

    $D_{12} = \{1,2,3,4,6,12\}$;

    $\Rightarrow GCD(18,12) = \mathbf{6}$.

    (ii) $M_{18} = \{18,36,54,72,90,\ldots\}$;

    $M_{12} = \{12,24,36,\ldots\}$;

    $\Rightarrow LCM(18,12) = \mathbf{36}$.

    (b) *(i)* $D_{20} = \{1,2,4,5,10,20\}$;

    $D_{36} = \{1,2,3,4,6,9,12,18,36\}$;

    $\Rightarrow GCD(20,36) = \mathbf{4}$.

(*ii*) $M_{20} = \{20, 40, 60, 80, 100, ..., 180, ...\}$;

$M_{36} = \{36, 72, 108, 144, 180, ...\}$;

$\Rightarrow LCM(24, 36) = \mathbf{180.}$

(c) (*i*) $D_8 = \{1, 2, 4, 8\}$;

$D_{24} = \{1, 2, 3, 4, 6, 8, 12, 24\}$;

$D_{64} = \{1, 2, 4, 8, 16, 32, 64\}$;

$\Rightarrow GCD(8, 24, 64) = \mathbf{8.}$

(*ii*) $M_8 = \{8, 16, 24, ..., 192, ...\}$;

$M_{24} = \{24, 48, 72, ..., 192, ...\}$;

$M_{64} = \{64, 128, 192, ...\}$;

$\Rightarrow LCM(8, 24, 64) = \mathbf{192.}$

(d) (*i*) $D_7 = \{1, 7\}$;

$D_9 = \{1, 3, 9\}$;

$\Rightarrow GCD(7, 9) = \mathbf{1.}$

(*ii*) $M_7 = \{7, 14, 21, ..., 63, ...\}$;

$M_9 = \{9, 18, 27, ..., 63, ...\}$;

$\Rightarrow LCM(7, 9) = \mathbf{63}.$

3. (a) $132 = 2 \cdot 2 \cdot 3 \cdot 11 = 2^2 \cdot 3 \cdot 11$;
$504 = 2 \cdot 2 \cdot 2 \cdot 3 \cdot 3 \cdot 7 =$
$2^3 \cdot 3^2 \cdot 7; \Rightarrow GCD(132, 504) =$
$2^2 \cdot 3 = \mathbf{12}.\ LCM(132, 504) =$
$2^3 \cdot 3^2 \cdot 7 \cdot 11 = \mathbf{5544}.$

(b) $65 = 5 \cdot 13$;

$1690 = 2 \cdot 5 \cdot 13 \cdot 13 = 2 \cdot 5 \cdot 13^2$;

$\Rightarrow GCD(65, 1690) = 5 \cdot 13 = \mathbf{65}.$

$LCM(65, 1690) = 2 \cdot 5 \cdot 13^2 = \mathbf{1690}.$

(c) $900 = 2 \cdot 2 \cdot 3 \cdot 3 \cdot 5 \cdot 5 = 2^2 \cdot 3^2 \cdot 5^2$;

$96 = 2 \cdot 2 \cdot 2 \cdot 2 \cdot 2 \cdot 3 = 2^5 \cdot 3$;

$630 = 2 \cdot 3 \cdot 3 \cdot 5 \cdot 7 = 2 \cdot 3^2 \cdot 5 \cdot 7$;

$\Rightarrow GCD(900, 96, 630) = 2 \cdot 3 = \mathbf{6}.$

$LCM(900, 96, 630) = 2^5 \cdot 3^2 \cdot 5^2 \cdot 7$
$= \mathbf{50,400}.$

(d) $108 = 2 \cdot 2 \cdot 3 \cdot 3 \cdot 3 = 2^2 \cdot 3^3$;

$360 = 2 \cdot 2 \cdot 2 \cdot 3 \cdot 3 \cdot 5 = 2^3 \cdot 3^2 \cdot 5$;

$\Rightarrow GCD(108, 360) = 2^2 \cdot 3^2 = \mathbf{36}.$

$LCM(108, 360) = 2^3 \cdot 3^3 \cdot 5 = \mathbf{1080}.$

4. $\rightarrow R$ symbolizes the reminder left after the indicated divisions.

(a) $GCD(2924, 220) = GCD(220, 64)$ since $2924 \div 220 \rightarrow R\,64$;

$GCD(220, 64) = GCD(64, 28)$ since $220 \div 64 \rightarrow R\,28$;

$GCD(64, 28) = GCD(28, 8)$ since $64 \div 28 \rightarrow R\,8$;

$GCD(28, 8) = GCD(8, 4)$ since $28 \div 8 \rightarrow R\,4$;

$GCD(8, 4) = GCD(4, 0)$ since $8 \div 4 \rightarrow R\,0$;

$\Rightarrow GCD(2924, 220) = \mathbf{4}.$

(b) $GCD(14595, 10856) = GCD(10856, 3739)$ since $14595 \div 10856 \rightarrow R\,3739$;

$GCD(10856, 3739) = GCD(3739, 3378)$ since $10856 \div 3739 \rightarrow R\,3378$;

$GCD(3739, 3378) = GCD(3378, 361)$ since $3739 \div 3378 \rightarrow R\,361$;

$GCD(3378, 361) = GCD(361, 129)$ since $3378 \div 361 \rightarrow R\,129$;

$GCD(361, 129) = GCD(129, 103)$ since $361 \div 129 \rightarrow R\,103$;

$GCD(129, 103) = GCD(103, 26)$ since $129 \div 103 \rightarrow R\,26$;

$GCD(103, 26) = GCD(26, 25)$ since $103 \div 26 \rightarrow R\,25$;

$GCD(26, 25) = GCD(25, 1)$ since $26 \div 25 \rightarrow R\,1$;

$GCD(25, 1) = GCD(1, 0)$ since $25 \div 1 \rightarrow R\,0$;

$\Rightarrow GCD(14595, 10856) = \mathbf{1}.$

5. (a) (*i*) Intersection of sets:

$M_{24} = \{24, 48, 72, 96, ...\}$;

$M_{36} = \{36, 72, 108, ...\}$;

$\Rightarrow LCM(24, 36) = \mathbf{72}.$

(*ii*) Prime factorization:

$24 = 2 \cdot 2 \cdot 2 \cdot 3 = 2^3 \cdot 3$;

$36 = 2 \cdot 2 \cdot 3 \cdot 3 = 2^2 \cdot 3^2$;

$\Rightarrow LCM(24, 36) = 2^3 \cdot 3^2 = \mathbf{72}.$

(b) ***(i)*** Intersection of sets:

$M_{72} = \{72, 144, 216, ..., 1440, ...\};$

$M_{90} = \{90, 180, 270, ..., 1440, ...\};$

$M_{96} = \{96, 192, 288, ..., 1440, ...\};$

$\Rightarrow LCM(72, 90, 96) = \mathbf{1440}.$

(ii) Prime factorization:

$72 = 2 \cdot 2 \cdot 2 \cdot 3 \cdot 3 = 2^3 \cdot 3^2;$

$90 = 2 \cdot 3 \cdot 3 \cdot 5 = 2 \cdot 3^2 \cdot 5;$

$96 = 2 \cdot 2 \cdot 2 \cdot 2 \cdot 2 \cdot 3$

$= 2^5 \cdot 3; \Rightarrow LCM(72, 90, 96)$

$= 2^5 \cdot 3^2 \cdot 5 = \mathbf{1440}.$

(c) ***(i)*** Intersection of sets:

$M_{90} = \{90, 180, 270, ..., 630, ...\};$

$M_{105} = \{105, 210, 315, ..., 630, ...\};$

$M_{315} = \{315, 630, 945, ...\};$

$\Rightarrow LCM(90, 105, 315) = \mathbf{630}.$

(ii) Prime factorization:

$90 = 2 \cdot 3 \cdot 3 \cdot 5 = 2 \cdot 3^2 \cdot 5;$

$105 = 3 \cdot 5 \cdot 7;$

$315 = 3 \cdot 3 \cdot 5 \cdot 7 = 3^2 \cdot 5 \cdot 7;$

$\Rightarrow LCM(90, 105, 315) = 2 \cdot 3^2 \cdot 5 \cdot 7$

$= \mathbf{630}$

(d) $9^{100} = (3^2)^{100} = 3^{200}$ and

$25^{100} = (5^2)^{100} = 5^{200}$

$\Rightarrow LCM(9^{100}, 25^{100}) = \mathbf{3^{200} \cdot 5^{200}}$

$= \mathbf{(15)^{200}}.$

6. If the GCD is 17 then the product of the two numbers divided by the GCD will result in the LCM. $1734 \div 17 = 102$, so **102** is their LCM.

7. If the GCD is 19 we can find the other factors of 57 by dividing 57 by 19. $57 \div 19 = 3$. The other number is the LCM divided by 3: $228 \div 3 = 76$. The **other number is 76**.

8. **(a)** $LCM(a, b) = \boldsymbol{ab}$. a and b have no common factors.

(b) $GCD(a, a) = \boldsymbol{a}$ and $LCM(a, a) = \boldsymbol{a}$. a has all factors in common with a.

(c) $GCD(a^2, a) = \boldsymbol{a}$ and $LCM(a^2, a) = \boldsymbol{a^2}$.

(d) $GCD(a, b) = \boldsymbol{a}$ and $LCM(a, b) = \boldsymbol{b}$, since $a|b$.

9. **(a)** **True**. If both a and b are even, then $GCD(a, b) \geq 2$.

(b) **True**. $GCD(a, b) = 2$ implies that $2|a$ and $2|b$.

(c) **False**. GCD could be any larger multiple of 2; e.g., $GCD(8, 20) = 4$.

10. **(a)** $GCD(120, 75) = GCD(75, 45) = GCD(45, 30) = GCD(30, 15) = GCD(15, 0);$

$GCD(120, 75) = 15.$

$GCD(105, 15) = GCD(15, 0);$

$GCD(105, 15) = 15.$

Thus $GCD(120, 75, 105) = \mathbf{15}.$

(b) $GCD(4618, 4619) = GCD(4618, 1) = 1;$

$GCD(34578, 1) = 1;$

Thus $GCD(34578, 4618, 4619) = \mathbf{1}.$

11. 2 is the only prime factor of 4 and $2 \nmid 97{,}219{,}988{,}751$, so 1 is their only common divisor; i.e., they are relatively prime.

12. **(a)** The Venn diagram below shows 5 as the only prime factor common to 10, 15, and 60; 2 and 5 as common prime factors of 10 and 60; 3 and 5 as common prime factors of 15 and 60; 2 as the remaining prime factor of 60:

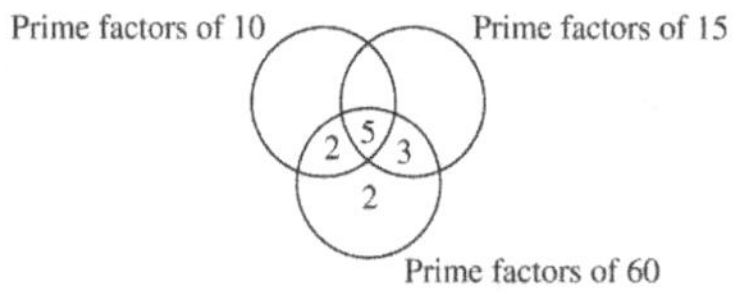

(b)

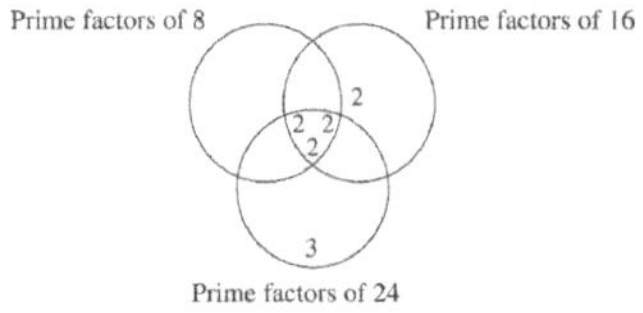

13. The prime factorization of 48 is $2^4 \cdot 3$. To find three pairs (a, b) such that their LCM is 48 is making sure that a and b have 2^4 and 3 in their LCM. For example: (1, 48), (2, 48), (3,48), etc.

14. This will be the set of all numbers between 1 and 49 that are relatively prime to 49. Since $49 = 7^2$, the set will contain all numbers between 1 and 49 **except** $1 \cdot 7, 2 \cdot 7, 3 \cdot 7, 4 \cdot 7, 5 \cdot 7, 6 \cdot 7,$ and $7 \cdot 7$.

15. **(a)** The real question is "what is $LCM(15, 40, 60)$?" since this is when the alarms will coincide: $LCM(15, 40, 60) = 120$, or **120 minutes;** i.e., 2 hours later, at 8:00 A.M.

(b) **No**. This would be equivalent to changing locations of clocks A and B in the room.

16. The **780th "like"**. $LCM(12,13,20) = 780$.

17. There are 15 children. The candy will come out evenly for $LCM(15,12) = 60$; 60 candies ÷ 12 candies per package = **5 packages**.

18. The question is really, "What is LCM (12, 18, 16)?" since that is when the times will coincide LCM (12, 18, 16) = **144 minutes**.

19. GCD (42, 54) = **6**. She can make 7 bags of chocolate chip cookies and 9 bags of sugar cookies. **Each bag has 6 cookies** in it.

20. LCM (100,18) = **900 inches or 75 feet** before points P and Q touch the sidewalk at the same time again.

21. GCD(72, 42) = 6. The largest pieces she can cut are **6 yards** long.

Mathematical Connections 4-3: Review Problems

19. (a) $17{,}496 = \mathbf{2^3 \cdot 3^7}$

(b) $32715 = \mathbf{3^2 \cdot 5 \cdot 727}$

(c) $2^4 \cdot 8^2 \cdot 27^3 = 2^4 \cdot (2^3)^2 \cdot (3^3)^3 = 2^4 \cdot 2^6 \cdot 3^9 = \mathbf{2^{10} \cdot 3^9}$

21. The question is really "What is the LCM (2, 4, 6, 8, 10)?". Their factors are $\{2, 2^2, 2 \cdot 3, 2^3, 2 \cdot 5\}$, thus LCM (2, 4, 6, 8, 10) $= 2^3 \cdot 3 \cdot 5 = \mathbf{120}$.

Chapter 4 Review

1. (a) **False**. $12 \nmid 4$ since $4 = 12x$ has no whole number solution for x.

(b) **False**. $0 \nmid 8$ since $8 = 0 \cdot x$ has no whole number solution for x.

(c) **True**. 0 divided by any natural number is 0.

(d) **No**. For example, 12. The test works only when the two numbers have no common factors and 4 and 6 have a common factor of 2.

(e) **No**. For example, 9 is not divisible by 12 but is divisible by 3.

2. (a) $m = 125{,}160$.

$2|m$ because $2|0$.

$3|m$ because $3|(1 + 2 + 5 + 1 + 6 + 0)$.

$4|m$ because $4|60$.

$5|m$ because $5|0$.

$6|m$ because $2|m$ and $3|m$.

$8|m$ because $8|160$.

$10|m$ because $10|0$.

(b) $m = 12{,}193$.

$2 \nmid m$ because $2 \nmid 3$.

$3 \nmid m$ because $3 \nmid (1+2+1+9+3)$.

$4 \nmid m$ because $4 \nmid 93$.

$5 \nmid m$ because $5 \nmid 3$.

$6 \nmid m$ because $2 \nmid m$ and $3 \nmid m$.

$8 \nmid m$ because $8 \nmid 193$.

$9 \nmid m$ because $9 \nmid (1+2+1+9+3)$.

$11 \nmid m$ because $11 \nmid (1 + 1 + 3) - (2 + 9)$.

3. (a) Write the number as $87a4$.

$6|87a4$ if $2|87a4$ and $3|87a4$;

$2|4$ so $2|87a4$, thus

$6|87a4$ if

$3|87a4 \Rightarrow 3|(8 + 7 + a + 4 = 19 + a)$;

$3|(21, 24, \text{or } 27)$ when $a = 2, 5, \text{or } 8$;

so $6|(87\underline{2}4, 87\underline{5}4, \text{or } 87\underline{8}4)$. **The greatest digit is 8.**

(b) Write the number as $4a856$.

$24|4a856$ if $3|4a856$ and $8|4a856$;

$8|856$ so $8|4a856$, thus

$24|4a856$ if

$3|(4 + a + 8 + 5 + 6) = 23 + a$;

$3|(24, 27, \text{or } 30)$ when $a = 1, 4, \text{or } 7$;

so $24|(4\underline{1}856, 4\underline{4}856, \text{or } 4\underline{7}856)$. **The greatest digit is 7.**

(c) Write the number as $87a4$.

$29|8700$ so $29|a4$.

$a4$ is a two-digit number; and, the only whole number to return 4 in the units position when multiplied by 9 is 6. However, $6 \cdot 29 = 174$, which is larger than any number $a4$ could be. Therefore, there is **no solution**; there is no number possible for a that will allow $87a4$ to be divisible by 29.

4. **(a)** The student's claim is true; examples may vary. For example.,

(i) $5|(3 + 4 + 5 + 6 + 7)$.

(ii) $5|(5 + 6 + 7 + 8 + 9)$.

(b) Let n be a whole number; then

$n + (n + 1) + (n + 2) + (n + 3) +$
$(n + 4) = 5n + 10 = 5(n + 2)$.

Thus the sum is divisible by 5.

5. *(i)* The number must be **divisible by 3 and 8.**

(ii) $3|4152$ and $8|4152$ so $24|4152$.

6. *(i)* Answers may vary; e.g., 16.

(ii) To obtain k divisors raise any prime to the $(k - 1)$st power; e.g., to obtain 5 divisors take $2^{5-1} = 16$; divisors are $2^0, 2^1, 2^2, 2^3$, and 2^4.

7. $144 = (2^2 \cdot 3)^2 = 2^4 \cdot 3^2$. There are thus $(4 + 1) \cdot (2 + 1) = 15$ divisors: **1, 2, 3, 4, 6, 8, 9, 12, 16, 18, 24, 36, 48, 72 and 144.**

8. $n = a \cdot 10^2 + b \cdot 10 + c$

$= a(99 + 1) + b(9 + 1) + c$

$= 99a + 9b + (a + b + c)$.

Since $9|99a$ and $9|9b$ then
$9 \mid [99a + 9b + (a + b + c)]$
if and only if $9|(a + b + c)$.

9. 1009 is prime, so $17 \nmid 1009$ but $17|17$.
$17 \nmid (1009 + 17)$ by Theorem 4-2(b).

10. **(a)** **Composite.** $3 \mid 147$.

(b) **Prime.** Primes through 19 do not divide 373; $23^2 > 373$.

11. **(a)** $7 \cdot 11 \cdot 13 \cdot 17 + 17 = 17(7 \cdot 11 \cdot 13 + 1)$, so 17 is a factor.

(b) $10! + k$ can be factored as in part (a), depending on the value of $k (2 \le k \le 10)$, thus composite.

12. First show that among any three consecutive odd whole numbers there is always one divisible by 3.

Suppose that the first whole number in the triplet is not divisible by 3. By the Division Algorithm that whole number can be written in the form $3n + 1$ or $3n + 2$ for some whole number n.

Then the three consecutive odd whole numbers are $(3n + 1, 3n + 3, 3n + 5)$ or $(3n + 2, 3n + 4, 3n + 6)$.

In the first triplet $3|(3n + 3)$; in the second $3|(3n + 6)$. If the first whole number is greater than 3 and not divisible by 3 then the second or third must be divisible by 3 and so cannot be prime.

13. **(a)**

$$\begin{array}{r|lll} 3 & 1 & 1 & 1 \\ \hline & & 3 & 7 \end{array} \Rightarrow 111 = \mathbf{3 \times 37}.$$

(b) $\Rightarrow 144 = \mathbf{2^4 \times 3^2}$.

$$\begin{array}{r|lll} 2 & 1 & 4 & 4 \\ \hline 2 & & 7 & 2 \\ \hline 2 & & 3 & 6 \\ \hline 2 & & 1 & 8 \\ \hline 3 & & & 9 \\ \hline & & & 3 \end{array}$$

(c)

$$\begin{array}{r|lll} 2 & 1 & 8 & 8 \\ \hline 2 & & 9 & 4 \\ \hline & & 4 & 7 \end{array} \Rightarrow 188 = \mathbf{2^2 \times 47}.$$

(d)

$$\begin{array}{r|lll} 2 & 5 & 2 & 0 \\ \hline 2 & 2 & 6 & 0 \\ \hline 2 & 1 & 3 & 0 \\ \hline 5 & & 6 & 5 \\ \hline & & 1 & 3 \end{array} \Rightarrow 520 = \mathbf{2^3 \times 5 \times 13}.$$

14. **(a)** $10^{10} = (2 \cdot 5)^{10} = \mathbf{2^{10} \cdot 5^{10}}$.

(b) $89^4 = \mathbf{89^4}$.

(c) $8^3 \cdot 6^4 \cdot 13^2 = (2^3)^3 \cdot (2 \cdot 3)^4 \cdot 13^2 =$
$2^9 \cdot 2^4 \cdot 3^4 \cdot 13^2 = \mathbf{2^{13} \cdot 3^4 \cdot 13^2}$

(d) $2^3 \cdot 3^2 + 2^4 \cdot 3^3 \cdot 7 =$

$2^3 \cdot 3^2(1 + 2 \cdot 3 \cdot 7) =$

$\mathbf{2^3 \cdot 3^2 \cdot 43}$.

(e) $2^4 \cdot 3 \cdot 5^7 + 2^4 \cdot 5^6 =$

$2^4 \cdot 5^6(3 \cdot 5 + 1) = 2^4 \cdot 5^6 \cdot 16 =$

$2^4 \cdot 5^6 \cdot 2^4 = \mathbf{2^8 \cdot 5^6}$.

15. Product (all whole numbers ≤ 10) $=$
$1 \cdot 2 \cdot 3 \cdot 4 \cdot 5 \cdot 6 \cdot 7 \cdot 8 \cdot 9 \cdot 10 =$
$1 \cdot 2 \cdot 3 \cdot 2^2 \cdot 5 \cdot (2 \cdot 3) \cdot 7 \cdot 2^3 \cdot 3^2 \cdot (2 \cdot 5)$.

Thus *LCM* (all whole numbers ≤ 10) $=$
$1 \cdot 2^3 \cdot 3^2 \cdot 5 \cdot 7 = \mathbf{2520}$.

16. **(a)** $24 = 2^3 \cdot 3$ and $52 = 2^2 \cdot 13$. Thus
$GCD(24, 52) = 2^2 = \mathbf{4}$.

(b) $GCD(5767, 4453) = GCD(4453, 1314)$
$= GCD(1314, 511) = GCD(511, 292)$
$= GCD(292, 219) = GCD(219, 73)$
$= GCD(73, 0) \Rightarrow GCD(5767, 4453) = \mathbf{73}$.

17. **(a)** $LCM(2^3 \cdot 5^2 \cdot 7^3, 2 \cdot 5^3 \cdot 7^2 \cdot 13,$
$2^4 \cdot 5 \cdot 7^4 \cdot 29) =$
$\mathbf{2^4 \cdot 5^3 \cdot 7^4 \cdot 13 \cdot 29}$.

(b) $GCD(277, 278) = \text{GCD}(277, 1) = 1$

Therefore no common factors in 277 and 278.

Thus $LCM(277, 278) \cdot 1 = 277 \cdot 278$
$= \mathbf{77{,}006}$.

18. **No**. *LCM* and *GCD* are the same if the numbers are equal.

19. $LCM(a, b, c) = LCM(m, c)$, where
$m = LCM(a, b)$. $LCM(a, b) = \frac{ab}{GCD(a,b)}$ and
$LCM(m, c) = \frac{mc}{GCD(m,c)}$. Each of the above
*GCD*s can be found using the Euclidean algorithm.

20. If $GCD(a, b) = 1 \Rightarrow \boldsymbol{LCM\,(a, b) = ab}$.

$GCD(a, b) \cdot LCM(a, b) = ab$; since $GCD\,(a, b) = 1$ then $LCM\,(a, b) = ab/1 = ab$.

21. **31¢**. The price must divide 3193¢; 31|3100 and 31|93 thus 31|3193. You could also argue the price is 1¢.

22. $9869 = 71 \cdot 139$ (both prime) implies **71 lattes** at \$1.39 each. She could not have sold 139 lattes at \$0.71 each, since she never sells for less than \$1.

23. LCM (45,30) = 90 minutes. 8:00 A.M. + 90 minutes = **9:30 A.M.**

24. The question is really "What is *GCD* (120, 144)?" since what is needed is the greatest number which will divide both 120 and 144: *GCD* (120, 144) = **24 coins**.

25. The runners will be at the starting place at the same time at *LCM* (3, 5) = **15 minutes**.

26. GCD(60,24) = 12. There will be 60 ÷12=5 bags of oranges and 24 ÷ 12 = 2 bags of apples. Each bag will have **12 pieces of fruit**.

27. The LCM (180,90,45) = 180 seconds. So if the lights all flash together at 8:00 a.m., they will all flash together 180 seconds later at **8:03 a.m. on Monday.**

28. $GCD(n + 2, n) = GCD(n + 2 - n, n)$
$= GCD(2, n)$; if n is even, then
$GCD(2, n) = 2$. $2 \mid n$ because n is even. If n is odd, then $GCD(2, n) = 1$. and $1 \mid n$.

CHAPTER 5

INTEGERS

Assessment 5-1A: Integers and the Operations of Addition and Subtraction

1. For every integer a, there exists a unique integer ^{-}a, such that $a + {}^{-}a = 0 = {}^{-}a + a$. ^{-}a is called the additive inverse of a.

   (a) The unique integer $\mathbf{^{-}2}$ is the additive inverse of 2 because $2 + {}^{-}2 = 0$.

   (b) The additive inverse of ^{-}a can be written as $^{-}(^{-}a) = a$, or the number of the opposite sign. Thus the additive inverse of $^{-}6$ is $^{-}(^{-}6) = \mathbf{6}$.

   (c) The additive inverse of m is $\boldsymbol{^{-}m}$.

   (d) The additive inverse of 0 is **0** because $0 + 0 = 0$.

   (e) The additive inverse of ^{-}m is $^{-}(^{-}m) = \boldsymbol{m}$.

   (f) The additive inverse of $(a + b)$ is $^{-}(a + b) = \boldsymbol{^{-}a + {}^{-}b}$.

2. **(a)** $^{-}(^{-}2) = \mathbf{2}$ is the additive inverse of $^{-}2$.

   (b) $^{-}(^{-}m) = \boldsymbol{m}$.

   (c) $^{-}0 = \mathbf{0}$ since $^{-}0 + 0 = 0$.

3. **(a)** Absolute value is the distance on a number line between the origin (0) and a specified number. Distance on the number line between 0 and $^{-}5$ is 5 units, so $|^{-}5| = \mathbf{5}$.

   (b) Distance on the number line between 0 and 10 is 10 units, so $|10| = \mathbf{10}$.

   (c) $^{-}|^{-}5|$ means the additive inverse of the absolute value of $^{-}5$, so $^{-}|^{-}5| = {}^{-}(5) = \mathbf{^{-}5}$.

   (d) $^{-}|5| = {}^{-}(5) = \mathbf{^{-}5}$.

4. **(a)** According to the Definition of Integer Addition Using Absolute Values,

$$7 + {}^{-}13 = {}^{-}(|^{-}13| - |7|) = {}^{-}(13 - 7) = {}^{-}(6) = {}^{-}6.$$

   (b) According to the Definition of Integer Addition Using Absolute Values,

$${}^{-}7 + {}^{-}13 = {}^{-}(|7| + |13|) = {}^{-}(7+13) = {}^{-}(20) = {}^{-}20.$$

5. **(a)**

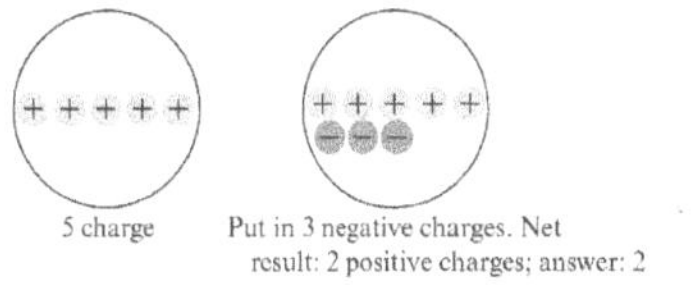

5 charge
Put in 3 negative charges. Net result: 2 positive charges; answer: 2

   (b)

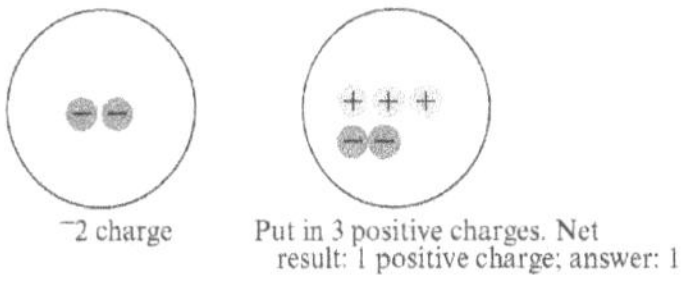

⁻2 charge
Put in 3 positive charges. Net result: 1 positive charge; answer: 1

   (c)

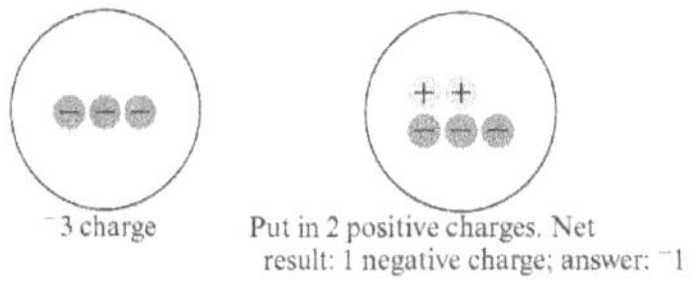

⁻3 charge
Put in 2 positive charges. Net result: 1 negative charge; answer: ⁻1

   (d)

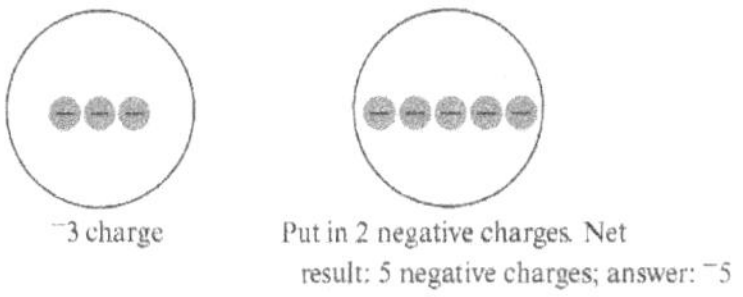

⁻3 charge
Put in 2 negative charges. Net result: 5 negative charges; answer: ⁻5

6. Movement on the number line in addition is always to the right for a positive number and to the left for a negative number.

   (a)

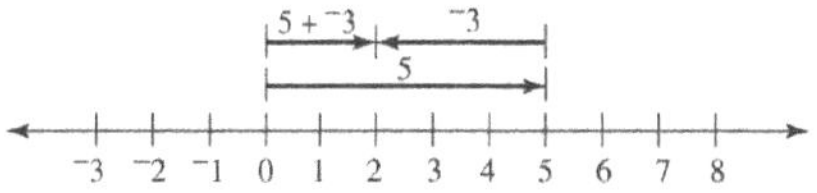

5 + ⁻3
⁻3
5
⁻3 ⁻2 ⁻1 0 1 2 3 4 5 6 7 8

(b)

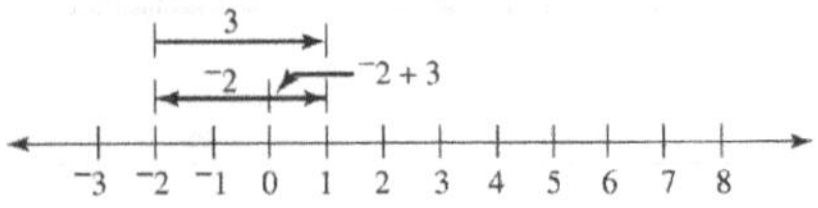

(c)

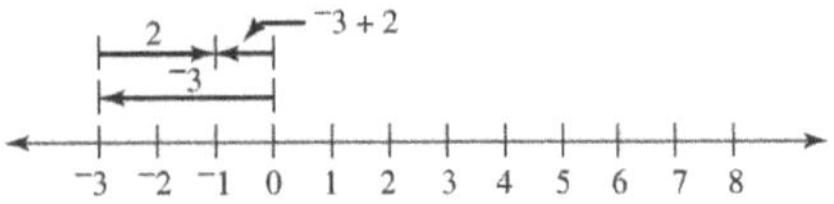

(d)

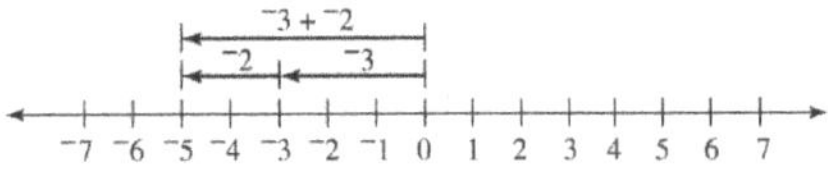

7. (a)

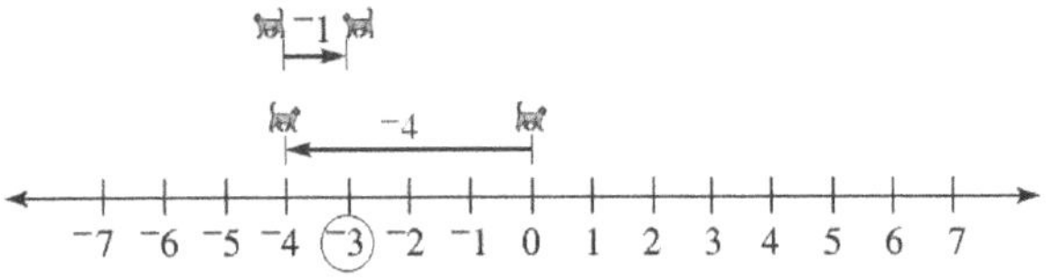

(b)

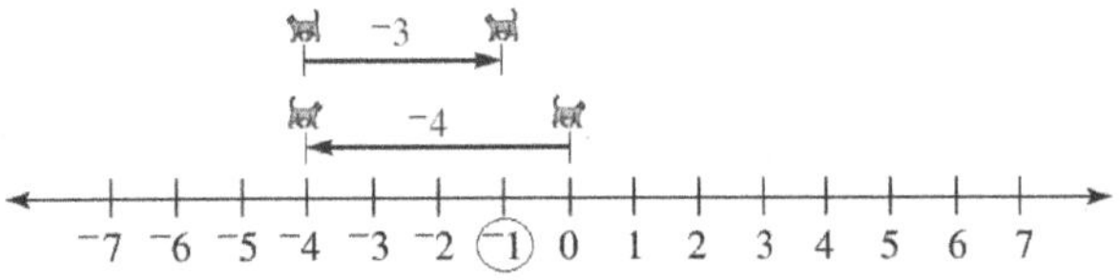

(c)

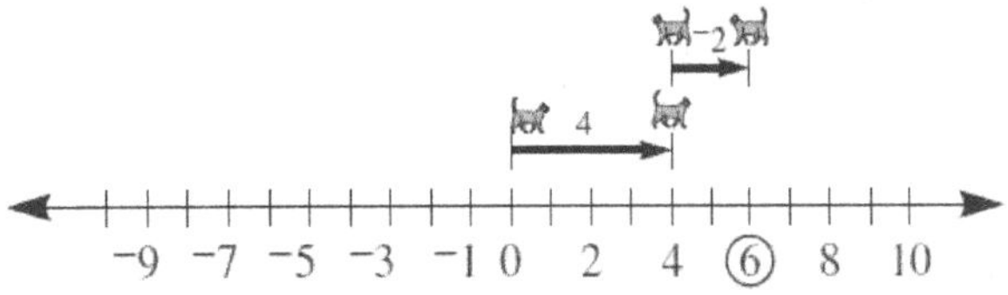

(d)

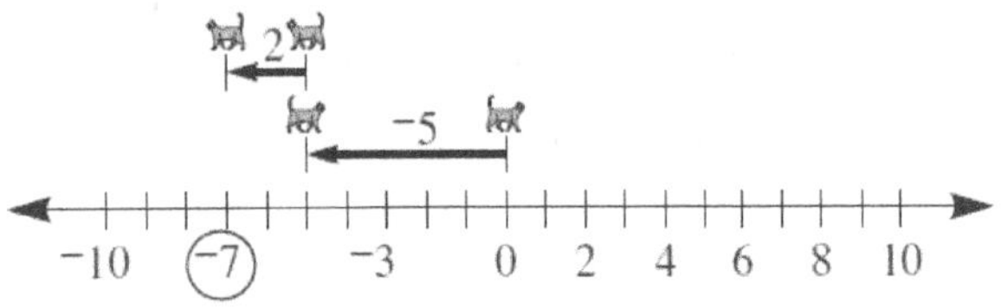

8. For all of these, we will use the "take away" model of subtraction.

(a)

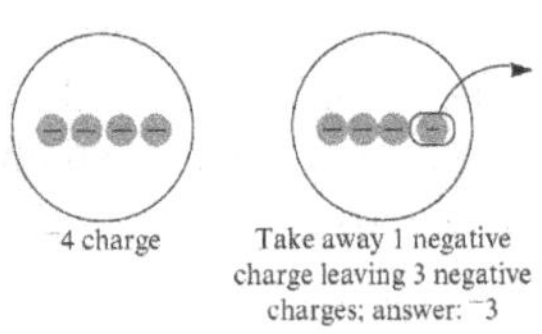

⁻4 charge
Take away 1 negative charge leaving 3 negative charges; answer: ⁻3

(b)

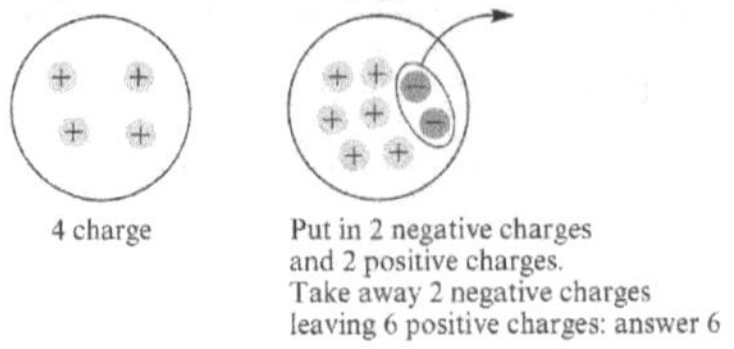

4 charge
Put in 2 negative charges and 2 positive charges. Take away 2 negative charges leaving 6 positive charges: answer 6

(c)

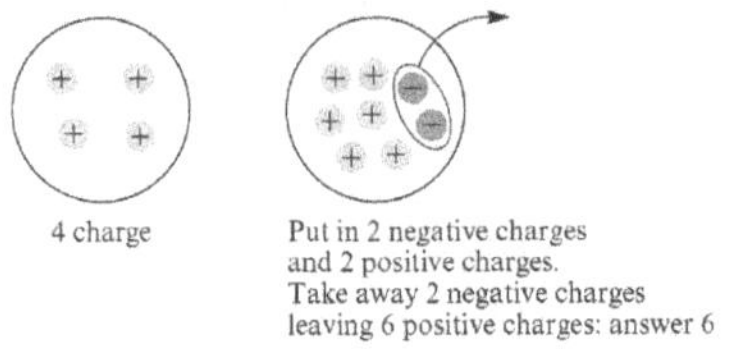

4 charge
Put in 2 negative charges and 2 positive charges. Take away 2 negative charges leaving 6 positive charges: answer 6

(d)

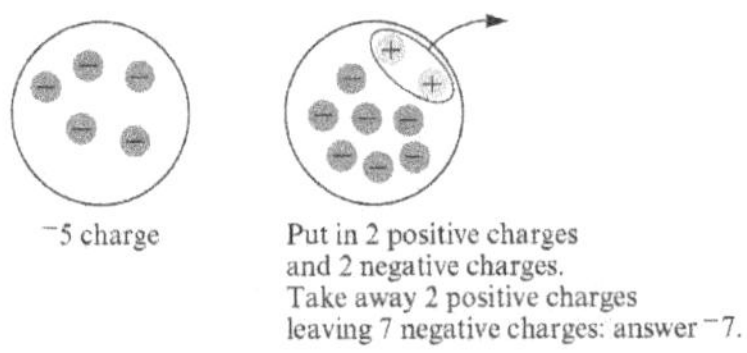

⁻5 charge
Put in 2 positive charges and 2 negative charges. Take away 2 positive charges leaving 7 negative charges: answer ⁻7.

9. (a) Start with $^-4 - n$ where n is a positive number that you know the difference, for example $^-4 - 2 = {}^-6$ then, continue to subtract smaller values as shown below:

$^-4 - 2 = {}^-6$

$^-4 - 1 = {}^-5$

$^-4 - 0 = {}^-4$

$\mathbf{^-4 - {}^-1 = {}^-3}$.

(b) Start with $n - 1$, where n is a positive number such that you know the difference; for example $2 - 1 = 1$ then, continue to subtract smaller values as shown below:

$2 - 1 = 1$

$1 - 1 = 0$

$0 - 1 = {}^-1$

$^-1 - 1 = {}^-2$

$\mathbf{^-2 - 1 = {}^-3}$

10. For problems (a) through (c), when trying to find a difference (D) between high (H) and low (L) temperatures, the equation is $H - L = D$.

(a) $121^{\circ}\text{F} - ({}^-60^{\circ}\text{F}) = \mathbf{181^{\circ}F}$.

(b) Let n equal the high temperature at 2:00 p.m. Then n°F – ($^-2$°)F = 19°F.

Adding $^-2$, to both sides of the equation, we get n°F = 17°F. After rising 19 degrees, the temperature at 2:00 p.m. is **17°F**.

(c) 45°F – ($^-4$)°F = **49°F**.

(d) 52°F + 13°F + 15°F = **80°F**.

11. For each problem there is a beginning amount of money, which could be positive or negative; then you either add deposits (which are positive) or add debts (which are negative) to that amount.

(a) ($^-$\$52) + \$150 = **\$98.**

(b) \$43 + ($^-$\$65) = **$^-$\$22.**

(c) \$80 + ($^-$\$59) + ($^-$\$43) = **$^-$\$22.**

(d) ($^-$\$56) + ($^-$\$234) = **$^-$\$290.**

12. **($^-$6) + 7 = 1.**

13. *score* = 5 + ($^-3$) + 0

score = 2; **2 above par.**

14. **(a)** $^-2 + (3 - 10) = {^-2} + {^-7} = {^-}\mathbf{9}$.

(b) $[8 - (^-5)] - 10 = [8 + {^-}(^-5)] - 10 = [8 + 5] - 10 = 13 - 10 = \mathbf{3}$.

(c) $(^-2 - 7) + 10 = (^-2 + {^-7}) + 10 = (^-9) + 10 = \mathbf{1}$.

15. **(a)** **(*i*)** 55 – 60.

(*ii*) 55 + ($^-60$).

(*iii*) T = 55 + ($^-60$) = **$^-5$°F.**

(b) **(*i*)** 200 – 220.

(*ii*) 200 + ($^-220$).

(*iii*) Balance = \$200 + $^-$\$220 = **$^-$\$20.**

16. **(a)** $3 - (2 - 4x) = 3 + {^-}(2 + {^-4x})$

$= 3 + {^-}(2) + {^-}(^-4x)$

$= 3 + {^-2} + 4x = \mathbf{1 + 4x}$.

(b) $x - (^-x - y) = x + {^-}(^-x + {^-y})$

$= x + {^-}(^-x) + {^-}(^-y)$

$= x + x + y = \mathbf{2x + y}$.

17. It can be shown that for all integers a, b, and c;

$a - b + c = a + {^-b} + c$ Theorem 5-4

$= a + {^-}(b + {^-c})$ Theorem 5-3 and Theorem 5-4

$= a - (b - c)$ Theorem 5-4

thus the original equation holds true for all integers.

18. Answers may vary. The key idea for the approach below is to make the sum across all rows, columns, and diagonals equal to zero.

$^-$3	**4**	**$^-$1**
2	**0**	**$^-$2**
1	**$^-$4**	**3**

19. If $y = {^-x} - 1$:

(a) $y = {^-}(^-1) - 1 = 1 + {^-1} = \mathbf{0}$,

when $x = {^-1}$

(b) $y = {^-}(100) + {^-1} = {^-100} + {^-1} = {^-}\mathbf{101}$, when $x = 100$

(c) $y = {^-}(^-2) - 1 = 2 + {^-1} = \mathbf{1}$,

when $x = {^-2}$

(d) $y = {^-}(^-a) - 1 = \mathbf{a - 1}$,

when $x = {^-a}$

(e) $3 = {^-x} - 1 \Rightarrow 3 + 1 = {^-x} - 1 + 1$

$\Rightarrow 4 = {^-x} \Rightarrow x = {^-}\mathbf{4}$.

20. **(a)** The common difference is 1. The nth term is

$n + (^-41)$. Set the nth term equal to the last term and solve for n.

$n + (^-41) = 40 \Rightarrow n = 81$. There are **81 terms** in the arithmetic sequence.

21. Answers vary. The east coast is 3 hours ahead of the west coast. So, if the east coast time is 0, then the west coast time is $^-3$; or if the west coast time is 0, then the east coast time is 3.

22. **(a)** **All negative integers**. If x is negative, then its additive inverse, ^-x, is positive, i.e.

$^-(^-x) = x$.

(b) **All positive integers**. If x is positive, then its additive inverse, ^-x, is negative.

(c) **All integers** $< {}^{-}\mathbf{1}$. If ${}^{-}x - 1 > 0 \Rightarrow$

${}^{-}x > 1 \Rightarrow x < {}^{-}1$.

(d) $x = \mathbf{2}$ **or** $x = {}^{-}\mathbf{2}$. If $|x| = 2$, then x is a value on the number line that is 2 units away, in either direction, from 0.

23. (a) If $|x - 6| = 6$, then $x - 6 = 6$ or $x - 6 = {}^{-}6$:

(*i*) If $x - 6 = 6 \Rightarrow x - 6 + 6 = 6 + 6 \Rightarrow \mathbf{x = 12}$, or

(*ii*) If $x - 6 = {}^{-}6 \Rightarrow x - 6 + 6 = {}^{-}6 + 6 \Rightarrow \mathbf{x = 0}$.

(b) $|x| + 2 = 10 \Rightarrow |x| + 2 - 2 = 10 - 2 \Rightarrow |x| = 8 \Rightarrow \mathbf{x = 8}$ **or** $\mathbf{x = {}^{-}8}$.

(c) $|{}^{-}x| = |x|$ is **true for all integers**, since the distance from 0 to x on a number line is the same as the distance from 0 to ${}^{-}x$.

24. To find the common difference (d) in an arithmetic sequence, subtract the first term from the second:

(a) $d = {}^{-}3 - 0 = {}^{-}\mathbf{3}$. The next two terms are ${}^{-}\mathbf{12}, {}^{-}\mathbf{15}$.

(b) $d = x - (x + y) = {}^{-}y$. The next two terms are $\mathbf{x - 2y}$, $\mathbf{x - 3y}$.

25. (a) **True**. Absolute value is always non-negative.

(b) **True**. Absolute value of the difference between the two elements is always non-negative.

(c) **True**. ${}^{-}x + {}^{-}y = {}^{-}(x + y)$, and $|{}^{-}(x + y)| = |x + y|$.

26. (a) $x + 7 = 3 \Rightarrow x + 7 - 7 = 3 - 7 \Rightarrow x = 3 + {}^{-}7 \Rightarrow \mathbf{x = {}^{-}4}$.

(b) ${}^{-}10 + x = {}^{-}7 \Rightarrow {}^{-}10 + 10 + x = {}^{-}7 + 10 \Rightarrow \mathbf{x = 3}$.

(c) ${}^{-}x = 5 \Rightarrow {}^{-}x + x = 5 + x \Rightarrow 0 = 5 + x \Rightarrow 0 + {}^{-}5 = 5 + {}^{-}5 + x \Rightarrow {}^{-}5 = x \Rightarrow \mathbf{x = {}^{-}5}$.

Assessment 5-2A: Multiplication and Division of Integers

1. $3 \cdot {}^{-}1 = {}^{-}1 + {}^{-}1 + {}^{-}1 = {}^{-}3$

$2 \cdot {}^{-}1 = {}^{-}1 + {}^{-}1 = {}^{-}2$

$1 \cdot {}^{-}1 = {}^{-}1$

$0 \cdot {}^{-}1 = 0$

${}^{-}1 \cdot {}^{-}1 = 1$.

2. (a) To find $({}^{-}4)({}^{-}2)$ using the chip model, the signs are interpreted as follows: ${}^{-}4$ is taken to mean *remove four groups of*; and ${}^{-}2$ is taken to mean *2 red chips*. So, begin with a set that has a zero charge, as shown below:

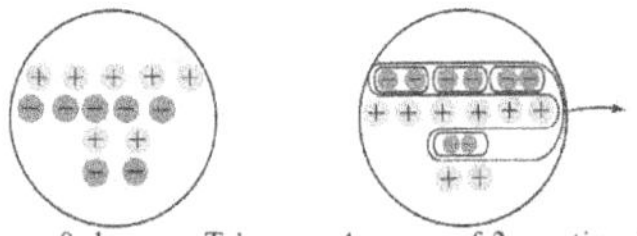

0 charge   Take away 4 groups of 2 negative charges. Net result is 8 positive charges; answer 8

Now, when four groups of 2 red chips are removed, the remaining positive chips are the solution. So $({}^{-}4)({}^{-}2) = \mathbf{8}$.

(b) $2 \cdot ({}^{-}5)$ is interpreted as creating two groups worth ${}^{-}5$ each, or creating *two groups of 5 red chips*, as shown below:

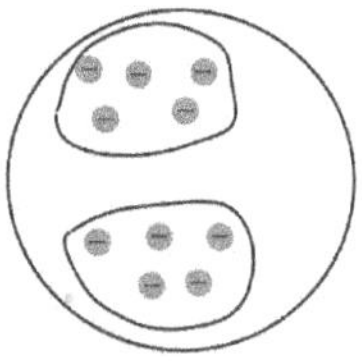

The solution is the value of the groups, which is negative 10. So $2 \cdot ({}^{-}5) = {}^{-}\mathbf{10}$.

(c) To find $^-12 \div {}^-4$, we will use the repeated subtraction model of division. This means we will begin with a set worth $^-12$, and repeatedly subtract from it sets worth $^-4$; our goal is to find out how many sets we create.

Two groups of $^-5 = {}^-10$

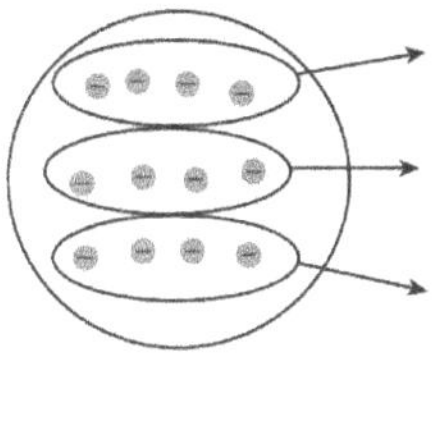

You can take away $^-4$ 3 times.

As shown above, three groups each worth $^-4$ can be created. So $^-12 \div {}^-4 = \mathbf{3}$.

(d) To find $^-24 \div 3$, the partition model of division will be used. This means we begin with a set worth $^-24$, and partition the chips into three groups. Our solution will be the value (worth) of each group. See below:

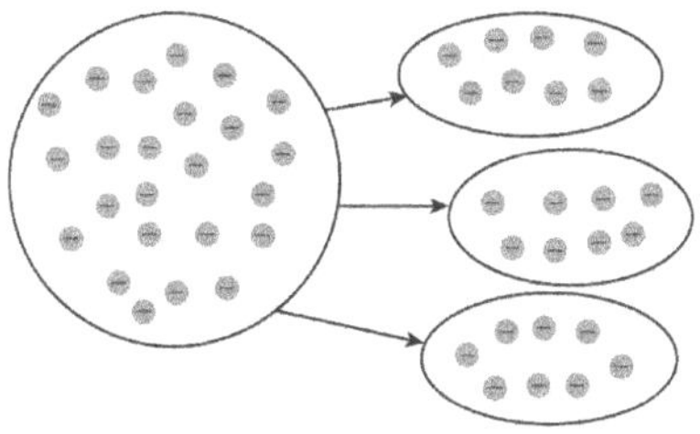

You can partition $^-24$ into 3 groups with $^-8$ in each group.

Each group is worth $\mathbf{^-8}$.

So $^-24 \div 3 = \mathbf{^-8}$.

3. Move four units to the left twice to arrive at the product $2(^-4) = {}^-8$:

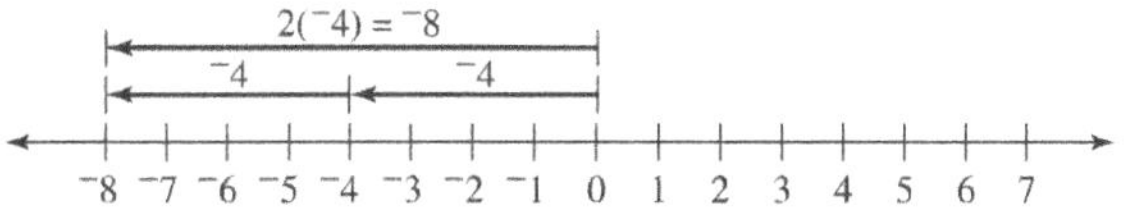

4. **(a)** Take away three groups of three negative charges, leaving nine positive charges; i.e., $\mathbf{(^-3) \cdot (^-3) = 9}$.

(b) Take away five groups of two positive charges, leaving ten negative charges; i.e., $\mathbf{(5)(^-2) = {}^-10}$.

5. **(a)** The number of students will decrease by 30 per year over the next four years, or $\mathbf{4(^-30) = {}^-120}$.

(b) There were 30 more students per year four years ago, or $\mathbf{(^-4)(^-30) = 120}$.

(c) The number of students will decrease by 30 per year over the next n years, or $n(^-30) = {}^-30n$ **students.**

(d) There were 30 more students per year during the past n years, or $\mathbf{(^-n)(^-30) = 30n}$ **students.**

6. The definition of integer division states that for all integers a and b, $a \div b$ is the *unique* integer c, if it exists, such that $a = bc$.

(a) By the definition of integer division, $^-40 \div {}^-5 = c \Rightarrow {}^-40 = {}^-5 \cdot c$. $c = 8$ makes this true; so $^-40 \div {}^-5 = \mathbf{8}$.

(b) By the definition of integer division, $^-143 \div 13 = c \Rightarrow {}^-143 = 13 \cdot c$. $c = {}^-11$ makes this true, so $^-143 \div 13 = \mathbf{^-11}$.

(c) By the definition of integer division, $^-5 \div 0 = c \Rightarrow {}^-5 = 0 \cdot c$. However, no integer c exists to satisfy the multiplication statement. Therefore, $^-5 \div 0$ is **undefined**.

7. **(a)** $(^-10 \div {}^-2)(^-2) = 5 \cdot {}^-2 = \mathbf{^-10}$.

(b) $(^-10 \cdot 5) \div 5 = {}^-50 \div 5 = \mathbf{^-10}$.

(c) $^-8 \div (^-8 + 8) = {}^-8 \div 0 \Rightarrow$ is **undefined** because there is no integer that would make the multiplication statement true.

(d) $(^-6 + 6) \div (^-2 + 2) = 0 \div 0 \Rightarrow$ is **undefined** because too many integers can make the multiplication equation true.

(e) $|^-24| \div [4 \cdot (9 - 15)] = 24 \div (4 \cdot {}^-6) = \mathbf{^-1}$.

Order of operations specifies that multiplication and division be performed from left to right after the operations within the absolute value and parentheses are performed.

8. For each of the following, if $a \cdot b = c \Rightarrow c \div b = a$ and $c \div a = b$ ($a \neq 0$ and $b \neq 0$):

(a) (*i*) $^{-}6 \cdot 5 = ^{-}\mathbf{30}$.

(*ii*) $^{-}30 \div 5 = ^{-}\mathbf{6}$.

(*iii*) $^{-}30 \div ^{-}6 = \mathbf{5}$.

(b) (*i*) $(^{-}5) \cdot (^{-}4) = \mathbf{20}$.

(*ii*) $20 \div ^{-}5 = ^{-}\mathbf{4}$.

(*iii*) $20 \div ^{-}4 = ^{-}\mathbf{5}$.

(c) (*i*) $^{-}3 \cdot 0 = \mathbf{0}$.

(*ii*) $0 \div ^{-}3 = \mathbf{0}$.

(*iii*) $0 \div 0$ **is undefined.**

9. **(a)** $4x \div 4 = n$ if and only if $4x = 4n$, where n is an integer. Then $4x = 4n$ if and only if $\boldsymbol{n = x}$, so $4x \div 4 = x$.

(b) $(^{-}xy) \div y = n$ if and only if $(^{-}xy) = yn$. Then $(^{-}xy) = yn$ if and only if $\boldsymbol{n = ^{-}x}$ $(y \neq 0)$.

10. **(a)** $32°C + (30 \text{ minutes} \times ^{-}1° \text{ per minute})$
$= 32° + ^{-}30° = \mathbf{2°C}$.

(b) $0°C + (^{-}25 \text{ minutes} \times ^{-}2° \text{ per minute})$
$= 0° + 50° = \mathbf{50° \, C}$.

(c) $^{-}20°C + (^{-}30 \text{ minutes} \times ^{-}2° \text{ per minute})$
$= ^{-}20° + 60° = \mathbf{40° \, C}$.

(d) $25°C + (^{-}20 \text{ minutes} \times 2° \text{ per minute})$
$= 25° + ^{-}40° = ^{-}\mathbf{15° \, C}$.

11. $(^{-}12{,}000 \text{ acres per year} \times 8 \text{ years}) =$
$^{-}96{,}000$ acres, or **96,000 acres lost.**

12.

$$a(b - c) = ab - ac$$
$$^{-}1(^{-}4 - ^{-}2) = (^{-}1)(^{-}4) - (^{-}1)(^{-}2)$$
$$^{-}1(^{-}2) = ^{-}4 - 2$$
$$2 = 2 \text{ check}$$

13. **(a)** $(^{-}2)^3 = ^{-}2 \cdot ^{-}2 \cdot ^{-}2 = (^{-}2 \cdot ^{-}2)\, ^{-}2 =$
$4 \cdot ^{-}2 = ^{-}\mathbf{8}$.

(b) $(^{-}2)^4 = (^{-}2 \cdot ^{-}2) \cdot (^{-}2 \cdot ^{-}2) =$
$4 \cdot 4 = \mathbf{16}$.

(c) $(^{-}10)^5 \div (^{-}10)^2 = ^{-}100{,}000 \div 100 =$
$^{-}\mathbf{1000}$.

(d) $(^{-}3)^5 \div (^{-}3) = ^{-}243 \div ^{-}3 = \mathbf{81}$.

(e) $(^{-}1)^{50} = \mathbf{1}$ (i.e., $^{-}1$ taken to an even power).

(f) $(^{-}1)^{151} = ^{-}\mathbf{1}$ (i.e., $^{-}1$ taken to an odd power).

(g) $^{-}2 + 3 \cdot 5 - 1 =$

$^{-}2 + 15 - 1 = 13 - 1 = \mathbf{12}$. Note the order of operation; multiplication before addition.

(h) $10 - 3 \cdot 7 - 4(^{-}2) + 3 =$

$10 - 21 - (^{-}8) + 3 =$

$(10 + 8 + 3) - 21 = \mathbf{0}$

(i) $(^{-}2)^{64} - 2^{64} = 2^{64} - 2^{64} = \mathbf{0}$.

(j) $^{-}2^8 + 2^8 =$

$^{-}(2^8) + 2^8 =$

$^{-}256 + 256 = \mathbf{0}$.

14. **(a)** Always **negative.** x^2 is always positive, so its additive inverse is always negative.

(b) Always **positive**. Any non-zero number taken to an even power is always positive.

(c) Always **positive**. Any non-zero number taken to an even power is always positive.

(d) **Positive when x is negative**; **negative when x is positive**. i.e., the additive inverse of x^3, which may be positive or negative.

(e) **Positive when x is negative** (the additive inverse of a negative integer is positive). **Negative when x is positive** (the additive inverse of a positive integer is negative).

15. **(a)** $^{-}x^2 = x^2$ **for x = 0.**

(b) $^{-}x^3 = (^{-}x)^3$ **for all integers.**

16. If $48 \div x$ is an integer and x is an integer, then x equals all positive and negative divisors of 48. In other words, x is equal to:
$\pm\mathbf{1}, \pm\mathbf{2}, \pm\mathbf{3}, \pm\mathbf{4}, \pm\mathbf{6}, \pm\mathbf{8}, \pm\mathbf{12}, \pm\mathbf{16}, \pm\mathbf{24},$
and $\pm\mathbf{48}$.

17. **(a)** **Commutative property of multiplication**, or $a \cdot b = b \cdot a$.

(b) **Closure property of addition**. The product of an integer and another integer is an integer.

(c) **Associative property of multiplication**, or $a(b \cdot c) = (a \cdot b)c$.

(d) **Distributive property of multiplication over addition**, or $a(b + c) = ab + ac$.

18. The pairs of integers (x, y) that multiply to $^{-}14$ are:

$(1, {}^{-}14)$; $(2, {}^{-}7)$; $({}^{-}1, 14)$; $({}^{-}2, 7)$;

$({}^{-}14, 1)$; $({}^{-}7, 2)$; $(14, {}^{-}1)$; $(7, {}^{-}2)$

19. **(a)** ${}^{-}2(x - y) = {}^{-}2x - {}^{-}2y = \mathbf{{}^{-}2x + 2y}$.

(b) $x(x - y) = x \cdot x - x \cdot y = \mathbf{x^2 - xy}$.

(c) ${}^{-}x(x - y) = {}^{-}x^2 - {}^{-}xy = \mathbf{{}^{-}x^2 + xy}$.

(d) ${}^{-}2(x + y - z) = {}^{-}2x + {}^{-}2y - {}^{-}2z =$

$\mathbf{{}^{-}2x - 2y + 2z}$.

20. **(a)** ${}^{-}3x = 6 \Rightarrow \frac{{}^{-}3x}{3} = \frac{6}{{}^{-}3} \Rightarrow \mathbf{x = {}^{-}2}$.

(b) ${}^{-}2x = 0 \Rightarrow \frac{{}^{-}2x}{{}^{-}2} = \frac{0}{{}^{-}2} \Rightarrow \mathbf{x = 0}$.

(c) $x \div 3 = {}^{-}12 \Rightarrow 3(x \div 3) = 3({}^{-}12)$

$\Rightarrow \mathbf{x = {}^{-}36}$.

(d) $x \div {}^{-}3 = {}^{-}2 \Rightarrow {}^{-}3(x \div {}^{-}3) = {}^{-}3({}^{-}2)$

$\Rightarrow \mathbf{x = 6}$.

(e) $x \div ({}^{-}x) = {}^{-}1 \Rightarrow {}^{-}x(x \div {}^{-}x) = {}^{-}x({}^{-}1)$

$\Rightarrow x = x$; i.e., the equation is satisfied by **all integers except 0** (since $0 \div 0$ is not defined).

(f) ${}^{-}3x - 8 = 7 \Rightarrow {}^{-}3x - 8 + 8 = 7 + 8$

$\Rightarrow {}^{-}3x = 15 \Rightarrow \frac{{}^{-}3}{{}^{-}3}x = \frac{15}{{}^{-}3} \Rightarrow \mathbf{x = {}^{-}5}$.

(g) ${}^{-}2(5x - 3) = 26 \Rightarrow {}^{-}10x + 6 = 26$

$\Rightarrow {}^{-}10x + 6 - 6 = 26 - 6 \Rightarrow {}^{-}10x = 20$

$\Rightarrow \frac{{}^{-}10}{{}^{-}10}x = \frac{20}{{}^{-}10} \Rightarrow \mathbf{x = {}^{-}2}$.

(h) $3x - x - 2x = 3 \Rightarrow 0 = 3 \Rightarrow$ **no solution**.

(i) ${}^{-}2(5x - 6) - 30 = {}^{-}x \Rightarrow {}^{-}10x + 12$

$-30 = {}^{-}x \Rightarrow {}^{-}10x - 18 = {}^{-}x \Rightarrow$

${}^{-}18 = 9x \Rightarrow \mathbf{x = {}^{-}2}$.

(j) $x^2 = 4 \Rightarrow \mathbf{x = 2}$ or $\mathbf{x = {}^{-}2}$, since

$2^2 = 4$ or $({}^{-}2)^2 = 4$.

(k) $(x - 1)^2 = 9 \Rightarrow$

(*i*) $(x - 1) = 3 \Rightarrow \mathbf{x = 4}$, or

(*ii*) $(x - 1) = {}^{-}3 \Rightarrow \mathbf{x = {}^{-}2}$.

(l) $(x - 1)^2 = (x + 3)^2 \Rightarrow$

(*i*) $(x - 1) = (x + 3) \Rightarrow 0 = 4 \Rightarrow$ not a viable solution.

(*ii*) $(x - 1) = {}^{-}(x + 3) \Rightarrow x - 1 = {}^{-}x - 3$

$\Rightarrow 2x = {}^{-}2 \Rightarrow \mathbf{x = {}^{-}1}$.

(m) $(x - 1)(x + 3) = 0 \Rightarrow$ either $(x - 1) = 0$ or $(x + 3) = 0$; i.e., the only way that the product of two numbers can be zero is that one or both must equal zero.

(*i*) **If** $(x - 1) = 0 \Rightarrow \mathbf{x = 1}$.

(*ii*) **If** $(x + 3) = 0 \Rightarrow \mathbf{x = {}^{-}3}$.

21. The difference-of-squares formula is:

$(a + b)(a - b) = a^2 - b^2$.

(a) $52 \cdot 48 = (50 + 2)(50 - 2) =$

$50^2 - 2^2 = 2500 - 4 = \mathbf{2496}$.

(b) $(5 - 100)(5 + 100) = 5^2 - 100^2 =$

$25 - 10{,}000 = \mathbf{{}^{-}9975}$.

(c) $({}^{-}x - y)({}^{-}x + y) = ({}^{-}x)^2 - y^2 =$

$\mathbf{x^2 - y^2}$.

22. To *factor* an expression means to find an equivalent expression that is a product; i.e., if $N = ab$, then a and b are factors of N. Factoring may be said to undo the distributive property of multiplication over addition or subtraction.

(a) $3x + 5x = x(3 + 5) = \mathbf{8x}$. The factor common to both terms, x, divides each and then multiplies their sum.

(b) $xy + x = x \cdot y + x \cdot 1 = \mathbf{x(y + 1)}$.

(c) $x^2 + xy = x \cdot x + x \cdot y = \mathbf{x(x + y)}$.

(d) $3xy + 2x - xz = \mathbf{x(3y + 2 - z)}$.

(e) $abc + ab - a = a(bc + b - 1) =$

$\mathbf{a[b(c + 1) - 1]}$.

(f) $16 - a^2 = 4^2 - a^2 = \mathbf{(4 + a)(4 - a)}$; i.e., the factorization of the difference-of-squares formula.

(g) $4x^2 - 25y^2 = (2x)^2 - (5y)^2$
$= \mathbf{(2x + 5y)(2x - 5y)}$.

23. $(a - b)^2 = (a - b)(a - b) = a(a - b)$
$+ {}^-b(a - b) = a^2 - ab + {}^-ba - {}^-b^2 =$
$a^2 - 2ab + b^2$.

24. **(a)** *(i)* Arithmetic sequence; **difference** $(d) = {}^-7 - {}^-10 = \mathbf{3}$.

(ii) The next two terms are: $5 + 3 = \mathbf{8}$ and $8 + 3 = \mathbf{11}$.

(iii) $a_n = a_1 + (n - 1)d = {}^-10 + (n - 1) \cdot 3 = {}^-10 + 3n - 3 = \mathbf{3n - 13}$.

(b) *(i)* Geometric sequence; **ratio** $(r) = {}^-4 \div {}^-2 = \mathbf{2}$

(ii) The next two terms are: ${}^-64 \cdot 2 = \mathbf{{}^-128}$ and ${}^-128 \cdot 2 = \mathbf{{}^-256}$.

(iii) $a_n = a_1(r)^{n-1} = {}^-2(2)^{n-1} = \mathbf{{}^-2^n}$.

25. **(a)** ${}^-3, {}^-9, \_, \_, {}^-243$ is a geometric sequence with a common ratio of 3. The missing terms are $\mathbf{{}^-27}$ and $\mathbf{{}^-81}$.

(b) $\_, 32, {}^-16, 8, \_$ is a geometric sequence.

The common ratio is ${}^-\frac{1}{2}$. The missing terms are $\mathbf{{}^-64}$ and $\mathbf{{}^-4}$.

26. Write an equation:
${}^-5(6) + 3(8) = {}^-30 + 24 = {}^-6$. The net result is a **descent of 6 feet.**

27. Let n be the number of cups. The height of each cup is 7 and ${}^-6$ represents the unseen part of each cup when they are nested together. There are (n-1) cups with unseen parts. The total height of n cups can then be written as:

$7n + ({}^-6)(n - 1) = 7n - 6n + 6 = n + 6$.

28. The scuba diver descended **450 feet /15 minutes** or **30 feet/minute (30 feet per minute).**

29. The temperature change is ${}^-3$ per hour, for 8 hours. ${}^-3(8) = \mathbf{{}^-24}$. The temperature will **drop 24 degrees in 8 hours**.

30. Let m equal the number of months you have paid the Sirius® radio bill out of this account. So the balance in your bank account will be $= 500 - 40m$. After one year, the balance will be $500 - 40(12) = 500 - 480 = \mathbf{\$20}$

31. Let d be the number of weeks the water level has dropped. Then the total distance the water level drops $= 2d$. After 4 weeks the drop $= 2(4) = \mathbf{8\ inches}$.

32. If the price loses \$2 per hour for six hours, the total price drop would be $2 \cdot 6 = \mathbf{\$12}$.

33. Since the debt is \$2000, and Bob has five siblings, each will pay $2000 \div 5 = \mathbf{\$400}$. Each sibling will pay \$400 to pay off Bob's debt. Bob has such nice siblings.

Mathematical Connections 5-2: Review Problems

21. **(a)** Since ${}^-5 + 5 = 0$, **5** is the additive inverse.

(b) Since $7 + {}^-7 = 0$, $\mathbf{{}^-7}$ is the additive inverse.

(c) Since $0 + 0 = 0$, **0** is the additive inverse.

23. Yes. When two integers are subtracted, their difference will always be another integer.

Chapter 5 Review

1. **(a)** The additive inverse of 3 is $\mathbf{{}^-3}$.

(b) The additive inverse of ${}^-a$ is ${}^-({}^-a) = \mathbf{a}$.

(c) The additive inverse of ${}^-2 + 3$ is ${}^-({}^-2 + 3) = \mathbf{{}^-1}$

(d) The additive inverse of $x + y$ is ${}^-(x + y) = {}^-x + {}^-y = \mathbf{{}^-x - y}$.

(e) The additive inverse of ${}^-x + y$ is ${}^-({}^-x + y) = x + {}^-y = \mathbf{x - y}$.

(f) The additive inverse of ${}^-x - y$ is ${}^-({}^-x - y) = x - {}^-y = \mathbf{x + y}$.

(g) $({}^-2)^5 = {}^-32$, thus the additive inverse is **32**.

(h) ${}^-2^5 = {}^-32$, thus the additive inverse is **32**.

2. **(a)** $(^{-}2 + {}^{-}8) + 3 = (^{-}10) + 3 = {}^{-}\mathbf{7}.$

(b) $^{-}2 - (^{-}5) + 5 = {}^{-}2 + 5 + 5 = \mathbf{8}.$

(c) $^{-}3(^{-}2) + 2 = 6 + 2 = \mathbf{8}.$

(d) $^{-}3(^{-}5 + 5) = {}^{-}3(0) = \mathbf{0}.$

(e) $^{-}40 \div {}^{-}5 = \mathbf{8}.$

(f) $(^{-}25 \div 5)(^{-}3) = (^{-}5)(^{-}3) = \mathbf{15}.$

3. **(a)** $^{-}x + 3 = 0 \Rightarrow {}^{-}x + 3 - 3 = 0 - 3 \Rightarrow$

$^{-}x = {}^{-}3 \Rightarrow {}^{-}x(^{-}1) = {}^{-}3(^{-}1) \Rightarrow \boldsymbol{x} = \mathbf{3}.$

(b) $^{-}2x = 10 \Rightarrow \frac{^{-}2x}{^{-}2} = \frac{10}{^{-}2} \Rightarrow \boldsymbol{x} = {}^{-}\mathbf{5}.$

(c) The definition of integer division states that for all integers a and b, $a \div b$ is the *unique* integer c, if it exists, such that $a = bc$. So $0 \div (^{-}x) = 0 \Rightarrow 0 = (^{-}x)(0) \Rightarrow x$ may be **any integer except 0**.

(d) $^{-}x \div 0 = {}^{-}1 \Rightarrow$ **no integer solution**. Division by 0 is undefined.

(e) $3x - 1 = {}^{-}124 \Rightarrow 3x = {}^{-}123 \Rightarrow x = \frac{^{-}123}{3}$

$= {}^{-}\mathbf{41}$

(f) $^{-}2x + 3x = x \Rightarrow x(^{-}2 + 3) = x$

$\Rightarrow x = x$; i.e., x may be **any integer**.

4. $2 \cdot {}^{-}3 = (^{-}3) + (^{-}3) = {}^{-}6;$

$1 \cdot {}^{-}3 = {}^{-}3;$

$0 \cdot {}^{-}3 = 0;$ and if the pattern continues:

$^{-}1 \cdot {}^{-}3 = 3;$

$^{-}2 \cdot {}^{-}3 = 6.$

5. **(a)** Create two sets, one worth negative 5, the other worth positive 5, as shown below:

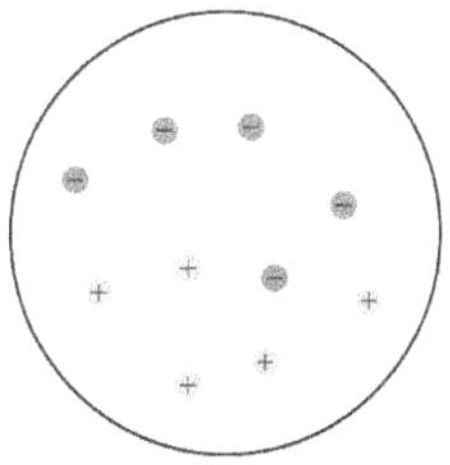

= 0

Make five zero pairs; all chips are paired, no chips are unpaired. So $^{-}5 + 5 = \mathbf{0}.$

(b) Create two sets, one worth positive 3, the other worth negative 7, as shown below:

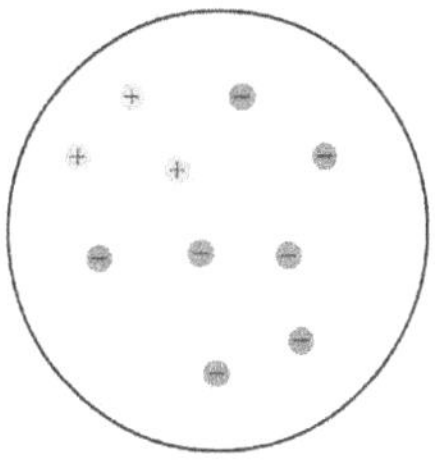

= $^{-}4$

make three zero pairs; the remaining negative chips are the solution. So $3 + {}^{-}7 = {}^{-}\mathbf{4}.$

(c) Create two sets, one worth negative 5, the other worth negative 6, as shown below:

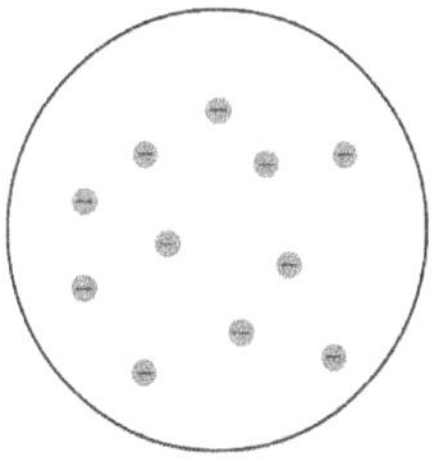

= $^{-}11$

Since the sets are same sign, combine the two sets into one group. The solution is what the group is worth.

So $^{-}5 + {}^{-}6 = {}^{-}\mathbf{11}.$

6. Movement on the number line: the cat walks forward when the number is positive, and the cat walks backwards when the number is negative.

(a)

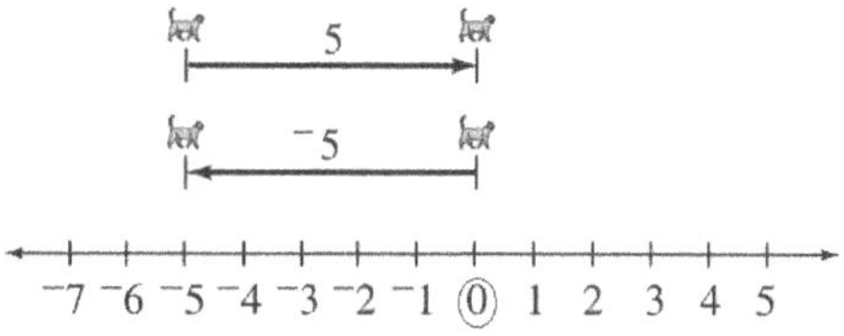

(b)

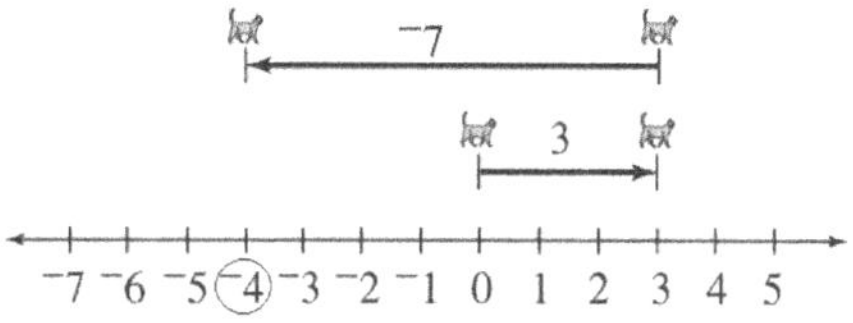

(c)

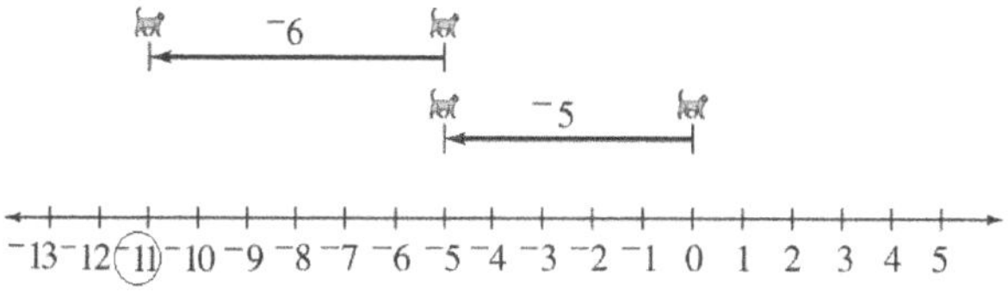

7. **(a)** Create one set, worth $^{-}5$. From this set, we need to take away a group worth positive 5. Since we don't have 5 positives to take away, we will add zero pairs until we do, as shown below:

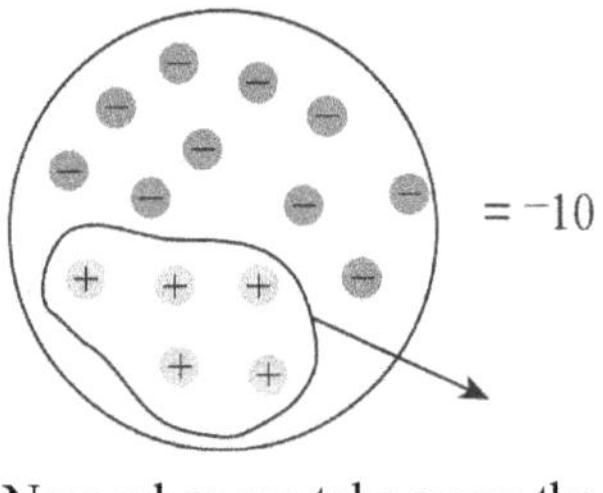

Now when we take away the five positives, what is left will be our answer. So

$^{-}5 - 5 = {}^{-}$**10.**

(b) Create one set worth 3. From this set, we need to take away a group worth positive 7. Since we don't have 7 positives to take away, we will add zero pairs until we do, as shown below:

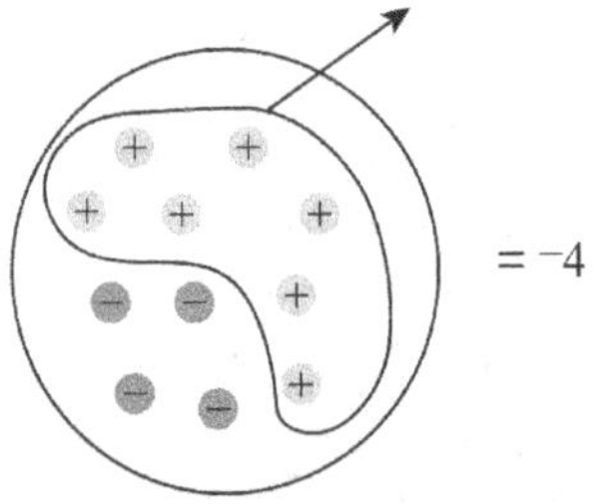

Now when we take away the seven positives, what is left will be our answer. So $3 - 7 = {}^{-}$**4.**

(c) Create one set worth $^{-}6$. Now we need to take away a group worth negative 5. Once that is done, the solution is the value of the chip(s) that are left. So $^{-}6 - {}^{-}5 = {}^{-}$**1.**

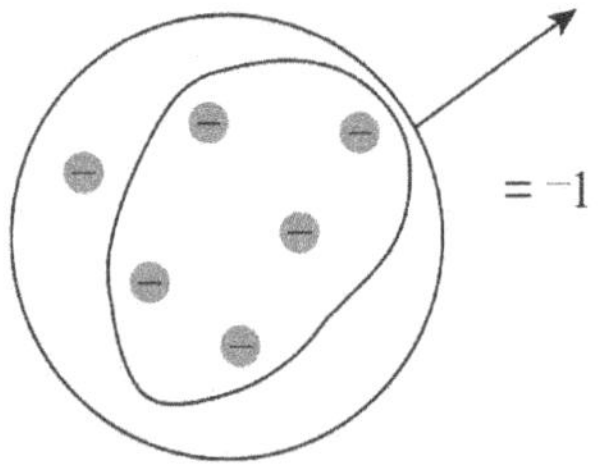

8. Movement on the number line: the cat walks forward when the number is positive, and the cat walks backwards when the number is negative.

(a)

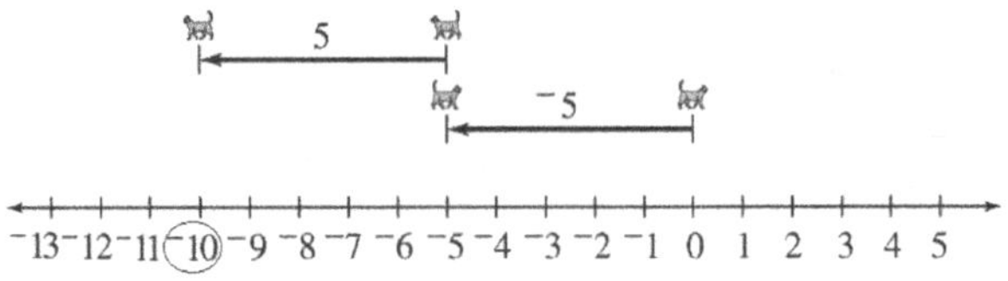

(b)

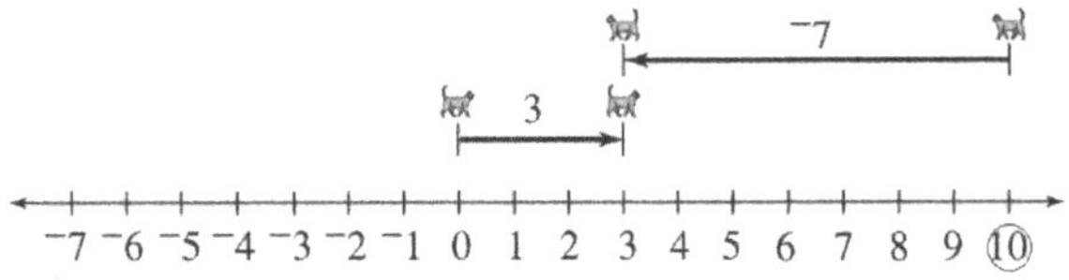

(c)

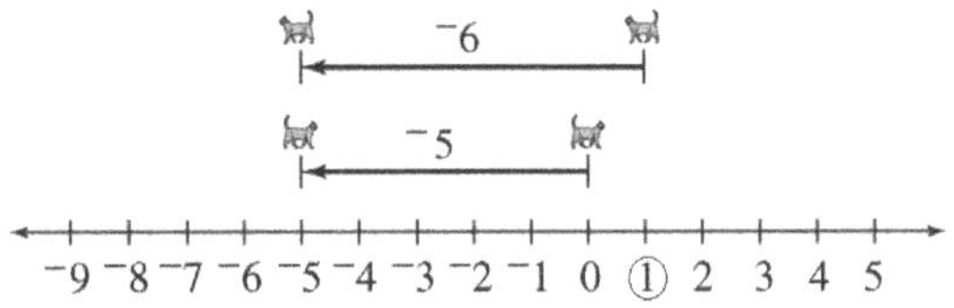

9. **(a)** $10 - 5 = 5.$

(b) $1 - ({}^{-}2) = 3.$

10. $4 + ({}^{-}3) + 18 + ({}^{-}5) + ({}^{-}6) + 21 =$ **29.** The football team gained **29 yards.**

11. $47 - ({}^{-}14) = 61.$ **61 degrees** is the change.

12. Let h be the height of Mount Everest. Then

$h - ({}^{-}1410) = 30439$

$h + 1410 = 30439$

$h = 29029.$

Mount Everest is **29,029 feet** tall.

13. **(a)** Let h be the ending height. Then

$300 - h = 450$

$300 - 450 = h$

$^{-}150 = h.$

Les ended her hike at an elevation of **$^{-}$150 feet**, or **150 feet below sea level.**

14. **(a)** $2+0+3+({}^{-}1)+1+({}^{-}2)+0+1+0.$

(b) $2+0+3+({}^{-}1)+1+({}^{-}2)+0+1+0=4.$ Brian's final score was **4 over par.**

15. $6-({}^{-}3)=9.$ The first golfer took **9 more shots.**

16. If we make years BCE negative, then the equation is $476-({}^{-}509)=985.$ However, there is no "year zero"; in other words, the calendar does not go from 1 BCE to 0 to 1 CE; it goes from 1 BCE to 1 CE. Therefore, the correct answer would be **984 years.**

17. If we make years BCE negative, then the equation is $1643 - (^-570) = 2213$. However, there was no "year zero" (see explanation on #16) so the correct answer would be **2212 years**.

18. $20 - 11 + 5 = 14$. The elevator is on the **14th floor.**

19. **(a)** $^-1x = {^-1} \cdot x = \mathbf{^-x}$.

(b) $(^-1)(x - y) = {^-1} \cdot x - {^-1} \cdot y = {^-x} - {^-y}$
$= {^-x} + y = \mathbf{y - x}$.

(c) $2x - (1 - x) = 2x - 1 - {^-x} = 2x - 1 + x$
$= \mathbf{3x - 1}$.

(d) $(^-x)^2 + x^2 = (^-x \cdot {^-x}) + x^2 = x^2 + x^2$
$= \mathbf{2x^2}$.

(e) $(^-x)^3 + x^3 = (^-x \cdot {^-x} \cdot {^-x}) + x^3$
$= {^-x^3} + x^3 = \mathbf{0}$.

(f) $(^-3 - x)(3 + x) = (^-3 - x)3 + (^-3 - x)x$
$= {^-9} - 3x + {^-3x} - x^2 = {^-9} - 6x - x^2$
$= \mathbf{^-x^2 - 6x - 9}$.

(g) $(^-2 - x)(^-2 + x) = (^-2)^2 - x^2 = \mathbf{4 - x^2}$.

20. **(a)** $x - 3x = x(1 - 3) = \mathbf{^-2x}$.

(b) $x^2 + x = \mathbf{x(x + 1)}$.

(c) $x^2 - 36 = x^2 - 6^2 = \mathbf{(x + 6)(x - 6)}$.

(d) $81y^4 - 16x^4 = (9y^2 + 4x^2)(9y^2 - 4x^2)$
$= \mathbf{(9y^2 + 4x^2)(3y + 2x)(3y - 2x)}$.

(e) $5 + 5x = \mathbf{5(1 + x)}$.

(f) $(x - y)(x + 1) - (x - y)$
$= (x - y)[(x + 1) - 1] = \mathbf{(x - y)x}$.

21. **(a)** **False.** $|x|$ is not positive when $x = 0$.

(b) **False.** $|x + y| \neq |x| + |y|$ when x and y are of opposite sign.

(c) **False.** Let $a = 3$ and $b = {^-4}$. Then $3 < {^-(^-4)}$ and $3 \nless 0$.

(d) **True.** $x - y$ and $y - x$ both represent the difference between x and y. Squaring that difference produces a positive number in either case.

22. Answer may vary:

(a) $2 \div 1 \neq 1 \div 2$.

(b) $3 - (4 - 5) \neq (3 - 4) - 5$.

(c) $1 \div 2 \notin I$.

(d) $8 \div (4 - 2) \neq (8 \div 4) - (8 \div 2)$.

23. **(a)** $5 \cdot (^-5)$ is interpreted as creating five groups worth $^-5$ each, or creating *five groups of 5 red chips*, as shown below:

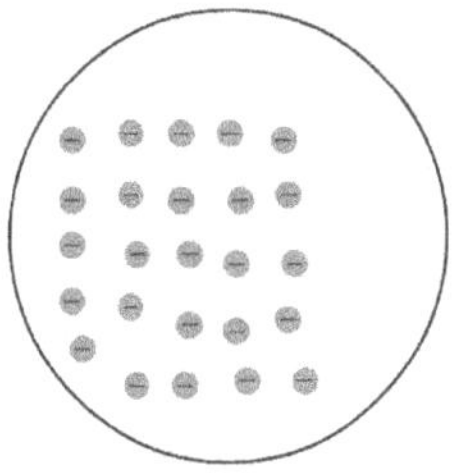

Five groups of $^-5 = {^-25}$

The solution is the value of the groups, which is negative 25. So $5 \cdot (^-5) = \mathbf{^-25}$.

(b) To find $^-12 \div {^-3}$ we will use the repeated subtraction model of division. This means we will begin with a set worth $^-12$, and repeatedly subtract from it sets worth $^-3$; our goal is to find out how many sets we create.

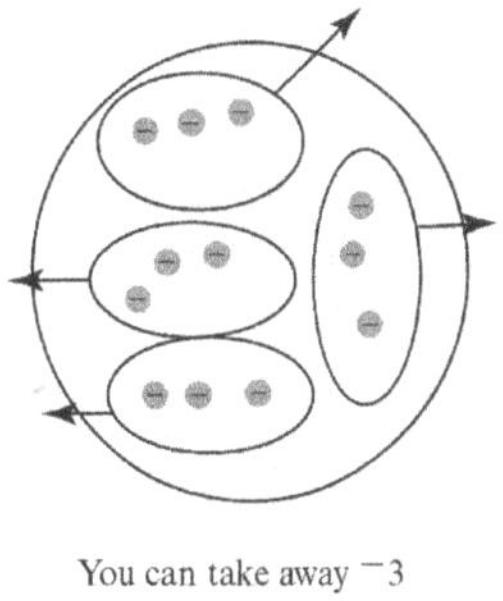

You can take away $^-3$
4 times.

As shown above, four groups each worth $^-3$ can be created. So $^-12 \div {^-3} = \mathbf{4}$.

(c) To find $(^-2)(^-3)$ using the chip model, the signs are interpreted as follows: $^-2$ is taken to mean *remove two groups of*; and $^-3$ is taken to mean *3 red chips*. So, begin with a set that has the value of zero, as shown below:

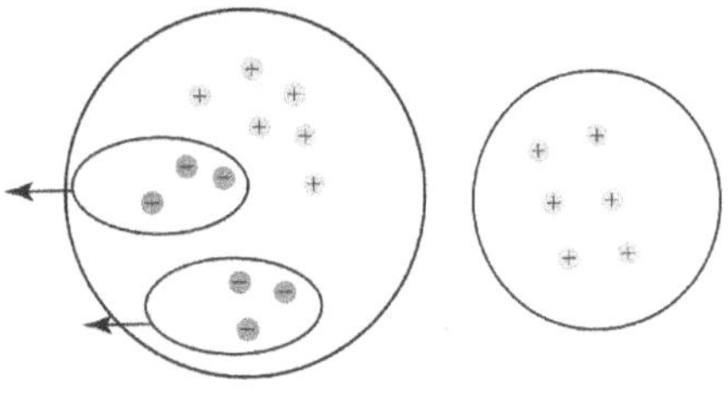

You add 2 groups of yellow chips and 2 groups of red chips. Take away two groups of $^-3$ amd you are left 6.

Now, when two groups of 3 red chips are removed, the remaining positive chips are the solution. So $(^-2)(^-3) = \mathbf{6}$.

24. Oregon, Nevada, Iowa, California, Arizona, Ohio, West Virginia, Alabama.

25. The questions answered correctly contribute $^+4$ points, and the questions answered incorrectly contribute $^-7$ points. Terry's score is $87(4) + 46(^-7) = 348 - 322 = \mathbf{26\ points}$

26. **(a)** $^-40 - (^-62) = -40 + 62 = \mathbf{22^{o}C}$.

(b) $8180 - 1100 = \mathbf{7080\ ft}$.

(c) The pattern above suggest that as one travels South from the North Pole elevations must be higher to achieve negative temperatures. Since $^-11 - (^-40) = 29$ degrees Celsius, we might guess that since Hawaii is closer to the equator than Arizona, the elevation where $^-11^\circ C$ was recorded was 7000 or 8000 feet higher than the elevation where $^-40^\circ C$ was recorded in Arizona. **A reasonable guess is that the elevation in Hawaii is 14,000 to 15,000 ft.**

27. **(a)** $^-8 - (^-282) = -8 + 282 = \mathbf{274\ feet.}$

(b) **Answers will vary**. The students should speculate that the states would be near an ocean (since the coastlines are 0 feet above sea level), or somewhere where it is known that the sea level is low due to some geographic formation. Louisiana is the state with a low point at 8 feet below sea level, and California (Death Valley) is the state with a low point of 282 feet below sea level

28. If sea level is considered 0 m elevation, then the average depth of the Pacific Ocean is $\mathbf{^-3963\ m}$.

29. **(a)** **degree Celsius $\approx$ kelvin $-273°$.**

(b) kelvin $\approx 100° + 273° = \mathbf{373°}$.

(c) kelvin $\approx {}^-40° + 273° = \mathbf{233°}$.

30. **(a)** If we consider the number of forces on the battlefield before the Union and Confederate force arrived as 0, we can describe the change in the number of forces engaged in the battle as $^+75,000,\ ^+82,289,\ ^-23,049,$ and $^-28,063$.

(b) $^-23,049 + {}^-28,063 = {}^-51,112$. Since the $^-$ denotes casualties, there were **51,112 casualties.**

31. $3(^-475) = {}^-1425$; the movie theater lost **$1425** in three days

32. $^-900 \div 2 = {}^-450$; Jared lost **$450** each month.

33. 30 centimeters = 300 millimeters; $^-300 \div {}^-3 = 100$; it will take **100 years**.

34. If we interpret 0 to be the ground floor, $^-2$ would mean that the elevator is below ground level, which might be described as a 2nd level basement.

35. **(a)** $^-2x < 14 \Rightarrow {}^-x < 7 \Rightarrow x > {}^-7$

$3x > {}^-15 \Rightarrow x > {}^-5$

The integer solutions are all integers x such that $\mathbf{x > {}^-5}$.

(b) $x^2 < 9 \Rightarrow {}^-3 < x < 3$

$x^2 \le 1 \Rightarrow {}^-1 \le x \le 1$

The integer solutions are $\mathbf{^-1, 0, 1}$.

(c) $|x| > {}^-4$ works for all integers x, since absolute values are always positive. However, $|x| < {}^-1$ doesn't work for any integer for the same reason. Therefore, there is **No solution**.

36. Write an equation:

$6 = -(-x+30-14)$

$6 = x-30+14$

$6 = x-16$

$22 = x$

37. **Answers will vary.** For any integer x to be a solution, it must be less than $^{-}3$, since $3+(^{-}3)=0$. So, three possible solutions are $^{-}\mathbf{4}, {}^{-}\mathbf{5}$, and $^{-}\mathbf{6}$.

All values of x that are solutions to $3+x<0$ are all x such that $\mathbf{x < {}^{-}3}$.

38. **(a)** $x+3 = {}^{-}x-17 \Rightarrow 2x = {}^{-}20$ $\Rightarrow x = \mathbf{{}^{-}10}$.

(b) $2x = {}^{-}2^{100} \Rightarrow x = \frac{-2^{100}}{2} \Rightarrow x = \mathbf{{}^{-}2^{99}}$.

(c) $2^{10}x = 2^{99} \Rightarrow x = \frac{2^{99}}{2^{10}} \Rightarrow x = \mathbf{2^{89}}$.

(d) ${}^{-}x = x \Rightarrow {}^{-}2x = 0 \Rightarrow x = \mathbf{0}$.

(e) $|{}^{-}x| = 3 \Rightarrow x = \mathbf{3}$ or $x = \mathbf{{}^{-}3}$.

(f) $|x| = {}^{-}x \Rightarrow x = 0, {}^{-}1, {}^{-}2, \ldots$; i.e., $\boldsymbol{x \le 0}$.

(g) $|x| > 3 \Rightarrow x = {}^{-}4, {}^{-}5, {}^{-}6, \ldots$, or $x = 4, 5, 6, \ldots$; i.e., $\boldsymbol{x > 3}$ **or** $\boldsymbol{x < {}^{-}3}$.

(h) $(x-1)^2 = 100 \Rightarrow$

(i) $x-1 = 10 \Rightarrow x = \mathbf{11}$, or

(ii) $x-1 = {}^{-}10 \Rightarrow x = \mathbf{{}^{-}9}$.

39. **(a)** If $a_n = ({}^{-}1)^n$:

$a_1 = ({}^{-}1)^1 = \mathbf{{}^{-}1}$;

$a_2 = ({}^{-}1)^2 = \mathbf{1}$;

$a_3 = ({}^{-}1)^3 = \mathbf{{}^{-}1}$;

$a_4 = ({}^{-}1)^4 = \mathbf{1}$;

$a_5 = ({}^{-}1)^5 = \mathbf{{}^{-}1}$;

$a_6 = ({}^{-}1)^6 = \mathbf{1}$.

(b) If $a_n = ({}^{-}2)^n$:

$a_1 = ({}^{-}2)^1 = \mathbf{{}^{-}2}$;

$a_2 = ({}^{-}2)^2 = \mathbf{4}$;

$a_3 = ({}^{-}2)^3 = \mathbf{{}^{-}8}$;

$a_4 = ({}^{-}2)^4 = \mathbf{16}$;

$a_5 = ({}^{-}2)^5 = \mathbf{{}^{-}32}$;

$a_6 = ({}^{-}2)^6 = \mathbf{64}$.

(c) If $a_n = {}^{-}2 - 3n$:

$a_1 = {}^{-}2 - 3 \cdot 1 = \mathbf{{}^{-}5}$;

$a_2 = {}^{-}2 - 3 \cdot 2 = \mathbf{{}^{-}8}$;

$a_3 = {}^{-}2 - 3 \cdot 3 = \mathbf{{}^{-}11}$;

$a_4 = {}^{-}2 - 3 \cdot 4 = \mathbf{{}^{-}14}$;

$a_5 = {}^{-}2 - 3 \cdot 5 = \mathbf{{}^{-}17}$;

$a_6 = {}^{-}2 - 3 \cdot 6 = \mathbf{{}^{-}20}$.

40. **(a)** Geometric; $r = {}^{-}1$.

(b) Geometric; $r = {}^{-}2$.

(c) Arithmetic; $d = {}^{-}3$.

41. **(a)** Fibonacci-type sequence. The missing terms are **1, 1, 2.**

(b) Geometric sequence with common ratio $r = {}^{-}2$. The missing terms are

392, $\mathbf{{}^{-}784}$, and 1568.

(c) Arithmetic sequence with common difference 147. The missing terms are **392, 539, 686.**

42. **(a)** ${}^{-}1+1+0+1+1+2+3 = \mathbf{7}$.

(b)

${}^{-}49+98+{}^{-}196+392+{}^{-}784+1568 = \mathbf{1029}$.

(c) ${}^{-}49+98+245+392+539+686 = \mathbf{1911}$.

43. **(a)** This is an arithmetic sequence with common difference -2. The nth term is - $2n$+103. We set the last term equal to the nth term and solve for n:

${}^{-}2n+103 = {}^{-}103 \Rightarrow {}^{-}2n = {}^{-}20$

$6 \Rightarrow n = 103$. There are 103 terms total. Subtracting the 5 given terms leaves **98 terms** missing by the indicated ellipse.

(b) Each term is the previous term multiplied by $^-5$. This gives:

$5, {}^-25, 125, {}^-625, 3125, {}^-15625$. There are 6 terms total. Subtracting the 4 given terms leaves **2 terms** missing by the indicated ellipse.

44. **(a)** The **common difference is $^-\mathbf{2}$**.

(b) The **common ratio is $^-\mathbf{5}$**.

45. The difference of squares formula is:

$a^2 - b^2 = (a - b)(a + b)$.

Applying this to $96 \cdot 104$:

$$(100-4)(100+4) = 100^2 - 4^2$$
$$= 10{,}000 - 16 = \mathbf{9984}.$$

46. **(a)** $1^2 - 4 = {}^-3$

$2^2 - 4 = 0$

$3^2 - 4 = 5$

$4^2 - 4 = 12$

$5^2 - 4 = 21$

(b) $^-5(1) - ({}^-3) = {}^-2$

$^-5(2) - ({}^-3) = {}^-7$

$^-5(3) - ({}^-3) = {}^-12$

$^-5(4) - ({}^-3) = {}^-17$

$^-5(5) - ({}^-3) = {}^-22$

(c) $^-(1)^2 - 4 = {}^-5$

$^-(2)^2 - 4 = {}^-8$

$^-(3)^2 - 4 = {}^-13$

$^-(4)^2 - 4 = {}^-20$

$^-(5)^2 - 4 = {}^-29$

47. **(a)** $({}^-1)^{2n}$ will always be equal to **1** for all integer values of n because $2n$ will always be even.

(b) $({}^-1)^{2n+1}$ will always be equal to $^-\mathbf{1}$ for all integer values of n because $2n+1$ will always be odd.

CHAPTER 6

RATIONAL NUMBERS AND PROPORTIONAL REASONING

Assessment 6-1A: The Set of Rational Numbers

1. (a) The solution to $8x = 7$ is $\mathbf{\frac{7}{8}}$.
   (b) Jane ate seven of the eight pizza slices.
   (c) The ratio of boys to girls in this math class is seven to eight.
2. Assume the shaded regions are symmetrical:
   (a) $\frac{1}{6}$. The ratio of shaded to unshaded areas is one to six.
   (b) $\frac{1}{4}$.. The ratio of shaded to unshaded areas is one to four.
   (c) $\frac{2}{6} = \frac{1}{3}$. The circle is in six parts; two are shaded.
   (d) $\frac{7}{12}$. Seven of the twelve dots are shaded.
3. Divide the whole into equal parts:

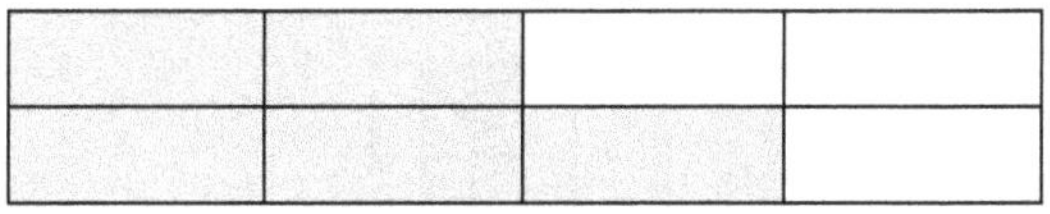

$\frac{5}{8}$ is shaded out of the whole.

4. The diagrams illustrate the Fundamental Law of Fractions; i.e., the value of a fraction does not change if its numerator and denominator are multiplied by the same nonzero number.
   (a) $\frac{2}{3}$. Two of the three parts are shaded.
   (b) $\frac{4}{6} = \frac{2}{3}$. Four of the six parts are shaded.
   (c) $\frac{6}{9} = \frac{2}{3}$. Six of the nine parts are shaded.
   (d) $\frac{8}{12} = \frac{2}{3}$. Eight of the twelve parts are shaded.
5. Assume the shaded regions are symmetrical:
   (a) **No**. The parts are not of equal area.
   (b) **Yes**. The sections are equal in area; three of the four are shaded.
   (c) **Yes**. The circle parts are equal in area; one of the two is shaded.
6. (a) One block of a square sectioned into four equal pieces is the same area as two blocks of a square sectioned into eight equal pieces; i.e., $\frac{1}{4} = \frac{2}{8}$:

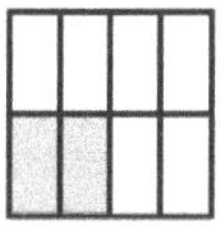

$\frac{2}{8}$

   (b) One arc of a circle sectioned into three equal pieces is the same area as three arcs of a circle sectioned into nine equal pieces; i.e., $\frac{1}{3} = \frac{3}{9}$:

$\frac{3}{9}$

   (c) One part of a hexagon sectioned into two equal pieces is the same area as three parts of a hexagon sectioned into six equal pieces; i.e., $\frac{1}{2} = \frac{3}{6}$:

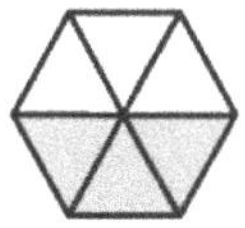

$\frac{3}{6}$

7. (a) $\frac{\text{Dots in circle}}{\text{Total dots}} = \frac{9}{24} = \mathbf{\frac{3}{8}}$.
   (b) $\frac{\text{Dots in rectangle}}{\text{Total dots}} = \frac{12}{24} = \mathbf{\frac{1}{2}}$.
   (c) $\frac{\text{Dots in intersection}}{\text{Total dots}} = \frac{4}{24} = \mathbf{\frac{1}{6}}$.
   (d) $\frac{\text{Dots in rectangle but outside circle}}{\text{Total dots}} = \frac{8}{24} = \mathbf{\frac{1}{3}}$.

8. **(a)** $A \cap B = \mathbf{\frac{4}{43}}$. There are 4 people in both an arts class and a botany class out of the 43 people in the universal set $(17 + 4 + 13 + 9 = 43)$.

(b) $A - B = \mathbf{\frac{17}{43}}$. There are 21 people in an arts class but 4 of those are also in a botany class out of 43 people in the universal set $(21 - 4 = 17)$.

(c) $U = \frac{43}{43} = \mathbf{1}$. There are 43 people in the universal set. As a fraction of the universal set it would be 43 out of 43.

(d) $\overline{A \cap B} = \frac{43-4}{43} = \mathbf{\frac{39}{43}}$. There are 43 people in the universal set and 4 people in set A and set B, so there are $43 - 4 = 39$ people who are not in A and B out of the the 43 people in the universal set.

9. **Answers may vary**, and are correct as long as the numerator and denominator of the given fractions are multiplied by the same number. Some possibilities are:

(a) $\frac{4}{18}, \frac{6}{27}, \frac{10}{45}$.

(b) $\frac{^-4}{10}, \frac{2}{^-5}, \frac{^-10}{25}$.

(c) $\frac{0}{6}, \frac{0}{9}, \frac{0}{12}$.

(d) $\frac{2a}{4}, \frac{3a}{6}, \frac{4a}{8}$.

10. **(a)** $\frac{156}{93} = \frac{3 \cdot 52}{3 \cdot 31} = \mathbf{\frac{52}{31}}$.

(b) $\frac{27}{45} = \frac{9 \cdot 3}{9 \cdot 5} = \mathbf{\frac{3}{5}}$.

(c) $\frac{^-65}{91} = \frac{^-5 \cdot 13}{7 \cdot 13} = \mathbf{\frac{^-5}{7}}$.

11. **(a)** **Undefined**. Division by 0 is undefined.

(b) **Undefined**. Division by 0.

(c) **0**. $\frac{0}{5} = 0$ because $0 \cdot 5 = 0$.

(d) **Cannot be simplified**. $(2 + a)$ and a have no common factors other than 1.

(e) **Cannot be simplified**. $(15 + x)$ and $3x$ have no common factors other than 1.

12. In both parts (a) and (b), a restriction must be maintained so that the denominator cannot be zero.

(a) $\frac{a^2-b^2}{3a+3b} = \frac{(a+b)(a-b)}{3(a+b)} = \mathbf{\frac{a-b}{3}} (a \neq {}^-b)$.

(b) $\frac{14x^2y}{63xy^2} = \frac{7 \cdot 2 \cdot x \cdot x \cdot y}{7 \cdot 9 \cdot x \cdot y \cdot y} = \mathbf{\frac{2x}{9y}} (x, y \neq 0)$.

13. **(a)** **Equal.** $\frac{375}{1000} = \frac{125 \cdot 3}{125 \cdot 8} = \frac{3}{8}$ or $375 \cdot 8 = 1000 \cdot 3$.

(b) **Equal.** $\frac{18}{54} = \frac{18}{3 \cdot 18} = \frac{1}{3}$ and $\frac{23}{69} = \frac{1 \cdot 23}{3 \cdot 23} = \frac{1}{3}$ or $18 \cdot 69 = 54 \cdot 23$.

14. **(a)** **Not equal.** $16 = 2^4; 18 = 2 \cdot 3^2 \Rightarrow$ $LCM(16,18) = 2^4 \cdot 3^2 = 144$. Then $\frac{10}{16} = \frac{10 \cdot 9}{16 \cdot 9} = \frac{90}{144}$ and $\frac{12}{18} = \frac{12 \cdot 8}{18 \cdot 8} = \frac{96}{144}$; $\frac{90}{144} \neq \frac{96}{144}$.

(b) **Not equal.** $86 = 2 \cdot 43; 215 = 5 \cdot 43 \Rightarrow$ $LCM(86, 215) = 2 \cdot 5 \cdot 43 = 430$. Then $\frac{^-21}{86} = \frac{^-21 \cdot 5}{86 \cdot 5} = \frac{^-105}{430}$ and $\frac{^-51}{215} = \frac{^-51 \cdot 2}{215 \cdot 2} = \frac{^-102}{430}$; $\frac{^-105}{430} \neq \frac{^-102}{430}$.

15. The shaded area takes in three of the four columns and six of the eight small rectangles. Since the area in each case is the same, $\frac{3}{4} = \frac{6}{8}$.

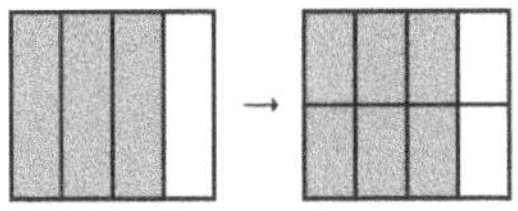

16. To obtain equal fractions, multiply numerator and denominator by the same number. All fractions equal to $\frac{3}{4}$ will then be of the form $\frac{3n}{4n}$. $3n + 4n = 84 \Rightarrow 7n = 84$; thus $n = 12$. $\frac{3 \cdot 12}{4 \cdot 12} = \mathbf{\frac{36}{48}}$.

17. Mr. Gomez used $\frac{7}{8}$ of the gas, so he had $\frac{1}{8}$ of a tank remaining; the needle points to the 1st division of 8 as shown:

(b) Mr. Gomez used 14 gallons: $\frac{14}{16} = \frac{7}{8}$.

18. **(a)** For equal fractions $\frac{a}{b} = \frac{c}{d}$, $ad = bc$.

If $\frac{2}{3} = \frac{x}{16}$, then $2 \cdot 16 = 3 \cdot x \Rightarrow$

$32 = 3x \Rightarrow \boldsymbol{x = \frac{32}{3}}$.

(b) $3 \cdot x = 4 \cdot {}^{-}27 \Rightarrow 3x = {}^{-}108 \Rightarrow$

$\boldsymbol{x = {}^{-}36}$.

19. **(a)** $\boldsymbol{\frac{7}{8} > \frac{5}{6}}$. The least common denominator of 6 and 8 is 24; $\frac{7}{8} = \frac{21}{24}$ and $\frac{5}{6} = \frac{20}{24}$.

(b) $\boldsymbol{\frac{-7}{8} < \frac{-4}{5}}$. The smallest common denominator of 8 and 5 is 40;

so $\frac{{}^{-}7}{8} = \frac{{}^{-}35}{40}$

and $\frac{{}^{-}4}{5} = \frac{{}^{-}32}{40}$ (note that ${}^{-}35 < {}^{-}32$).

20. Answers may vary; one method is to convert the given fractions into equal fractions having larger common denominators, thus creating spaces between them. For example:

(a) $\frac{3}{7} = \frac{9}{21}$ and $\frac{4}{7} = \frac{12}{21}$. Two rational numbers between them could then be $\frac{10}{21}$ and $\frac{11}{21}$.

(b) $\frac{{}^{-}7}{9} = \frac{{}^{-}28}{36}$ and $\frac{{}^{-}8}{9} = \frac{{}^{-}32}{36}$. Two rational numbers between them could then be $\frac{{}^{-}30}{36}$ and $\frac{{}^{-}31}{36}$.

21. **(a)** ***(i)*** $6 \text{ oz}\left(\frac{1 \text{ lb}}{16 \text{ oz}}\right) = \frac{6}{16} \text{ lb} = \boldsymbol{\frac{3}{8}}$ **pound.**

(ii) $6 \text{ oz}\left(\frac{1 \text{ lb}}{16 \text{ oz}}\right)\left(\frac{1 \text{ ton}}{2000 \text{ lb}}\right) = \frac{6}{32{,}000} \text{ ton} =$

$\boldsymbol{\frac{3}{16{,}000}}$ **ton.**

(b) $1 \text{ dime}\left(\frac{1 \text{ dollar}}{10 \text{ dime}}\right) = \boldsymbol{\frac{1}{10}}$ **dollar.**

(c) $15 \text{ min}\left(\frac{1 \text{ hr}}{60 \text{ min}}\right) = \frac{15}{60} \text{ hr} = \boldsymbol{\frac{1}{4}}$ **hour.**

(d) $8 \text{ hr}\left(\frac{1 \text{ day}}{24 \text{ hr}}\right) = \frac{8}{24} \text{ day} = \boldsymbol{\frac{1}{3}}$ **day.**

22. Even though the figure is divided in an inconsistent manner, we could redraw the figure as follows.

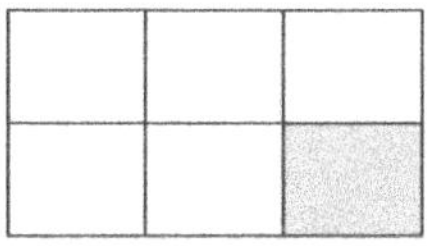

The **shaded region is $\boldsymbol{\frac{1}{6}}$** of the entire figure.

23. Note that decreasing means "largest to smallest".

(a) When two or more fractions have the same numerator, then the fraction that has the smallest denominator will be the largest fraction. Therefore

$\boldsymbol{\frac{11}{13} > \frac{11}{16} > \frac{11}{22}}$.

(b) First, note that $-\frac{1}{5}$ is between 0 and $-\frac{1}{2}$, while the other two fractions are between $-\frac{1}{2}$ and 1. That means $-\frac{1}{5}$ is larger than the other two fractions. To compare $-\frac{19}{36}$ and $-\frac{17}{30}$, use the least common denominator of 36 and 30, which is 180. So $-\frac{19}{36} = \frac{-95}{180}$ and $-\frac{17}{30} = -\frac{102}{180}$. Therefore

$\boldsymbol{\frac{-1}{5} > \frac{-19}{36} > \frac{-17}{30}}$.

24. **False**. For example, let $a = 1$ and $c = 2$. Since $b < 0$, let $b = -3$. So $\frac{1}{-3} > \frac{2}{-3}$ is true, but a is not larger than c.

25. **(a)** **Answers vary**. For example

$\frac{1}{3} < \frac{2}{4} < \frac{3}{5} < \frac{4}{6}$.

The terms in this sequence go in ascending order: $\frac{20}{60} < \frac{30}{60} < \frac{36}{60} < \frac{40}{60}$.

(b) **Answers vary**. For example

$\frac{{}^{-}3}{3} < \frac{{}^{-}4}{5} < \frac{{}^{-}5}{7} < \frac{{}^{-}6}{9}$. The terms in this sequence go in ascending order:

$\frac{{}^{-}315}{315} < \frac{{}^{-}252}{315} < \frac{{}^{-}225}{315} < \frac{{}^{-}210}{315}$.

26. $n = \frac{n}{1}$ for all integers n, so every integer can be written as a fraction.

27. 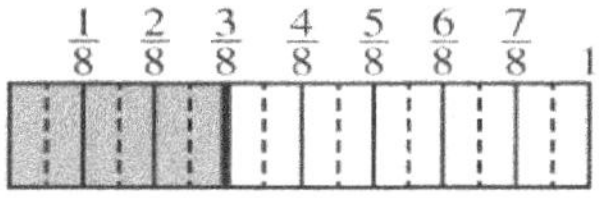

The drawing is not to scale.

28. $\frac{1}{5}$ of $10 = \frac{1}{5} \cdot 10 = 2$. **Two light bulbs** were not shining.

29. $\frac{2}{5}$ were men so $\boldsymbol{\frac{3}{5}}$ **were women.**

Assessment 6-2A: Addition and Subtraction, and Estimation with Rational Numbers

1. **(a)** Three possible methods are illustrated below:

   (i) $\frac{1}{2}+\frac{2}{3}=\frac{1\cdot 3+2\cdot 2}{2\cdot 3}=\frac{3+4}{6}=\mathbf{\frac{7}{6}}$;

   or

   (ii) $LCD(2,3)=6\Rightarrow$

   $\frac{1}{2}+\frac{2}{3}=\frac{1}{2}\cdot\frac{3}{3}+\frac{2}{3}\cdot\frac{2}{2}$

   $=\frac{3}{6}+\frac{4}{6}=\mathbf{\frac{7}{6}}$; or

   (iii)

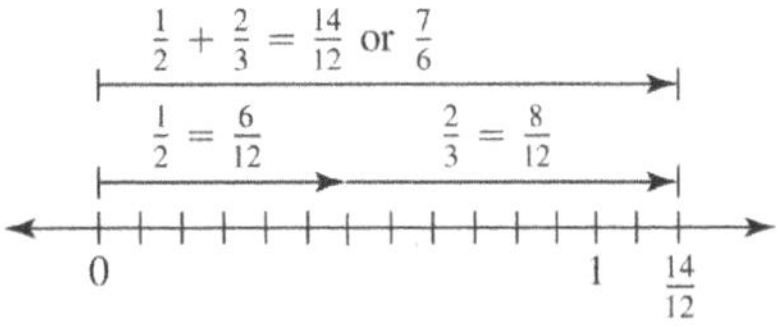

   (b) $LCD(12,3)=12\Rightarrow\frac{4}{12}-\frac{2}{3}=$

   $\frac{4}{12}-\frac{2}{3}\cdot\frac{4}{4}=\frac{4}{12}-\frac{8}{12}=\frac{^{-}4}{12}=\mathbf{\frac{^{-}1}{3}}$.

   (c) $\frac{5}{x}+\frac{-3}{y}=\frac{5\cdot y-x\cdot 3}{x\cdot y}=\mathbf{\frac{5y-3x}{xy}}$.

   (d) $LCD(2x^2y,2xy^2,x^2)=2x^2y^2\Rightarrow$

   $\frac{^{-}3}{2x^2y}+\frac{5}{2xy^2}+\frac{7}{x^2}=\frac{^{-}3}{2x^2y}\cdot\frac{y}{y}+$

   $\frac{5}{2xy^2}\cdot\frac{x}{x}+\frac{7}{x^2}\cdot\frac{2y^2}{2y^2}=\mathbf{\frac{^{-}3y+5x+14y^2}{2x^2y^2}}$.

   (e) $\frac{5}{6}+2\frac{1}{8}=\frac{5}{6}+\frac{17}{8}=\frac{5\cdot 8+6\cdot 17}{6\cdot 8}=$

   $\frac{40+102}{48}=\frac{142}{48}=\frac{71}{24}=\mathbf{2\frac{23}{24}}$.

   (f) $^{-}4\frac{1}{2}-3\frac{1}{6}={}^{-}4\frac{3}{6}-3\frac{1}{6}={}^{-}7\frac{4}{6}$

   $=\mathbf{^{-}7\frac{2}{3}}\left(\text{or }\mathbf{\frac{^{-}23}{3}}\right)$.

   (g) $7\frac{1}{4}+3\frac{5}{12}-2\frac{1}{3}=\frac{29}{4}+\frac{41}{12}-\frac{7}{3}=$

   $\frac{87}{12}+\frac{41}{12}-\frac{28}{12}=\frac{128}{12}-\frac{28}{12}=\frac{100}{12}=8\frac{4}{12}$

   $=\mathbf{8\frac{1}{3}}$

2. **(a)** Two possible methods are illustrated below:

   (i) $\frac{56}{3}=\frac{3\cdot 18+2}{3}=\frac{3\cdot 18}{3}+\frac{2}{3}=$

   $18+\frac{2}{3}=\mathbf{18\frac{2}{3}}$; or

   (ii) $56\div 3=18$, remainder

   $2\Rightarrow\frac{56}{3}=\mathbf{18\frac{2}{3}}$.

   (b) $-\frac{293}{100}=-\left(\frac{2\cdot 100+93}{100}\right)={}^{-}\mathbf{2\frac{93}{100}}$.

3. **(a)** $6\frac{3}{4}=\frac{6}{1}+\frac{3}{4}=\frac{6\cdot 4+1\cdot 3}{1\cdot 4}=\frac{24+3}{4}$

   $=\mathbf{\frac{27}{4}}$.

   (b) $^{-}3\frac{5}{8}={}^{-}\left(\frac{3}{1}+\frac{5}{8}\right)={}^{-}\left(\frac{3\cdot 8+1\cdot 5}{1\cdot 8}\right)=$

   $-\left(\frac{24+5}{8}\right)=\mathbf{\frac{^{-}29}{8}}$.

4. **(a)** $\frac{15}{46}\approx\frac{15}{45}=\mathbf{\frac{1}{3}}$. 45 is smaller than the actual denominator so the estimate is **high**.

   (b) $\frac{7}{41}\approx\frac{7}{42}=\mathbf{\frac{1}{6}}$. 42 is larger than the actual denominator so the estimate is **low**.

   (c) $\frac{62}{80}\approx\frac{60}{80}=\mathbf{\frac{3}{4}}$. 60 is smaller than the actual numerator so the estimate is **low**.

   (d) $\frac{9}{19}\approx\frac{9}{18}=\mathbf{\frac{1}{2}}$. The denominator in the estimate is smaller than the actual denominator, so the estimate is **high**.

5. **(a)** **Beavers.** $\frac{10}{19}>\frac{1}{2}$.

   (b) **Ducks.** $\frac{10}{22}<\frac{1}{2}$ by the least margin.

   (c) **Bears.** $\frac{8}{23}>\frac{1}{3}$ by the least margin.

6.

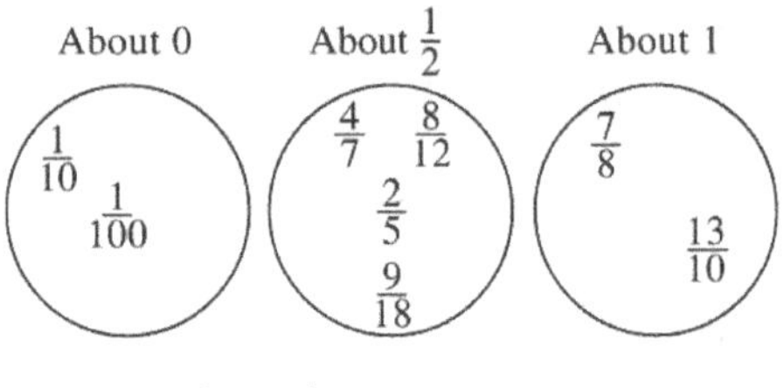

   Note that $\frac{8}{12}=\frac{3}{4}$ so it could be placed in either the "about $\frac{1}{2}$" or the "about 1" oval.

7. **(a)** $\mathbf{\frac{1}{2}}$**; high.** $\frac{19}{38}=\frac{1}{2}\Rightarrow\frac{19}{39}<\frac{1}{2}$.

   (b) **0; low.** $\frac{3}{197}>0$.

   (c) $\mathbf{\frac{3}{4}}$**; high.** $\frac{150}{200}=\frac{3}{4}\Rightarrow\frac{150}{201}<\frac{3}{4}$.

   (d) **1; high.** $\frac{8}{9}<1$.

8. **(a)** **2**. Each addend is about $\frac{1}{2}$; thus the best approximation would be $4\cdot\frac{1}{2}=2$.

(b) $\mathbf{\frac{3}{4}}$. $\frac{30}{41}$ is about $\frac{3}{4}$; the other two addends are negligible compared to $\frac{3}{4}$.

9. Possible thought processes could be:

(a) $\frac{8}{4} - \frac{3}{4} = \mathbf{\frac{5}{4}}$.

(b) $\left(3 + 2 + \frac{3}{8} + \frac{2}{8}\right) - 5\frac{5}{8} = 5\frac{5}{8} - 5\frac{5}{8} = \mathbf{0}$.

10. **(a)** Region ***A***. $\frac{20}{8} = \frac{10}{4}$ is between 2 and 3.

(b) Region ***H***. $\frac{36}{8} = \frac{18}{4}$ is between 4 and 5.

(c) Region ***T***. $\frac{60}{16} = \frac{15}{4}$ is between 3 and 4.

(d) Region ***H***. $\frac{18}{4}$ is between 4 and 5.

11. About $4 \cdot 3 = \mathbf{12}$. The numbers cluster around 3.

12. The entire student population is represented by 1; subtract to obtain the senior's fraction; i.e., seniors make up $1 - \frac{2}{5} - \frac{1}{4} - \frac{1}{10}$ of the class. Thus
$1 - \frac{2}{5} - \frac{1}{4} - \frac{1}{10} = \frac{20}{20} - \frac{8}{20} - \frac{5}{20} - \frac{2}{20} = \frac{5}{20} = \mathbf{\frac{1}{4}}$.

13. $\frac{1}{3} + 2\frac{3}{4} + 3\frac{1}{2} = \frac{4}{12} + 2\frac{9}{12} + 3\frac{6}{12} = 5\frac{19}{12} = 5 + 1\frac{7}{12} = \mathbf{6\frac{7}{12}}$ **yards**.

14. The amount of fabric to be used is
$1\frac{7}{8} + 2\frac{3}{8} + 1\frac{2}{3} = 1\frac{21}{24} + 2\frac{9}{24} + 1\frac{16}{24} = 4\frac{46}{24} = 5\frac{22}{24}$ yards. The amount left over is
$8\frac{3}{4} - 5\frac{22}{24} = 8\frac{18}{24} - 5\frac{22}{24} = 7\frac{42}{24} - 5\frac{22}{24} = 2\frac{20}{24} = \mathbf{2\frac{5}{6}}$ **yards**.

15. Examples may vary:

(a) Closure property: If two rational numbers are added, the sum should also be rational; e.g., $\frac{1}{2} + \frac{3}{4} = \frac{5}{4}$, which is rational.

(b) Commutative property:
$\frac{a}{b} + \frac{c}{d} = \frac{c}{d} + \frac{a}{b}$; e.g.,
$\frac{1}{2} + \frac{3}{4} = \frac{5}{4} = \frac{3}{4} + \frac{1}{2}$.

(c) Associative property: $\frac{a}{b} + \left(\frac{c}{d} + \frac{c}{f}\right) = \left(\frac{a}{b} + \frac{c}{d}\right) + \frac{c}{f}$; e.g.,
$\frac{1}{2} + \left(\frac{1}{3} + \frac{1}{4}\right) = \frac{13}{12} = \left(\frac{1}{2} + \frac{1}{3}\right) + \frac{1}{4}$.

16. **(a)** (*i*) $\frac{1}{4} + \frac{1}{3 \cdot 4} = \frac{1}{4} + \frac{1}{12} = \frac{3}{12} + \frac{1}{12} = \frac{4}{12} = \frac{1}{3}$.

(***ii***) $\frac{1}{5} + \frac{1}{4 \cdot 5} = \frac{1}{5} + \frac{1}{20} = \frac{4}{20} + \frac{1}{20} = \frac{5}{20} = \frac{1}{4}$.

(***iii***) $\frac{1}{6} + \frac{1}{5 \cdot 6} = \frac{1}{6} + \frac{1}{30} = \frac{5}{30} + \frac{1}{30} = \frac{6}{30} = \frac{1}{5}$.

(b) $\mathbf{\frac{1}{n} = \frac{1}{n+1} + \frac{1}{n(n+1)}}$.

17. **(a)** $x + 2\frac{1}{2} = 3\frac{1}{3} \Rightarrow x = 3\frac{1}{3} - 2\frac{1}{2} = 2\frac{8}{6} - 2\frac{3}{6}$. $\mathbf{x = \frac{5}{6}}$.

(b) $x - 2\frac{2}{3} = \frac{5}{6} \Rightarrow x = \frac{5}{6} + \frac{16}{6}$.
$\mathbf{x = \frac{21}{6} = 3\frac{1}{2}}$.

18. In 10 minutes, the difference of the two runners' distances is $\frac{7}{8} - \frac{5}{8} = \frac{2}{8} = \frac{1}{4}$ mile. In 20 minutes the difference will be twice as large. Answer: $2\left(\frac{1}{4}\right) = \frac{2}{4} = \mathbf{\frac{1}{2}}$ **mile**.

Alternative:

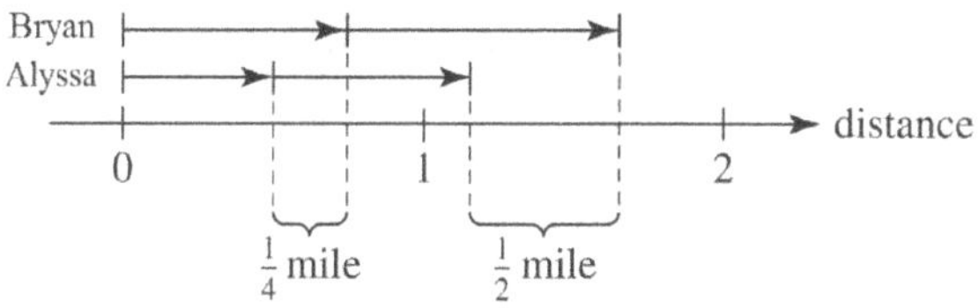

19. **No**. To make both recipes, you need $1\frac{3}{4} + 1\frac{1}{2} = 1 + \frac{3}{4} + 1 + \frac{1}{2} = 2 + \frac{3}{4} + \frac{1}{2} = 2 + 1\frac{1}{4} = 3\frac{1}{4}$ cups. $3\frac{1}{4} - 3 = \frac{1}{4}$ cups less than what is needed.

20. **Answers may vary**. Some possible solutions:

(a) $\frac{76,432 + 76,503 + 78,672}{150,155 + 149,904 + 154,144} = \frac{231,607}{454,203} \approx \frac{23}{45}$

(b) $\frac{73,723 + 73,401 + 75,472}{150,155 + 149,904 + 154,144} = \frac{222,596}{454,203} \approx \frac{22}{45}$

(c) $\dfrac{231,607}{12,742,886} \approx \dfrac{2}{127}$

21. *i.* **Answers may vary**. Some possible solutions are
$\frac{1}{2}+\frac{1}{2}=1; \frac{1}{4}+\frac{3}{4}=1; \frac{1}{3}+\frac{2}{3}=1;$
$\frac{2}{5}+\frac{3}{5}=1; \frac{1}{5}+\frac{4}{5}=1; \frac{5}{6}+\frac{1}{6}=1;$ etc

ii. $\frac{a}{b}+\frac{b-a}{b}=\frac{a+b-a}{b}=\frac{b}{b}=1$; the other fraction is $\mathbf{\frac{b-a}{b}}$

22. $40^\circ + 5^\circ = 45^\circ$ and $\dfrac{45^\circ}{360^\circ} = \dfrac{1}{8}$ so rent and saving represent $\mathbf{\dfrac{1}{8}}$ of the total.

Mathematical Connections 6-2: Review Problems

19. (a) **Equal.** $\frac{a^2b}{b^2}=\frac{a^2\cdot b}{b\cdot b}=\frac{a^2}{b}\ (b \neq 0)$.

(b) **Not equal.** $377 \cdot 401 = 151{,}177 \neq 400 \cdot 378 = 151{,}200$.

(c) **Equal.** $\frac{0}{10}=0=\frac{0}{^-10}$.

(d) **Not equal.** $a(b+1)=ab+a \neq b(a+1)=ab+b$.

21. There are infinitely many fractions equivalent to $\dfrac{3}{5}$ because the equivalent fractions can be written in the form $\dfrac{3n}{5n}$ with n an integer.Since the set of integers is infinite, the set of fractions equivalent to $\dfrac{3}{5}$ is infinite.

23. $\dfrac{-1}{10}=\dfrac{^-10}{100}$ and $\dfrac{^-10}{100}$ is $\dfrac{10}{100}$ ths away from 0 and $\dfrac{-1}{100}$ is $\dfrac{1}{100}$ ths away from 0 so $\dfrac{^-1}{100}$ is greater than $\dfrac{^-1}{10}$.

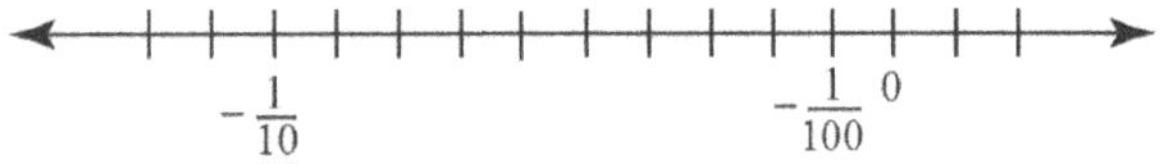

Assessment 6-3A: Multiplication and Division, and Estimation with Rational Numbers

1. (a) The blue-shaded vertical region represents $\frac{1}{3}$ of the total area; the yellow-shaded horizontal region represents $\frac{1}{4}$ of the total area. The blue-yellow (green) region therefore represents $\frac{1}{4}$ of $\frac{1}{3}$, or the product of the two fractions. Since one of the twelve blocks is blue-yellow it represents $\mathbf{\frac{1}{4}\cdot\frac{1}{3}=\frac{1}{12}}$.

(b) The blue-shaded vertical region represents $\frac{3}{5}$ of the total area; the yellow-shaded horizontal region represents $\frac{2}{4}$ of the total area. The blue-yellow (green) region therefore represents $\frac{2}{4}$ of $\frac{3}{5}$, or the product of the two fractions. Since six of the twenty blocks are blue-yellow it represents $\mathbf{\frac{2}{4}\cdot\frac{3}{5}=\frac{6}{20}=\frac{3}{10}}$.

2. (a) There are 12 boxes in the figure below. The dark-shaded region represents $\frac{9}{12}$ of $\frac{4}{12}$, or $\frac{3}{4}\cdot\frac{1}{3}=\mathbf{\frac{3}{12}}=\mathbf{\frac{1}{4}}$:

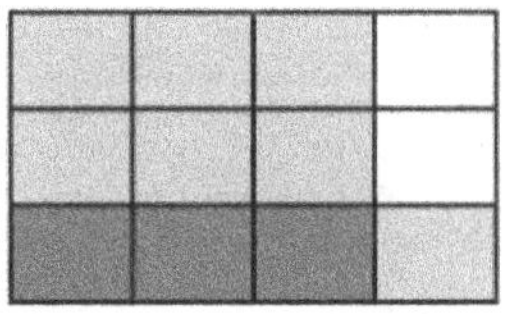

(b) There are 15 boxes in the figure below. The dark-shaded region represents $\frac{3}{15}$ of $\frac{10}{15}$, or $\frac{1}{5}\cdot\frac{2}{3}=\mathbf{\frac{2}{15}}$:

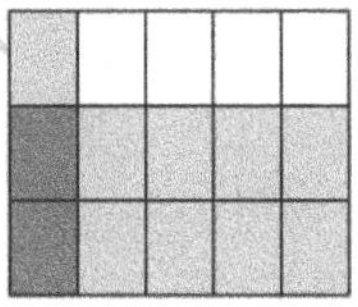

3. (a) $\frac{49}{65}\cdot\frac{26}{98}=\frac{1274}{6370}=\frac{1\cdot1274}{5\cdot1274}=\mathbf{\frac{1}{5}}$.

(b) $\frac{a}{b}\cdot\frac{b^2}{a^2}=\frac{ab^2}{a^2b}=\frac{b\cdot ab}{a\cdot ab}=\mathbf{\frac{b}{a}}$.

(c) $\frac{xy}{z}\cdot\frac{z^2a}{x^3y^2}=\frac{axyz^2}{x^3y^2z}=\frac{az\cdot xyz}{x^2y\cdot xyz}=\mathbf{\frac{az}{x^2y}}$.

4. **(a)** $4\frac{1}{2} \cdot 2\frac{1}{3} = \left(4 + \frac{1}{2}\right) \cdot \left(2 + \frac{1}{3}\right)$
$= 4\left(2 + \frac{1}{3}\right) + \frac{1}{2}\left(2 + \frac{1}{3}\right)$
$= 8 + \frac{4}{3} + 1 + \frac{1}{6}$
$= 9 + \frac{8}{6} + \frac{1}{6} = \mathbf{10\frac{1}{2}}.$

(b) $3\frac{1}{3} \cdot 2\frac{1}{2} = \left(3 + \frac{1}{3}\right) \cdot \left(2 + \frac{1}{2}\right)$
$= 3\left(2 + \frac{1}{2}\right) + \frac{1}{3}\left(2 + \frac{1}{2}\right)$
$= 6 + \frac{3}{2} + \frac{2}{3} + \frac{1}{6}$
$= 6 + \frac{9}{6} + \frac{4}{6} + \frac{1}{6} = \mathbf{8\frac{1}{3}}.$

5. **(a)** Multiplicative inverse of $\frac{^-1}{3}$ is $\frac{3}{^-1} = \mathbf{^-3}.$

(b) Multiplicative inverse of $3\frac{1}{3} = \frac{10}{3}$ is $\mathbf{\frac{3}{10}}.$

(c) Multiplicative inverse of $\frac{x}{y}$ is $\mathbf{\frac{y}{x}}(x, y \neq 0).$

(d) Multiplicative inverse of $^-7 = \frac{^-7}{1}$ is $\frac{1}{^-7} = \mathbf{\frac{^-1}{7}}.$

6. **(a)** $\frac{2}{3}x = \frac{7}{6} \Rightarrow x = \frac{7}{6} \div \frac{2}{3} = \frac{7}{6} \cdot \frac{3}{2} \Rightarrow \mathbf{x = \frac{7}{4}}.$

(b) $\frac{3}{4} \div x = \frac{1}{2} \Rightarrow \frac{3 \cdot 1}{4 \cdot x} = \frac{1}{2} \Rightarrow$
$3 \cdot 1 \cdot 2 = 4 \cdot x \cdot 1 \Rightarrow 4x = 6 \Rightarrow$
$\mathbf{x = \frac{6}{4} = \frac{3}{2}}.$

(c) $\frac{5}{6} + \frac{2}{3}x = \frac{3}{4} \Rightarrow \frac{2}{3}x = \frac{3}{4} - \frac{5}{6} = \frac{^-1}{12} \Rightarrow$
$x = \frac{^-1}{12} \div \frac{2}{3} = \frac{^-1}{12} \cdot \frac{3}{2} \Rightarrow \mathbf{x = \frac{^-1}{8}}.$

(d) $\frac{2x}{3} - \frac{1}{4} = \frac{x}{6} + \frac{1}{2} \Rightarrow$
$\frac{2}{3}x - \frac{1}{6}x = \frac{1}{2} + \frac{1}{4} \Rightarrow$
$\frac{1}{2}x = \frac{3}{4} \Rightarrow$
$x = \frac{3}{4} \div \frac{1}{2} = \frac{3}{4} \cdot \frac{2}{1} \Rightarrow \mathbf{x = \frac{3}{2}}.$

7. Answers may vary; e.g.,

(a) $\frac{1}{2} \div \frac{1}{4} \neq \frac{1}{4} \div \frac{1}{2}.$

(b) $\left(\frac{2}{3} \div \frac{1}{2}\right) \div \frac{3}{5} \neq \frac{2}{3} \div \left(\frac{1}{2} \div \frac{3}{5}\right)$

8. Possible thought processes are described; all provide exact answers:

(a) $3 \cdot 8 = 24, \frac{1}{4} \cdot 8 = 2,$ and $24 + 2 = \mathbf{26}.$

(b) $7 \cdot 4 = 28, \frac{1}{4} \cdot 4 = 1,$ and $28 + 1 = \mathbf{29}.$

(c) $9 \cdot 10 = 90, \frac{1}{5} \cdot 10 = 2,$ and $90 + 2 = \mathbf{92}.$

(d) $8 \cdot 2 = 16, 8 \cdot \frac{1}{4} = 2,$ and $16 + 2 = \mathbf{18}.$

9. **(a)** **20.** $3\frac{11}{12} \cdot 5\frac{3}{100} \approx 4 \cdot 5 = \mathbf{20}.$

(b) **16.** $2\frac{1}{10} \cdot 7\frac{7}{8} \approx 2 \cdot 8 = \mathbf{16}.$

(c) **1.** $\frac{1}{101}$ and $\frac{1}{103}$ are approximately equal.

10. **(a)** **Less than 1.** $\frac{13}{14} \cdot \frac{17}{19}$ is the product of two proper fractions; i.e., each is less than 1. The product of two positive proper fractions will be smaller than either, and therefore will always be less than 1.

(b) **Less than 1.** $3\frac{2}{7} \div 5\frac{1}{9}$ has a positive number divided by a larger positive number. Their quotient, a proper fraction, would always be less than 1.

(c) **Greater than 2.** If the quotient was 2, then checking would show that $2\left(2\frac{3}{100}\right) = 4\frac{6}{100}$, which is less than $4\frac{1}{3}$. Thus the true quotient must be greater than 2.

11. **(d)**, between \$33 and \$40. Estimation gives a cost of approximately $6 \cdot \$4.00 + 3 \cdot \$3.00 = \$24.00 + \$9.00 = \$33.00$. All rounding was down, so the estimate is low.

12. Let p be the student population. The 6000 students living in dorms are $\frac{5}{8}$ of p; i.e.,

$6000 = \frac{5}{8}p \Rightarrow$

$p = \frac{8}{5}(6000) = $ **9600 students.**

13. Alberto has $\frac{5}{9}$ of the stock; Renatta has $\frac{1}{2} \cdot \frac{5}{9} = \frac{5}{18}$ of the stock. Thus $1 - \frac{5}{9} - \frac{5}{18} = \mathbf{\frac{1}{6}}$ **of the stock** is not owned by them.

14. Let p be the original price. Then $p - \frac{1}{4}p = 180 \Rightarrow \frac{3}{4}p = 180 \Rightarrow p = \frac{4}{3}(180) = \mathbf{\$240}$.

15. Let a be the amount of money in the account. After spending \$50 there was $a - 50$ remaining. He spent $\frac{3}{5}$ of that, or $\frac{3}{5}(a - 50)$, leaving him $\frac{2}{5}(a - 50)$; half goes back into the bank, or $\frac{1}{2} \cdot \frac{2}{5}(a - 50) = \frac{1}{5}(a - 50)$. The other half was \$35; or
$\frac{1}{5}(a - 50) = 35 \Rightarrow \frac{1}{5}a - 10 = 35 \Rightarrow$
$\frac{1}{5}a = 45 \Rightarrow a = 5 \cdot 45 = \mathbf{\$225}$.

16. Al's marbles are halved three times in the process of Dani receiving 4 marbles, i.e., if m is the number of marbles Al had originally, then
$\frac{1}{2} \cdot \frac{1}{2} \cdot \frac{1}{2} \cdot m = 4 \Rightarrow$
$\frac{1}{8}M = 4 \Rightarrow M = 8 \cdot 4 = \mathbf{32\ marbles}$.

17. If a is any number and m and n are natural numbers, then $a^m \cdot a^n = a^{m+n}$, $\frac{a^m}{a^n} = a^{m-n}$, and $a^{-m} = \frac{1}{a^m}$:

(a) $3^{-7} \cdot 3^{-6} = 3^{-7+-6} = 3^{-13} = \frac{1}{3^{13}} = \left(\frac{1}{3}\right)^{13}$.

(b) $3^7 \cdot 3^6 = 3^{7+6} = \mathbf{3^{13}}$.

(c) $5^{15} \div 5^4 = 5^{15-4} = \mathbf{5^{11}}$.

(d) $5^{15} \div 5^{-4} = 5^{15-{-4}} = 5^{15+4} = \mathbf{5^{19}}$.

(e) $(^-5)^{-2} = \frac{1}{(^-5)^2} = \frac{1}{5^2} = \left(\frac{1}{5}\right)^2$.

(f) $\frac{a^2}{a^{-3}} = a^{2-{-3}} = a^{2+3} = \mathbf{a^5}$.

18. **(a)** $\left(\frac{1}{2}\right)^3 \cdot \left(\frac{1}{2}\right)^7 = \left(\frac{1}{2}\right)^{3+7} = \left(\frac{1}{2}\right)^{10}$.

(b) $\left(\frac{1}{2}\right)^9 \div \left(\frac{1}{2}\right)^6 = \left(\frac{1}{2}\right)^{9-6} = \left(\frac{1}{2}\right)^3$.

(c) $\left(\frac{2}{3}\right)^5 \cdot \left(\frac{4}{9}\right)^2 = \left(\frac{2}{3}\right)^5 \cdot \left[\left(\frac{2}{3}\right)^2\right]^2 = \left(\frac{2}{3}\right)^{5+4} = \left(\frac{2}{3}\right)^9$.

(d) $\left(\frac{3}{5}\right)^7 \div \left(\frac{3}{5}\right)^7 = \left(\frac{3}{5}\right)^{7-7} = \left(\frac{3}{5}\right)^0 = \mathbf{1}$.

19. Counterexamples may vary:

(a) **False.** $2^3 \cdot 3^2 = 8 \cdot 9 = 72 \neq (2 \cdot 3)^{3+2} = 6^5 = 7776$.

(b) **False.** $2^3 \cdot 3^2 = 72 \neq (2 \cdot 3)^{2 \cdot 3} = 6^6 = 46{,}656$.

(c) **False.** $2^3 \cdot 3^3 = (2 \cdot 3)^3 = 216 \neq (2 \cdot 3)^{2 \cdot 3} = 46{,}656$.

(d) **True**. Since $a \neq 0$ and $b \neq 0$, $(ab)^0 = 1$ for all rational numbers a and b.

(e) **False.**
$(2 + 3)^2 = (2 + 3)(2 + 3) = 25 \neq 2^2 + 3^2 = 13$.

(f) **False.** $(2 + 3)^{-2} = \frac{1}{(2+3)^2} = \frac{1}{25} \neq \frac{1}{2^2} + \frac{1}{3^2} = \frac{13}{36}$.

20. **(a)** $2^n = 32 \Rightarrow 2^n = 2^5 \Rightarrow \mathbf{n = 5}$.

(b) $n^2 = 36 \Rightarrow n^2 = (\pm 6)^2 \Rightarrow \mathbf{n = 6}$ or $\mathbf{^-6}$.

(c) $2^n \cdot 2^7 = 2^5 \Rightarrow 2^{n+7} = 2^5 \Rightarrow n + 7 = 5 \Rightarrow \mathbf{n = {^-2}}$.

(d) $2^n \cdot 2^7 = 8 \Rightarrow 2^{n+7} = 2^3 \Rightarrow n + 7 = 3 \Rightarrow \mathbf{n = {^-4}}$.

21. **(a)** $3^x \leq 9 \Rightarrow 3^x \leq 3^2 \Rightarrow \mathbf{x \leq 2}$, where x is an integer.

(b) $25^x < 125 \Rightarrow (5^2)^x < 5^3 \Rightarrow 5^{2x} < 5^3 \Rightarrow 2x < 3 \Rightarrow \mathbf{x < \frac{3}{2}}$,
$\Rightarrow \mathbf{x < 2}$
where x is an integer.

(c) $3^{2x} > 27 \Rightarrow 3^{2x} > 3^3 \Rightarrow 2x > 3 \Rightarrow \mathbf{x > \frac{3}{2}} \Rightarrow \mathbf{x \geq 2}$ where x is an integer.

(d) $4^x > 1 \Rightarrow 4^x > 4^0 \Rightarrow x > 0 \Rightarrow \mathbf{x \geq 1}$, where x is an integer.

22. **(a)** $\left(\frac{1}{2}\right)^3 > \left(\frac{1}{2}\right)^4$; $\left(\frac{1}{2}\right)^3 = \frac{1}{8}$ while $\left(\frac{1}{2}\right)^4 = \frac{1}{16}$; $\frac{1}{8}$ is greater.

(b) $\left(\frac{3}{4}\right)^8 > \left(\frac{3}{4}\right)^{10}$; $\left(\frac{3}{4}\right)^8 \div \left(\frac{3}{4}\right)^{10} = \left(\frac{3}{4}\right)^{-2}$ $= \left(\frac{4}{3}\right)^2$, which is greater than 1 $\Rightarrow$ an improper fraction $\Rightarrow \left(\frac{3}{4}\right)^8$ is greater.

(c) $\left(\frac{4}{3}\right)^{10} > \left(\frac{4}{3}\right)^8$; $\left(\frac{4}{3}\right)^{10} \div \left(\frac{4}{3}\right)^8 = \left(\frac{4}{3}\right)^2$, which is greater than 1 $\Rightarrow \left(\frac{4}{3}\right)^{10}$ is greater.

(d) $\left(\frac{4}{5}\right)^{10} > \left(\frac{3}{4}\right)^{10}$; $\frac{4}{5} \div \frac{3}{4} > 1 \Rightarrow \frac{4}{5} > \frac{3}{4}$ $\Rightarrow \left(\frac{4}{5}\right)^{10} > \left(\frac{3}{4}\right)^{10}$.

(e) $\mathbf{32^{50} > 4^{100}} \Rightarrow 32^{50} = (2^5)^{50} = 2^{250}$

$\Rightarrow 4^{100} = (2^2)^{100} = 2^{200}$

$2^{250} > 2^{200}$

$\mathbf{(^-3)^{^-75} > (^-27)^{^-15}}$

(f) $(^-27)^{^-15} = (^-3^3)^{^-15} = (^-3)^{^-45}$

$(^-3)^{^-75} > (^-3)^{^-45}$

23. If either $\frac{a}{b}$ and $\frac{c}{d}$ is negative, it can be written with the negative signs associated with the numerators. Thus, $\frac{a}{b} < \frac{c}{d} \Rightarrow ad < bc$. Adding ad to both sides of the inequalities results in $2ad < ad + bc$. Adding bc to both sides of the inequalities results in $ad + bc < 2bc$. Dividing both sides of the inequalities by $2bd$ gives us $\frac{a}{b} < \frac{1}{2}\left(\frac{a}{b} + \frac{c}{d}\right) < \frac{c}{d}$.

24. **(a)** $1 - \frac{17}{25} = \frac{25}{25} - \frac{17}{25} = \frac{8}{25}$. There were $\mathbf{\frac{8}{25}}$ of the students who were female.

(b) **No**, the number of male students depends on the total number of students. If there were 25 students total then there would be 17 male students.

25. One batch of green paint consists of $\frac{3}{4} + \frac{3}{4} = \frac{6}{4} = \frac{3}{2}$ cups. $12 \times \frac{3}{2} = \frac{36}{2} =$ **18 cups.**

26. $2 \div \frac{2}{3} = 2 \times \frac{3}{2} = \frac{6}{2} =$ **3 friends** went to the movies.

27. $\frac{1}{33} + \frac{2}{7} = \frac{7}{231} + \frac{66}{231} = \mathbf{\frac{73}{231}}$ is the fractional part of the deer population that is in Mississippi and in Montana.

28. $\frac{1}{4} \div 8 = \mathbf{\frac{1}{32}}$ of the entire estate is what each cousin receives.

Mathematical Connections 6-3: Review Problems

21. The portion of students that take Spanish, French, or German is $\frac{2}{3} + \frac{1}{9} + \frac{1}{18} = \frac{5}{6}$. The portion of students not taking one of these languages is $1 - \frac{5}{6} = \frac{1}{6}$. That number is $\frac{1}{6} \cdot 720 =$ **120 students.**

23. **(a)** This is **not correct**. For example, $\frac{2 \cdot 3 + 5}{3} \neq 2 + 5$ because $\frac{11}{3} \neq 7$.

(b) This is **not correct**. For example, because $\frac{5}{7} \neq \frac{3}{5}$.

(c) This **is correct** as long as a≠0. For example, $\frac{2 \cdot 3 + 2 \cdot 5}{2 \cdot 5} = \frac{3+5}{5}$ because $\frac{6+10}{10} = \frac{3+5}{5} \Rightarrow \frac{16}{10} = \frac{8}{5}$.

Assessment 6-4A: Proportional Reasoning

1. **(a)** There are five vowels and 21 consonants. Their ratio is $\frac{5 \text{ vowels}}{21 \text{ consonants}} = \frac{5}{21}$, or **5:21.**

(b) $\frac{21 \text{ consonants}}{5 \text{ vowels}} = \frac{21}{5}$, or **21:5.**

(c) $\frac{21 \text{ consonants}}{26 \text{ alphabet letters}} = \frac{21}{26}$, or **21:26.**

(d) **Answers may vary**. "Broke" (2 vowels, 3 consonants) or "minor" (2 vowels, 3 consonants) or "reteaching" (4 vowels, 6 consonants) are three examples.

2. $\frac{a}{b} = \frac{c}{d}$ only if $ad = bc$:

(a) $\frac{12}{x} = \frac{18}{45} \Rightarrow 18x = 12 \cdot 45 \Rightarrow$ $x = \frac{12 \cdot 45}{18} =$ **30.**

(b) $\frac{x}{7} = \frac{^{-}10}{21} \Rightarrow 21x = 7 \cdot {^{-}10} \Rightarrow$

$x = \frac{7 \cdot {^{-}10}}{21} = \frac{^{-}10}{3} = \mathbf{^{-}3\frac{1}{3}}.$

(c) $\frac{5}{7} = \frac{3x}{98} \Rightarrow 7 \cdot 3x = 5 \cdot 98 \Rightarrow$

$x = \frac{5 \cdot 98}{7 \cdot 3} = \frac{70}{3} = \mathbf{23\frac{1}{3}}.$

(d) $3\frac{1}{2}$ is to 5 as x is to 15 $\Rightarrow \frac{3\frac{1}{2}}{5} = \frac{x}{15} \Rightarrow$

$5 \cdot x = 3\frac{1}{2} \cdot 15 \Rightarrow x = \frac{3\frac{1}{2} \cdot 15}{5} = \frac{21}{2} = \mathbf{10\frac{1}{2}}.$

3. **(a)** Because the ratio is 2:3, there are $2x$ boys and $3x$ girls. The ratio of boys to all students is then $\frac{2x}{2x+3x} = \frac{2x}{5x} = \frac{2}{5} =$ **2:5**.

(b) $\mathbf{m:(m+n)}$. See part (a) above.

(c) Because the ratio of girls to all students is $\frac{3}{5}$, there are 3 girls to every 2 boys, or a ratio of girls to boys of **3:2**.

4. $\frac{\text{2 pounds muscle}}{\text{5 pounds body weight}} = \frac{x \text{ pounds muscle}}{\text{90 pounds body weight}} \Rightarrow$

$5 \cdot x = 2 \cdot 90 \Rightarrow x = \frac{2 \cdot 90}{5} =$ **36 pounds.**

5. $\frac{\text{4 grapefruit}}{80¢}$ and $\frac{\text{12 grapefruit}}{\$2.00}$ $\Rightarrow \frac{\text{1 grapefruit}}{20¢}$ and $\frac{\text{1 grapefruit}}{16.\overline{6}¢}$ so **\$2.00 for 12** is a better buy.

6. $\frac{\frac{1}{3}\text{ inch}}{\text{5 miles}} = \frac{\text{18 inches}}{x \text{ miles}} \Rightarrow \frac{1}{3} \cdot x = 5 \cdot 18 \Rightarrow x =$

$\frac{5 \cdot 18}{\frac{1}{3}} =$ **270 miles.**

7. $\frac{\text{40 pages}}{\text{50 minutes}} = \frac{x \text{ pages}}{\text{80 minutes}} \Rightarrow 50 \cdot x$

$= 40 \cdot 80 \Rightarrow$

$x = \frac{40 \cdot 80}{50} =$ **64 pages.**

8. **(a)** Let u be a unit value. Then $3u + 4u = 98 \Rightarrow 7u = 98 \Rightarrow u = 14$. $3u =$ **42** and $4u =$ **56**.

(b) Let u be a unit value. $3u \cdot 4u = 768 \Rightarrow 12u^2 = 768 \Rightarrow u^2 = 64 \Rightarrow u = 8$ or

$u = {^{-}8}$

[since $8^2 = 64$ and $({^{-}8})^2 = 64$]. Then $3u =$ **24** and $4u =$ **32 or** $3u =$ **$^{-}$24** and $4u =$ **$^{-}$32.**

9. **(*i*)** There are a total of $2 + 3 + 5 = 10$ shares. Using a proportion, and letting G represent the amount of money Gary would receive, we write:

$\frac{2}{10} = \frac{G}{82{,}000} \Rightarrow 10G =$

$2 \cdot 82{,}000 \Rightarrow G =$ **\$16,400.**

Alternative thinking: Common Core State Standards suggests we consider using rates to solve this type of problem. Gary's rate is 2 to 10. Thus,

$\frac{2}{10} \cdot 82{,}000 = \frac{2 \cdot 82{,}000}{10} =$ **\$16,400.**

(*ii*) $\frac{3}{10} = \frac{\text{Bill's amount}}{82{,}000} \Rightarrow 10(\text{Bill's}) =$

$3 \cdot 82{,}000 \Rightarrow$ Bill's amount $=$ **\$24,600.**

(*iii*) $\frac{5}{10} = \frac{\text{Carmella's amount}}{82{,}000} \Rightarrow$

$10 \cdot$ Carmella's $= 5 \cdot 82{,}000 \Rightarrow$

Carmella's amount $=$ **\$41,000.**

10. The ratio of Sheila's hours to Dora's is $3\frac{1}{2}$ to $4\frac{1}{2}$. Then

$3\frac{1}{2}x + 4\frac{1}{2}x = \$176 \Rightarrow 8x = 176 \Rightarrow$

$x = 22$. Thus **Sheila's** earnings are

$3\frac{1}{2} \cdot 22 =$ **\$77** and **Dora's** are

$4\frac{1}{2} \cdot 22 =$ **\$99.**

11. Success:failure $= 5{:}4 \Rightarrow \frac{\text{5 successes}}{\text{4 failures}} =$

$\frac{\text{75 successes}}{x \text{ failures}} \Rightarrow 5 \cdot x = 4 \cdot 75 \Rightarrow x = \frac{4 \cdot 75}{5} =$

60 failures. 75 successes + 60 failures = **135 attempts.**

12. **(a)** $\frac{1}{6}{:}1 \Rightarrow (6)\frac{1}{6}{:}(6)1 \Rightarrow 1{:}6 = \mathbf{\frac{1}{6}}.$

(b) $\frac{1}{3}{:}\frac{1}{3} \Rightarrow (3)\frac{1}{3}{:}(3)\frac{1}{3} \Rightarrow 1{:}1 = \mathbf{\frac{1}{1}}.$

(c) $\frac{1}{6}{:}\frac{2}{7} \Rightarrow (42)\frac{1}{6}{:}(42)\frac{2}{7} \Rightarrow 7{:}12 = \mathbf{\frac{7}{12}}.$

13. The proportion implies $12¢ \cdot 48\,\text{oz} = 16¢ \cdot 36\,\text{oz}$. Other equivalent proportions are thus:

(*i*) $\frac{12¢}{16¢} = \frac{\text{36 ounces}}{\text{48 ounces}}.$

(*ii*) $\frac{\text{48 ounces}}{16¢} = \frac{\text{36 ounces}}{12¢}.$

(*iii*) $\frac{16¢}{12¢} = \frac{\text{48 oz}}{\text{36 oz}}.$

14. **(a)** $\frac{\text{Rise}}{\text{half-span}} = \frac{10}{14} = \frac{5}{7}$ or **5:7**.

(b) $\frac{\text{Rise}}{\text{half-span}} = \text{pitch}.\ \frac{\text{Rise}}{8} = \frac{3}{4} \Rightarrow \text{rise} \cdot 4 = 8 \cdot 3 \Rightarrow \text{rise} = \frac{8 \cdot 3}{4} =$ **6 feet.**

15. **(a)** $\frac{4 \text{ rpm on large gear}}{6 \text{ rpm on small gear}} = \frac{18 \text{ teeth on small gear}}{x \text{ teeth on large gear}}$

$\Rightarrow 4 \cdot x = 6 \cdot 18$

$\Rightarrow x = \frac{6 \cdot 18}{4} =$ **27 teeth.**

(b) $\frac{200 \text{ rpm on large gear}}{600 \text{ rpm on small gear}} = \frac{x \text{ teeth on small gear}}{60 \text{ teeth on large gear}}$

$\Rightarrow 600 \cdot x = 200 \cdot 60$

$\Rightarrow x = \frac{200 \cdot 60}{600} =$ **20 teeth.**

16. $\frac{230 \text{ feet length}}{195 \text{ feet wingspan}} = \frac{40 \text{ cm length}}{x \text{ cm wingspan}} \Rightarrow$

$230 \cdot x = 195 \cdot 40 \Rightarrow$

$x = \frac{195 \cdot 40}{230} = 33\frac{21}{23}$, or **about 34 cm.**

17. **(a)** $\frac{3 \text{ cups tomato sauce}}{2 \text{ cups tomato sauce}} = \frac{3}{2}$, so:

(i) $\frac{1 \text{ tsp mustard seed}}{x \text{ tsp mustard seed}} = \frac{3}{2} \Rightarrow 3 \cdot x = 1 \cdot 2 \Rightarrow x = \frac{2}{3}$ **tsp mustard seed.**

(ii) $\frac{1\frac{1}{2} \text{ cups scallions}}{x \text{ cups scallions}} = \frac{3}{2} \Rightarrow 3 \cdot x = 1\frac{1}{2} \cdot 2 \Rightarrow x = \frac{1\frac{1}{2} \cdot 2}{3} =$

1 cup scallions.

(iii) $\frac{3\frac{1}{4} \text{ cups beans}}{x \text{ cups beans}} = \frac{3}{2} \Rightarrow 3 \cdot x = 3\frac{1}{4} \cdot 2 \Rightarrow x = \frac{3\frac{1}{4} \cdot 2}{3} =$

$2\frac{1}{6}$ **cups beans.**

(b) $\frac{1\frac{1}{2} \text{ cups sacallions}}{1 \text{ cup sacallions}} = \frac{3}{2}$, which is the same ratio as in (a) above, or $\frac{2}{3}$ **tsp mustard seed, 2 cups tomato sauce,** $2\frac{1}{6}$ **cups beans.**

(c) $\frac{3\frac{1}{4} \text{ cups beans}}{1\frac{3}{4} \text{ cups beans}} = \frac{13}{7}$, so:

(i) $\frac{1 \text{ tsp mustard seed}}{x \text{ tsp mustard seed}} = \frac{13}{7} \Rightarrow 13 \cdot x = 1 \cdot 7 \Rightarrow x = \frac{1 \cdot 7}{13} =$

$\frac{7}{13}$ **tsp mustard seed.**

(ii) $\frac{3 \text{ cups tomato sauce}}{x \text{ cups tomato sauce}} = \frac{13}{7} \Rightarrow 13 \cdot x = 3 \cdot 7 \Rightarrow x = \frac{3 \cdot 7}{13} =$

$1\frac{8}{13}$ **cups tomato sauce.**

(iii) $\frac{1\frac{1}{2} \text{ cups scallions}}{x \text{ cups scallions}} = \frac{13}{7} \Rightarrow 13 \cdot x = 1\frac{1}{2} \cdot 7 \Rightarrow x = \frac{1\frac{1}{2} \cdot 7}{13} =$

$\frac{21}{26}$ **cups scallions.**

18. $\frac{4 \text{ ohms}}{5 \text{ feet}} = \frac{x \text{ ohms}}{20 \text{ feet}} \Rightarrow 5 \cdot x = 4 \cdot 20 \Rightarrow x = \frac{4 \cdot 20}{5} =$ **16 ohms.**

19. $\frac{2 \text{ cm (daughter)}}{6 \text{ cm (father)}} = \frac{x \text{ cm (daughter)}}{183 \text{ cm (father)}} \Rightarrow 6x = 2 \cdot 183 \Rightarrow x = \frac{2 \cdot 183}{6} =$ **61 cm.**

20. The ratio between the mass of the gold in the ring to the mass of the ring is 18:24. If x is the number of ounces of pure gold in a ring that weighs 0.4 ounces, then $\frac{18 \text{ ounces of gold}}{24 \text{ ounces of ring}} = \frac{x \text{ ounces of gold}}{4 \text{ ounces of ring}} \Rightarrow 24x = 18 \cdot 4 \Rightarrow$

$x = \frac{18 \cdot 4}{24} = 3$ ounces of gold in the ring.

3 ounces at \$1800 per ounce of = **\$5400.**

21. **(a)** $\frac{5 \text{ hours}}{\$40} = \frac{40 \text{ hours}}{\$x} \Rightarrow 5x = 40 \cdot 40$

$\Rightarrow x = \frac{40 \cdot 40}{5} \Rightarrow x =$ **\$320.** The same result could have been obtained by using any of the other ratios; e.g., \$16 for 2 hours.

(b) The constant of proportionality is **8**, representing \$8 dollars per hour.

22. **(a)** The total number of men in all three rooms is $1 + 2 + 5 = 8$; the total number of women in all three rooms is $2 + 4 + 10 = 16$. The ratio of men to women is $\frac{8}{16} = \frac{1}{2}$, or **1:2.**

(b) Let $\frac{a}{b} = \frac{c}{d} = \frac{e}{f} = r$.

Then $a = br$;

$c = dr$;

$e = fr$.

So $a + c + e = br + dr + fr \Rightarrow$

$a + c + e = r(b + d + f) \Rightarrow$

$r = \frac{a+c+e}{b+d+f}$

Thus $r = \frac{a}{b} = \frac{c}{d} = \frac{e}{f} = \frac{a+c+e}{b+d+f}$.

23. $\frac{26 \text{ miles}}{1 \text{ gallon}}$ for the car and

$\frac{250 \text{ miles}}{14 \text{ gallons}} \approx \frac{17.85 \text{ miles}}{1 \text{ gallon}}$ for the truck. The **car gets the better gas mileage** because it travels 26 miles on 1 gallon and the truck only travels 17.85 miles on 1 gallon.

24. $\frac{20 \text{ miles}}{2 \text{ hours}} = \frac{10 \text{ miles}}{1 \text{ hour}}$ for Susan and

$\frac{32 \text{ miles}}{3 \text{ hours}} \approx \frac{10.\bar{6} \text{ miles}}{1 \text{ hour}}$ for Nick. So **Nick travels faster.**

25. The desired ratio of students to computers is 3:1. If there are 40 students then the ratio should be 40:13.3. So there would need to be **14 computers.**

26. The ratio of oblong tables to round tables is 5:1. The ratio of oblong tables to total tables is then 5:6. If there are 102 total tables then the ratio of oblong tables to total tables is $\frac{5}{6} = \frac{85}{102}$. There are **85 oblong tables.** 102- 85 = 17, so there are **17 round tables.**

27. $1\frac{1}{2}$ cups for 2 dozen cookies

3 cups for 4 dozen cookies

$4\frac{1}{2}$ for 6 dozen cookies

4 cups would not be enough for 6 dozen cookies because it would take $4\frac{1}{2}$ cups for 6 dozen cookies.

28. **(a)** For 360 to be divided in the ratio of 1:5, there will be a total of six sections. $360 \div 6 = 60$, so each section will be size 60, as shown below.

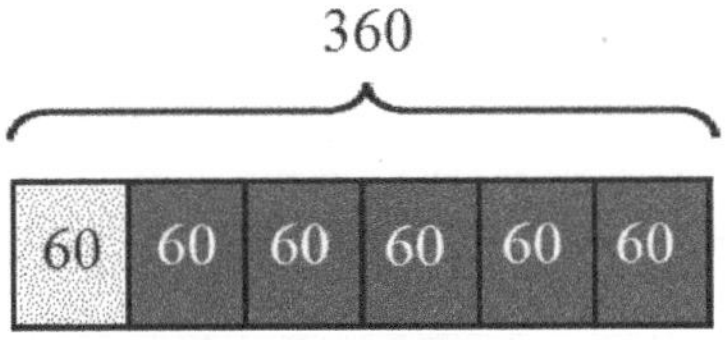

Note that the ratio 60:300 = 1:5.

(b) The bar model below models the situation of the ratio of boys to girls being 6:5, and their being 90 total girls:

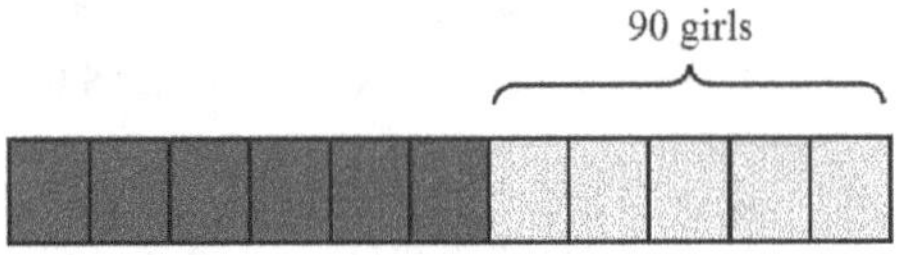

Since $90 \div 5 = 18$, each section represents 18 students at the school. This means there are $18 \times 6 = $ **108 boys** at the school.

29. As seen in the drawing, $\frac{1}{2}$ of stick A is $\frac{2}{3}$ of stick B, and stick B is 18 cm shorter than stick A.

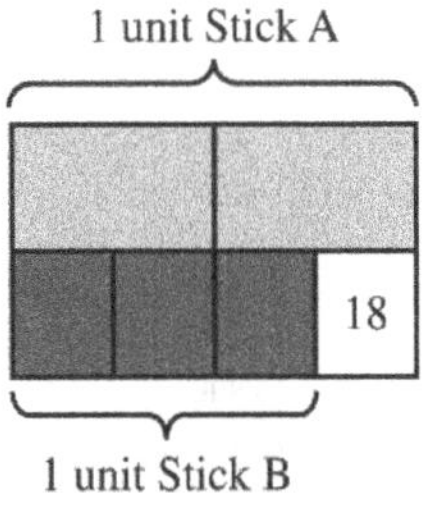

We have the following: Stick B is 3 • 18 unit sections while stick A is 4 • 18 unit sections. Thus the lenths of the sticks are **A = 72 cm, and B = 54 cm.**

Mathematical Connections 6-4: Review Problems

15. $\frac{3}{4}$ is a proper fraction because the numerator and denominator are both positive with the numerator less than the denominator.

17. The statement is not true in general; for example, $\frac{25}{35} \neq \frac{2}{3}$ because $75 \neq 70$.

19. **No,** $\frac{2}{3}$ of a class is not equivalent to $\frac{4}{5}$ of the school being absent because 2 • 5 ≠ 3 • 4, 10 ≠ 12.

Chapter 6 Review

1. **(a)** Answers may vary. Shade three of four parts:

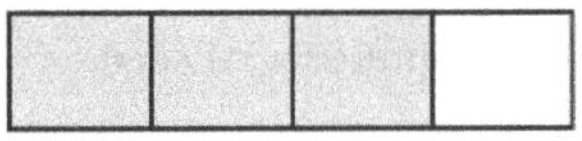

(b) Two of three blocks shaded:

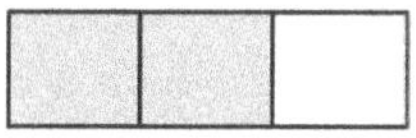

(c) Three of four horizontal bars are shaded, meshed with two of three vertical bars. Six of the twelve bars are dark-shaded; i.e., $\frac{3}{4}\cdot\frac{2}{3}=\frac{6}{12}$:

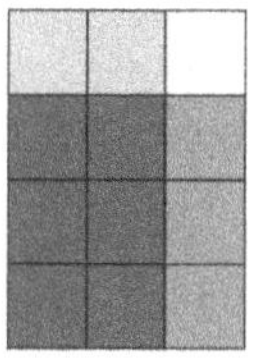

2. Answers may vary; three are $\frac{10}{12}, \frac{15}{18}$, and $\frac{50}{60}$.

3. **(a)** $\frac{24}{28}=\frac{6\cdot 4}{7\cdot 4}=\mathbf{\frac{6}{7}}$.

(b) $\frac{ax^2}{bx}=\frac{ax\cdot x}{b\cdot x}=\mathbf{\frac{ax}{b}}$.

(c) $\frac{0}{17}=\frac{0\cdot 17}{1\cdot 17}=\mathbf{\frac{0}{1}}=0$.

(d) $\frac{45}{81}=\frac{5\cdot 9}{9\cdot 9}=\mathbf{\frac{5}{9}}$.

(e) $\frac{bx^2+bx}{b+x}=\frac{b\cdot(b+x)}{1\cdot(b+x)}=\mathbf{\frac{b}{1}}=b$.

(f) $\frac{16}{216}=\frac{2\cdot 8}{27\cdot 8}=\mathbf{\frac{2}{27}}$.

(g) $\frac{x+a}{x-a}$ **cannot be further reduced**. There are no factors common to both numerator and denominator (x and a are terms).

(h) $\frac{xa}{x+a}$ **cannot be further reduced**.

4. **(a)** $=.\ \frac{6\cdot 20}{10\cdot 20}=\frac{120}{200}$.

(b) $>.\ \frac{^-3}{4}=\frac{^-18}{24}>\frac{^-5}{6}=\frac{^-20}{24}$.

(c) $>.\ \left(\frac{4}{5}\right)^{10}>\left(\frac{4}{5}\right)^{20}$.

(d)

$$<.\ \left(1+\frac{1}{3}\right)^2=\left(\frac{4}{3}\right)^2<\left(1+\frac{1}{3}\right)^3=\left(\frac{4}{3}\right)^3.$$

5. Additive inverse: $n+$ inverse $=0$.
Multiplicative inverse: $n\cdot$ inverse $=1$.

	n	**Additive Inverse**	**Multiplicative Inverse**
(a)	3	$^-3$	$\frac{1}{3}$
(b)	$3\frac{1}{7}=\frac{22}{7}$	$^-3\frac{1}{7}$	$\frac{7}{22}$
(c)	$\frac{5}{6}$	$\frac{^-5}{6}$	$\frac{6}{5}$
(d)	$\frac{^-3}{4}$	$\frac{3}{4}$	$\frac{^-4}{3}$

6. $^-2\frac{1}{3}<{}^-1\frac{7}{8}<0<\left(\frac{71}{140}\right)^{300}<\frac{69}{140}<\frac{1}{2}<\frac{71}{140}<\left(\frac{74}{73}\right)^{300}$.

7. **Yes**. Apply the laws of multiplication and the commutative and associative laws of multiplication to find: $\frac{4}{5}\cdot\frac{7}{8}\cdot\frac{5}{14}=\frac{4\cdot 7\cdot 5}{5\cdot 8\cdot 14}=\frac{4\cdot 7\cdot 5}{8\cdot 14\cdot 5}=\frac{4}{8}\cdot\frac{7}{14}\cdot\frac{5}{5}$.

8. Methods may vary:

(a) $\frac{1}{3}\cdot(8\cdot 9)=\left(\frac{1}{3}\cdot 9\right)\cdot 8=3\cdot 8=\mathbf{24}$.

(b) $36\cdot 1\frac{5}{6}=36\cdot\frac{11}{6}=6\cdot 11=\mathbf{66}$.

9. **(*a*)** Assuming no waste, $54\frac{1}{4}\div 3\frac{1}{12}=17\frac{22}{37}$, so **17 pieces** can be cut.

(b) $\frac{22}{37}\cdot 3\frac{1}{12}=\mathbf{\frac{11}{6}}=\mathbf{1\frac{5}{6}}$ **yards left over**.

10. **(a)** **15.**
$\frac{30\frac{3}{8}}{4\frac{1}{9}}\cdot\frac{8\frac{1}{3}}{3\frac{8}{9}}\approx\frac{30\frac{2}{8}}{4\frac{1}{8}}\cdot\frac{8}{4}=\frac{30}{4}\cdot 2=7\frac{1}{2}\cdot 2$.

(b) **15.** $\left(\frac{3}{800}+\frac{4}{5000}+\frac{15}{6}\right)\cdot 6\approx\frac{15}{6}\cdot 6$.

(c) **4.** $\frac{1}{407}\div\frac{1}{1609}\approx\frac{1}{400}\div\frac{1}{1600}=\frac{1}{400}\cdot\frac{1600}{1}$.

11. **Answers may vary**; for instance: A carpenter has a board measuring $4\frac{5}{8}$ feet. How many 6-inch pieces can be cut from it (disregarding the width of the cuts)? The solution is then $4\frac{5}{8}\div\frac{1}{2}$; a diagram would show four foot-long pieces plus $\frac{5}{8}$ of another, with $\frac{1}{2}$-foot size pieces shaded.

12. **Answers may vary.** $\frac{3}{4}=\frac{60}{80}$ and $\frac{4}{5}=\frac{64}{80}$ so $\frac{61}{80}$ and $\frac{62}{80}$ are between $\frac{3}{4}$ and $\frac{4}{5}$.

13. Think of $504792 \div 23$ as $504792 \cdot \frac{1}{23}$ and enter:

5	0	4	7	9	2	×	2	3	$\frac{1}{x}$	=

14. Lucas ate $\frac{1}{3} \cdot \frac{1}{2} = \frac{1}{6}$ pizza.

$\frac{1}{6} \cdot 2000 = \mathbf{333\frac{1}{3}}$ **calories.**

15. Let t be the number of times the coin was flipped. Then $\frac{1}{2}t = 376 \Rightarrow t = 376 \div \frac{1}{2} =$ **752 times.**

16. $\frac{240 \text{ heads}}{1000 \text{ flips}} = \frac{6 \cdot 40}{25 \cdot 40} = \mathbf{\frac{6}{25}}$ **of the time.**

17. **Not reasonable**. The statement implies that the University won $\frac{3}{4} + \frac{5}{8} = \frac{11}{8}$ of its games, or that it won more games than it played.

18. We could convert to the same units (for example inches to feet); however, this is not necessary when using proportional reasoning provide corresponding measures are in the same units.

Let $x =$ the distance between pupils on the carring of George Washington's head.

$\frac{2\frac{1}{2} \text{ in}}{9 \text{ in}} = \frac{x \text{ ft}}{60 \text{ ft}} \Rightarrow 9x = 2\frac{1}{2} \cdot 60 \Rightarrow x =$

$\frac{2\frac{1}{2} \cdot 60}{9} = \frac{\frac{5}{2} \cdot 60}{9} = \frac{150}{9} = \frac{50}{3} = \mathbf{16\frac{2}{3}}$ **ft.**

19. $\frac{\left(\frac{2}{3}\right)}{\left(\frac{3}{4}\right)} = \frac{\frac{2}{3} \cdot \frac{4}{3}}{\frac{3}{4} \cdot \frac{4}{3}} = \frac{\left(\frac{8}{9}\right)}{1} = \frac{8}{9}$, which is the quotient of two integers.

20. $12 \text{ acres} \cdot 9\frac{1}{3}$ bags per acre $= 108 + 4 =$ **112 bags.**

21. $\frac{2}{3}$ female $\cdot \frac{2}{5}$ blond $= \mathbf{\frac{4}{15}}$ **blond females**; i.e., a fraction of a fraction of the student population.

22. $\frac{^-12}{10}$ is greater than $\frac{^-11}{9}$ because

$^-12 \cdot 9 > {}^-11 \cdot 10$. Alternatively,

$\frac{^-12}{10} - \frac{^-11}{9} = \frac{^-108}{90} - \frac{^-110}{90} = \frac{2}{90}$, which is positive; therefore $-\frac{^-12}{10} > \frac{^-11}{9}$.

23. **(a)** $7^x = 343 \Rightarrow 7^x = 7^3 \Rightarrow \mathbf{x = 3}$.

(b) $2^{^-3x} = \frac{1}{512} \Rightarrow 2^{-3x} = 2^{-9}$

$\Rightarrow {}^-3x = {}^-9 \Rightarrow \mathbf{x = 3}$.

(c) $2x - \frac{5}{3} = \frac{5}{6} \Rightarrow 2x = \frac{5}{6} + \frac{5}{3} = \frac{15}{6} \Rightarrow$

$x = \frac{15}{6} \div 2 = \frac{15}{6} \cdot \frac{1}{2} \Rightarrow \mathbf{x = \frac{5}{4}}$.

(d) $x + 2\frac{1}{2} = 5\frac{2}{3} \Rightarrow x = 5\frac{2}{3} - 2\frac{1}{2} \Rightarrow$

$\mathbf{x = 3\frac{1}{6}}$.

(e) $\frac{20+x}{x} = \frac{4}{5} \Rightarrow 5(20 + x) = 4x \Rightarrow$

$100 + 5x = 4x \Rightarrow 100 = {}^-x \Rightarrow$

$\mathbf{x = {}^-100}$.

(f) $2x + 4 = 3x - \frac{1}{3} \Rightarrow 4 + \frac{1}{3} =$

$3x - 2x \Rightarrow \mathbf{x = 4\frac{1}{3}}$.

24. **(a)** $\frac{(x^3a^{^-1})^{^-2}}{xa^{^-1}} = \frac{x^{^-6}a^2}{xa^{^-1}} = \frac{\left(\frac{a^2}{x^6}\right)}{\left(\frac{x}{a}\right)} = \mathbf{\frac{a^3}{x^7}}$.

(b) $\left(\frac{x^2y^{^-2}}{x^{^-3}y^2}\right)^{^-2} = \left(\frac{x^{^-4}y^4}{x^6y^{^-4}}\right) = \frac{\left(\frac{y^4}{x^4}\right)}{\left(\frac{x^6}{y^4}\right)} = \mathbf{\frac{y^8}{x^{10}}}$.

25. **(a)** $\frac{3a}{xy^2} + \frac{b}{x^2y^2} = \frac{3a}{xy^2} \cdot \frac{x}{x} + \frac{b}{x^2y^2}$

$= \mathbf{\frac{3ax + b}{x^2y^2}}$.

(b) $\frac{5}{xy^2} - \frac{2}{3x} = \frac{5}{xy^2} \cdot \frac{3}{3} - \frac{2}{3x} \cdot \frac{y^2}{y^2}$

$= \mathbf{\frac{15 - 2y^2}{3xy^2}}$.

(c) $\frac{7}{2^3 3^2} + \frac{5}{2^2 3^3} = \frac{7}{2^3 3^2} \cdot \frac{3}{3} + \frac{5}{2^2 3^3} \cdot \frac{2}{2} =$

$\frac{21 + 10}{2^3 3^3} = \mathbf{\frac{31}{216}}$.

26. **Answers may vary**. The problem is to find how many $\frac{1}{2}$-yard pieces can be cut from a $1\frac{3}{4}$-yard ribbon. The method is to divide $1\frac{3}{4}$ by $\frac{1}{2} \Rightarrow \frac{7}{4} \div \frac{1}{2} = \frac{14}{4}$ or 3 pieces $\left(1\frac{1}{2} \text{ yards}\right)$ with $1\frac{3}{4} - 1\frac{1}{2} = \frac{1}{4}$ yard left over. This agrees with his answer obtained by his drawing. His algorithm concludes that he will have $3\frac{1}{2}$ half-yard pieces, and $\frac{1}{2}$ of a half-yard piece would be $\frac{1}{4}$ yard. He mistook a half-piece for a half-yard.

27. **(a)** $\frac{\text{Number of heads}}{\text{Total tosses}} = \frac{17}{30} = \mathbf{17{:}30}$.

(b) If 17 heads were recorded, there were 13 tails $\Rightarrow \frac{\text{Number of heads}}{\text{Number of tails}} = \frac{17}{13} = \mathbf{17{:}13}$.

(c) $\frac{\text{Number of tails}}{\text{Number of heads}} = \frac{13}{17} = \mathbf{13{:}17}$.

28. **(*i*)** 48 fl oz for \$3 $\Rightarrow \frac{300¢}{48 \text{ fl oz}} \approx$ 6.25¢ per fl oz;

(*ii*) 64 fl oz for \$4 $\Rightarrow \frac{400¢}{64 \text{ fl oz}} \approx$ 6.25¢ per fl oz;

Based on cost per ounce, neither is a better buy because they are the same price.

29. $\frac{18 \text{ parts gold}}{6 \text{ parts other metals}} \neq \frac{12 \text{ parts gold}}{3 \text{ parts other metals}}$ because $18 \cdot 3 \neq 12 \cdot 6$. Her ring may be more or less, but it is **not 18 karat gold**.

30. 4 people = 3 oranges

1 person = $\frac{3}{4}$ orange

11 people = $11 \cdot \frac{3}{4}$ oranges

$= \frac{33}{4} = \mathbf{8\frac{1}{4}}$ **oranges.**

4 people = 16 grapes

And 1 person = $\frac{16}{4}$ = 4 grapes

11 people = $11 \cdot 4$ grapes

= **44 grapes.**

31. $\frac{1 \text{ cm}}{2.5 \text{ m}} = \frac{3 \text{ cm}}{x \text{ m}} \Rightarrow x = 3 \cdot 2.5 = \mathbf{7.5\,m}$.

32. If the ratio of O weight to H_2 weight is 8:1 then $8x + x = 16$(ounces). $9x = 16 \Rightarrow$ $x = \frac{16}{9}$ and $1x = \frac{16}{9} = \mathbf{1\frac{7}{9}}$ **ounces.**

33. The ratio **cannot be determined exactly**, but it will always be between 12:100 and 15:100 as a ratio of defective to non-defective chips. 12:100 is $\frac{12}{112} = 10\frac{5}{7}\%$ defective; 15:100 is $\frac{15}{115} = 13\frac{1}{23}\%$ defective. If the observed defective percentage is closer to $13\frac{1}{23}\%$ then more chips came from the first plant. If the percentage is closer to $10\frac{5}{7}\%$, then more chips came from the second.

34. The area of a square is the square of the length of its side. If the ratio of the sides is $1{:}r$, then the ratio of their areas is $1^2{:}r^2$, or $\mathbf{1{:}r^2}$.

35. **(a)** Games won to games lost was $\frac{18}{7}$ or **18:7**.

(b) 25 games were played. Games won to games played was $\frac{18}{25}$ or **18:25**.

36. **(a)** $\frac{1}{5}{:}1 \Rightarrow 5\left(\frac{1}{5}\right){:}5(1) = \mathbf{1{:}5}$.

(b) $\frac{2}{5}{:}\frac{3}{4} \Rightarrow 20\left(\frac{2}{5}\right){:}20\left(\frac{3}{4}\right) = \mathbf{8{:}15}$.

37. $\frac{\text{boys}}{\text{girls}} = \frac{3}{5} = \frac{x \text{ boys}}{15 \text{ girls}} \Rightarrow 5 \cdot x = 3 \cdot 15 \Rightarrow$ $x = \frac{3 \cdot 15}{5} = \mathbf{9}$ **boys.**

38. There are a total of 66 slots on this roulette wheel.

(a) $\mathbf{\frac{32}{66} = \frac{16}{33}}$

(b) $\mathbf{\frac{2}{66} = \frac{1}{33}}$

(c) $\mathbf{\frac{34}{66} = \frac{17}{33}}$

(d) $\mathbf{\frac{64}{66} = \frac{32}{33}}$

39. **(a)** $\frac{1}{2} = \frac{10}{20}$

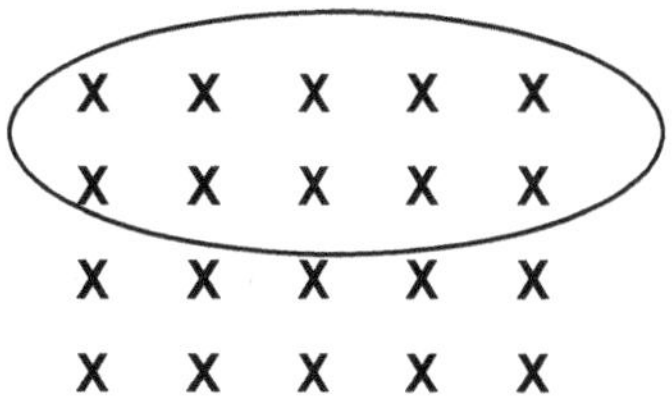

(b) $\frac{1}{4} = \frac{5}{20}$

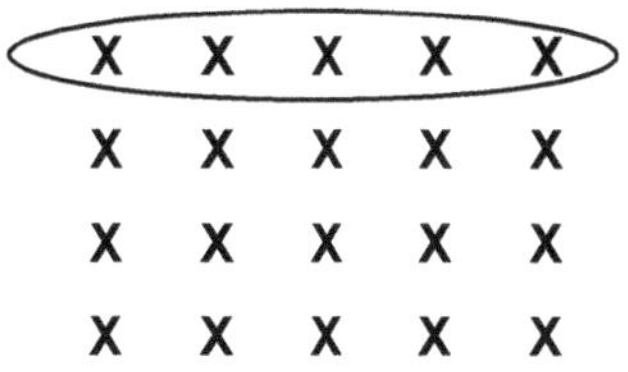

(c) $\frac{3}{4}=\frac{15}{20}$

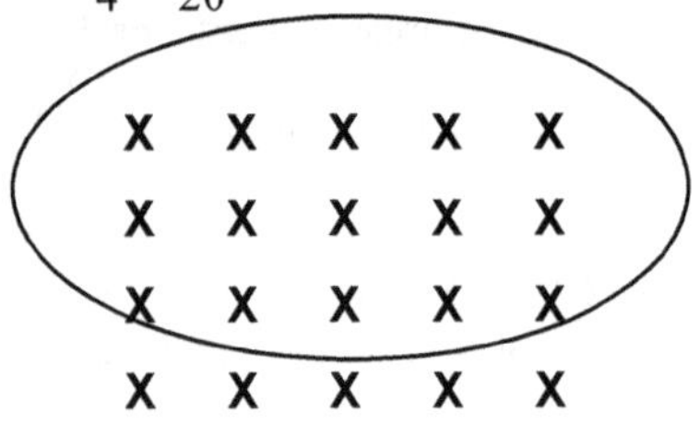

(d) $\frac{4}{5}=\frac{16}{20}$

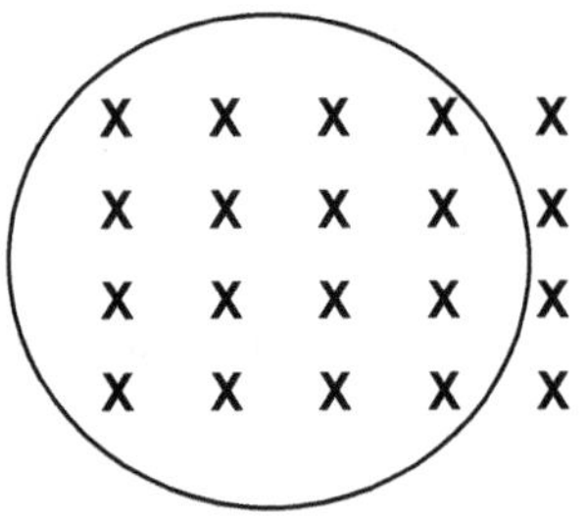

40. **1:5; 2:3; 3:4; 5:6**. There are several different ways to approach this. One method involves converting each to a fraction and comparing fractions. Another method would be to determine the percentages of each (for example, 1:5 ratio is 20%; 2:3 ratio is about 66%, and so on).

41. Since the ratio of two numbers is 5:9, the two numbers themselves must be $5n$ and $9n$, where n is an integer. So, $5n+9n=154$ Therefore $14n=154$, so $n=11$. The two numbers are **55 and 99**.

42. **2**. 2+2 = 4, and 2+3 = 5.

43. There are a total of $2+3+5=10$ shares in sharing the marbles. Since Tom gets 30 marbles, and he has two shares, each share must be worth 15 marbles. Therefore, **Dick gets 45 marbles** and **Mary gets 75 marbles**.

44. **(a)** setting up a proportion, $\frac{4(\text{boys})}{3(\text{girls})}=\frac{x(\text{boys})}{15(\text{girls})}$. Solving for x, we get there are **20 boys** in the classroom

(b) 20+15 = **35 total students**

45. The ratio of perimeters is 6 + 6 + 6 = 18 to 10 + 10 + 10 = 30. **The ratio 18:30 is equal to 3:5.**

46. The **cup that has 1 oz cream** because 1oz > $\frac{9}{10}$ oz.

47. $\frac{45}{99}\cdot 121=\frac{5}{11}\cdot 121=5\cdot 11=55$ bulbs are expected to bloom and 121 – 55 = **66** are not expected to bloom.

48. **Answers will vary**. Technically, the will was not followed; but who wants to take (for example) one half of 17 cats? The judge's solution was probably the best way of handling the situation without (literally) dividing the cats.

49. Work the problem backwards. After Juan went over the fourth bridge, he had only 1 bag left. Before he crossed that bridge, he had four bags…because he gave the guard half his bags plus one more, which means he gave the guard 3 bags (half of 4 is 2, plus one more). Similarly, before he crossed the 3rd bridge he had 10 bags (then he gave that guard 5 bags plus one more); before he crossed the second bridge, he had 22 bags; and before he crossed the first bridge, he had **46 bags of gold**.

CHAPTER 7

DECIMALS, PERCENTS, AND REAL NUMBERS

Assessment 7-1A: Introduction to Decimals

1. (a) $0.023 = \mathbf{0 \cdot 10^0 + 0 \cdot 10^{-1} + 2 \cdot 10^{-2} + 3 \cdot 10^{-3}}$.
   (b) $206.06 = \mathbf{2 \cdot 10^2 + 0 \cdot 10^1 + 6 \cdot 10^0 + 0 \cdot 10^{-1} + 6 \cdot 10^{-2}}$.
   (c) $312.0103 = \mathbf{3 \cdot 10^2 + 1 \cdot 10^1 + 2 \cdot 10^0 + 0 \cdot 10^{-1} + 1 \cdot 10^{-2} + 0 \cdot 10^{-3} + 3 \cdot 10^{-4}}$.
   (d) $0.000132 = \mathbf{0 \cdot 10^0 + 0 \cdot 10^{-1} + 0 \cdot 10^{-2} + 0 \cdot 10^{-3} + 1 \cdot 10^{-4} + 3 \cdot 10^{-5} + 2 \cdot 10^{-6}}$.
2. (a) $4000 + 300 + 50 + 6 + 0.7 + 0.08 = \mathbf{4356.78}$.
   (b) $4000 + 0.6 + 0.008 = \mathbf{4000.608}$.
   (c) $40{,}000 + 0.03 = \mathbf{40{,}000.03}$.
   (d) $0.2 + 0.0004 + 0.0000007 = \mathbf{0.2004007}$.
3. (a) **536.0076**
   (b) **3.008**
   (c) **0.000436**
   (d) **5,000,000.2**
4. (a) 0.34 = **thirty-four hundredths**.
   (b) 20.34 = **twenty and thirty-four hundredths**.
   (c) 2.034 = **two and thirty-four thousandths.**
   (d) 0.000034 = **thirty-four mllionths; i.e.,** $34 \cdot 10^{-6}$.
5. (a) $0.436 = \frac{436}{1000} = \frac{109 \cdot 4}{250 \cdot 4} = \mathbf{\frac{109}{250}}$.
   (b) $25.16 = 25\frac{16}{100} = 25 + \frac{16}{100} = \frac{2500}{100} + \frac{16}{100} = \frac{2516}{100} = \mathbf{\frac{629}{25}}$.
   (c) $^{-}316.027 = {}^{-}316\frac{27}{1000} = \mathbf{\frac{^{-}316{,}027}{1000}}$.
   (d) $28.1902 = 28\frac{1902}{10{,}000} = \frac{281{,}902}{10{,}000} = \mathbf{\frac{140{,}951}{5000}}$.
   (e) $^{-}4.3 = {}^{-}4\frac{3}{10} = \mathbf{\frac{^{-}43}{10}}$.
   (f) $^{-}62.01 = {}^{-}62\frac{1}{100} = \mathbf{\frac{^{-}6201}{100}}$.
6. A rational number in simplest form can be written as a terminating decimal if and only if the prime factorization of the denominator contains no primes other than 2 or 5:
   (a) **Terminating**. The denominator contains only 5 as a prime factor.
   (b) **Terminating**. The denominator contains no prime factors other than 2 and 5.
   (c) **Terminating**. The denominator in reduced form contains only 2 as a prime factor.
   (d) **Terminating**. The denominator contains only 2 as a prime factor.
   (e) **Not terminating**. The denominator contains 2 and 17 as prime factors.
   (f) **Terminating**. The denominator contains only 5 as a prime factor.
7. (a) $\frac{4}{5} = \frac{4}{5} \cdot \frac{2}{2} = \frac{8}{10} = \mathbf{0.8}$.
   (b) $\frac{61}{2^2 \cdot 5} = \frac{61}{2^2 \cdot 5} \cdot \frac{5}{5} = \frac{305}{2^2 \cdot 5^2} = \frac{305}{100} = \mathbf{3.05}$.
   (c) $\frac{3}{6} = \frac{1}{2} = \frac{1}{2} \cdot \frac{5}{5} = \frac{5}{10} = \mathbf{0.5}$.
   (d) $\frac{1}{2^5} = \frac{1}{2^5} \cdot \frac{5^5}{5^5} = \frac{3125}{10^5} = \frac{3125}{100{,}000} = \mathbf{0.03125}$.
   (e) Not a terminating decimal.
   (f) $\frac{133}{625} = \frac{133}{5^4} \cdot \frac{2^4}{2^4} = \frac{2128}{10^4} = \frac{2128}{10{,}000} = \mathbf{0.2128}$.
8. One hour is 60 minutes, so 7 minutes would be $\frac{7}{60}$ of an hour. This is **nonterminating** decimal because $\frac{7}{60}$ is in simplest form and 3 is a factor of the denominator.

9. **(a)** If one long represents $\frac{1}{10}$, then one cube represents $\frac{1}{10} \cdot \frac{1}{10} = \frac{1}{100} = \mathbf{0.01}$.

(b) If one long represents $\frac{1}{10}$, then one flat represents $10 \cdot \frac{1}{10} = \mathbf{1}$.

(c) If one long represents $\frac{1}{10}$, then one block represents $100 \cdot \frac{1}{10} = \mathbf{10}$.

(d) If one long represents $\frac{1}{10}$, then one block, 2 flats, 3 longs, and 5 cubes represent

$$1 \cdot 10 + 2 \cdot 1 + 3 \cdot \frac{1}{10} + 5 \cdot \frac{1}{100}$$
$$= 10 + 2 + 0.3 + 0.05$$
$$= \mathbf{12.35}$$

10. **(a)** Line up by place value:

13.49190
13.49200
13.49183
13.49199

From greatest to least is:
13.492 > 13.49199 > 13.4919 > 13.49183.

(b) Line up by place value:

$^-1.45300$
$^-1.45000$
$^-1.40530$
$^-1.49300$

From greatest (i.e., closest to 0) to least is:

$\mathbf{^-1.4053 > {}^-1.45 > {}^-1.453 > {}^-1.493}$.

11. **(a)** Fourteen thousandths inch $= \frac{14}{1000}$ inch $=$ **0.014 inch.**

(b) Twenty-four hundredths $= \frac{24}{100} = 0.24$ days. The rotational period is thus **365.24 days**.

12. There are 32 of 100 squares shaded, representing $\frac{32}{100}$ of the whole grid $= 0.32$ of the grid.

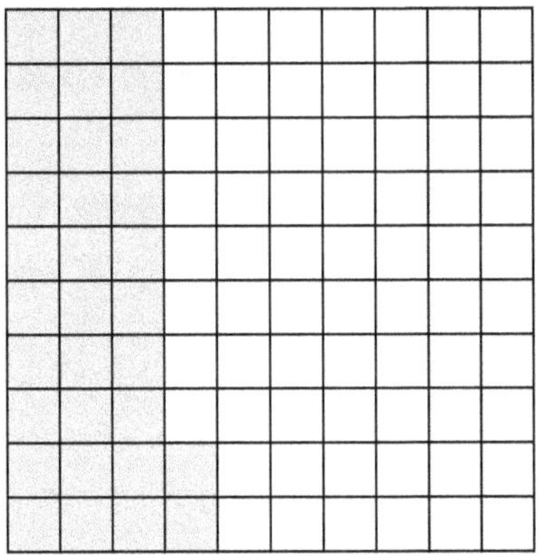

13. The largest number is furthest to the right on the number line. Thus $0.804 < 0.8399 < \mathbf{0.84}$.

14. Answers may vary. A decimal carried to the ten-thousandths place would have four digits to the right of the decimal point, and a number between 8.3400 and 8.3410 might be 8.3401 or 8.3405.

15. **(a)** Answers may vary. One method could be to find the difference, not matter how slight, between the two decimal numbers and add some fraction of that to the smaller.

(b) Part (a) is a recursive process; no matter how small the difference between the terminating decimals, that differcence can be divided.

16. **(i)** No. On a centimeter ruler, 8cm would represent 8. 0.8 would be between 0 and 1 at 8mm.

(ii) Yes. On a numberline, 0.8 is $\frac{8}{10}$ between 0 and 1.

(iii) No. 8 out of 20 are shaded:
$\frac{8}{20} = \frac{4}{10} = 0.4 \neq 0.8$.

17. **Rhonda, Martha, Kathy, Molly, Emily,** because 63.54 (Rhonda) < 63.59 (Martha) < 64.02 (Kathy) < 64.46 (Molly) < 64.54 (Emily).

18. Using our knowledge of mixed number from the previous chapters, we can write: $\frac{1\frac{1}{2}}{16} = \frac{1.5}{16} = \frac{1.5 \cdot 2}{16 \cdot 2} = \frac{3}{32} \cdot \frac{3}{32} = \frac{3 \cdot 5^5}{2^5 \cdot 5^5} = \mathbf{0.09375}$.

19. Label the marks between 24.0 and 25.0. The arrow points at **24.3.**

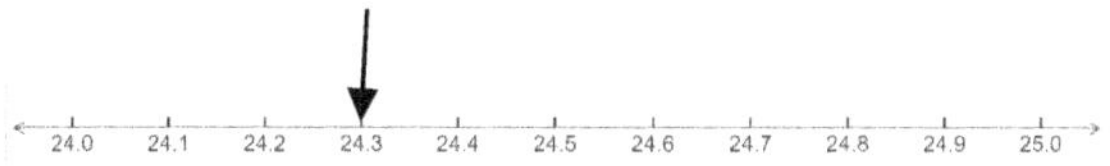

Assessment 7-2A: Operations on Decimals

1.

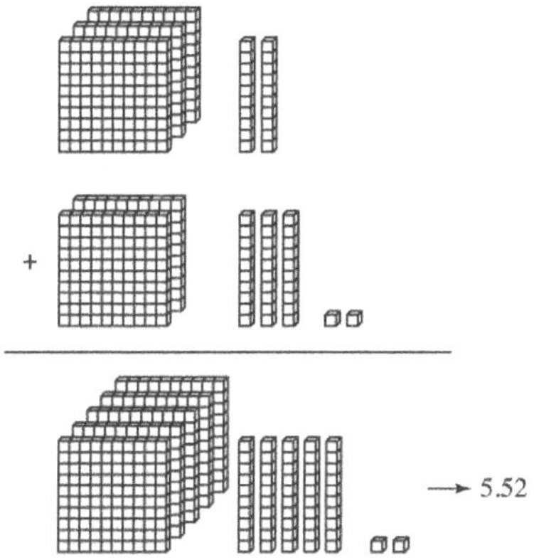

2. Shade 0.56 by shading 5 tenths and 6 hundredths. Take away 0.4 and you are left with 0.16.

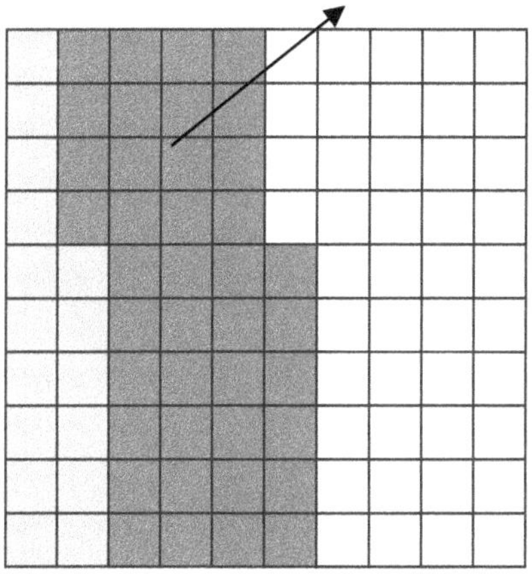

3. **(a)** $\frac{86}{10}+\frac{231}{10}+\frac{92}{100}=\frac{860}{100}+\frac{2310}{100}+\frac{92}{100}$

$=\frac{3262}{100}=\mathbf{32.62}.$

(b) $\frac{232}{100}+\frac{21008}{1000}=\frac{2320}{1000}+\frac{21008}{1000}$.

$=\frac{23328}{1000}=\mathbf{23.328}$

4. To find the greatest possible number use the largest digits in the highest place values:

4 [9] [7] 3 [6] · [5] [2] 8 [1].

5. She bought a total of:

$$\begin{array}{r} \$17.95 \\ 13.59 \\ 14.86 \\ 179.98 \\ 2.43 \\ \underline{2.43} \\ \mathbf{\$231.24} \end{array}$$

in her shopping (excluding sales tax)

6. The bank is **not correct**. The total of outstanding checks is:

$$\begin{array}{r} \$3.21 \\ 14.56 \\ 12.44 \\ 6.98 \\ 9.51 \\ \underline{7.49} \\ \$54.19 \end{array}$$

Adding the total of outstanding checks to the checkbook balance gives $\$54.19 + 21.69 = \$75.88$, which differs from the bank statement by \$7.74. The bank balance is too high.

7. Let p be the price of the stock. Then $p + 0.24 = 73.245 \Rightarrow p = 73.245 - 0.24$, or

$$\begin{array}{r} 73.245 \\ \underline{-0.240} \\ \mathbf{73.005} \end{array}$$

8. **(a)** There is a difference of 0.9 between each element of the sequence, thus it is arithmetic. The next three elements are: $4.5 + 0.9 =$ **5.4**, $5.4 + 0.9 =$ **6.3**, $6.3 + 0.9 =$ **7.2.**

(b) There is a difference of 0.2 between each element of the sequence, thus it is arithmetic. The next three elements are: $1.1 + 0.2 =$ **1.3**, $1.3 + 0.2 =$ **1.5**, $1.5 + 0.2 =$ **1.7**.

9. **(a)** Sum along the diagonal, yielding 16.5. In rows and columns with two figures subtract their sum from 16.5 to obtain the missing element:

8.2	**1.9**	**6.4**
3.7	5.5	**7.3**
4.6	9.1	2.8

(b) **Yes.**

(c) Each row, column, and diagonal will have $3 \cdot 0.85 = 2.55$ added to it, for a sum of $16.5 + 2.55 = \mathbf{19.05}$.

10. **(a)** $0.5 \bullet 0.6 = \frac{5}{10} \bullet \frac{6}{10} = \frac{30}{100} = \mathbf{0.30}$.

(b) $203 \bullet 0.03 = 203 \bullet \frac{3}{100} = \frac{609}{100} = \mathbf{6.09}$.

(c) $0.003 \bullet 0.006 = \frac{3}{1000} \bullet \frac{6}{1000} = \frac{18}{1{,}000{,}000}$

$= \mathbf{0.000018}$.

11. To find the greatest products use the largest digits in the highest place values:

(i) Least: $\boxed{2}.\boxed{3} \times \boxed{1} = \mathbf{2.3}$.

(ii) Greatest: $\boxed{8}.\boxed{7} \times \boxed{9} = \mathbf{78.3}$.

12. **(a)** When you multiply a natural number by a decimal the result is less than the natural number if the decimal is less than 1. For example $5 \bullet 0.5 = 2.5$ and $2.5 < 5$.

(b) When you multiply a natural number by a decimal the result is greater than the natural number when the decimal is greater than 1. For example $5 \bullet 1.5 = 7.5$ and $7.5 > 5$.

13. If: $a_1 = 0.9$;

$a_2 = 0.9 \cdot 0.2 = 0.18$;

$a_3 = 0.18 \cdot 0.2 = 0.036$;

$a_4 = 0.036 \cdot 0.2 = 0.0072$; and

$a_5 = 0.0072 \cdot 0.2 = 0.00144$:

Their sum is **1.12464**.

14. $\frac{7}{0.25} = \frac{70}{2.5} = \frac{700}{25}$. Thus, ***(i)*, *(ii)*, and *(iv)*** have equal quotients.

15. $1776 \text{ ft} \bullet 0.005 = 8.88 \text{ ft}$ The height of the model will be **8.88 ft.**

16. $102.3 \text{ cm} \bullet 1.5 = 153.45 \text{ cm}$ If Allie is 102.3 cm now she will be **153.45 cm** when she is twelve years old.

17. **(a)** $\$150 \bullet 109.415 = 16{,}412.25$ yen. If \$1 is worth 109.415 yen then \$150 is worth **16,412.25 yen.**

(b) If \$1 is worth 109.415 yen then 1 yen is worth \$0.0091395147 and 3500 yen is worth approximately **\$31.99.**

18. $\frac{243 \text{ miles}}{12 \text{ gallons}} = \mathbf{20.25 \text{ mpg}}$.

19. Let 63 is 30% of $n \Rightarrow 63 = \frac{30}{100} \cdot n \Rightarrow 30n$ the price per pound of the other type of fruit. The total cost of the fruit is $25 \cdot 4 + 15 \cdot 2 + 10x = 130 + 10x$ dollars. The total weight of the fruit is $25 + 15 + 10 = 50$ lbs.

To find the average cost per pound divide the total cost by the total weight. This will equal \$3.50 per lb.

$$\frac{130 + 10x}{50} = 3.50 \Rightarrow \frac{130 + 10x}{50} = \frac{350}{100}$$

$$\Rightarrow 20.5n = 6150 \Rightarrow n = \frac{6150}{20.5} = 300,$$

$$\Rightarrow 10x = 45 \Rightarrow x = \mathbf{\$4.5/\ lb}.$$

20. If one U.S. dollar is valued at 1.256 Canadian, then \$27.32 U.S. would be valued at $27.32 \times 1.256 =$ \$34.31392. which rounds to **\$34.31** Canadian.

21. Bumble Bee is ten 5 oz. cans for \$11.27 so $10 \bullet 5$ oz = 50 oz for \$11.27 or

$\frac{50 \text{ oz}}{50}$ for $\frac{\$11.27}{50}$

1 oz for \$0.2254

Starkist is eight 5 oz. cans for \$7.88 so $8 \bullet 5$ oz = 40 oz for \$7.88 or

$\frac{40 \text{ oz}}{40}$ for $\frac{\$7.88}{40}$

1 oz for \$0.197

Safe Catch is twelve 5 oz. cans for \$49.32 so $12 \bullet 5$ oz = 60 oz for \$49.32 or

$\frac{60 \text{ oz}}{60}$ for $\frac{\$49.32}{60}$

1 oz for \$0.822

Great Value is four 5 oz. cans for \$2.96 so $4 \bullet 5$ oz = 20 oz for \$2.96 or

$\frac{20 \text{ oz}}{20}$ for $\frac{\$2.96}{20}$

1 oz for \$0.148

The best buy based solely on cost per ounce is Great Value at \$0.148 per ounce.

22. Divide 8.46 into 6 equal pieces: $8.46 \div 6 = \mathbf{1.41\ meters}$.

23. The baker starts with 8.6 pounds. When he is done pouring he has 0.35 pound left, so he poured $8.6 - 0.35 = 8.25$ pounds of sugar into 5 bags. Divide 8.25 by 5: $8.25 \div 5 = \mathbf{1.65}$ pounds of sugar in each bag.

24. **(a)** The question is asking for a whole number of bunches of balloons, so the remainder is not necessary as part of the answer. Divide \$75 by \$6 = 12 R 3, so she can buy **12 bunches of balloons.**

(b) The question is asking how long each piece is, so an exact answer is necessary, because there will be no leftover rope, so no remainder. Divide 93 inches into 12 pieces: $93 \div 12 = 7.75$, but it is custom to write the decimal as a fraction, so since $0.75 = \frac{75}{100} = \frac{3}{4}$ the answer would be $\mathbf{7\frac{3}{4}}$ **inches.**

(c) The question is asking how much each ticket costs, so an exact answer is necessary, because there is no money left over, so no remainder. Divide \$50 by 8: $\$50 \div 8 = \$6.25$, so the **cost per ticket is \$6.25.**

(d) The question is asking for a whole number of baskets. Divide 75 by 9: $75 \div 9 = 8R3$ but we need to have enough baskets to store all 75 DVDs, so the answer is **9 baskets** even if not all baskets are full.

(e) The questions is asking for a whole number of penny sets and a remainder. Divide 348 by 16: $348 \div 16 = 21R12$ so she can make **21 sets of 16 pennies each with 12 pennies left over.**

25. **(a)** $3.2 \div 10^9 = \mathbf{0.0000000032}$; i.e., move the decimal point 9 places to the left.

(b) $3.2 \cdot 10^9 = \mathbf{3{,}200{,}000{,}000}$; i.e., move the decimal point 9 places to the right.

(c) $4.2 \div 10^1 = \mathbf{0.42}$; i.e., move the decimal point 1 place to the left.

(d) $6.2 \cdot 10^5 = \mathbf{620{,}000}$; i.e., move the decimal point 5 places to the right.

26. **(a)** Move the decimal point 7 places to the left to obtain the product of a number between 1 and 10 and an integer power of 10. I.e., 1.27 multiplied by 10^7, or about $\mathbf{1.27 \cdot 10^7\,m}$.

(b) Move the decimal point 9 places to the left and multiply the result by 10^9 to obtain about $\mathbf{4.486 \cdot 10^9\ km}$.

27. $4 \cdot 10^{12} = 4 \cdot 1{,}000{,}000{,}000{,}000$ which is four trillion and also $4e12$. The answers are **i., iii., iv.**

28. **(a)** Move the decimal point ten places to the left to obtain **0.000000000753 g.**

(b) Move the decimal point five places to the right to obtain **298,000 km per sec.**

(c) Move the decimal point eight places to the right to obtain **778,570,000 km.**

29. **(a)** $(8 \cdot 10^{12}) \cdot (6 \cdot 10^{15}) = 8 \cdot 6 \cdot 10^{12} \cdot 10^{15}$
$= 48 \cdot 10^{27} = \mathbf{4.8 \cdot 10^{28}}$.

(b) $(16 \cdot 10^{12}) \div (4 \cdot 10^5) = \frac{16}{4} \cdot \frac{10^{12}}{10^5} =$
$4 \cdot 10^{12} \cdot 10^{-5} = \mathbf{4 \cdot 10^7}$.

(c) $(5 \cdot 10^8) \cdot (6 \cdot 10^9) \div (15 \cdot 10^{15}) =$
$\frac{5 \cdot 6}{15} \cdot 10^8 \cdot 10^9 \cdot 10^{-15} = \mathbf{2 \cdot 10^2}$.

30. Dividing each number by 0.01 is equal to multiplying each number by $\frac{1}{100}$ which is equal to dividing each number by 100.

(a) $\frac{2586}{100} = \mathbf{25.86}$

(b) $\frac{34.79}{100} = \mathbf{0.3479}$

(c) $\frac{0.24}{100} = \mathbf{0.0024}$

(d) $\frac{0.0037}{100} = \mathbf{0.000037}$.

31. Answers vary. For example:
$40 \cdot \$8 + 40 \cdot \$\left(\frac{1}{4}\right) = \$320 + \$10 = \mathbf{\$330}$.

32. **(a)** $8.4 \cdot 6 = 4.2 \cdot \underline{\mathbf{12}}\left[\frac{8.4}{\cancel{2}} \cdot (6 \cdot \cancel{2}) = 8.4 \cdot 6\right]$.

(b) $10.2 \div 0.3 = 20.4 \div \underline{\mathbf{0.6}}\left(\frac{10.2 \cdot \cancel{2}}{0.3 \cdot \cancel{2}} = \frac{10.2}{0.3}\right)$.

(c) $a \cdot b = \frac{a}{2} \cdot \underline{\mathbf{2b}}\left(\frac{a}{\cancel{2}} \cdot \frac{\cancel{2}b}{1} = a \cdot b\right)$.

(d) $a \div b = (2a) \div \underline{\mathbf{(2b)}}\left(\frac{\cancel{2}a}{\cancel{2}b} = \frac{a}{b}\right)$.

33. **(a)** $\underline{2}03.651$ is between 200 and 300 and is closer to 200, so round down to **200.**

(b) 203.651 is between 200 and 210 and is closer to 200, so round down to **200**.

(c) 203.651 is between 203 and 204 and is closer to 204, so round up to **204**.

(d) 203.651 is between 203.6 and 203.7 and is closer to 203.7, so round up to **203.7**.

(e) 203.651 is between 203.65 and 203.66 and is closer to 203.65, so round down to **203.65**.

34. **(a)** $0.3 \div 0.31$ is close to 1.

(b) $0.3 \cdot 0.31$ is close to 0.09.

Thus on the numbers line:

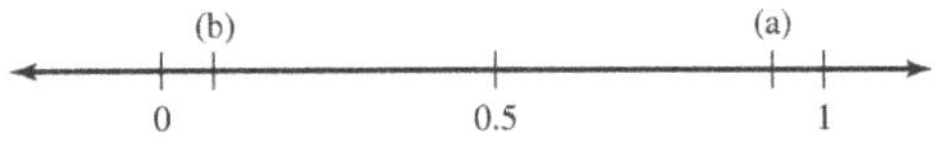

35. **(a)** $3.6 \cdot 85.5 (\approx 4 \cdot 85 \approx (4 \cdot 80) + (4 \cdot 5) \approx 320 + 20 \approx 340) = 307.8$

(b) $\frac{137.025}{1.75} (\approx \frac{140}{2} \approx 70) = 78.3$.

36. Camera rounds to \$55, film rounds to \$5, and case rounds to \$18. Total estimated cost is $55 + 5 + 18 =$ **\$78**.

37. Answers vary:

(a) **(*i*)** 65.84 rounds to 66;
24.29 rounds to 24;
12.18 rounds to 12;
19.75 rounds to 20;
Rounded estimate $= 66 + 24 + 12 + 20 =$ **122**.

(*ii*) Lead digits sum to $60 + 20 + 10 + 10 = 100$; adjustments are about $6 + 4 + 2 + 10 = 22$. Front-end estimate $= 100 + 22 =$ **122**.

(*iii*) Actual sum $=$ **122.06**.

(b) **(*i*)** 89.47 rounds to 89;
32.16 rounds to 32.
Rounded estimate $= 89 - 32 =$ **57**.

(*ii*) Lead digit difference is $80 - 30 = 50$; adjustments are about $9 - 2 = 7$. Front-end estimate $= 50 + 7 =$ **57**.

(*iii*) Actual difference $=$ **57.31**.

(c) **(*i*)** 5.85 rounds to 6;
6.13 rounds to 6;
9.10 rounds to 9;
4.32 rounds to 4.
Rounded estimate $= 6 + 6 + 9 + 4 =$ **25**.

(*ii*) Lead digits sum to $5 + 6 + 9 + 4 = 24$; adjustments are about $1 + 0 + 0 + 0 = 1$.
Front-end estimate $= 24 + 1 =$ **25**.

(*iii*) Actual sum $=$ **25.40**.

(d) **(*i*)** 223.75 rounds to 224;
87.60 rounds to 88.
Rounded estimate $= 224 - 88 =$ **136**.

(*ii*) Lead digit difference $= 200 - 80 = 120$; adjustments are about $24 - 8 = 16$. Front-end estimate $= 120 + 16 =$ **136**.

(*iii*) Actual difference $=$ **136.15**.

Mathematical Connections 7-2: Review Problems

27. **(a)** **Nonterminating**. The reduced denominator has only 3 as a factor.

(b) **Terminating**. The simplified fraction has a denominator with only 2 as a factor.

29. $\frac{35}{56} = \frac{5}{8}$ in simplest form, and 8 in the denominator may be written as 2^3.

Assessment 7-3A: Nonterminating Decimals

1. The fraction is in simplest form and if the prime factorization of the denominator contains only 2 and/or 5's then the fraction will be a terminating decimal. Since the denominator ends on 9 it is not divisible by 2 or by 5 it will be a non-terminating decimal.

2. **(a)** $0.45\overline{777}, 0.45\overline{77}$, and $0.45\overline{7}$ are all **the same number**. The only difference is that the repeating block of 7's is indicated in different ways.

(b) The preferred part is the shortest repeating part: $\mathbf{0.45\overline{7}}$.

(c) The shortest period in $0.45\overline{7}$ is **1**.

3. **(a)** The first ten decimal places of $0.\overline{123} =$ **0.1231231231**

(b) The first ten decimal places of $0.1\overline{23} = \mathbf{0.1232323232}$

(c) The first ten decimal places of $0.12\overline{3} = \mathbf{0.1233333333}$

(d) The first ten decimal places of $0.123\overline{43} = \mathbf{0.1234343434}$

4. We can write out the first ten decimal to compare:

$0.36\overline{25} = 0.3625252525$

$0.3\overline{625} = 0.3625625625$

and in the fifth decimal place 6>2 so $\mathbf{0.3\overline{625}}$ **is greater.**

5. **(a)** $0.4444... = \mathbf{0.\overline{4}}$ **and the period is 1.**

(b) $0.36454545... = \mathbf{0.36\overline{45}}$ **and the period is 2.**

(c) $0.18273273273... = \mathbf{0.18\overline{273}}$ **and the period is 3.**

6. In each of the following, divide the numerator by the denominator either with a calculator or by use of the division algorithm. The overline in the quotient indicates that the block of digits underneath is repeated an infinite number of times.

(a) $\frac{4}{9} = 4 \div 9 = \mathbf{0.\overline{4}}.$

(b) $\frac{2}{7} = 2 \div 7 = \mathbf{0.\overline{285714}}.$

(c) $\frac{3}{11} = 3 \div 11 = \mathbf{0.\overline{27}}.$

(d) $\frac{1}{15} = 1 \div 15 = \mathbf{0.0\overline{6}}.$

(e) $\frac{2}{75} = 2 \div 75 = \mathbf{0.02\overline{6}}.$

(f) $\frac{1}{99} = 1 \div 99 = \mathbf{0.\overline{01}}.$

(g) $\frac{5}{6} = 5 \div 6 = \mathbf{0.8\overline{3}}.$

(h) $\frac{1}{13} = 1 \div 13 = \mathbf{0.\overline{076923}}.$

7. **(a)** $n = 0.\overline{4}$; a one-digit repetend:

$$\begin{aligned} 10n &= 4.4444\ldots \\ {}^-n &= {}^-0.4444\ldots \\ \hline 9n &= 4 \end{aligned}$$

$\Rightarrow n = \mathbf{\frac{4}{9}}.$

(b) $n = 0.\overline{61}$; a two-digit repetend:

$$\begin{aligned} 100n &= 61.6161\ldots \\ {}^-n &= {}^-0.6161\ldots \\ \hline 99n &= 61 \end{aligned}$$

$\Rightarrow n = \mathbf{\frac{61}{99}}.$

(c) $n = 1.3\overline{96}$; a two-digit repetend:

$$\begin{aligned} 1000n &= 1396.9696\ldots \\ {}^-10n &= {}^-13.9696\ldots \\ \hline 990n &= 1383 \end{aligned}$$

$\Rightarrow n = \frac{1383}{990} = \mathbf{\frac{461}{330}}.$

(d) $n = 0.\overline{55} = 0.\overline{5}$; a one-digit repetend:

$$\begin{aligned} 10n &= 5.5555\ldots \\ {}^-n &= {}^-0.5555\ldots \\ \hline 9n &= 5 \end{aligned}$$

$\Rightarrow n = \mathbf{\frac{5}{9}}.$

(e) $n = {}^-2.3\overline{4}$; $10n = {}^-23.\overline{4}$; now a one-digit repetend:

$$\begin{aligned} 100n &= {}^-234.4444\ldots \\ {}^-10n &= 23.4444\ldots \\ \hline 90n &= {}^-211 \end{aligned}$$

$\Rightarrow n = \mathbf{\frac{{}^-211}{90}}.$

(f) $n = {}^-0.0\overline{2}$; a one-digit repetend:

$$\begin{aligned} 100n &= {}^-2.2222\ldots \\ {}^-10n &= 0.2222\ldots \\ \hline 90n &= {}^-2 \end{aligned}$$

$\Rightarrow n = \frac{{}^-2}{90} = \mathbf{\frac{{}^-1}{45}}.$

8. 1 minute $= \frac{1}{60}$ hour, and

$1 \div 60 = \mathbf{0.01\overline{6}}$ **hour.**

9. Line up the decimal points:

$$^{-}1.4\overline{54} = {}^{-}1.4545454\ldots$$
$$^{-}1.\overline{454} = {}^{-}1.4544544\ldots$$
$$^{-}1.\overline{45} = {}^{-}1.4545454\ldots$$
$$^{-}1.45\overline{4} = {}^{-}1.4544444\ldots$$
$$^{-}1.454 = {}^{-}1.4540000\ldots$$

Ordering from greatest (i.e., closest to 0) yields:
$\mathbf{^{-}1.454 > {}^{-}1.45\overline{4} > {}^{-}1.\overline{454} > {}^{-}1.4\overline{54} = {}^{-}1.\overline{45}}$.

10. Each element could be a decimal representation of $\frac{n}{n+1}$, beginning at $n = 0$. If so, then the next three elements would be:

$$\tfrac{6}{7} = \mathbf{0.\overline{857142}},$$
$$\tfrac{7}{8} = \mathbf{0.875}, \text{ and}$$
$$\tfrac{8}{9} = \mathbf{0.\overline{8}}.$$

11. The sum is $0.4 + 0.4(0.5) + 0.4(0.5)^2 + 0.4(0.5)^3 + 0.4(0.5)^4 = 0.4 + 0.2 + 0.1 + 0.05 + 0.025 = \mathbf{0.775}$.

Another method by which to attack this problem would be to find the sum of a finite geometric sequence:

$$S_n = a_1 + a_1r + a_1r^2 + \cdots + a_1r^{n-1}$$
$$rS_n = a_1r + a_1r^2 + a_1r^3 + \cdots + a_1r^n$$
$$rS_n - S_n = a_1r^n - a_1$$
$$(r-1)S_n = a_1(r^n - 1)$$
$$S_n = \frac{a_1(r^n - 1)}{r - 1} = \frac{a_1(1 - r^n)}{1 - r}$$

Thus if $a_1 = 0.4$, $r = 0.5$, and $n = 5$,

$$S_5 = \frac{0.4[(1-0.5^5)]}{1-0.5} = \frac{0.4 \cdot 0.96875}{0.5} = \mathbf{0.775}.$$

12. If we sum $0.\overline{235}$ and $0.\overline{2356}$, the digits in each place value will not realign until we reach the least common multiple of the length of each number's repetend. Thus, we expect the length of the repetend to be $LCM(3,4) = \mathbf{12}$. To see this, consider the follwing.

$$\begin{array}{r} 0.235\ 235\ 235\ 235\ 235\ldots \\ +\ 0.235\ 623\ 562\ 356\ 2356\ldots \\ \hline \end{array}$$

At the 10^{-13} place the digits realign.

13. (*i*) $a + b =$

$$\begin{array}{r} 0.323232323232\ldots \\ +\ 0.123123123123\ldots \\ \hline \mathbf{0.446355446355\ldots} \end{array}$$

(*ii*) there are **six digits** in the repetend.

14. **Yes**. Zeros after the last non-zero digit of the decimal can be repeated.

15. Answers may vary; e.g.,

(a) Write $3.\overline{2}$ as 3.222... and write 3.22 as 3.220.

$3.220 < 3.221 < 3.2211 < 3.22111 < 3.222\ldots$

(b) Write $462.\overline{24}$ as $464.2424\ldots$and write 462.243 as 462.2430. $462.2424\ldots < 462.2425 < 462.2426 < 462.2427 < 462.2430$.

16. To find a number halfway between any two others, find their average; i.e., add the original numbers and divide by 2.

$(0.5 + 0.\overline{4}) \div 2 = (0.9444\ldots) \div 2 =$ $\mathbf{0.47222\cdots} = \mathbf{0.47\overline{2}}$.

17. Answers may vary; e.g.,

(a) Write $\frac{3}{4}$ as 0.750 and write $0.\overline{75}$ as 0.757...;
$\frac{3}{4} < 0.751 < 0.752 < 0.753 < 0.\overline{75}$.

(b) Write $\frac{1}{3}$ as 0.333... and write $0.\overline{34}$ as $0.343\ldots$; $\frac{1}{3} < 0.334 < 0.335 < 0.336 < 0.\overline{34}$.

18. The repeating decimal part is determined by the 7 in the denominator of $3\frac{1}{7} = \frac{22}{7}$. Seven has no prime factors of 2 or 5 and so is a repeating decimal.

19. **(a)** (*i*) Use the repeating block strategy:
$n = 0.\overline{1}$, a one-digit repetend.

$$\begin{array}{rcl} 10n & = & 1.1111\ldots \\ ^{-}n & = & ^{-}0.1111\ldots \\ \hline 9n & = & 1 \end{array}$$

$\Rightarrow n = \frac{1}{9}$.

Observe that when the repetend is immediately to the right of the decimal point in the repeating decimal, a in the fraction $\frac{a}{b}$ will be the digit(s) in the repetend and b will be one or more 9's. The number of 9's will be the same as the number of digits in the repetend.

(*ii*) $0.\overline{01} = \mathbf{\frac{1}{99}}$.

(*iii*) $0.\overline{001} = \mathbf{\frac{1}{999}}$.

(b) $0.\overline{0001} = \mathbf{\frac{1}{9999}}$.

(c) $0.\bar{1} = \frac{1}{9}$ is true, so $(0.\bar{1})\left(\frac{1}{10}\right) = \frac{1}{9}\left(\frac{1}{10}\right) \Rightarrow$

$\mathbf{0.0\bar{1}} = \frac{1}{90}$.

20. **(a)** Since $0.\bar{9} = 1$, $0.0\bar{9} = \frac{1}{10} \cdot (0.\bar{9}) = \frac{1}{10}(1) = \mathbf{\frac{1}{10}} = \mathbf{0.1}$.

(b) $0.3\bar{9} = 0.3 + 0.0\bar{9} = 0.3 + 0.1 = \mathbf{0.4}$.

(c) $9.\bar{9} = 9 + 0.\bar{9} = 9 + 1 = \mathbf{10}$.

21. See problem 19(a)(*i*) above and observe that when the repetend is immediately to the right of the decimal point in the repeating decimal, *a* in the fraction $\frac{a}{b}$ will be the digit(s) in the repetend and *b* will be one or more 9's. The number of 9's will be the same as the number of digits in the repetend.

(a) $0.\bar{2} = \mathbf{\frac{2}{9}}$.

(b) $0.\bar{3} = \frac{3}{9} = \mathbf{\frac{1}{3}}$.

(c) $9.\bar{9} = 9\frac{9}{9} = \mathbf{10}$.

22. **(a)** $0.\overline{05} = \mathbf{\frac{5}{99}}$. In the fraction $\frac{a}{b}$, $a = 5$ and b is two 9's.

(b) $0.\overline{003} = \frac{3}{999} = \mathbf{\frac{1}{333}}$. In the fraction $\frac{a}{b}$, $a = 3$ and b is three 9's.

23. If $\dfrac{1}{3} = 0.\bar{3}$ and $\dfrac{1}{33} = 0.\overline{03}$,

then $\dfrac{1}{333} = 0.\overline{003}$, $\dfrac{1}{3333} = 0.\overline{0003}$,

$\dfrac{1}{33{,}333} = 0.\overline{00003}$ and

$\dfrac{1}{333{,}333} = \mathbf{0.\overline{000003}}$

24. To find the sum of an infinite geometric series with ratio $0 < r < 1$ (if r were greater than 1, the series could not have a finite sum; if $r = 0$ the series would be just the first term):

$$\begin{aligned} S_\infty &= a + ar + ar^2 + ar^3 + \cdots \\ {}^-rS_\infty &= {}^-(ar + ar^2 + ar^3 + \cdots) \\ S_\infty - rS_\infty &= a \\ S_\infty(1 - r) &= a \\ S_\infty &= \frac{a}{1 - r} \end{aligned}$$

(a) $0.2\bar{9} = 0.29999\ldots = 0.2 + 0.9999\ldots = 0.2 + 0.09 + 0.09(0.1) + 0.09(0.01) + \cdots$.

Thus $0.2\bar{9}$ is 0.2 plus an infinite geometric sequence where $a_1 = 0.09$ and $r = 0.1$.

$S_\infty = 0.2 + \frac{0.09}{1-0.1} = 0.2 + \frac{0.09}{0.9}$

$= 0.2 + 0.1 = \mathbf{\frac{3}{10}}$.

(b) $2.0\overline{29} = 2 + 0.029 + 0.00029 + \cdots$.

Thus $2.0\overline{29}$ is 2 plus an infinite geometric sequence where $a_1 = 0.029$ and $r = 0.01$.

$S_\infty = 2 + \frac{0.029}{1-0.01} = 2\frac{29}{990} = \mathbf{\frac{2009}{990}}$.

25. $n = 0.\overline{abc}$; a three-digit repetend:

$$\begin{aligned} 1000n &= abc.abcabc\ldots \\ {}^-n &= {}^-0.abcabc\ldots \\ \hline 999n &= abc \end{aligned}$$

$\Rightarrow n = \frac{abc}{999}$.

$0.\overline{abc}$ written as a fraction is $\mathbf{\dfrac{abc}{999}}$.

26. **(a)** $1 - 3x = 8 \Rightarrow -3x = 7 \Rightarrow$

$x = \frac{^-7}{3} = \mathbf{^-2.\bar{3}}$.

(b) $1 = 3x + 8 \Rightarrow {}^-7 = 3x \Rightarrow 3x = {}^-7$

$\Rightarrow x = \frac{^-7}{3} = \mathbf{^-2.\bar{3}}$.

(c) $1 = 8 - 3x \Rightarrow {}^-7 = -3x \Rightarrow 3x = 7 \Rightarrow x = \frac{7}{3} = \mathbf{2.\bar{3}}$.

Mathematical Connections 7-3: Review Problems

19. Total deductions were \$1520.63 + \$723.30 + \$2843.62 = \$5087.55. Gross pay less deductions was \$27,849.50 − \$5087.55 = **\$22,761.95**.

21. The rule states that decimal point placement should be four places, or 0.0770. Because 0.077 as found on the calculator equals 0.0770; i.e., trailing zeros add no value; the rule still applies.

Assessment 7-4A: Percents and Interest

1. **(a)** Divide the shape into 16 equal size parts. One out of 16 is shaded.

$\frac{1}{16} = 0.0625 = \mathbf{6.25\%}$

(b) Divide the shape into 6 equal size parts. One out of 6 is shaded.

$\frac{1}{6} = 0.1\overline{6} = 16.\overline{6}\% = \mathbf{16\frac{2}{3}\%}$

2. 8 out of 12 circles are black.

$\frac{8}{12} = \frac{2}{3} = 0.\overline{6} = 66.\overline{6}\% = \mathbf{66\frac{2}{3}\%}$

3. **(a)** $7.89 = 100\left[\frac{7.89}{100}\right] = \frac{789}{100} = \mathbf{789\%}.$

(b) $193.1 = 100\left[\frac{193.1}{100}\right] = \frac{19310}{100} = \mathbf{19,310\%}.$

(c) $\frac{5}{6} = 100\left[\frac{\frac{5}{6}}{100}\right] = \frac{100\left[\frac{5}{6}\right]}{100}.$

$= \frac{\frac{50}{1}\cdot\frac{5}{3}}{100} = \frac{\frac{250}{3}}{100} = \frac{83\frac{1}{3}}{100} = \mathbf{83\frac{1}{3}\%}.$

(d) $\frac{1}{8} = 100\left[\frac{\left(\frac{1}{8}\right)}{100}\right] = \frac{\left(\frac{100}{8}\right)}{100} = \frac{12.5}{100} = \mathbf{12.5\%}.$

(e) $\frac{5}{8} = 100\left[\frac{\left(\frac{5}{8}\right)}{100}\right] = \frac{\left(\frac{500}{8}\right)}{100} = \frac{62.5}{100}$

$= \mathbf{62.5\%}.$

(f) $\frac{4}{5} = 100\left[\frac{\left(\frac{4}{5}\right)}{100}\right] = \frac{\left(\frac{400}{5}\right)}{100} = \frac{80}{100} = \mathbf{80\%}.$

4. **(a)** $16\% = \frac{16}{100} = \mathbf{0.16}.$

(b) $\frac{1}{5}\% = 0.2\% = \frac{0.2}{100} = \frac{2}{1000} = \mathbf{0.002}.$

(c) $13\frac{2}{3}\% = 13.\overline{6}\% = \frac{13.\overline{6}}{100} = \mathbf{0.13\overline{6}}.$

(d) $\frac{1}{3}\% = 0.\overline{3}\% = \frac{0.\overline{3}}{100} = \mathbf{0.00\overline{3}}.$

5. **(a)** <u>4</u> for every 100. $4\% = \frac{4}{100}.$

(b) <u>2</u> for every 50. $4\% = \frac{4}{100} = \frac{2}{50}.$

(c) 1 for every <u>**25**</u>. $4\% = \frac{4}{100} = \frac{1}{25}.$

(d) 8 for every <u>**200**</u>. $4\% = \frac{4}{100} = \frac{8}{200}.$

(e) 0.5 for every <u>**12.5**</u>. $4\% = \frac{4}{100} = \frac{0.5}{12.5}.$

6.

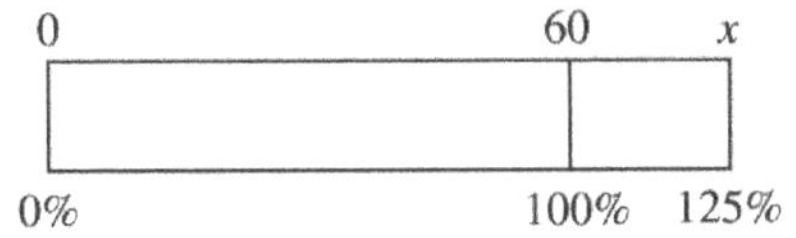

$\frac{60}{100} = \frac{x}{125} \Rightarrow 100x = 6000 \Rightarrow x = 75.$

7. **(a)** $6\% \text{ of } 34 \Rightarrow \frac{6}{100}\cdot 34 = \frac{204}{100} = \mathbf{2.04}.$

(b) 17 is $n\%$ of 34 $\Rightarrow 17 = \frac{n}{100}\cdot 34$

$\Rightarrow 34n = 1700 \Rightarrow n = \frac{1700}{34} =$

50, or 17 is **50%** of 34.

(c) 18 is 30% of $n \Rightarrow 18 = \frac{30}{100}\cdot n$

$\Rightarrow 30n = 1800 \Rightarrow n = \frac{1800}{30} = \mathbf{60}.$

(d) $7\% \text{ of } 49 \Rightarrow \frac{7}{100}\cdot 49 = \frac{343}{100} = \mathbf{3.43}.$

8.

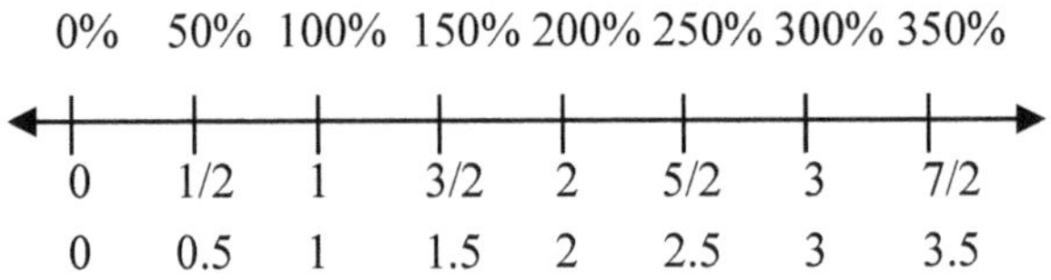

9. $\frac{325}{500}\cdot\frac{325}{500} = \frac{650}{1000} = \frac{65}{100} = 65\%;\ \frac{600}{1000} =$

$\frac{60}{100} = 60\%.$

$\mathbf{\frac{325}{500}}$ **represents the greater fraction.**

10. 75% of 84 boxes $\Rightarrow \frac{75}{100}\cdot 84 =$ **63 boxes.**

11. (a) **Bill**. Zake sold 180 papers; Bill sold $\frac{85}{100} \cdot 260 = \mathbf{221}$ **papers;** Maksim sold 212 papers.

(b) **Zake**. Zake sold $100\left[\frac{\left(\frac{180}{200}\right)}{100}\right] = \mathbf{90\%}$ of his papers; Bill sold 85% of his papers; Ron sold 80% of his papers.

(c) **Maksim**. Zake started with 200 papers; Bill started with 260 papers; Maksim started with $\frac{212}{0.80} = \mathbf{265}$ **papers.**

12. $66\frac{2}{3}\%$ of 1800 employees means $\frac{66\frac{2}{3}}{100} = \frac{\left(\frac{200}{3}\right)}{100} = \frac{200}{300} = \frac{2}{3}$ of 1800. $\frac{2}{3} \cdot 1800 =$ **1200 employees**.

13. (a) $\frac{\text{20 math majors}}{\text{500 incoming students}} = \frac{4}{100} = \mathbf{4\%}$ math majors.

(b) (*i*) There were 480 non-math-major students originally. 5% of $480 = 0.05 \cdot 480 = 24$ who switched.
20 original + 24 switches = **44** math majors now.

(*ii*) $\frac{\text{44 math majors}}{\text{500 students}} = 0.088 = 100\left(\frac{0.088}{100}\right) = \frac{8.8}{100} =$ **8.8%** math majors.

14. Sale price was regular price – 20% of regular price. $\$28.00 - 20\%$ of $\$28.00 = \$28 - (0.2 \cdot \$28) = \$28.00 - \$5.60 = \mathbf{\$22.40}$.

15. (a) 15% of $\$30 = 0.15 \cdot 30 = \mathbf{\$4.50}$.

(b) $100\left[\frac{\left(\frac{1}{2}\right)}{100}\right] = \mathbf{50\%}$.

(c) $100\left(\frac{1}{100}\right) = \mathbf{100\%}$.

16. The amount of the tax is 5% of $320, or $0.05 \cdot 320 = \$16$. The total cost is $\$320 + \$16 = \mathbf{\$336}$.

17. A journeyman makes 200% of an apprentice's pay and a master makes 150% of a journeyman's pay.
150% of 200% = $1.5 \cdot 2 = 3 = 300\%$ of an apprentice's pay.
The $4200 must have $1 + 2 + 3 = 6$ shares, or $\frac{\$4200}{6} = \$700$ per share. The **apprentice earns $700**. The **journeyman** earns 200% of $\$700 = 2 \cdot \$700 = \mathbf{\$1400}$. The **master** earns 300% of $\$700 = 3 \cdot \$700 = \mathbf{\$2100}$.

18. **No**. Since 56% is more than double 25%, $950 should be more than double $500, but it is not.

19. Louis answered $80 - 52 = 28$ questions incorrectly. 28 as percentage of 80 is $\frac{28}{80} = 0.35 = 100\left(\frac{0.35}{100}\right) = \frac{35}{100} = \mathbf{35\%}$.

20. 30 out of 45 students passed their physics test, that is $\frac{30}{45} = \frac{2}{3} = 0.\bar{6} = 66.\bar{6}\% = \mathbf{66\frac{2}{3}\%}$.

21. 25% of the girls are going, so $0.25 \cdot 12 = 3$ girls. 25% of the boys are going, so $0.25 \cdot 8 = 2$ boys. 3 girls and 2 boys is 5 students out of 20 total students (12 girls + 8 boys), that is $\frac{5}{20} = \frac{1}{4} = 0.25 = \mathbf{25\%}$.

22. 68% of all students is equal to 374 students, so $0.68 \cdot$ total students = 374 and total students = $374 \div 0.68 = 550$. **There is a total of 550 students in the school.**

23. 8 questions represent 40% of all the questions. 4 questions would represent 20% and $5 \cdot 20\% = 100\%$, so $5 \cdot 4 = 20$ questions. **There were 20 questions on his math test.**

24. **(a)** If there are 100 students, then 25 play sports and 60% of these play football. $0.60 \cdot 25 = 15$ students play football. 15 is 15% of 100, so **15% play football**. Or, 60% of 25% is $0.60 \cdot 0.25 = 0.15 = 15\%$.

(b) 180 students play football, which is 15% of the students enrolled at Riverdale High. So 180 = $0.15 \cdot$ total students and total students = $\frac{180}{0.15} = 1200$. **There are 1200 students enrolled at Riverdale High.**

25. (a) $\frac{14}{15} = 0.9\bar{3} = \mathbf{93\frac{1}{3}\%}$.

(b) $80\% = \frac{80}{100} = \mathbf{\frac{4}{5}}$ of the sun's UV rays are blocked.

(c) The SPF is **5**.

(d) **Yes.** A sunblock with SPF 30 blocks $\frac{29}{30} = 0.9\bar{6} = 96\frac{2}{3}\%$. The label is correct.

26. Let s be her last salary. 7% of $s + 100\%$ of $s =$ $\$27{,}285 \Rightarrow 107\% \cdot s = \$27{,}285 \Rightarrow$ $1.07 \cdot s = 27{,}285$.

Thus $s = \frac{27{,}285}{1.07} = \mathbf{\$25{,}500}$.

27. Let s be the salary of the previous year. 100% of $s + 10\%$ of $s = 110\%$ of $s = 1.1 \cdot s$ is current salary. $1.1s = \$100{,}000$ (this year), so

$s = \frac{\$100.000}{1.1} \approx \$90{,}909.09$ (last year).

$1.1s = \$90{,}909.09$ (last year), so

$s = \frac{\$90{,}909{,}09}{1.1}$, or about **\$82,644.63** (two years ago).

28. The amount of the discount is $\$35 - \$28 = \$7$. The percent of the discount is \$7 as percent of $\$35 \Rightarrow \frac{7}{35} = 0.2 = 100\left(\frac{0.2}{100}\right) = \frac{20}{100} = \mathbf{20\%}$.

29. The amount of decrease in value was $\$359{,}000 - \$195{,}000 = \$164{,}000$. The decrease as a percentage of the original value was $\frac{164{,}000}{359{,}000} \approx 0.46 = 100\left(\frac{0.46}{100}\right) = \frac{46}{100}$, or **about 46%**.

30. $\$1200 + 0.20 \cdot \$1200 = \$1440$

$\$1440 - 0.20 \cdot \$1440 = \mathbf{\$1152}$

31. **(a)** **(*i*)** $8 \cdot \$9.50 = \mathbf{\$76.00}$ for 8 items.

(*ii*) 10 items at \$9.50 each $= \$95$.

$\$95 - 20\%$ of $\$95$

$= 95 - 0.2 \cdot 95$

$= \$95.00 - \$19.00$

$=$ **\$76** for 10 items.

(b) **10 items**. You get two more items for the same price.

32. The amount of Shaun's 20% profit is $0.20 \cdot \$330 = \$66$. Sale price of the bike after a 10% discount must be $\$330 + \$66 = \$396$. Let p be the list price; then $p - 10\%$ of p is the sale price. $p - 10\%$ of $p = \$396 \Rightarrow 0.9p = 396 \Rightarrow p = \frac{396}{0.9} = \$440$. Thus if Shaun prices the bike at **\$440** he can offer a 10% discount of \$44 and still realize his \$66 profit.

33. Techniques may vary:

(a) 10% of $\$22 = \$2.20$. 5% of \$22 is half 10%, or \$1.10. $\$2.20 + \$1.10 = \mathbf{\$3.30}$.

(b) 10% of $\$120 = \$12$. 20% is twice 10%, or twice $\$12 = \mathbf{\$24}$.

(c) 10% of $\$38 = \$3.80$. 5% is half of 10%, or half of $\$3.80 = \mathbf{\$1.90}$.

(d) 25% is $\frac{1}{4}$; $\frac{1}{4}$ of $\$98 = \$98 \div 4$, (To or **\$24.50**.
divide 98 by 4, think of it as $100 \div 4 = 25$. Then subtract $2 \div 4 = 0.50$.)

34. Where I is interest, P is principal, r is rate expressed as a decimal, and t is time in years: $I = Prt = \$42{,}000 \cdot 0.0875 \cdot 1 = \mathbf{\$3675.00}$.

35. $0.34 \cdot 100\% = 34\%$ and $\frac{34}{100} = \frac{17}{50}$

$29\frac{1}{5}\% = 29.2\% = 0.292$

$0.292 = \frac{292}{1000} = \frac{73}{250}$

$\frac{5}{6} = 0.8\overline{3}$

$0.8\overline{3} = 83.\overline{3}\% = \mathbf{83\frac{1}{3}\%}$

Mathematical Connections 7-4: Review Problems

31. **(a)** $0.18(120) = \mathbf{21.6}$ **lb.**

(b) $0.4(120) = \mathbf{48}$ **lb.**

33. **(a)** $5 = 4 + 1 = 4 + 0.\overline{9} = \mathbf{4.\overline{9}}$.

(b) $5.1 = 5 + 0.1 = 5 + \frac{1}{10}0.\overline{9} = \mathbf{5.0\overline{9}}$.

(c) $\frac{1}{2} = 0.5 = 0.4\overline{9}$.

35. $0.\overline{24} = \frac{24}{99} = \frac{8.3}{3.33} = \mathbf{\frac{8}{33}}$.

Assessment 7-5A: Real Numbers

1. **Answers vary**. For example: one such number could be 0.232233222333…, continuing the pattern of adding a 2 and a 3 to each succeeding group.

2. **(a)** $x^2 = 2^2 + (\sqrt{2})^2 = 4 + 2 = 6$.

$x = \mathbf{\sqrt{6}}$.

(b) $x^2 = (2)^2 + (2)^2 = 4 + 4 = 8.$

$x = \sqrt{8} = \mathbf{2\sqrt{2}}.$

(c) $5^2 = 2^2 + x^2 \Rightarrow x^2 = 5^2 - 2^2$

$= 25 - 4 = 21$

$x = \mathbf{\sqrt{21}}.$

3. Line up the decimal points:

$0.9 = 0.90000000\ldots$

$0.\overline{9} = 0.99999999\ldots$

$0.\overline{98} = 0.98989898\ldots$

$0.9\overline{88} = 0.98888888\ldots$

$0.9\overline{98} = 0.99898989\ldots$

$0.\overline{898} = 0.89889889\ldots$

$\sqrt{0.98} = 0.98994949\ldots$

Ordering from greatest to least:

$\mathbf{0.\overline{9} > 0.9\overline{98} > \sqrt{0.98} > 0.\overline{98} > 0.9\overline{88} > 0.9}$

$\mathbf{> 0.\overline{898}}.$

4. **(a)** **Irrational**. There is no rational number s such that $s^2 = 51$.

(b) **Rational**. $8^2 = 64$.

(c) **Rational**. $18^2 = 324$.

(d) **Irrational**. There is no rational number s such that $s^2 = 325$.

(e) **Irrational**. The sum of a rational number and an irrational number is irrational.

(f) **Irrational**. The quotient of any non-zero rational number and any irrational number is irrational.

5. **(a)** $15 \cdot 15 = 225 \Rightarrow \sqrt{225} = \mathbf{15}.$

(b) $13 \cdot 13 = 169 \Rightarrow \sqrt{169} = \mathbf{13}.$

(c) $^-1 \cdot 9 \cdot 9 = {}^-81 \Rightarrow {}^-\sqrt{81} = \mathbf{^-9}.$

(d) $25 \cdot 25 = 625 \Rightarrow \sqrt{625} = \mathbf{25}.$

(e) $\frac{1}{2} \cdot \frac{1}{2} = \frac{1}{4} \Rightarrow \sqrt{\frac{1}{4}} = \mathbf{\frac{1}{2}}.$

(f) $0.01 \cdot 0.01 = 0.0001 \Rightarrow \sqrt{0.0001} = \mathbf{0.01}.$

6. **(a)** $2 < \sqrt{7} < 3.$

$(2.6)^2 = 6.76$ and $(2.7)^2 = 7.29$

$\Rightarrow 2.6 < \sqrt{7} < 2.7.$

$(2.64)^2 = 6.9696$ and $(2.65)^2 = 7.0225$

$\Rightarrow 2.64 < \sqrt{7} < 2.65.$

7 is closer to 7.0225 than to 6.9696 $\Rightarrow$

$\sqrt{7} \approx \mathbf{2.65}.$

(b) $0.1 < \sqrt{0.0120} < 0.2.$

$(0.10)^2 = 0.0100$ and $(0.11)^2 = 0.0121$

$\Rightarrow 0.10 < \sqrt{0.0120} < 0.11.$

0.0120 is closer to 0.0121 than to 0.0100

$\Rightarrow \sqrt{0.0120} \approx \mathbf{0.11}.$

7. Answers vary.

(a) **False**. $0 + \sqrt{2}$ is irrational.

(b) **False**. $^-\sqrt{2} + \sqrt{2} = 0$, which is rational.

(c) **False**. $\sqrt{2} \cdot \sqrt{2} = 2$, which is rational.

(d) **True**. $\sqrt{2} - \sqrt{2} = 0$, which is rational.

8. **Answers vary**.

(a) $\sqrt{2}, \sqrt{3}$, and $\sqrt{5}$ are three such irrational numbers.

(b) $0.\overline{54} = 0.54545454\ldots$

$< 0.546010010001\ldots$

$< 0.547010010001\ldots$

$< 0.548010010001\ldots$

$< 0.\overline{55} = 0.55555555\ldots.$

(c) $\frac{1}{3} = 0.\overline{3} = 0.333\ldots.$

$< 0.34010010001\ldots$

$< 0.35010010001\ldots$

$< 0.36010010001\ldots < 0.5.$

9. Consider 8(c). We form an irrational number between $0.\overline{3}$ and 0.5 by placing digits in lesser place-values so that the decimal does not terminate or repeat, e.g., 0.3434434443…. A similar approach can be taken for any two rational numbers. For example, 1.7171171117… is between 1.7 and 1.8.

10. **(a)** $Q \cup S = \boldsymbol{R}$. The set of real numbers contains any number which is either rational or irrational.

(b) $Q \cap S = \varnothing$. A rational number cannot be an irrational number.

(c) $Q \cap R = \mathbf{Q}$. $Q \subset R$.

(d) $S \cap W = \varnothing$. No whole number can be irrational.

(e) $W \cup R = \mathbf{R}$. $W \subset R$.

(f) $Q \cup R = \mathbf{R}$. $Q \subset R$.

11. In the tables of 11. and 12. below: N is the set of natural (or counting) numbers; I is the set of integers; Q is the set of rational numbers; R is the set of real numbers; and S is the set of irrational numbers.

$N \subset I \subset Q \subset R$; $R = Q \cup S$.

		N	I	Q	R	S
(a)	6.7			√	√	
(b)	5	√	√	√	√	
(c)	$\sqrt{2}$				√	√
(d)	$^-5$		√	√	√	
(e)	$3\frac{1}{7}$			√	√	

12.

		$x =$	N	I	Q	R	S
(a)	$x^2 + 1 = 5$	$2, {}^-2$		√	√	√	
(b)	$2x - 1 = 32$	$\frac{33}{2}$			√	√	
(c)	$x^2 = 3$	$\sqrt{3}, {}^-\sqrt{3}$				√	√
(d)	$\sqrt{x} = {}^-1$	no solution					
(e)	$\frac{3}{4}x = 0.\overline{4}$	$\frac{16}{27}$			√	√	

13. (a) $x = \mathbf{64}$. $\sqrt{64} = 8$.

(b) **No real values.** $\sqrt{x}$ is the principal square root of x.

(c) $x = \mathbf{{}^-64}$. $\sqrt{{}^-({}^-64)} = \sqrt{64} = 8$.

(d) **No real values.** $\sqrt{{}^-x}$ is the principal square root of x, if $x < 0$.

(e) All real numbers > 0. If $x = 0$ then $\sqrt{x} \not> 0$.

(f) **No real values.** $\sqrt{x}$ is the principal square root of x.

14. (a) $\sqrt{180} = \sqrt{36 \cdot 5} = \sqrt{36} \cdot \sqrt{5} = \mathbf{6\sqrt{5}}$.

(b) $\sqrt{363} = \sqrt{121 \cdot 3} = \sqrt{121} \cdot \sqrt{3} = \mathbf{11\sqrt{3}}$.

(c) $\sqrt{252} = \sqrt{36 \cdot 7} = \sqrt{36} \cdot \sqrt{7} = \mathbf{6\sqrt{7}}$.

15. (a) $\sqrt[3]{{}^-54} = \sqrt[3]{{}^-27 \cdot 2} = \sqrt[3]{{}^-27} \cdot \sqrt[3]{2}$
$= \mathbf{{}^-3\sqrt[3]{2}}$.

(b) $\sqrt[5]{96} = \sqrt[5]{32 \cdot 3} = \mathbf{2\sqrt[5]{3}}$.

(c) $\sqrt[3]{250} = \sqrt[3]{125 \cdot 2} = \mathbf{5\sqrt[3]{2}}$.

(d) $\sqrt[5]{{}^-243} = {}^-3\sqrt[5]{1} = {}^-3 \times 1 = \mathbf{{}^-3}$.

16. (a) $a_1 = 5$; $a_4 = 10$; $n = 4$. Thus $10 = 5(r)^{4-1} \Rightarrow 2 = r^3 \Rightarrow r = \sqrt[3]{2}$. The terms of the sequence are $5, \underline{\mathbf{5\sqrt[3]{2}}}, \underline{\mathbf{5\sqrt[3]{4}}}, 5\sqrt[3]{8} = 10$.

(b) $a_1 = 2$; $a_5 = 1$; $n = 5$. Thus $1 = 2(r)^{5-1}$ $\Rightarrow r = \sqrt[4]{\frac{1}{2}}$ or $r = {}^-\sqrt[4]{\frac{1}{2}}$. The terms of the sequence are then
$2 = \sqrt[4]{16}, \underline{\mathbf{\sqrt[4]{8}}}, \underline{\mathbf{\sqrt[4]{4}}}, \underline{\mathbf{\sqrt[4]{2}}},$
$\sqrt[4]{1} = 1$ or $2 = \sqrt[4]{16}, \underline{\mathbf{{}^-\sqrt[4]{8}}}, \underline{\mathbf{\sqrt[4]{4}}}, \underline{\mathbf{{}^-\sqrt[4]{2}}},$
$\sqrt[4]{1} = 1$.

(c) $a_2 = 5$ and $a_4 = 3$ Thus $5 = a_1(r)^{2-1}$ and $3 = a_1(r)^{4-1}$, so $5 = a_1 r$ and $3 = a_1 r^3$ Thus

$$a_1 = \frac{5}{r} \text{ and } a_1 = \frac{3}{r^3} \Rightarrow \frac{5}{r} = \frac{3}{r^3} \Rightarrow$$

$$5r^3 = 3r \Rightarrow 5r^3 - 3r = 0 \Rightarrow r(5r^2 - 3) = 0$$

$$\Rightarrow r = 0 \text{ and } 5r^2 - 3 = 0 \Rightarrow 5r^2 = 3$$

$$\Rightarrow r^2 = \frac{3}{5} \Rightarrow r = \pm\sqrt{\frac{3}{5}} \Rightarrow r = \pm\frac{\sqrt{15}}{5}$$

The terms of the sequence are

$$\frac{5\sqrt{15}}{3}, 5, \sqrt{15}, 3 \text{ or } {}^-\frac{5\sqrt{15}}{3}, 5, {}^-\sqrt{15}, 3$$

(d) $a_2 = {}^-2$ and $a_4 = {}^-3$ Thus ${}^-2 = a_1(r)^{2-1}$ and ${}^-3 = a_1(r)^{4-1}$, so ${}^-2 = a_1 r$ and ${}^-3 = a_1 r^3$

Thus

$$a_1 = \frac{{}^-2}{r} \text{ and } a_1 = \frac{{}^-3}{r^3} \Rightarrow \frac{{}^-2}{r} = \frac{{}^-3}{r^3} \Rightarrow {}^-2r^3 = {}^-3r$$

$$\Rightarrow {}^-2r^3 + 3r = 0 \Rightarrow {}^-r(2r^2 - 3) = 0 \Rightarrow {}^-r = 0$$

and $2r^2 - 3 = 0 \Rightarrow 2r^2 = 3 \Rightarrow r^2 = \frac{3}{2}$

$$\Rightarrow r = \pm\sqrt{\frac{3}{2}} \Rightarrow r = \pm\frac{\sqrt{6}}{2}$$

The terms of the sequence are

$$\frac{2\sqrt{6}}{3}, {}^-2, \sqrt{6}, {}^-3 \text{ or } \frac{{}^-2\sqrt{6}}{3}, {}^-2, {}^-\sqrt{6}, {}^-3$$

17. **(a)** $E(0) = 2^{10} \cdot 16^0 = \mathbf{2^{10}}$ **bacteria.**

(b) $E\left(\frac{1}{4}\right) = 2^{10} \cdot 16^{1/4} = 2^{10} \cdot 2$

$= \mathbf{2^{11}}$ **bacteria.**

(c) $E\left(\frac{1}{2}\right) = 2^{10} \cdot 16^{1/2} = 2^{10} \cdot 2^2$

$= \mathbf{2^{12}}$ **bacteria.**

18. **(a) True,** odd root of any real number exists.

(b) True, $x^2 = x \cdot x$ and $x^3 = x \cdot x \cdot x$, so $x^2 \cdot x^3 = x \cdot x \cdot x \cdot x \cdot x = x^5$.

(c) False, for example: $3^2 + 3^3 = 9 + 27 = 36$; $3^{2\cdot3} = 3^6 = 729 \neq 36$

(d) False, for example:

$$\sqrt[4]{({}^-4)^2} = \sqrt[4]{16} = 2 \neq ({}^-4)^{\frac{1}{2}}$$

(e) False, for example: $0^3 \div 0^2 \neq 0$, the division is undefined.

(f) False, $a^3a^2 = a \cdot a \cdot a \cdot a \cdot a = a^5$

$$a^{3\cdot2} = a^6 \neq a^5$$

19. **(a)** $3^x = 243 \Rightarrow 3^x = 3^5 \Rightarrow x = \mathbf{5}$.

(b) $9^{{}^-x} = 27 \Rightarrow 3^{{}^-2x} = 3^3 \Rightarrow {}^-2x = 3 \Rightarrow$

$$x = \mathbf{\frac{{}^-3}{2}}$$

(c) $\left(\frac{9}{4}\right)^{3x} = \frac{32}{243} \Rightarrow \left(\frac{3^2}{2^2}\right)^{3x} = \frac{2^5}{3^5}$

$$\Rightarrow \left(\frac{3}{2}\right)^{6x} = \left(\frac{2}{3}\right)^{5} \Rightarrow \left(\frac{3}{2}\right)^{6x} = \left(\frac{3}{2}\right)^{{}^-5}$$

$$\Rightarrow 6x = {}^-5 \Rightarrow x = \mathbf{\frac{{}^-5}{6}}$$

(d) $\sqrt{{}^-x} = 3\sqrt{2} \Rightarrow \sqrt{{}^-x} = \sqrt{18} \Rightarrow$

${}^-x = 18 \Rightarrow x = \mathbf{{}^-18}$

(e) $x^{\frac{{}^-3}{4}} = 2 \Rightarrow \left(x^{\frac{{}^-3}{4}}\right)^{\frac{{}^-4}{3}} = (2)^{\frac{{}^-4}{3}} \Rightarrow x = \mathbf{(2)^{\frac{{}^-4}{3}}}$

(f) $(x-1)^2 = 2 \Rightarrow x - 1 = \pm\sqrt{2} \Rightarrow x = \mathbf{1 \pm \sqrt{2}}$

20. Answers vary.

(a) $\sqrt[5]{20} \Rightarrow x^5 - 20 = 0$

(b) $\sqrt[3]{{}^-2} \Rightarrow x^3 + 2 = 0$

(c) $\sqrt[3]{10} - 1 \Rightarrow (x+1)^3 - 10 = 0$

(d) $\frac{\sqrt{2}}{\sqrt{3}} \Rightarrow 3x^2 - 2 = 0$

21. Using the "squeeze" strategy

$4^3 = 64$

$4.6^3 = 97.3$

$4.7^3 = 103.8$

$5^3 = 125$

So, an integer approximation of $\sqrt[3]{103}$ is 5.

22. **(a)** $\frac{\sqrt{500}}{\sqrt{20}} = \frac{10\sqrt{5}}{2\sqrt{5}} = 5$ so it is rational.

(b) $8^{\frac{1}{3}} + 8^{\frac{{}^-1}{3}} = (2^3)^{\frac{1}{3}} + (2^3)^{\frac{{}^-1}{3}} = 2 + \frac{1}{2} = 2\frac{1}{2}$ so it is rational.

(c) $\frac{2}{\sqrt{2}} - \sqrt{2} = \frac{2}{\sqrt{2}} - \frac{2}{\sqrt{2}} = 0$ so it is rational.

(d) $\sqrt{1000} = 31.6227766...$ so it is irrational.

23. **(a) True,** any two natural numbers added together equal another natural number.

(b) False, counterexample: $3 - 2 = {}^-1$ which is not a whole number.

(c) False, counterexample: ${}^-3 \div 2 = {}^-1.5$ wich is not an integer.

(d) True, the set of integers goes to infinity, so there's no largest positive integer.

(e) **False**, the associative property does not work for subtraction:

$5-(4-2) \neq (5-4)-2$

$5-2 \neq 1-2$

$3 \neq {}^{-}1.$

It works for addition and multiplication.

24. (a) $\frac{1}{\sqrt{2}} = \frac{\sqrt{2}}{x}$

$1 \cdot x = \sqrt{2} \cdot \sqrt{2}$

$x = 2$

(b) $\frac{5}{x} = x$ or $\frac{5}{x} = \frac{x}{1}$

$5 \cdot 1 = x \cdot x$

$5 = x^2$

$\pm\sqrt{5} = x$

25. $\pi^3 = 31.006276680...$

$30.\overline{9} = 30.99999999...$

$\sqrt[3]{29,791} = 31.00000$

$31\frac{1}{9} = 31.11111111...$

$30.5 = 30.50000000$

So $30.5 < 30.\overline{9} < \sqrt[3]{29791} < \pi^3 < 31\frac{1}{9}$.

Mathematical Connections 7-5: Review Problems

27. The number of repossessed homes in 2006 plus 51% to get 405,000 repossessed homes in 2007. So the 2006home + 0.51 •2006 #homes = 405.000. Solving for 2006 homes:

$(2006\,\text{homes}) + 0.51 \cdot (2006\,\text{homes}) = 405{,}000$

$1.51(2006\,\text{homes}) = 405{,}000$

$2006\,\text{homes} = \frac{405{,}000}{1.51} = \mathbf{268{,}212}$

29. $5\% = 0.05 = \frac{5}{100} = \mathbf{\frac{1}{20}}$

$56\% = 0.56 = \frac{56}{100} = \mathbf{\frac{14}{25}}$

$\frac{7}{8} = 0.875 = \mathbf{87.5\%}$

$0.25 = \frac{25}{100} = \frac{1}{4} = \mathbf{25\%}$

$33\frac{1}{3}\% = 0.\overline{3} = \mathbf{\frac{1}{3}}$

$\frac{5}{12} = 0.41\overline{6} = 41.\overline{6}\% = \mathbf{41\frac{2}{3}\%}$

$0.08 = 8\% = \frac{8}{100} = \mathbf{\frac{2}{25}}$

Chapter 7 Review

1. Each division on the number line corresponds to 0.01.

(a) (*i*) Point A is two divisions to the right of point 0, or **0.02**.

(*ii*) Point B is five divisions to the right of point 0, or **0.05**.

(*iii*) Point C is one division to the right of point 0.1, or **0.11**.

(b) (*i*) Point D (0.09) is one division to the left of point 0.1; i.e., $0.1 - 0.01 = 0.09$.

(*ii*) Point E (0.15) is five divisions to the right of point 0.1; i.e., $0.1 + 0.05 = 0.15$.

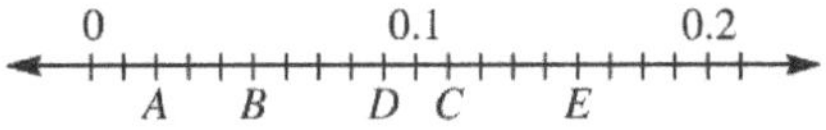

2. (a) $32.012 = 32\frac{12}{1000} = \frac{32.012}{1000} = \mathbf{\frac{8003}{250}}$.

(b) $0.00103 = \mathbf{\frac{103}{100{,}000}}$.

3. $8000 + 500 + 6 + 0.0008$
$= \mathbf{8506.0008}$

4. $442.4\text{ cm} \div 55.3\text{ cm} =$ **8 shelves.**

5. (a) $\frac{4}{7} = 4 \div 7 = \mathbf{0.\overline{571428}}$.

(b) $\frac{1}{8} = 1 \div 8 = \mathbf{0.125}$.

(c) $\frac{2}{3} = 2 \div 3 = \mathbf{0.\overline{6}}$.

(d) $\frac{5}{8} = 5 \div 8 = \mathbf{0.625}$.

6. (a) $0.28 = \frac{28}{100} = \mathbf{\frac{7}{25}}$.

(b) ${}^{-}6.07 = {}^{-}6\frac{7}{100} = \mathbf{\frac{{}^{-}607}{100}}$.

(c) $0.\overline{3} = \frac{3}{9} = \mathbf{\frac{1}{3}}$.

(d) $2.0\overline{8} = 2\frac{80}{900} = 2\frac{4}{45} = \mathbf{\frac{94}{45}}$.

7. **(a)** **307.63**. $307.6\underline{2}5$ is between 307.62 and 307.63; when the number to be rounded is at the mid-point, by convention it is commonly rounded up.

(b) **307.6**. $307.\underline{6}25$ is between 307.6 and 307.7 and is closer to 307.6.

(c) **308**. $30\underline{7}.625$ is between 307 and 308 and is closer to 308.

(d) **300**. $\underline{3}07.625$ is between 300 and 400 and is closer to 300.

8. **(a)** Move the decimal point five units to the left and multiply by 10^5, or $\mathbf{4.26 \cdot 10^5}$.

(b) $324 \cdot 10^{-6} = 3.24 \cdot 10^2 \cdot 10^{-6} =$ $\mathbf{3.24 \cdot 10^{-4}}$.

(c) Move the decimal point six units to the right and multiply by 10^{-6}, or $0.00000237 = \mathbf{2.37 \cdot 10^{-6}}$.

(d) $^{-}0.325 = \mathbf{^{-}3.25 \cdot 10^{-1}}$.

9. $\frac{^{-}2}{3} = {}^{-}0.666666...$

$0.76 = 0.760000$

$\frac{25}{33} = 0.75757575...$

$1.232 = 1.23200000$

$^{-}0.666 = {}^{-}0.666000$

$1.\overline{23} = 1.23232323...$

So $\mathbf{\frac{^{-}2}{3}} < 0.666 < \mathbf{\frac{25}{33}} < 0.76 < 1.232 < \mathbf{1.\overline{23}}$

10. **(a)** Move the decimal point six units to the right and multiply by 10^6, or $\mathbf{1.78341156 \cdot 10^6}$.

(b) $\frac{347}{10^8} = 347 \cdot 10^{-8} = 3.47 \cdot 10^2 \cdot 10^{-8} =$ $\mathbf{3.47 \cdot 10^{-6}}$.

(c) $49.3 \cdot 10^8 = 4.93 \cdot 10^1 \cdot 10^8$ $= \mathbf{4.93 \cdot 10^9}$.

(d) $29.4 \cdot \frac{10^{12}}{10^1} = 2.94 \cdot 10^1 \cdot 10^{12} \cdot 10^4 =$ $\mathbf{2.94 \cdot 10^{17}}$.

(e) $0.47 \cdot 1000^{12} = 4.7 \cdot 10^{-1} \cdot (10^3)^{12} =$ $4.7 \cdot 10^{-1} \cdot 10^{36} = \mathbf{4.7 \cdot 10^{35}}$.

(f) $\frac{3}{5^9} = \frac{3 \cdot 2^9}{5^9 \cdot 2^9} = \frac{1536}{10^9} = 1536 \cdot 10^{-9}$ $= 1.536 \cdot 10^3 \cdot 10^{-9} = \mathbf{1.536 \cdot 10^{-6}}$.

11.

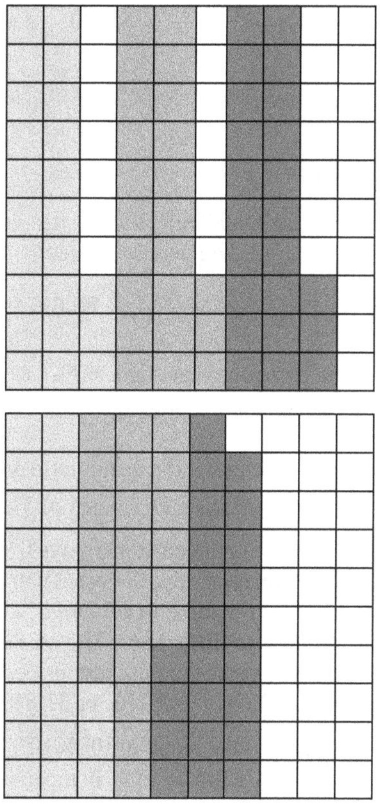

= 0.69

12. **(a)** $6 = n\%$ of $24 \Rightarrow n\% = \frac{6}{24}$ $\Rightarrow \frac{n}{100} = \frac{1}{4} \Rightarrow 4n = 100 \Rightarrow n = 25$, or $6 =$ **25%** of 24.

(b) $n = 320\%$ of $60 \Rightarrow n =$ $\frac{320}{100} \cdot 60 = \mathbf{192}$.

(c) $17 = 30\%$ of $n \Rightarrow n =$ $\frac{17}{\left(\frac{30}{100}\right)} = \frac{1700}{30} = \mathbf{56.\overline{6}}$.

(d) $0.2 = n\%$ of $1 \Rightarrow \frac{n}{100} = \frac{0.2}{1}$ $\Rightarrow n = 20$, or $0.2 =$ **20%** of 1.

13. $\frac{5}{8} = 0.625 = 62.5\%$

$0.02 = 2\% = \frac{2}{100} = \frac{1}{50}$

$0.5\% = 0.005 = \frac{5}{1000} = \frac{1}{200}$

14. $\$7.08 - \$0.91 = \$6.17$. $\frac{6.17}{7.08} \approx 0.87146$ or a 87.15% decrease

15. $11\% \cdot \text{investment} = \$1020.80$
$\Rightarrow \text{investment} = \frac{\$1020.80}{0.11} = \mathbf{\$9280.}$

16. Percent defective $= 100\left[\frac{\left(\frac{5}{150}\right)}{100}\right] = \frac{\left(\frac{500}{150}\right)}{100} =$
$\frac{3.\bar{3}}{100} = \mathbf{3.\bar{3}\%}$ or $\mathbf{3\frac{1}{3}\%}$.

17. Assume "nearest tenth" refers to the nearest tenth of a percent. Then percent correct
$= 100\left(\frac{\frac{62}{70}}{100}\right) = \frac{\left(\frac{6200}{70}\right)}{100} \approx \frac{88.6}{100} = \mathbf{88.6\%.}$

18. Let C be the cost four years ago. $60\% \cdot C = \$3450 \Rightarrow C = \frac{\$3450}{0.60} = \mathbf{\$5750.}$

19. A discount of $d\%$ means the customer pays $1 - \frac{d}{100}$ for the purchase. Discounts of 5%, 10%, and 20% mean the customer pays 0.95, 0.90, and 0.80, respectively, of cost. Their product, 0.648, is the same in any order, therefore there is **no difference**. The customer would pay 0.684 times the purchase price, for a discount of $1 - 0.684 = 0.316$, or 31.6%.

20. Let c be the cost of the bicycle. 100% of $c + 30\%$ of $c = \$104 \Rightarrow c + 0.3c = \$104 \Rightarrow 1.3c = \$104$. $c = \frac{\$104}{1.3} = \mathbf{\$80.}$

21. The price difference was $\$89.95 - \$62.00 = \$27.95$. The percentage difference was
$100\left[\frac{\left(\frac{27.95}{89.95}\right)}{100}\right] \approx \mathbf{31.1\%.}$

22. Answers may vary. If the dress was originally priced at \$100, 60% off would result in a sale price of \$40. Then the 40% off coupon would result in a final price of $\$40 - (0.40 \cdot \$40) = \$24$. The reasoning could be applied to the actual list price of the dress.

23. If the item is \$100 then the markup is \$130. The sales price is then $\$130 - (0.30)\$130 = \$91$ which is \$9 less than the wholesale price. The store would be losing 9%.

24. $I = Prt$, where $P = \$30{,}000$, $r = 12.5\% = 0.125$, and $t = 4$ years. $I = \$30{,}000 \cdot 0.125 \cdot 4 = \mathbf{\$15{,}000.}$

25. To reduce the crust from 25% to 20% the crust is decreased by 5% which is 25% of 20%. So **it should be reduced by 25%.**

26. (a) **Irrational**. The pattern never repeats in blocks of the same length.
(b) **Irrational**. Any non-zero rational number divided by any irrational number is irrational.
(c) **Rational**. The ratio of two integers.
(d) **Rational**. The pattern repeats in blocks of 0011.
(e) **Irrational**. The pattern does not repeat.

27. (a) $\sqrt{484} = \mathbf{22}$
(b) $\sqrt{288} = \sqrt{144 \cdot 2} = \mathbf{12\sqrt{2}.}$
(c) $\sqrt{180} = \sqrt{36 \cdot 5} = \mathbf{6\sqrt{5}.}$
(d) $\sqrt[3]{162} = \sqrt[3]{27 \cdot 6} = \mathbf{3\sqrt[3]{6}.}$

28. (a) **No.** $^{-}\sqrt{2} + \sqrt{2} = 0$ is rational.
(b) **No.** $\sqrt{2} - \sqrt{2} = 0$ is rational.
(c) **No.** $\sqrt{2} \cdot \sqrt{2} = 2$ is rational.
(d) **No.** $\frac{\sqrt{2}}{\sqrt{2}} = 1$ is rational.

29. $4 < \sqrt{23} < 5$.
$(4.7)^2 = 22.09$ and $(4.8)^2 = 23.04 \Rightarrow$
$4.7 < \sqrt{23} < 4.8$;
$(4.79)^2 = 22.9441$ and $(4.80)^2 = 23.04 \Rightarrow$
$4.79 < \sqrt{23} < 4.80$;
$(4.795)^2 = 22.992025$ and $(4.796)^2 = 23.001616 \Rightarrow 4.795 < \sqrt{23} < 4.796$;
23 is closer to 23.001616 than to 22.992025, thus $\sqrt{23} \approx \mathbf{4.796}$.

30.

Number	Whole	Natural	Integer	Rational	Irrational
100	X	X	X	X	
$\sqrt{3}$					X
0	X		X	X	
$\frac{3}{8}$				X	
$^{-}17$			X	X	

CHAPTER 8

ALGEBRAIC THINKING

Assessment 8-1A: Variables

1. **(a)** Use the format for an arithmetic sequence: $a_n = a_1 + (n-1)d$, if $a_1 = 10$. Then $a_4 = 10 + (4-1)d = \mathbf{10 + 3d}$

   (b) If n is the number, then twice the number is $2n$ and 15 less than that is $\mathbf{2n - 15}$.

   (c) If n is the number, then its square is n^2 and 15 times that is $\mathbf{15n^2}$.

   (d) If n is the number, its square is n^2 and twice the number is $2n$. Their difference is $\mathbf{n^2 - 2n}$.

2. **(a)** Let n be the number:

   (*i*) Adding $\sqrt{3} \Rightarrow n + \sqrt{3}$

   (*ii*) Multiplying the sum by $7 \Rightarrow 7(n + \sqrt{3})$

   (*iii*) Subtracting $14 \Rightarrow 7(n + \sqrt{3}) - 14$

   (*iv*) Dividing the difference by $7 \Rightarrow \frac{7(n+\sqrt{3})-14}{7}$

   (*v*) Subtracting the original number $\Rightarrow \mathbf{\frac{7(n+\sqrt{3})-14}{7} - n}$.

   (b) $\frac{7n+7\sqrt{3}-14}{7} - n = n + \sqrt{3} - 2 - n = \mathbf{\sqrt{3} - 2}$.

3. **(a)** There are 4, 6, 8, and 10 shaded tiles, respectively, in the four figures. Assume and arithmetic sequence with $a_1 = 4$ and $d = 2$. Thus $a_n = 4 + (n-1)2 = \mathbf{2n + 2}$ or $\mathbf{2(n + 1)}$.

   (b) There are a total of $(n+2)^2$ squares in each figure. Assume the pattern continues; then the number of white squares is $(n+2)^2$ – the number of shaded squares, or $(n+2)^2 - (2n+2) = n^2 + 4n + 4 - 2n - 2 = \mathbf{n^2 + 2n + 2}$.

4. **(a)** Cost is **\$(50 + 60*h*)**

   (b) Let n, d, and q be the number of nickels, dimes, and quarters, respectively. It is given that $n = 3d$ and $q = 3n = 9d$. Then the value of dimes + Value of nickels + value of quarters is $10d + 5(3d) + 25(9d) = \mathbf{250d}$ **cents**

   (c) The sum of the three numbers is $x + (x+1) + (x+2) = \mathbf{3x + 3}$.

   (d) Make a table of the number of bacteria in terms of n minutes:

n minutes	Number of bacteria
1	$q \cdot 2$
2	$(q \cdot 2) \cdot 2 = q \cdot 2^2$
3	$(q \cdot 2^2) \cdot 2 = q \cdot 2^3$
⋮	⋮
n	$\mathbf{q \cdot 2^n}$

   (e) The temperature after t hours is $\mathbf{(40 - 3t)°F}$.

   (f) Pawel's salary is \$$s$ the first years; \$$(s + 5000)$ the second year; and \$$2(s + 5000)$ = \$$(2s + 10{,}000)$ the third year, for a total of **\$(4*s* + 15,000)**.

   (g) The sum of the three numbers is $x + (x+2) + (x+4) = \mathbf{3x + 6}$.

   (h) The sum of the three numbers is $(m-1) + m + (m+1) = \mathbf{3m}$.

5. If the number of students is 45 times the number of professors, then $\mathbf{S = 45P}$

6. If there are five more girls than boys, then $\mathbf{g = b + 5}$.

7. Let m be the number of matchsticks Ryan uses. Then $m_1 = 13, m_2 = 19, m_3 = 25, \ldots$. Assume an arithmetic sequence with $m_1 = 13$ and $d = 6$. Thus $m_n = 13 + (n-1) \cdot 6 = \mathbf{6n + 7}$ **matchsticks**.

8. **(a)** $P = \$8$ per hour $\times$ t hours, or $\boldsymbol{P = \$(8t)}$.

(b) \$15 is paid as a flat fee for the first hour. Afterwards $P = \$15 + \$10$ per hour. So combining these ideas,
$P = \$15 + \$10$ per hour $\times$ $(t - 1)$ hours,
or $\boldsymbol{P = \$10(t - 1) + \$15}$

9. Let r be total revenue. Then
$r = 5x + 13(100) = \boldsymbol{\$5(x) + \$1300}$.

10. **(a)** If the youngest receives \$$x$:
The eldest receives **\$3*x***;
The middle receives $\frac{1}{2}(3x) = \boldsymbol{\$\frac{3}{2}x}$.

(b) If the middle receives \$$y$:
The eldest receives **\$2*y***;
The youngest receives $\frac{1}{3}(2y) = \boldsymbol{\$\frac{2}{3}y}$.

(c) If the eldest receives \$$z$:
The middle receives $\boldsymbol{\$\frac{1}{2}}z$;
The youngest receives $\boldsymbol{\$\frac{1}{3}}z$.

11. **(a)** **(*i*)** The pattern appears to be subtracting 3 from each term to obtain its successor. At the 100th term, 3 will be subtracted 99 times. So the 100th term is
$0 - 3(99) = \boldsymbol{{}^{-}297}$.

(*ii*) The nth term can be found using the expression $a_n = a_1 + d(n - 1)$. So
$a_n = 0 - 3(n - 1) = \boldsymbol{{}^{-}3(n - 1) =}$
$\boldsymbol{3 - 3n}$.

(b) **(*i*)** The pattern appears to be adding $\frac{1}{4}$ to each term to obtain its successor. At the 100th term, $\frac{1}{4}$ will be added to $\frac{1}{2}$ 99 times. The 100th term is $\boldsymbol{25\frac{1}{4}}$.

(*ii*) The nth term can be found using the expression $a_n = a_1 + d(n - 1)$. So

$$a_n = \frac{1}{2} + \frac{1}{4}(n - 1)$$

$$a_n = \frac{1}{2} + \frac{1}{4}n - \frac{1}{4}$$

$$a_n = \frac{1}{4} + \frac{1}{4}n$$

$$a_n = \boldsymbol{\frac{1}{4}(n + 1)}.$$

(c) **(*i*)** The pattern appears to be subtracting $2\frac{1}{2}$ from each term to obtain its successor. At the 100th term, $2\frac{1}{2}$ will be subtracted from 17 99 times. The 100th term is $\boldsymbol{{}^{-}230\frac{1}{2}}$.

(*ii*)
$$a_n = 17 + \left({}^{-}2\frac{1}{2}\right)(n - 1) \Rightarrow$$

$$a_n = 17 - 2\frac{1}{2}n + 2\frac{1}{2} \Rightarrow$$

$$a_n = \boldsymbol{19\frac{1}{2} - 2\frac{1}{2}n}$$

12. **(a)** **(*i*)** Multiply each term by $\frac{3}{2}$ to obtain its successor. The 7th term is given by

$$a_1 \cdot r^6 = \frac{1}{2} \cdot \left(\frac{3}{2}\right)^6 = \frac{1}{2} \cdot \frac{729}{64} = \boldsymbol{\frac{729}{128}}.$$

(*ii*) The nth term is given by

$$\boldsymbol{a_n = \frac{1}{2}\left(\frac{3}{2}\right)^{n-1}}$$

(b) **(*i*)** Multiply each term by $\sqrt{3}$ to obtain its successor. The 7th term is given by

$$a_1 \cdot r^6 = \sqrt{3} \cdot \left(\sqrt{3}\right)^6 = \sqrt{3} \cdot \left(\sqrt{3^2}\right)^3$$

$$= \sqrt{3} \cdot 3^3 = \sqrt{3} \cdot 27 = \boldsymbol{27\sqrt{3}}.$$

(*ii*) The nth term is given by

$$a_n = \sqrt{3} \cdot \left(\sqrt{3}\right)^{n-1} = \boldsymbol{\left(\sqrt{3}\right)^n}$$

(c) **(*i*)** Multiply each term by $-\frac{3}{2}$ to obtain its successor. The 7th term is given by

$$a_1 \cdot r^6 = -\frac{1}{2} \cdot \left(-\frac{3}{2}\right)^6 = -\frac{1}{2} \cdot \frac{729}{64} = \boldsymbol{-\frac{729}{128}}.$$

(*ii*) The nth term is given by

$$\boldsymbol{a_n = -\frac{1}{2}\left(-\frac{3}{2}\right)^{n-1}}$$

13. If the 100th term is 27 and the 200th term is 98 and the sequence is arithmetic, then the difference is:

$a_1 + d(100 - 1) = 27 \Rightarrow a_1 = 27 - 99d$

$a_1 + d(200 - 1) = 98 \Rightarrow a_1 = 98 - 199d$

$\Rightarrow 27 - 99d = 98 - 199d$

$\Rightarrow 100d = 98 - 27$

$\Rightarrow d = \dfrac{71}{100}$

Now, use this result to find the first term:

$a_1 + 99\left(\dfrac{71}{100}\right) = 27$

$a_1 = 27 - \dfrac{7029}{100} \Rightarrow a_1 = \dfrac{2700 - 7029}{100}$.

$a_1 = -\dfrac{4329}{100}$

14. If the 11th term is $^-128$ and the ratio is $\dfrac{^-1}{3}$
then the first term is

$^-128 = a_1\left(\dfrac{^-1}{3}\right)^{10} \Rightarrow {}^-128 = a_1\left(\dfrac{1}{59049}\right)$

$a_1 = {}^-128(59,049) \Rightarrow \boldsymbol{a_1 - 7{,}558{,}272}$

15. $10 - 0.20(10) = 8;\ 8 - 0.20(8) = 6.4;$

$6.4 - 0.20(6.4) = 5.12;\ 5.12 - 0.20(5.12)$
$= 4.096;$

$4.096 - 0.20(4.096) = 3.2768$ or about **3.3 feet.**

16. Starting with two hours and increasing by 20 minutes each day can be represented with an arithmetic sequence with $a_1 = 2$ and $d = \dfrac{1}{3}$.

The nth term is $2 + \dfrac{1}{3}(n - 1)$. Set the nth term equal to 12 hours and solve for n.

$2 + \dfrac{1}{3}(n - 1) = 12 \Rightarrow 2 + \dfrac{1}{3}n - \dfrac{1}{3} = 12$

$\Rightarrow 6 + n - 1 = 36 \Rightarrow 5 + n = 36 \Rightarrow \boldsymbol{n = 31}$

After 31 days Jake will be able to wear his contacts for 12 hours.

17. Answers will vary

18. (a) The Fibonacci sequence is 1, -1, 0, -1, -1, -2, -3, -5, -8, -13, -21, -34, …

(b) $F_4 = F_3 + F_2$; $F_5 = F_4 + F_3$; and so on; so

$\boldsymbol{F_n = F_{n-1} + F_{n-2}}$, **for all $n = 3, 4, 5, …$**

Assessment 8.2A: Equals Relation and Equations

1. The balance scales imply addition, so:

$\triangle + \square = 12.$

$O + O + \triangle + \square = 18;$

$2O + 12 = 18;$ so

$O = 3$.

$2O + \square = 10;$

$6 + \square = 10;$ so

$\square = 4.$

$\triangle + \square = 12;$

$\triangle + 4 = 12;$ so

$\triangle = 8.$

2. (a) The bar diagram below illustrates

$3x - 11 = 13$

X	X	X
13		11

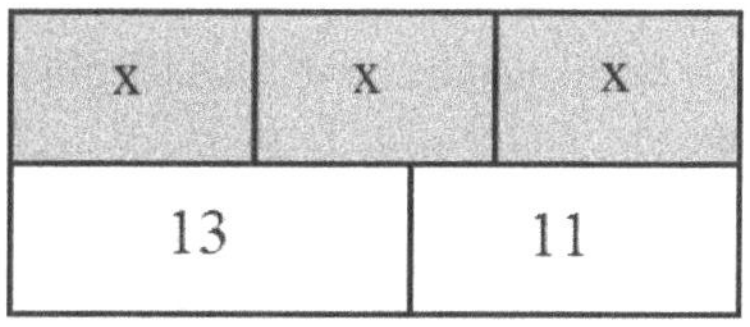

So $3x = 13 + 11 \Rightarrow 3x = 24 \Rightarrow \boldsymbol{x = 8}$.

(b) The bar diagram below illustrates

X	X	5
13		

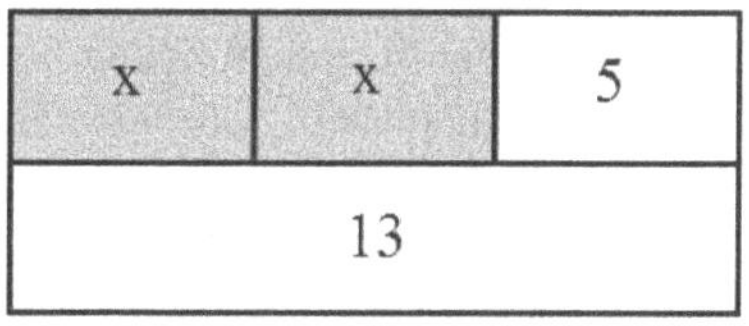

Note we can redraw the bar graph as follows:

X	X	5
8		5

13

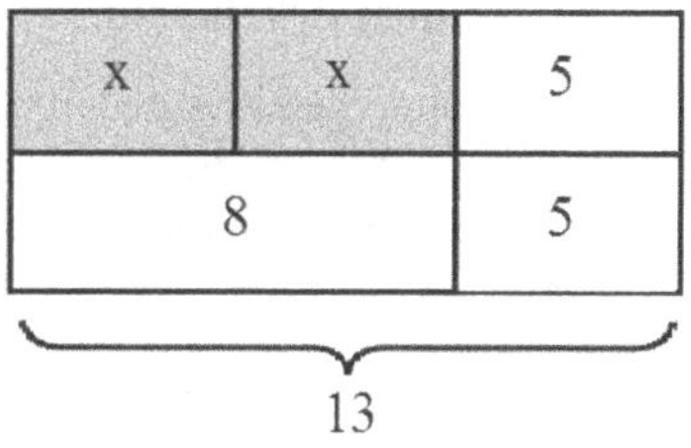

So $2x = 8 \Rightarrow x = 4$.

3. (a) $x + 3 = 2 - x \Rightarrow 2x = -1 \Rightarrow \boldsymbol{x = -\dfrac{1}{2}}$

(b) $\dfrac{3}{2}x - 3 = x + 2 \Rightarrow \dfrac{3}{2}x - x = 2 + 3 \Rightarrow$

$\dfrac{1}{2}x = 5 \Rightarrow \boldsymbol{x = 10}$

(c) $5\left(2x+\sqrt{2}\right)+7\left(2x+\sqrt{2}\right)=12\sqrt{2} \Rightarrow$

$12(2x+\sqrt{2}) = 12\sqrt{2} \Rightarrow$

$2x+\sqrt{2} = \sqrt{2} \Rightarrow$

$2x = 0 \Rightarrow \mathbf{x = 0}$.

(d) $3(x-3) = 5(x-3)$

$3(x-3) - 5(x-3) = 0 \Rightarrow$

$^{-}2(x-3) = 0 \Rightarrow$

$x - 3 = 0 \Rightarrow \mathbf{x = 3}$

(e) $(x-2)^2 = 9$

$(x-2)(x-2) = 9 \Rightarrow$

$x^2 - 4x + 4 = 9 \Rightarrow$

$x^2 - 4x - 5 = 0 \Rightarrow$

$(x-5)(x+1) = 0 \Rightarrow$

$\mathbf{x = 5}$ **or** $\mathbf{x = {}^{-}1}$

4. Let m be the number of matchsticks Ryan uses. Then $m_1 = 4, m_2 = 7, m_3 = 10, ..., m_n = 67$. Assume an arithmetic series with $m_1 = 4, d = 3$, and $m_n = 67$.

Thus $67 = 4 + (n-1)\cdot 3 \Rightarrow 67 = 4 + 3n - 3 \Rightarrow 66 = 3n \Rightarrow n =$ **22 squares.**

5. If s is the number of student tickets sold, then $812 - s$ is the number of nonstudent tickets. Adding their values gives the total amount taken in:

$\$2\cdot s + \$3\cdot(812 - s) = \$1912$

$2s + 2436 - 3s = 1912$

$\Rightarrow s =$ **524 student tickets sold.**

6. Let e be the amount the eldest receives,
m be the amount the middle sibling receives,
y be the amount the youngest receives.

Then $e = 3y$ and $m = y + 14{,}000$

So
$$e + m + y = 486{,}000$$
$$3y + (y + 14{,}000) + y = 486{,}000$$
$$5y + 14{,}000 = 486{,}000$$
$$5y = 472{,}000$$

Thus $\mathbf{y = \$94{,}400}$

$\mathbf{m = \$108{,}400}$

$\mathbf{e = \$283{,}200}$

7. Let e be the length of the equal length pieces.

Then $2e + (e-3) = 120$ inches

$3e - 3 = 120$

$3e = 123 \Rightarrow e = 41$

So the equal length pieces are **41 inches** and the short piece is **38 inches** (disregarding the width of the saw cuts).

8. Let d be the number of dimes; then $67 - d$ is the number of nickels. $0.10d$ is the amount of money in dimes and $0.05(67 - d)$ is the amount of money in nickels.

Thus $0.10d + 0.05(67 - d) = 4.20$, or

$10d + 5(67 - d) = 420$

$10d + 335 - 5d = 420$

$5d = 85 \Rightarrow \quad d =$ **17 dimes**

$67 - d =$ **50 nickels.**

9. Let m be Miriam's age now; then $m - 10$ is Ricardo's age now and $m - 2$ was Miriam's age two years ago. Thus

$m - 2 = 3(m - 10) \Rightarrow m - 2 =$

$m - 2 = 3m - 30 \Rightarrow 28 = 2m$, or

Miriam is 14

Ricardo is 4.

10. Let g be the number of graduate students; $20g$ is the number of undergraduates. Then

$g + 20g = 21{,}000$

$21g = 21{,}000 \Rightarrow$

1000 graduate students.

11. Let the perpendicular sides to the river be of length a; then the parallel side is of length $b = 2a$. Thus $a + 2a + a = 1200$, $4a = 1200 \Rightarrow a = 300$. The perpendicular sides are then **300 yards**; the parallel side is **600 yards.**

12. In the given sequence $d = 3$, thus

$n + (n+3) + (n+6) = 903$

$3n + 9 = 903$

$3n = 894$

$n = 298$

The three terms are 298, 301, and 304.

13. Since he used 360 feet of fencing and the backyard is a square, all sides are the same length. $360 \div 4 =$ **90 feet.**

14. Let the smallest of the four consecutive numbers be represented by the variable x.

Therefore
$x + (x + 1) + (x + 2) + (x + 3) = 598.$

$\Rightarrow 4x + 6 = 598 \Rightarrow 4x = 592 \Rightarrow$

$x = 148.$

Therefore the four numbers are **148, 149, 150, 151**.

Review Problems

17. If the middle even number is $2n$, then the next two are $2n + 2$ and $2n + 4$; the previous two are $2n - 4$ and $2n - 2$. Their sum is $2n - 4 +$
$2n - 2 + 2n + 2n + 2 + 2n + 4 = \mathbf{10n}$.

19. **(a)** $P = 30 + (30 + d) + ((30 + d) + d)$
$= \mathbf{\$90 + 3d}.$

(b) $\$(d + 2d + 4d + 8d) = \mathbf{\$15d}$.

21. The sixth term is given by

$$\sqrt{5}(0.5)^5 = \sqrt{5} \cdot \left(\frac{1}{2}\right)^5 = \sqrt{5}\left(\frac{1}{32}\right) = \mathbf{\frac{\sqrt{5}}{32}}.$$

Assessment 8-3A: Functions

1. Where x is the first element in each ordered pair and $f(x)$ is the second:

(a) **Multiply** the input number by $^-2$;
i.e., $f(x) = {}^-2x$.

(b) **Add 6** to the input number;
i.e., $f(x) = x + 6$.

(c) **Square the input; i.e.** $f(x) = x^2$

2. **(a)** **Not a function**. The element 1 is paired with both a and d.

(b) **A function**. Each element in the first set is paired with a unique element from the second.

(c) **A function**. Each distinct element in the first set is paired with an element from the second.

(d) **A function**. Each distinct element in the first set is paired with an element from the second.

3. **(a)** Answers may vary; e.g.:

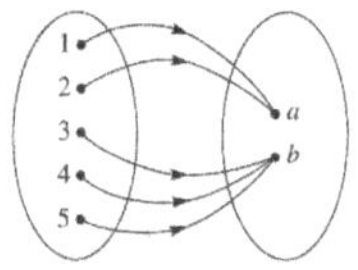

(b) Each of the five elements in the domain have two choices for a pairing $\Rightarrow 2^5 = 32$ possible functions. However, two of these 32 do not use both range items: ((1,a), (2,a), (3,a) and (1,b), (2,b), (3,b)) only use a or b. Therefore, there are **30 possible functions**.

4. **(a)**

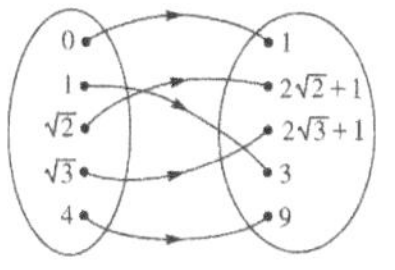

(b) $\{(0, 1), (1, 3), (\sqrt{2}, 2\sqrt{2} + 1), (\sqrt{3}, 2\sqrt{3} + 1), (4, 9)\}$.

(c)

x	$f(x)$
0	1
1	3
$\sqrt{2}$	$2\sqrt{2} + 1$
$\sqrt{3}$	$2\sqrt{3} + 1$
4	9

(d)

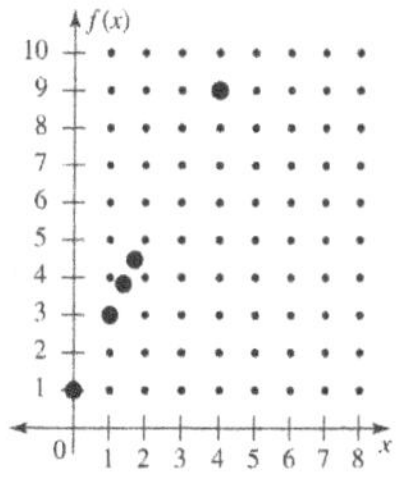

5. **(a)** **Function from $\mathbb{R}$ to {2}**, a subset of R.

$f(x) = 2$ is called a constant function and its only output is 2.

(b) **Function from $\mathbb{R}$ to $\{x \mid x \geq 0\}$**, a subset of R. $f(x) = \sqrt{x}$ has output of nonnegative numbers only because $\sqrt{x}$ is the principal square root.

(c) **Not a function** because $(4, {}^-2)$ and $(4, 2)$ both satisfy the relation, but inputs can have only one associated output.

(d) Function from $\mathbb{R}$ to $\mathbb{R}$.

$x = y^3 \Rightarrow y = \sqrt[3]{x}$ has output of all real numbers only because $\sqrt[3]{x}$ is the cubed root of all real numbers.

6. (a) (*i*)

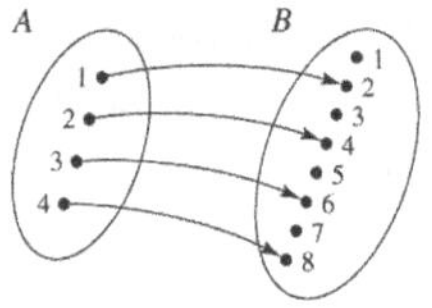

(*ii*)

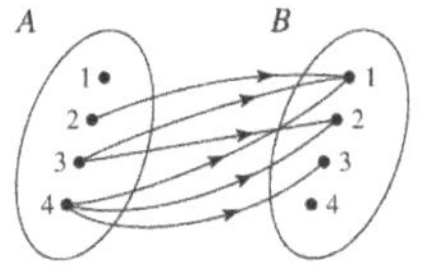

(b) (*i*) A function from A to B because each element in A corresponds to one and only one element in B. The range is $\{2, 4, 6, 8\}$.

(*ii*) Not a function. Elements 3 and 4 in A correspond to more than one element in B; element 1 in A does not correspond to any element in B.

7. This is an example of **step function**; note in the following graph that a child weighing exactly 30, 32, 34,... pounds uses the lower dosage at the break point:

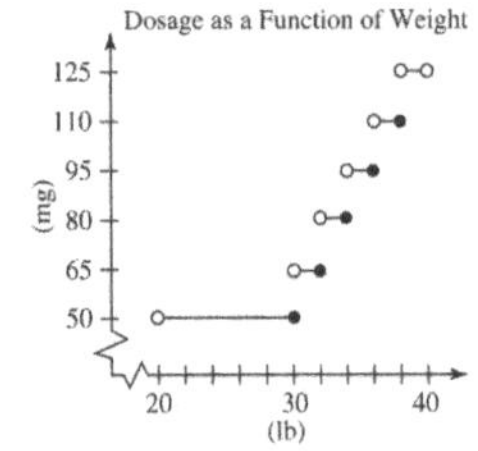

8. Answers may vary.

(a) $L(n)$ could be an arithmetic sequence with $a_3 = 8$ and $d = 3$. Then $8 = a_1 + (3 - 1) \cdot 3 \Rightarrow a_1 = 2$.

$L(n) = 2 + (n - 1) \cdot 3 = \mathbf{3n - 1}$.

(b) $L(n)$ could be obtained by squaring n and adding 1, or $\boldsymbol{L(n) = n^2 + 1}$.

9. (a)

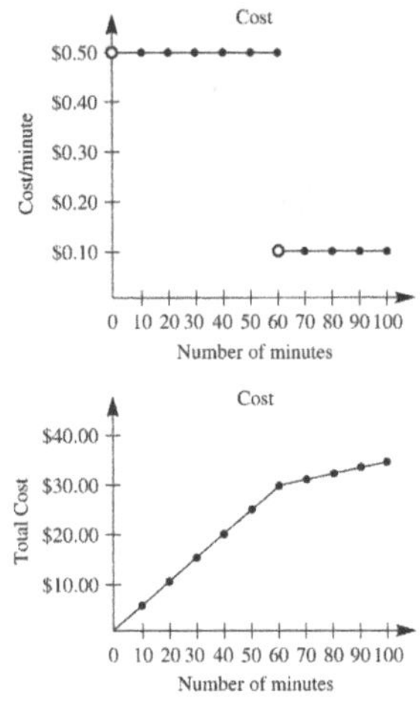

(b) It must be assumed that the company charges only for the **exact fraction of minutes used.**

(c) The two segments represent the **two different rates** per minute; the steeper one comes from the 50¢ per minute charge for the first
60 minutes.

(d) If $c(t)$ is the cost in dollars as a function of time t in minutes, then

$$c(t) = \begin{cases} \mathbf{0.50}t & \textbf{if } t \leq \mathbf{60} \\ \mathbf{30 + 0.10}(t - \mathbf{60}) & \textbf{if } t > \mathbf{60} \end{cases}$$

Where $30 is the cost of the first 60 minutes.

10. Let n be the elements of the domain:

(a) (*i*) Assume an arithmetic sequence with $a_1 = 3$ and $d = 5$. Thus

$f(n) = 3 + (n - 1) \times 5 = \mathbf{5n - 2}$.

(*ii*) $a_{20} = f(20) = 5(20) - 2 \Rightarrow f(20) = 98$.

So the sum of the first 20 terms in the sequence is

$$S(n) = \frac{n(a_1 + a_n)}{2} \Rightarrow S(20) = \frac{20(3 + 98)}{2} \Rightarrow$$

$$S(20) = \frac{20(101)}{2} = 10(101) = \mathbf{1010}.$$

(b) (*i*) The output is obtained if 3 is raised to the *nth* power, so $f(n) = 3^n$.

(*ii*) $S(n) = a_1\dfrac{1-r^n}{1-r} \Rightarrow S(20) = 3\dfrac{1-3^{20}}{1-3} \Rightarrow$

$S(20) = \dfrac{3-3^{21}}{-2} = \dfrac{3^{21}-3}{2}$.

(c) (*i*) A constant function, so $f(n) = 3$.

(*ii*) $S(20) = \dfrac{20(3+3)}{2} = 10(3+3) = \mathbf{60}$.

(d) (*i*) Assume a geometric sequence with

$a_1 = 3$ and $r = 3^{\frac{1}{3}}$. Thus

$f(n) = 3\left(3^{\frac{1}{3}}\right)^{n-1} = 3^1 3^{\frac{1}{3}n-\frac{1}{3}} = 3^{\frac{1}{3}n+\frac{2}{3}}$

$= 3^{\frac{n+2}{3}}$

(*ii*) $S(20) = 3\dfrac{1-\left(3^{\frac{1}{3}}\right)^{20}}{1-3^{\frac{1}{3}}} = \dfrac{3\left(1-3^{\frac{20}{3}}\right)}{1-3^{\frac{1}{3}}}$

11. **(a)** $(g \circ f)(6) = g[f(6)] = f(6) - 5 = 7(6) - 5 = \mathbf{37}$

(b) $(g \circ f)(10) = g[f(10)] = f(10) - 5 = 7(10) - 5 = \mathbf{65}$.

(c) $(g \circ f)(\sqrt{10}) = g[f(\sqrt{10})] = f(\sqrt{10}) - 5 = \mathbf{7\sqrt{10} - 5}$.

(d) $(g \circ f)(0) = g[f(0)] = f(0) - 5 = {}^{-}\mathbf{5}$.

(e) $(g \circ f)(\frac{5}{8}) = g[f(\frac{5}{8})] = f(\frac{5}{8}) - 5 = 7(\frac{5}{8}) - 5 = \frac{35}{8} - \frac{40}{8} = \mathbf{\frac{^{-}5}{8}}$.

(f) $(g \circ f)(n) = g[f(n)] = f(7n) - 5 = \mathbf{7n - 5}$.

12. $7n - 5 = 2n \Rightarrow 5n = 5 \Rightarrow \mathbf{n = 1}$.

13. **(a)** **(i)** $(f \circ g)(6) = f[g(6)] = f[6 - 5] = f(1) = 7(1) = 7$.

(ii) $(f \circ g)(10) = f[g(10)] = f(10 - 5) = f(5) = 7(5) = \mathbf{35}$.

(iii) $(f \circ g)(\sqrt{10}) = f[g(\sqrt{10})] = f(\sqrt{10} - 5) = 7(\sqrt{10} - 5) = \mathbf{7\sqrt{10} - 35}$.

(iv) $(f \circ g)(0) = f[g(0)] = f(0 - 5) = f({}^{-}5) = ({}^{-}5)(7) = {}^{-}\mathbf{35}$.

(v) $(f \circ g)(\frac{5}{8}) = f[g(\frac{5}{8})] = f(\frac{5}{8} - 5) = f(\frac{5}{8} - \frac{40}{8}) = f(\frac{^{-}35}{8}) = \frac{^{-}35}{8} \cdot 7 = -\mathbf{\frac{245}{8}}$.

(vi) $(f \circ g)(n) = f[g(n)] = f(n - 5) = 7(n - 5) = \mathbf{7n - 35}$.

(b) $(f \circ g)(x) = x$

$7x - 35 = x \Rightarrow 6x = 35$

$x = \mathbf{\frac{35}{6}}$.

14. **(a)** $f(f(0)) = f({}^{-}0 + b) = {}^{-}({}^{-}0 + b) + b = \mathbf{0}$.

(b) $f(f(2) = f({}^{-}2 + b) = {}^{-}({}^{-}2 + b) + b = \mathbf{2}$.

(c) $f(f(x)) = f({}^{-}x + b) = {}^{-}({}^{-}x + b) + b = x$. This works for **all real numbers**.

15. **(a)** (*i*) **Yes**. $4n - 3 = 1 \Rightarrow t(1) = 1$.

(*ii*) **Yes**. $4n - 3 = 385 \Rightarrow t(97) = 385$.

(*iii*) **Yes**. $4n - 3 = 389 \Rightarrow t(98) = 389$.

(*iv*) **No**. If $4n - 3 = 392$, $n = \frac{395}{4}$ which is not a natural number.

(b) (*i*) **No**. 0 is not a natural number.

(*ii*) **Yes**. $n^2 = 25 \Rightarrow t(5) = 25$.

(*iii*) **Yes**. $n^2 = 625 \Rightarrow t(25) = 625$.

(*iv*) **No**. If $n^2 = 1000$, $n = \sqrt{1000}$ which is not a natural number.

(*v*) **No**. If $n^2 = 90$, $n = \sqrt{90}$, which is not a natural number.

(c) (*i*) **Yes**. $n(n - 1) = 0 \Rightarrow t(1) = 0$.

(*ii*) **Yes**. $n(n - 1) = 2 \Rightarrow t(2) = 2$.

(*iii*) **Yes**. $n(n-1) = 5 \Rightarrow t(5) = 20$.

(*iv*) **No**. If $n(n-1) = 999 \Rightarrow t(32.11) \approx 999$, but 32.11 is not a natural number.

16. **(a)** $f(x) = x^2 \Rightarrow x^2 = 2 \Rightarrow x = \pm\sqrt{2}$.

(b) $f(x) = x^3 \Rightarrow x^3 = 2 \Rightarrow x = \sqrt[3]{2}$.

17. **(a)** (*i*) $(1,7) \Rightarrow 2 \cdot 1 + 2 \cdot 7 = \mathbf{16}$.

(*ii*) $(2,6) \Rightarrow 2 \cdot 2 + 2 \cdot 6 = \mathbf{16}$.

(*iii*) $(6,2) \Rightarrow 2 \cdot 6 + 2 \cdot 2 = \mathbf{16}$.

(b) If output (or perimeter) $= 20$ then $2\ell + 2w = 20 \Rightarrow 2(\ell + w) = 20$
$\Rightarrow \ell + w = 10$ or $w = 10 - \ell$

The possibilities are any ordered pair **$(l, 10 - l)$** with $0 < l < 10$, where l is a natural number: {(1,9), (2,8), (3,7), (4,6), (5,5) (6,4), (7,3), (8,2), (9,1)}.

(c) Domain: ℓ and w are any natural number

Range: any even natural number greater than or equal to 4.

18. **(a)** There were 600 cars on the road at 6:30; there were (as nearly as can be determined from the graph) 650 cars at 7:00 $\Rightarrow 650 - 600 =$ **50 cars** increase.

(b) The graph rises most steeply **between 6:00 and 6:30**; i.e., that is the period in which the increase in number of cars was greatest (by 200 cars).

(c) The graph is flat between 7:30 and 8:30; i.e., there was **no increase** in the number of cars.

(d) Only one period of the graph shows a drop: 700 cars at **8:30,** decreasing to 600 at **9:00,** or a decrease of $700 - 600 =$ **100 cars.**

(e) Segments are used because the data are continuous rather than discrete; e.g., there is a unique number of cars at 5:47 A.M. The assumption of a linear increase or decrease in traffic between hours may, however, be invalid and could lead to erroneous conclusions.

19. **(a)** (*i*) $H(2) = 128(2) - 16(2)^2$
$=$ **192 feet.**

(*ii*) $H(6) = 128(6) - 16(6)^2$
$=$ **192 feet.**

(*iii*) $H(3) = 128(3) - 16(3)^2$
$=$ **240 feet.**

(*iv*) $H(5) = 128(5) - 16(5)^2$
$=$ **240 feet.**

Some of the heights correspond to the height of the ball as it rises; some to its height as it falls.

(b) Table:

t	$H(t) = 128t - 16t^2$
0	0
1	112
2	192
3	240
4	256
5	240
6	192
7	112
8	0

Graph:

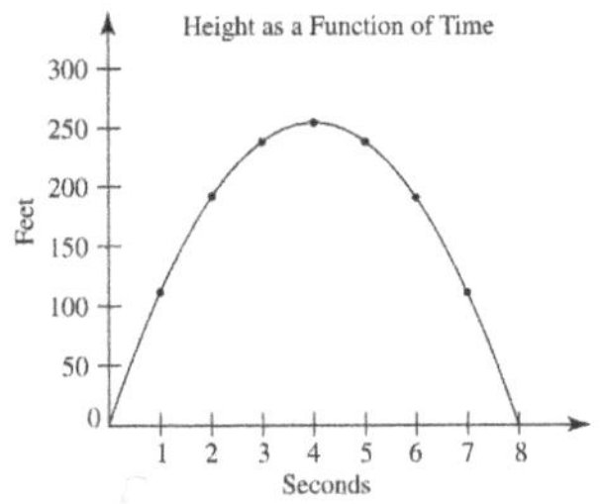

$H(t) = 0$ at $t = 8$; i.e., the ball will hit the ground in **8 seconds**. Note that $H(t) = 0$ also at $t = 0$.

(c) Domain: $\{t \mid 0 \le t \le 8\}$ **seconds.**

20. The total number of matchsticks are 7, 17, 31,... In a table write the number of matchsticks vertically and horizontally:

n	vertical	horizontal	total
1	4	3	7
2	9	8	17
3	16	15	31

The number of vertical matchsticks in each shape are square numbers: 4, 9, 16, … = $(n+1)^2$. The horizontal match sticks are $3\cdot 1, 4\cdot 2, 5\cdot 3, \ldots : (n+2)n$ So $T(n) = \mathbf{(n+1)^2 + (n+2)n}$.

21. 1, 4, 16. …It is a geometric sequence with $a_1 = 1$ and $r = 4$. Then $a_n = \mathbf{4^{n-1}}$.

22. **(a)** **A function**. $x + y = 2 \Rightarrow y = 2 - x$; for any input x, y is unique.

(b) **Not a function.** $x - y < 2 \Rightarrow y > x - 2$; for any input x, y may be any value greater than $x - 2$.

(c) **A function**. y is unique for any input x.

(d) **A function**. $xy = 2 \Rightarrow y = \frac{2}{x}$; for any input x (except $x = 0$, for which y is undefined), y is unique.

(e) **A function**. y is unique for any input x.

(f) $|y| = x$ is **not a function** from x to y, $x = 1 \Rightarrow y = \pm 1$ which is not unique.

(g) $x^2 + y^2 = 1$ is **not a function** from x to y, $x = 0 \Rightarrow y = \pm 1$ which is not unique.

(h) **A function**. y is unique for any input x.

23. **(a)** **A function**. There is only one value of y for each value of x.

(b) **Not a function**. When $x = 1$ there are many values of y.

(c) **A function**. There is only one value of y for each value of x.

Review Problems

21. The variable is time, denoted t. Since distance $=$ rate $\times$ time, the fast car will travel $70t$ miles in the same time as the slow car travels $60t$ miles. Then $70t - 40 = 60t \Rightarrow 10t = 40 \Rightarrow \mathbf{t = 4}$ **hours**.

23. Since it is a Fibonacci-like sequence, the terms are found by adding the two previous terms. S the first ten terms are **-7, -2, -9, -11, -20, -31, -51, -82, -133 and -215**

Assessment 8-4A: Equations in a Cartesian Coordinate System

1. The graph of $y = mx + 3$ contains the point (0, 3) and is parallel to the line $y = mx$. Similarly, the graph of $y = mx - 3$ contains the point $(0, {}^{-}3)$ and is parallel to $y = mx$ (see below).

(a) Parallel line; y-intercept $= 3$.

(b) Parallel line; y-intercept $= {}^{-}3$.

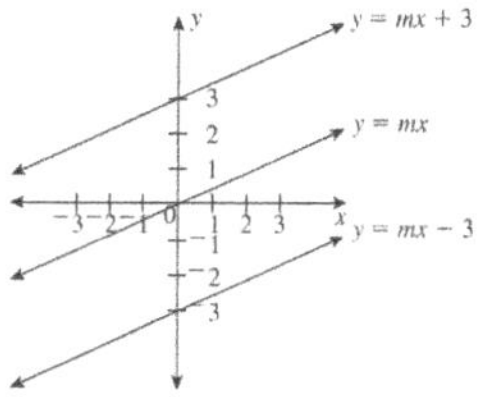

(c) The graph of $y = 3mx$ has a slope three times m and the same y-intercept as $y = mx$.

(d) The graph of $y = {}^{-}3mx$ has a slope negative three times m and the same y-intercept as $y = mx$.

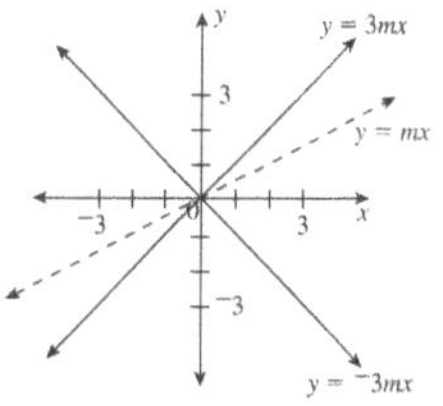

2. Each of the following is in the form $y = mx + b$, where m represents the slope of the line and b represents the y-coordinate of the y-intercept. The x-intercept in each case can be found by setting y equal to 0 and solving for x.

(a) Given $y = \frac{{}^{-}3}{4}x + 3$:

(i) $m = \frac{{}^{-}3}{4}$ and $b = 3$.

(ii) If $y = 0 = \frac{{}^{-}3}{4}x + 3 \Rightarrow x = 4$; the x-intercept is at (4, 0).

(iii) Draw a line through (0, 3) and (4, 0).

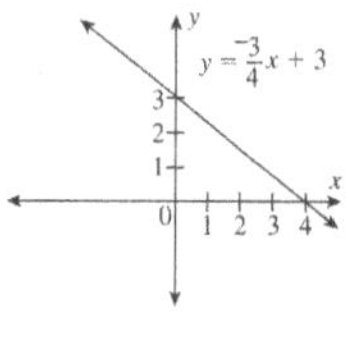

(b) Given $y = {}^{-}3$:

(i) $m = 0$ and $b = {}^{-}3$.

(ii) There is no x-intercept; i.e., $y = {}^{-}3$ is a horizontal line.

(iii) Draw a line through $(0, {}^{-}3)$ parallel to the x-axis.

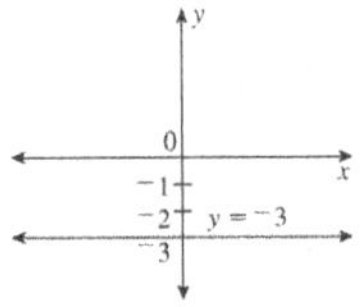

(c) Given $y = 15x - 30$:

(i) $m = 15$ and $b = {}^{-}30$.

(ii) If $y = 0 = 15x - 30 \Rightarrow x = 2$.

(iii) Draw a line through $(0, {}^{-}30)$ and $(2, 0)$.

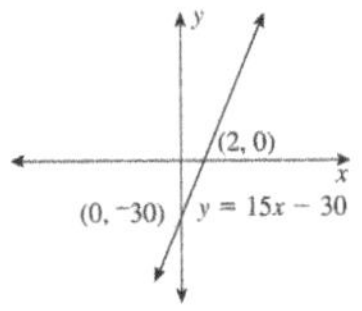

3. When the equations in Problem 2 are in the form $y = mx + b$, b is the y-intercept and is located at $(0, b)$. The x-intercept may be found by setting y equal to 0 and solving for x. If x is some value a, the x-intercept is located at $(a, 0)$.

(a) $y = \frac{^{-}3}{4}x + 3$:

(i) $b = 3$. y-intercept is at **(0, 3)**.

(ii) If $y = 0 = \frac{^{-}3}{4}x + 3 \Rightarrow x = 4$.
x-intercept is at **(4, 0)**.

(b) $y = {}^{-}3$:

(i) $b = {}^{-}30$. y-intercept is at **$(0, {}^{-}3)$**.

(ii) There is **no x-intercept**; i.e., $y = {}^{-}3$ is a horizontal line parallel to the x-axis.

(c) $y = 15x - 30$.

(i) $b = {}^{-}30$. y-intercept is at **$(0, {}^{-}30)$**.

(ii) If $y = 0 = 15x - 30 \Rightarrow x = 2$.
x-intercept is at **(2, 0)**.

4. From example 20, we have the equation $F = \frac{9}{5}C + 32$. To find out when $F = C$, substitute C for F:

$$C = \frac{9}{5}C + 32$$
$$5C = 5\left(\frac{9}{5}C + 32\right)$$
$$5C = 9C + 160$$
$$-4C = 160$$
$$C = -40.$$

The ordered-pair **(−40, −40)** represents where degrees Celsius and degrees Fahrenheit are the same. Note you could have used the other formula from example 20, $C = \frac{5}{9}(F - 32)$, substituted F in for C, and obtained the same result.

5. When the equations are in slope-intercept form, $y = mx + b$, m is the slope of the line and b is the y-intercept, located at $(0, b)$.

(a) *(i)* $3y - x = 0 \Rightarrow 3y = x \Rightarrow$
$\boldsymbol{y = \frac{1}{3}x}$.

(ii) $m = \frac{1}{3}$; y-intercept is at **(0, 0)**.

(b) *(i)* $x + y = 3 \Rightarrow \boldsymbol{y = {}^{-}x + 3}$.

(ii) $m = {}^{-}\mathbf{1}$; y-intercept is at **(0, 3)**.

(c) *(i)* $x = 3y \Rightarrow \boldsymbol{y = \frac{1}{3}x}$.

(ii) $m = \frac{1}{3}$; y-intercept is at **(0, 0)**.

6. **(a)** $m = \frac{^{-}2-3}{1-^{-}4} = \frac{^{-}5}{5} = {}^{-}1$, thus
$y = {}^{-}x + b$.

Substituting x and y from $({}^{-}4, 3)$: $(3) = {}^{-}({}^{-}4) + b$, thus $b = {}^{-}1$.

So $\boldsymbol{y = {}^{-}x - 1}$.

(b) $m = \frac{1-0}{2-0} = \frac{1}{2}$, thus $y = \frac{1}{2}x + b$. The line goes through the origin, thus $b = 0$.
So $\boldsymbol{y = \frac{1}{2}x}$.

(c) $m = \dfrac{0 - \frac{^{-}1}{2}}{\frac{1}{2} - 0} = \dfrac{\left(\frac{1}{2}\right)}{\left(\frac{1}{2}\right)} = 1$, thus
$y = x + b$.

Substituting x and y from $\left(\frac{1}{2}, 0\right)$: $(0) = \left(\frac{1}{2}\right) + b$, thus $b = \frac{^-1}{2}$.

So $\mathbf{y = x - \frac{1}{2}}$.

(d) $m = \dfrac{^-3 - {}^-3}{^-1 - 0} = 0$, thus $y = 0x + b$.

$^-3 = 0x + b \Rightarrow b = {}^-3$

So $\mathbf{y = -3}$.

7. Answers may vary.

(a) Both points include the coordinate $y = 2$, thus the equation of the line is $y = 2$. Other points could be $(^-3, 2), (3, 2), (5, 2), \ldots$.

(b) Both points include the coordinate $x = 0$, thus the line is the y-axis. Other points could be $(0, {}^-1), (0, 2), (0, 3), \ldots$.

8. **(a)** $(^-2, 2), (^-2, 1)$, and (x, y) are collinear, thus the value of x at each point is $^-2$.

(i) $\mathbf{x = {}^-2}$ **is the equation of the line.**

(ii) y may be **any real number**.

(b) ***(i)*** The fourth quadrant implies that $\mathbf{x > 0}$ and $\mathbf{y < 0}$.

(ii) x is any real number greater than zero, and y is any real number less than zero.

9. The rectangle is shown below:

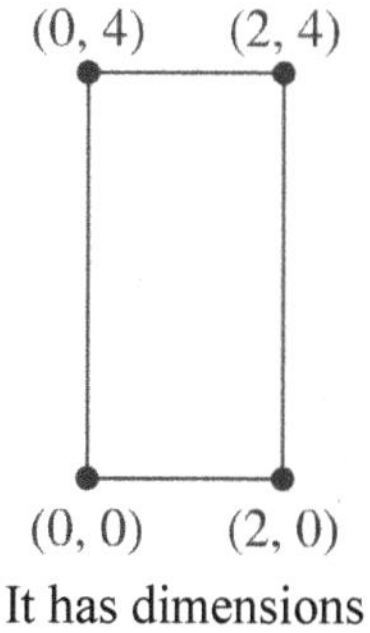

It has dimensions of $2 - 0 = 2$ and $4 - 0 = 4$.

(i) Area $= 2 \cdot 4 =$ **8 square units.**

(ii) Perimeter $= 2 \cdot 2 + 2 \cdot 4 =$ **12 units.**

10. **(a)** $\mathbf{x = 3}$, a vertical line through $(3, 0)$.

(b) $\mathbf{y = 5}$, a horizontal line through $(^-4, 5)$.

11. In each case, slope $(m) = \frac{\text{rise}}{\text{run}} = \frac{y_2 - y_1}{x_2 - x_1}$.

(a) $m = \frac{0-3}{^-5-4} = \frac{^-3}{-9} = \mathbf{\frac{1}{3}}$.

(b) $m = \frac{2-2}{1-\sqrt{5}} = \frac{0}{1-\sqrt{5}} = \mathbf{0}$.

(c) $m = \frac{b-a}{b-a} = \mathbf{1}$ (if $b \neq a$).

(d) $m = \frac{^-2-2}{\sqrt{5}-\sqrt{5}} = \frac{^-4}{0}$ does not exist.

12. In each case where m is known, substitute x and y values into the equation $y = mx + b$ and solve for b.

(a) Use the point (4,3). $(3) = \frac{1}{3}(4) + b \Rightarrow b = \frac{5}{3}$, thus $\mathbf{y = \frac{1}{3}x + \frac{5}{3}}$.

(b) $\mathbf{y = 2}$ (a horizontal line).

(c) Use the point $(a, a) \Rightarrow (a) = 1(a) + b \Rightarrow b = 0$, thus $\mathbf{y = x}$ (if $b \neq a$).

(d) $\mathbf{x = \sqrt{5}}$.

13. Answers may vary slightly from those below, depending on estimates from the fitted line; e.g.:

(a) From the fitted line, estimate point coordinates of $(50, 10)$ and $(80, 40)$. $m = \frac{40-10}{80-50} = \frac{30}{30} = 1$.

Use the point $(50, 10)$ and substitute $T = 50$ and $C = 10$ into $C = 1T + b$ (i.e., an equation of the form $y = mx + b$). $(10) = 1(50) + b \Rightarrow b = {}^-40$. Therefore, based on these points the equation is $\mathbf{C = T - 40}$.

(b) **If** $T = 90°$, then $C = (90) - 40 =$ **50 chirps** per 15-second interval.

(c) $N = 4C$; i.e., there are four 15-second intervals in one minute. $N = 4(T - 40)$, or $\mathbf{N = 4T - 160}$.

14. **(a)** $y = 2x - 20$ **is line** $\overleftrightarrow{\mathbf{BC}}$ because the y-intercept is $(0, {}^-20)$ and it has positive slope. So $y = 4 - 2x$ **is line** $\overleftrightarrow{\mathbf{AB}}$, as its y-intercept is (0,4) and it has negative slope.

(b) Point D has the same x-coordinate as point C (Point C is the x-intercept of line $\overleftrightarrow{BC}$), which is 10. Point D lies on $\overleftrightarrow{AB}$, so $y = 4 - 2(10) = {}^-16$. Point D has coordinates $\mathbf{(10, -16)}$.

15. **(a)** The lines have the same x-intercept, $^-\mathbf{3}$.

(b) The lines have the same x-intercept, 1.

(c) The lines have the same slope, $^-\mathbf{2}$.

(d) The lines have the same slope, $^-\mathbf{1}$.

16. **(a)** Graphs may vary.

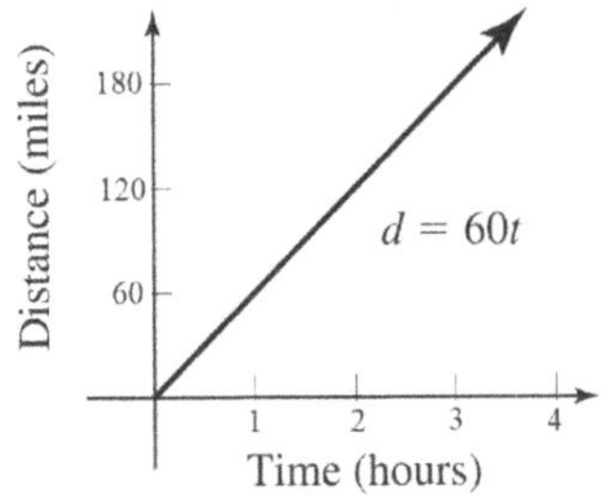

(b) Use the points $(0,0)$ and $(1,60)$. Then

$m = \frac{60-0}{1-0} = \mathbf{60}$.

17. **(a)** If $y = 3x - 1$ and $y = x + 3$, then

$3x - 1 = x + 3 \Rightarrow 2x = 4 \Rightarrow x = 2$.

Substituting $x = 2$ into $y = x + 3 \Rightarrow y = 5$; or a **unique solution** of **(2, 5)**.

(b) Multiply $3x + 4y = {}^-17$ by $2 \Rightarrow$

$6x + 8y = {}^-34$. Multiply $2x + 3y = {}^-13$ by $^-3 \Rightarrow {}^-6x + {}^-9y = 39$.

Add the equations:

$$\begin{array}{rcrcr} 6x & + & 8y & = & ^-34 \\ ^-6x & + & ^-9y & = & 39 \\ \hline & & ^-y & = & 5 \end{array}$$

or $y = {}^-5$.

Substitute $y = {}^-5$ into $3x + 4y = {}^-17$.

$3x + 4({}^-5) = {}^-17 \Rightarrow x = 1$; or a **unique solution** of **(1, $^-$5)**.

(c) Multiply $\frac{^-2}{3}x + y = \frac{1}{3}$ by $3 \Rightarrow {}^-2x + 3y = 1$

and add the equations:

$$\begin{array}{r} ^-2x + 3y = 1 \\ 2x - 3y = {}^-1 \\ \hline 0 = 0 \end{array}$$

There are **infinitely many solutions** of the form $\left(x, \frac{2}{3}x + \frac{1}{3}\right)$.

(d) Substitute $y = 1 - x$ in $y = x - 1$:

$1 - x = x - 1 \Rightarrow 2 = 2x \Rightarrow x = 1$. Substitute 1 for x: $y = 1 - 1 \Rightarrow y = 0$. There is a **unique solution (1, 0)**.

18. Let g be the number of gallons of gasoline and let k be the number of gallons of kerosene on the truck.

Then: $g + k = 5000 \Rightarrow g = 5000 - k$.

$\$0.13g + \$0.12k = \$640 \Rightarrow 13g + 12k = 64000$.

Substitute $g = 5000 - k$ into $13g + 12k = 64000 \Rightarrow 13(5000 - k) + 12k = 64000 \Rightarrow$ $k =$ **1000 gallons of kerosene.**

Substitute $k = 1000$ into $g = 5000 - k \Rightarrow$ $g =$ **4000 gallons of gasoline.**

19. Let d be the number of dimes and let q be the number of quarters. Then:

$d + q = 27 \Rightarrow d = 27 - q$.

$\$0.10d + \$0.25q = \$5.25 \Rightarrow 10d + 25q = 525$.

Substitute $d = 27 - q$ into $10d + 25q = 525 \Rightarrow$

$10(27 - q) + 25q = 525$. $q =$ **17 quarters.**

Substitute $q = 17$ into $d = 27 - q$. $d =$ **10 dimes.**

8-4 Review Problems

17. **(a)** $x\sqrt{2} = {}^-3x\sqrt{2} + 2 \Rightarrow$

$x\sqrt{2} + 3x\sqrt{2} = 2 \Rightarrow$

$4x\sqrt{2} = 2 \Rightarrow$

$x\sqrt{2} = \frac{1}{2} \Rightarrow \mathbf{x} = \frac{1}{2\sqrt{2}} = \frac{\sqrt{2}}{4}$.

(b) $x^2 - 81 = 0 \Rightarrow (x - 9)(x + 9) = 0$

$\Rightarrow x - 9 = 0$ or $x + 9 = 0$

$\Rightarrow \mathbf{x = 9}$ or $\mathbf{x = {}^-9}$.

(c) $3x < {}^-\sqrt{7} \Rightarrow \mathbf{x < \frac{^-\sqrt{7}}{3}}$.

19. $12 = 3x - \sqrt{2} \Rightarrow 12 + \sqrt{2} = 3x$

$\Rightarrow x = \frac{\mathbf{12+\sqrt{2}}}{\mathbf{3}}$.

Chapter 8 Review

1. (a) Since it is a geometric sequence, 5 multiplied by some number x yields the middle term; then, that middle term multiplied by x yields 10. Calling the middle term a, the equations are:

$5x = a$ and $ax = 10$. Substituting $5x$ in for a in the second equation yields

$(5x)x = 10 \Rightarrow 5x^2 = 10 \Rightarrow$

$x^2 = 2 \Rightarrow x = \pm\sqrt{2}$.

If $x = \sqrt{2}$, then $\boldsymbol{a} = \mathbf{5\sqrt{2}}$; if $x = {}^{-}\sqrt{2}$,

then $\boldsymbol{a} = \mathbf{{}^{-}5\sqrt{2}}$;

(b) Since it is a geometric sequence, 1 multiplied by some number x yields the second term; in this case, the second term is x. It follows, then, that the third term is x^2, the fourth term is x^3, and the fifth term, $81 = x^3x = x^4$. So,

$x^4 = 81 \Rightarrow x = \sqrt[4]{81} \Rightarrow x = 3$ or $x = {}^{-}3$.

When $x = 3$, the missing terms are **3, 9, and 27**; when $x = {}^{-}3$, the missing terms are **⁻3, 9, and ⁻27**.

2. (*i*) $\sqrt{7} \times ({}^{-}1)^{10} = \sqrt{7} \times 1 = \sqrt{7}$.

(*ii*) ratio $=$ **⁻1**.

3. (a) The difference is $a_2 - a_1 = \sqrt{2} - 1$.

Thus, $\boldsymbol{a_n = 1 + (n - 1)(\sqrt{2} - 1)}$.

(b) The sum of the first 100 terms is **20,100**.

4. $\boldsymbol{S = 13P}$.

5. **Answers may vary.** One possibility: There are 103 times as many girls as boys.

6. $\boldsymbol{f = 3y}$.

7. Let $S = n_1 + n_2 + n_3 + \cdots$. Then $S_{new} = (10n_1 - 10) + (10n_2 - 10) + (10n_3 - 10) + \cdots = 10[(n_1 - 10) + (n_2 - 10) + (n_3 - 10) + \cdots] = \mathbf{10S - 10n}$.

8. Let n be the whole number. Then $12\left(\frac{n}{13}\right) - 20 + 89 = 93 \Rightarrow 12\left(\frac{n}{13}\right) = 24 \Rightarrow \frac{n}{13} = 2 \Rightarrow \boldsymbol{n = 26}$.

9. (a) Let n be the number. Then:

(*i*) $n + 17$

(*ii*) $2(n + 17) = 2n + 34$

(*iii*) $(2n + 34) - 4 = 2n + 30$

(*iv*) $2(2n + 30) = 4n + 60$

(*v*) $(4n + 60) + 20 = 4n + 80$

(*vi*) $\frac{4n+80}{4} = n + 20$

(*vii*) $n + 20 - 20 = \boldsymbol{n}$.

(b) **Answers may vary**; the first three steps yield

$4(n + 18) - 7 = 4n + 72 - 28 = 4n + 44$.

So for example, the next two lines could be to divide by 4, and subtract 11. That would mean $\frac{[4(n+18)-7]}{4} - 11 = n$.

10. (a) $4(7x - 21) = 14(7x - 21) \Rightarrow$

$0 = 10(7x - 21) \Rightarrow 0 = 70x - 210 \Rightarrow$

$210 = 70x \Rightarrow \boldsymbol{x = 3}$.

(b) $3(\sqrt{x} - 1) = \sqrt{9x} + 5 \Rightarrow 3\sqrt{x} - 3 =$

$3\sqrt{x} + 5 \Rightarrow 3\sqrt{x} - 3\sqrt{x} = {}^{-}8 \Rightarrow 0 \neq {}^{-}8$

so there is **no solution**.

(c) $2(3x + 5) = 6x + 11 \Rightarrow$

$6x + 10 = 6x + 11 \Rightarrow 10 = 11$.

so there is **no solution.**

(d) $2(x + \sqrt{3}) = 3(x - \sqrt{3}) \Rightarrow 2x + 2\sqrt{3} =$

$3x - 3\sqrt{3} \Rightarrow \mathbf{5\sqrt{3}} = x.$.

11. Let p be the number of Paige's cards. Then Jordan has $2p$ cards and Mike has $6p$ cards.

So

$p + 2p + 6p = 999$ so

$9p = 999 \Rightarrow \boldsymbol{p = 111}$.

So Paige has **111 cards**

Jordan has **222 cards**

Mike has **666 cards.**

12. Let s be the number of science book overdue days and c be the number of children's book overdue days. It is given that $c = s - 14$.

Then $8(0.20)(s - 14) + 2(0.20)s = 11.60 \Rightarrow$

$1.6s - 22.4 + 0.4s = 11.60 \Rightarrow 2s = 34 \Rightarrow$

$s = 17$.

Each science book was overdue by **17 days**

Each children's book was overdue by $17 - 14 =$ **3 days.**

13. Let j be the number of papers delivered by Jacobo, d be the number of papers delivered by Dahlia, and r be the number of papers delivered by Rashid.

It is given that $d = r + 100$ and $j = 2d = 2(r + 100)$.

Then if $j + d + r = 500 \Rightarrow r + (r + 100) + 2(r + 100) = 500 \Rightarrow 4r + 300 = 500 \Rightarrow r = 50$.

So Rashid delivered **50 papers**

Dahlia delivered $50 + 100 =$ **150 papers**

Jacobo delivered $2(50 + 100) =$ **300 papers**.

14. **(a)** **A function**. Each component of the domain corresponds to a unique component of the range.

(b) **Not a function**. a and b both correspond to two components of the range.

(c) **A function**. a and b correspond to unique components of the range.

(d) **A function**. Each component of the domain corresponds to a value of 2 in the range.

15. **(a)** $\{0, 1, 2, 3\} + 3 \Rightarrow$ Range is $\{x + 3 | x \in \{0, 1, 2, 3\}\} = \mathbf{\{3, 4, 5, 6\}}$.

(b) Range is $\{3x - 1 | x \in \mathbb{R}\} = \mathbb{R}$

(c) The domain is $\{x^2 | x \in \mathbb{R}\}$ so the range is $\{x | x \geq 0\}$.

(d) $\{0, 1, 2\}^2 + 3\{0, 1, 2\} + 5 \Rightarrow$ Range is $\{x^2 + 3x + 5 \mid x \in \{0, 1, 2\}\} = \mathbf{\{5, 9, 15\}}$.

16. **(a)** **Not a function**. A student may have more than one major.

(b) **A function**. The range is the subset of the natural numbers that includes the number of pages in each book in the library.

(c) **A function**. The range is $\{x | x \geq 6$ and x is even$\}$.

(d) **A function**. The range is $\{0, 1\}$.

(e) **A function**. The range is the set of all natural numbers N.

17. **(a)** $\mathbf{C(x) = \$[200 + 55(x - 1)]}$ where x is the number of months.

(b) Plot:

x	$C(x)$
1	200
2	255
3	310
4	365
5	420
6	475
7	530
8	585
9	640
10	695
11	750
12	805

(c) The cost will exceed \$600 beginning with the **ninth month**.

(d) If $C(x) = 200 + 55(x - 1) = 6000$ then $55x - 55 = 5800 \Rightarrow x = 106.5$.

Or the cost will exceed \$6000 in the **107th month**.

18. **(a)** $4x - 5 = 15 \Rightarrow 4x = 20 \Rightarrow \mathbf{x = 5}$.

(b) $x^2 - 1 = 2 \Rightarrow x^2 = 3 \Rightarrow \mathbf{x = \pm\sqrt{3}}$.

(c) $\dfrac{x+1}{x+2} = {}^-1 \Rightarrow x + 1 = ({}^-1)(x + 2)$

$x + 1 = {}^-x - 2 \Rightarrow 2x = {}^-3 \Rightarrow \mathbf{x = \dfrac{{}^-3}{2}}$.

(d) $\dfrac{x+1}{x+2} = 1 \Rightarrow x + 1 = x + 2 \Rightarrow 0x = 1 \Rightarrow 0 \neq 1$

no solution.

19. **(a)** **A function**. Each value on the x-axis corresponds to a unique value on the y-axis.

(b) **Not a function**. 4 and 5 on the x-axis each correspond to two values on the y-axis.

(c) **Not a function**. The x-value 5 corresponds to two values on the y-axis.

20. Assume Jilly starts with unpainted blocks each time:

(a)

Number of cubes	Number of squares to paint
1	6
2	10
3	14
4	18
5	22
6	26

(b)

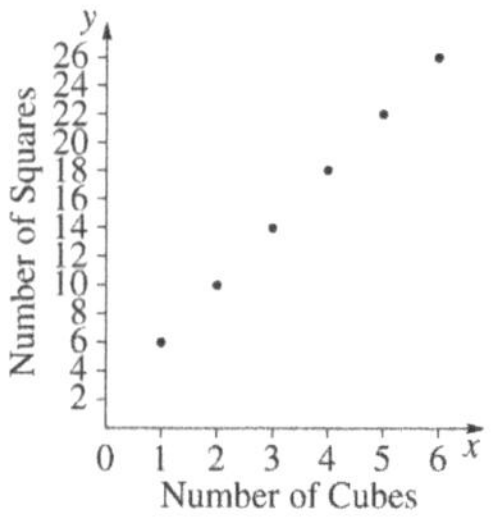

(c) This is a function which is an arithmetic sequence with $a_1 = 6,\ d = 4,$ and $a_6 = 26 \Rightarrow a_n = 6 + (n-1)4 \Rightarrow 4n + 2$. The function is $y = \mathbf{4x + 2}$ for $x = 1, 2, ..., 6$.

(d) No. The graph does represent a function, but not a straight line because values of x cannot assume anything but natural numbers.

21. **(a)** $y = {}^{-}2x + 5$ Some values of x and y are in the following table:

x	y
$^{-}1$	7
0	5
1	3
2	1
⋮	⋮

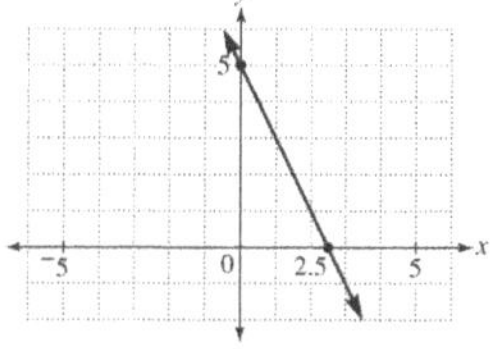

(b) $\dfrac{{}^{-}x}{3} + \dfrac{y}{5} = 1 \Rightarrow {}^{-}5x + 3y = 15 \Rightarrow y = \dfrac{5}{3}x + 5$

(values of x are multiples of 3):

x	y
$^{-}6$	$^{-}5$
$^{-}3$	0
0	5
3	10
6	15
⋮	⋮

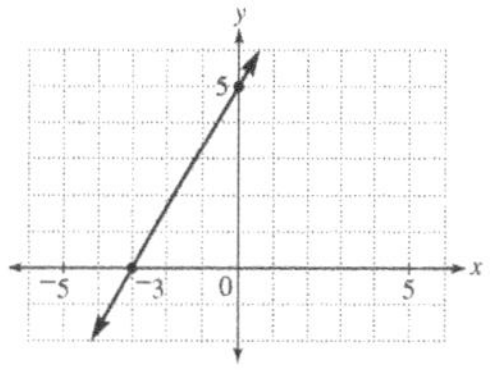

(c) $y = {}^{-}x\sqrt{2}$:

x	y
${}^{-}\sqrt{2} \approx {}^{-}1.4$	2
0	0
$\sqrt{2} \approx 1.4$	$^{-}2$
$\sqrt{8} \approx 2.8$	$^{-}4$
⋮	⋮

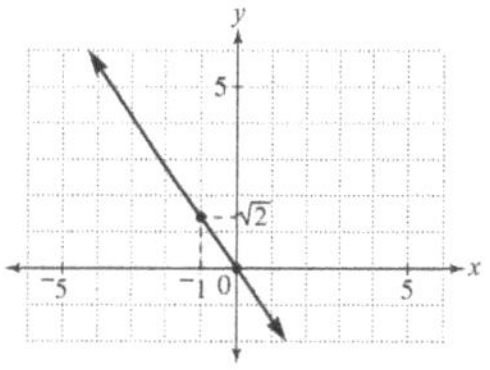

22. **(a)** $4x + 3y - 12 = 0 \Rightarrow 3y = {}^{-}4x + 12 \Rightarrow$

$y = -\dfrac{4}{3}x + 4$ is the graph of $\overleftrightarrow{CD}$.

(b) $x - y - 1 = 0 \Rightarrow y = x - 1$

$4x + 3y - 12 = 0 \Rightarrow y = 4 - \dfrac{4}{3}x$

$x - 1 = 4 - \dfrac{4}{3}x \Rightarrow 3x - 3 = 12 - 4x$

$7x = 15 \Rightarrow x = \dfrac{15}{7}$

$y = \dfrac{15}{7} - 1 \Rightarrow y = \dfrac{8}{7}$

The coordinates of point E are $\left(\mathbf{0, \dfrac{8}{7}}\right)$.

(c) $x - y - 1 = 0 \Rightarrow 0 - y - 1 = 0 \Rightarrow y = {}^{-}1$, the coordinates of point B are (0,⁻1). Point F has the same y-coordinate. Using the equation for $\overleftrightarrow{CD}$ substitute ⁻1 for y and solve for x.

$4x + 3({}^{-}1) - 12 = 0 \Rightarrow 4x - 15 = 0 \Rightarrow x = \frac{15}{4}$

. The distance of $\overleftrightarrow{BF}$ is $\frac{15}{4} - 0 = \mathbf{\frac{15}{4}}$.

23. **(a)** ${}^{-}3x + 12 = 23 \Rightarrow {}^{-}3x = 11 \Rightarrow x = \mathbf{{}^{-}\frac{11}{3}}$.

(b) $2x - 5 = 8x + 1 \Rightarrow {}^{-}6 = 6x \Rightarrow$

$\mathbf{{}^{-}1 = x}$

(c) $(2x + 1)^2 = 9 \Rightarrow (2x + 1)(2x + 1) = 9$

$4x^2 + 4x + 1 = 9 \Rightarrow 4x^2 + 4x - 8 = 0$

$4(x^2 + x - 2) = 0 \Rightarrow 4(x + 2)(x - 1) = 0$

so

$4 = 0$ (nonsense)

$x + 2 = 0 \Rightarrow \mathbf{x = {}^{-}2}$

$x - 1 = 0 \Rightarrow \mathbf{x = 1}$

(d) Set each factor equal to zero and solve.

So $x - 1 = 0 \Rightarrow \mathbf{x = 1}$
or $2x + 1 = 0 \Rightarrow \mathbf{x = \frac{1}{2}}$.

CHAPTER 9

PROBABILITY

Assessment 9-1A: Determining Probabilities

1. (a) **No.** There are fewer face cards than there are not face cards.

   (b) **Yes.** Each suit has the same number of cards.

   (c) **Yes.** There are equal numbers of black and red cards.

   (d) **No.** There are 4 each kings, queens, jacks, and aces; 20 even-numbered cards; and 16 odd-numbered cards in a standard deck.

2. $P(4) = \frac{24}{100}$, the theoretical probability is

   $P(4) = \frac{1}{6}$

   The difference is $\frac{24}{100} - \frac{1}{6} = \frac{36}{150} - \frac{25}{150} = \mathbf{\frac{11}{150}}$.

3. (a) $P(vowel) = \frac{n(vowel)}{n(S)} = \mathbf{\frac{5}{26}}$.

   (b) $P(consonant) = 1 - P(vowel)$
   $= 1 - \frac{5}{26} = \mathbf{\frac{21}{26}}$.

   (c) $P(\text{vowel or letter from "probability"}) = P(vowel) + P(letter) - P(vowel \cap letter) = \frac{5}{26} + \frac{9}{26} - \frac{3}{26} = \mathbf{\frac{11}{26}}$. There are 11 letters in the word "probability," but *b* and *i* are repeated. There are only nine unique choice of letters; the probabilities of a vowel and a letter from "probability" are not mutually exclusive.

4. (a) $P(1, 5 \text{ or } 7) = \frac{1}{8} + \frac{1}{8} + \frac{1}{8} = \mathbf{\frac{3}{8}}$. Factors of 35 are 1, 5, or 7 and their probabilities are mutually exclusive.

   (b) $P(3 \text{ or } 6) = \frac{1}{8} + \frac{1}{8} = \frac{2}{8} = \mathbf{\frac{1}{4}}$. **Multiples of 3 are 3 or 6 and their probabilities are mutually exclusive.**

   (c) $P(2,4,6 \text{ or } 8) = \frac{1}{8} + \frac{1}{8} + \frac{1}{8} + \frac{1}{8}$
   $= \frac{4}{8} = \mathbf{\frac{1}{2}}$.

   (d) $P(6 \text{ or } 2) = \frac{1}{8} + \frac{1}{8} = \frac{2}{8} = \mathbf{\frac{1}{4}}$.

   (e) $P(11) = \mathbf{0}$. The probability of an impossible event is 0.

   (f) $P(4, 6 \text{ or } 8) = \frac{1}{8} + \frac{1}{8} + \frac{1}{8} = \mathbf{\frac{3}{8}}$. 4, 6, and 8 are the only composite numbers on the spinner.

   (g) $P(1) = \mathbf{\frac{1}{8}}$. 1 is the only number on the spinner which is neither prime nor composite.

5. (a) $P(red\ card) = \frac{n(red)}{n(S)} = \frac{26}{52} = \mathbf{\frac{1}{2}}$.

   (b) $P(face\ card) = \frac{n(face\ card)}{n(S)} = \frac{12}{52} = \mathbf{\frac{3}{13}}$.

   (c) $P(red\ card\ or\ 10) = \frac{n(red)}{n(S)} + \frac{n(10)}{n(S)} - \frac{n(red\ and\ 10)}{n(S)} = \frac{26}{52} + \frac{4}{52} - \frac{2}{52} = \frac{28}{52}$
   $= \mathbf{\frac{7}{13}}$. Red and 10 are not mutually exclusive.

   (d) $P(queen) = \frac{n(queen)}{n(S)} = \frac{4}{52} = \mathbf{\frac{1}{13}}$.

   (e) $P(not\ a\ queen) = 1 - P(queen) = 1 - \frac{1}{13} = \mathbf{\frac{12}{13}}$.

   (f) $P(face\ card\ or\ club)$
   $= \frac{n(face\ card)}{n(S)} + \frac{n(club)}{n(S)} - \frac{n(face\ card\ and\ club)}{n(S)} = \frac{12}{52} + \frac{13}{52} - \frac{3}{52} = \frac{22}{52}$
   $= \mathbf{\frac{11}{26}}$. Face card and club are not mutually exclusive.

   (g) $P(face\ card\ and\ club)$
   $= \frac{n(face\ card\ and\ club)}{n(S)} = \mathbf{\frac{3}{52}}$.

(h) *P*(*not a face card and not a club*) $=$
$1 - P(\textit{face card or club}) = 1 - \frac{11}{26} = \mathbf{\frac{15}{26}}$.
Use a Venn diagram to verify this use of the complementary property.

6. Assume the socks are randomly mixed in the drawer.

(a) $P(brown) = \frac{n(brown)}{n(S)} = \frac{4}{12} = \mathbf{\frac{1}{3}}$. There are four brown socks out of the twelve in the drawer.

(b) $P(\textit{black or green}) = \frac{n(black)}{n(S)} + \frac{n(green)}{n(S)} =$
$\frac{6}{12} + \frac{2}{12} = \frac{8}{12} = \mathbf{\frac{2}{3}}$. The two events are mutually exclusive.

(c) $P(red) = \mathbf{0}$. There are no red socks in the drawer, so this is an impossible event.

(d) *P*(*not black*)
$= 1 - P(black) = 1 - \frac{n(black)}{n(S)} =$
$1 - \frac{6}{12} = \frac{6}{12} = \mathbf{\frac{1}{2}}$.

or

$P(\textit{not black}) = \frac{n(broun)}{n(S)} + \frac{n(green)}{n(S)} =$
$\frac{4}{12} + \frac{2}{12} = \frac{6}{12} = \mathbf{\frac{1}{2}}$.

(e) $P(\textit{pair of same color}) = \mathbf{1}$. Since there are only three colors, if four socks are pulled out at least two of them must be of the same color.

7. (a) Neither Unlikely or Likely. There are 3 odd numbers and 3 even numbers on a die, so the probability of landing on an odd number or an even number is the same.

(b) Unlikely. In a standard deck of cards there are 4 aces and 48 non-aces, so the probability of drawing an ace is unlikely.

(c) Certain. Every number on a die is less than or equal to six, so the probability of landing on a number less than or equal to six is 1 and it is certain to happen.

(d) Impossible. There is no card numbered "11" in a standard deck of cards.

(e) Neither Unlikely or Likely. Each flip is independent from the other flip and each flip has a probability of $\frac{1}{2}$.

8. (a) Answers will vary. For example, choose one cell of the grid at random.

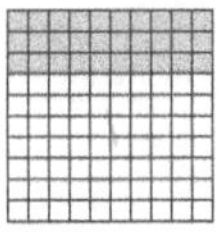

(b) *P*(*no rain*)
$= 1 - P(rain) = 1 - 0.30 = 0.70$,
or **70%**.

9. There are $17 + 13 + 12 = 42$ dolls. The probability of drawing a doll with no red hair is
$P(\text{no red}) = 1 - P(\text{red}) = 1 - \frac{12}{42} = \frac{30}{42} = \mathbf{\frac{5}{7}}$.

10. There are 10 plays that are considered tragedies.

(a) $P(\text{"King"}) = \mathbf{\frac{1}{10}}$. *King Lear* is the only title that contains the word "King".

(b) $P(\text{Starts with "T"}) = \frac{2}{10} = \mathbf{\frac{1}{5}}$. *Timon of Athens* and *Titus Andronicus* are the two titles that start with the letter "T".

(c) $P(\text{Contains "Q"}) = \frac{0}{10} = \mathbf{0}$. No titles contain the letter "Q".

(d) $P(\text{two names}) = \frac{2}{10} = \mathbf{\frac{1}{5}}$. *Antony and Cleopatra* and *Romeo and Juliet* are the two titles that contain the names of two people.

11. $P(o) = \mathbf{\frac{1}{5}}$. There is 1 letter "o" in "*e-i-e-i-o*".

12. There are $350 + 320 + 310 + 400 = 1380$ students in all. $P(\textit{freshman}) = \frac{n(\textit{freshman})}{n(S)} =$
$\frac{350}{1380} = \mathbf{\frac{35}{138}}$.

13. (a) Mutually Exclusive. There are no cards that are both red and black cards.

(b) Not Mutually Exclusive. 2 is both even and prime.

(c) Not Mutually Exclusive. There are 13 cards that are black club cards.

(d) Mutually Exclusive. There are no cards that are both a face card and an odd number card.

(e) Not Mutually Exclusive. There are 3 cards that are heart face cards: Jack of hearts, queen of hearts, and king of hearts.

(f) Mutually Exclusive. There are no cards that are both spades and diamonds.

14. $n(S) = n(men) + N(women)$
$= 35 + 45 = 80$.

(a) $P(female) = \frac{n(female)}{n(S)} = \frac{45}{80} = \mathbf{\frac{9}{16}}.$

(b) $P(computer\ science\ major) = \frac{n(computer\ science)}{n(S)} = \frac{10}{80} = \mathbf{\frac{1}{8}}.$

(c) $P(not\ math\ major) = 1 - P(math\ major) = 1 - \frac{n(math)}{n(S)} = 1 - \frac{20}{80} = \frac{60}{80} = \mathbf{\frac{3}{4}}.$ It is easier to use the complementary property in this case than to calculate the sum of the non-math majors.

(d) $P(computer\ science\ or\ math\ major) = \frac{n(computer\ science)}{n(s)} + \frac{n(math)}{n(S)} = \frac{10}{80} + \frac{20}{80} = \frac{30}{80} = \mathbf{\frac{3}{8}}.$ Since there are no double majors, the events are mutually exclusive.

15. (a) If r red balls are added then $P(red\ ball) = \frac{2+r}{10+r} = \frac{3}{4} \Rightarrow 4(2+r) = 3(10+r).$
$r = $ **22 red balls.**

(b) There are originally $\frac{5\text{ white}}{10\text{ total}}$ balls. b black balls are added so that $P(white) = \frac{5}{10+b} = \frac{1}{4} \Rightarrow 20 = 10 + b.\ b = $ **10 black balls** added.

16. (a) P(red)=P(hearts or diamonds)

$=\frac{12}{48}+\frac{12}{48}=\frac{24}{48}=\mathbf{\frac{1}{2}}$

(b) P(face)=P(jack or queen or king)

$=\frac{8}{48}+\frac{8}{48}+\frac{8}{48}+\frac{8}{48}=\frac{24}{48}=\mathbf{\frac{1}{2}}$

(c) P(red card or 10)

$=\frac{24}{48}+\frac{8}{48}-\frac{4}{48}=\frac{28}{48}=\mathbf{\frac{7}{12}}$

(d) P(black queen)

$=\frac{2}{48}+\frac{2}{48}=\frac{4}{48}=\mathbf{\frac{1}{12}}$

(e) P(not a black queen)

$=1-\frac{1}{12}=\frac{12}{12}-\frac{1}{12}=\mathbf{\frac{11}{12}}$

(f) P(numbered card)=P(9 or 10)

$=\frac{8}{48}+\frac{8}{48}=\frac{16}{48}=\mathbf{\frac{1}{3}}$

(g) P(face card or heart)

$=\frac{24}{48}+\frac{12}{48}-\frac{6}{48}=\frac{30}{48}=\mathbf{\frac{5}{8}}$

(h) P(not a face card and not a heart)

$=1-\frac{5}{8}=\frac{8}{8}-\frac{5}{8}=\mathbf{\frac{3}{8}}$

17. A fair coin will always exhibit equal probabilities of a head or a tail when flipped, regardless of past history. Thus the probability of a head on toss 16 is $\mathbf{\frac{1}{2}}$.

18. (a) P(vowel)=$\frac{6}{15}=\mathbf{\frac{2}{5}}$

(b) P(consonant)=$\frac{9}{15}=\mathbf{\frac{3}{5}}$

(c) P(letter after m)=$\mathbf{\frac{7}{15}}$

(d) Answers vary. For example: if your name is Barbara, then P(a letter in Barbara)=$\mathbf{\frac{2}{15}}$.

19. $P(\text{correct digit}) = \mathbf{\frac{1}{10}}$. There are ten digits: 0, 1, 2, 3, 4, 5, 6, 7, 8, 9 and only one can be the correct digit.

20. $P(\text{first letter is a vowel}) = \mathbf{\frac{1}{9}}$. There are nine letters and one vowel.

21. Because A and B are mutually exclusive, $P(A \cup B) = P(A) + P(B) = 0.3 + 0.4 = \mathbf{0.7}.$

22. $P(4) = \frac{1}{6} \approx \frac{167}{1000}$. So **about 167 times.**

Assessment 9-2A: Multistage Experiments and Modeling Games

1. (a) P(sum of 6)=$\mathbf{\frac{5}{36}}$

(b) P(sum multiple of 3)

=P(3)+P(6)+P(9)+P(12)

$=\frac{2}{36}+\frac{5}{36}+\frac{4}{36}+\frac{1}{36}=\frac{12}{36}=\mathbf{\frac{1}{3}}$

(c) P(sum of 6 or 7)=$\frac{5}{36}+\frac{6}{36}=\mathbf{\frac{11}{36}}$

(d) P(sum of 5 or multiple of 2)

$=P(5)+P(2)+P(4)+P(6)+P(8)+P(10)+P(12)$

$=\frac{4}{36}+\frac{1}{36}+\frac{3}{36}+\frac{5}{36}+\frac{5}{36}+\frac{3}{36}+\frac{1}{36}$

$=\frac{22}{36}=\mathbf{\frac{11}{18}}$

(e) P(sum of 5 and multiple of 2)$=\frac{0}{36}=\mathbf{0}$

(f) P(sum of 6 given that the first roll is 2)

$=\mathbf{\frac{1}{6}}$

2. **(a)** P(both are aces)$=\frac{4}{52}\cdot\frac{3}{51}=\frac{12}{2652}=\mathbf{\frac{1}{221}}$

(b) P(at least one card is black)

= P(1 black)+P(2 black)

$=1-P(0\text{ black})=1-\left(\frac{26}{52}\cdot\frac{25}{51}\right)$

$=1-\frac{650}{2652}=1-\frac{25}{102}=\mathbf{\frac{77}{102}}$

(c) P(2nd is black given that 1st is red)

$=\mathbf{\frac{26}{51}}$

3. **(a)** A tree diagram showing paths for each possible outcome is shown below. Paths are defined by finding the probability of each event in draw one followed by the probability of each event in draw two; for example, the probability of drawing a white ball first is $\frac{3}{5}$, and the probability of drawing another is then $\frac{2}{4}$, since there will be two white balls of a total of four:

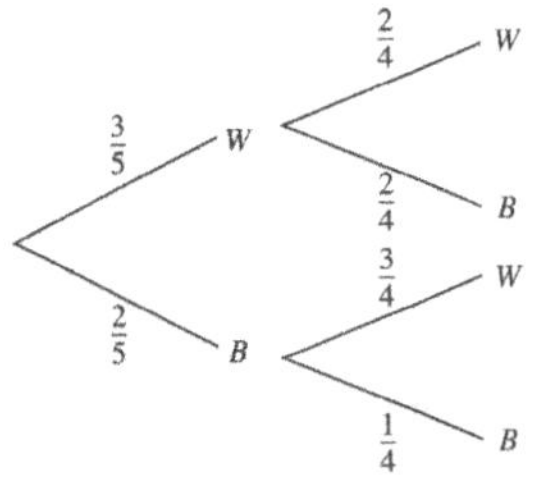

The possible outcomes to obtain different colors are white and black or black and white. *P*(*different colors*)

$=\frac{3}{5}\cdot\frac{2}{4}+\frac{2}{5}\cdot\frac{3}{4}=\frac{6}{20}+\frac{6}{20}=\frac{12}{20}=\mathbf{\frac{3}{5}}.$

(b) A tree diagram is shown below. Probabilities in each path differ from (a) above in that balls are replaced; thus the probability of drawing a white or black ball is the same in each path:

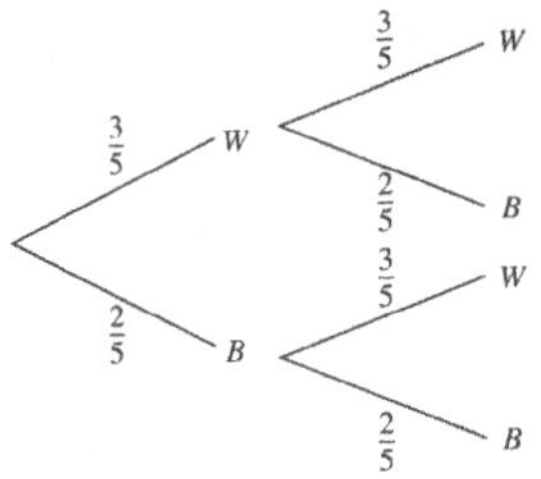

$P(\text{different colors})=\frac{3}{5}\cdot\frac{2}{5}+\frac{2}{5}\cdot\frac{3}{5}$

$=\mathbf{\frac{12}{25}}.$

4. **(a)** This is a multistage event with replacement. $P(DAN)=P(D)\cdot P(A)\cdot P(N)=\frac{1}{6}\cdot\frac{1}{6}\cdot\frac{1}{6}=\mathbf{\frac{1}{216}}.$

(b) Without replacement $P(DAN)=P(D)\cdot P(A)\cdot P(N)=\frac{1}{6}\cdot\frac{1}{5}\cdot\frac{1}{4}=\mathbf{\frac{1}{120}}.$

5. This is a multistage event without replacement. $P(\text{three women})=\frac{4}{10}\cdot\frac{3}{9}\cdot\frac{2}{8}=\frac{24}{720}=\mathbf{\frac{1}{30}}.$

6. *P*(*all boys*) = *P* (*first child boy*) · *P*(*second child boy*) · *P*(*third child boy*) · *P*(*fourth child boy*) $=\frac{1}{2}\cdot\frac{1}{2}\cdot\frac{1}{2}\cdot\frac{1}{2}=\mathbf{\frac{1}{16}}.$

7. Draw a tree diagram with two stages of branches:

(*i*) One stage with each of the numbers 4, 6, 7, 8, or 9 as branches, each with probability $\mathbf{\frac{1}{5}}$.

(*ii*) A second stage: for each branch of (*i*) extend four branches with the remaining numbers as choices, each with probability $\mathbf{\frac{1}{4}}$.

(*iii*) There are $5\cdot 4=20$ branches, each with probability $\frac{1}{5}\cdot\frac{1}{4}=\mathbf{\frac{1}{20}}.$

(a) Add the two numbers in each branch of the tree diagram; there are eight even sums. $P(\text{even sum; win})=\frac{8}{20}=\mathbf{\frac{2}{5}}.$

(b) **Yes**. Only two of the 20 products are odd ($7\cdot 9$ and $9\cdot 7$). *P*(*even product; win*) $=\frac{18}{20}=\mathbf{\frac{9}{10}}.$

8. **(a)**

-	1	2	3	4	5	6
1	0	1	2	3	4	5
2	$^-1$	0	1	2	3	4
3	$^-2$	$^-1$	0	1	2	3
4	$^-3$	$^-2$	$^-1$	0	1	2
5	$^-4$	$^-3$	$^-2$	$^-1$	0	1
6	$^-5$	$^-4$	$^-3$	$^-2$	$^-1$	0

(b) P(zero)=$\frac{6}{36}=\mathbf{\frac{1}{6}}$

(c) P(positive prime number)

=P(2)+P(3)+P(5)

$=\frac{4}{36}+\frac{3}{36}+\frac{1}{36}=\frac{8}{36}=\mathbf{\frac{2}{9}}$

(d) P(negative)=$\frac{15}{36}=\mathbf{\frac{5}{12}}$

9. **(a)** **Spinner A wins with the following combinations:**

Outcome on spinner A	Outcome on spinner B
4	3
6	3 or 5
8	3 or 5

Assuming each spinner is divided evenly into thirds,

$P(A > B;\ win) = \frac{1}{3}\cdot\frac{1}{3}+\frac{1}{3}\cdot\frac{2}{3}+\frac{1}{3}\cdot\frac{2}{3}=\frac{1}{9}+\frac{2}{9}+\frac{2}{9}=\frac{5}{9};\ P(B > A;\ win) = 1-\frac{5}{9}=\mathbf{\frac{4}{9}}$. Therefore, choose **spinner A**.

(b) Answers vary; one example is shown below:

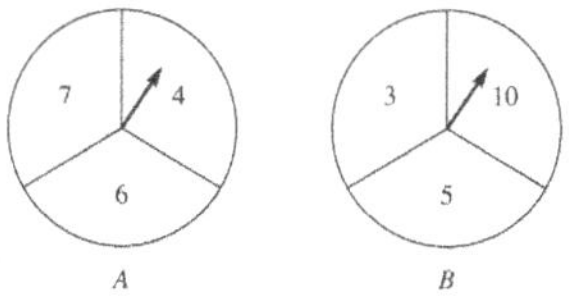

$P(A > B) = \frac{5}{9};\ P(B > A) = \frac{4}{9}.$

10. Each question has a probability of $\frac{1}{2}$ of being right, and the result of each question has no effect on subsequent questions.
$P(100\%) = \frac{1}{2}\cdot\frac{1}{2}\cdot\frac{1}{2}\cdot\frac{1}{2}\cdot\frac{1}{2}=\mathbf{\frac{1}{32}}.$

11. **(a)** $P(Paxson\ loses) = P(Rattlesnake\ wins).$
$P(4\ Paxson\ losses) = \frac{2}{3}\cdot\frac{2}{3}\cdot\frac{2}{3}\cdot\frac{2}{3}=\mathbf{\frac{16}{81}}.$

(b) There are six ways in which each school wins two games:
$S = \{(PPRR), (PRPR), (PRRP), (RPPR), (RPRP), (RRPP)\}$; each occurs with probability $\frac{2}{3}\cdot\frac{2}{3}\cdot\frac{1}{3}\cdot\frac{1}{3}=\frac{4}{81}.$
$P(draw) = 6\cdot\frac{4}{81}=\frac{24}{81}=\mathbf{\frac{8}{27}}.$

12. The total area of the dart board $= 5x$ by $5x = 25x^2$.

(a) The area of $A = x$ by $x = x^2$. $P(section\ A) = \frac{x^2}{25x^2}=\mathbf{\frac{1}{25}}.$

(b) The area of $B = (3x)^2 - x^2 = 8x^2$.
$P(section\ B) = \frac{8x^2}{25x^2}=\mathbf{\frac{8}{25}}.$

(c) The area of $C = (5x)^2 - (3x)^2 = 16x^2$.
$P(section\ C) = \frac{16x^2}{25x^2}=\mathbf{\frac{16}{25}}.$

or

$P(section\ C)$
$=1-[P(section\ A + P(section\ B)] = 1-\left(\frac{1}{25}+\frac{8}{25}\right)=1-\frac{9}{25}=\mathbf{\frac{16}{25}}.$

13. **(a)** P(1 green or 1 blue)

$=\frac{5}{50}+\frac{6}{50}=\mathbf{\frac{11}{50}}$

(b) P(not yellow or orange)

1 − P(yellow or orange)

$=1-\left(\frac{4}{50}+\frac{8}{50}\right)=1-\frac{12}{50}=\frac{38}{50}=\mathbf{\frac{19}{25}}$

(c) P(2 green)=$\frac{5}{50}\cdot\frac{5}{50}=\frac{1}{10}\cdot\frac{1}{10}=\mathbf{\frac{1}{100}}$

(d) P(2 brown)=$\frac{15}{50}\cdot\frac{15}{50}=\frac{3}{10}\cdot\frac{3}{10}=\mathbf{\frac{9}{100}}$

(e) P(2 brown)=$\frac{15}{50}\cdot\frac{14}{49}=\frac{3}{10}\cdot\frac{14}{49}$

$=\frac{42}{490}=\mathbf{\frac{3}{35}}$

(f) P(2 red)=$\frac{12}{50}\cdot\frac{12}{50}=\frac{6}{25}\cdot\frac{6}{25}=\mathbf{\frac{36}{625}}$

(g) P(3 red)=$\frac{12}{50}\cdot\frac{11}{49}\cdot\frac{10}{48}=\frac{1320}{117{,}600}=\mathbf{\frac{11}{980}}$

(h) P(2 red)$=\frac{11}{44}\cdot\frac{11}{44}=\frac{121}{1936}=\mathbf{\frac{1}{16}}$

(i) Answers vary. For example: drawing an M&M.

(j) Answers vary. For example: drawing a purple M&M.

14. Of the ten secretaries, two are male. $P(\textit{secretary given male}) = \frac{n(\textit{male and secretary})}{n(\textit{male})} = \frac{2}{28} = \mathbf{\frac{1}{14}}$.

15. (a)

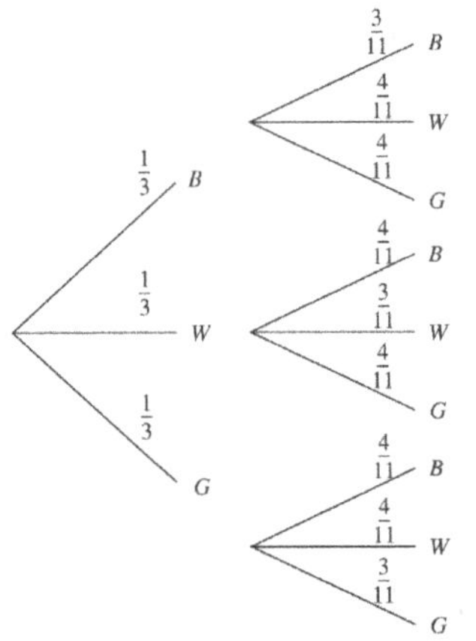

(b) The probability of two the same color is composed of three mutually exclusive events: two blues, two whites, or two grays.
$P(\textit{same color})$
$= P(\textit{two blues}) + P(\textit{two whites}) + P(\textit{two grays}) =$
$\frac{1}{3}\cdot\frac{3}{11}+\frac{1}{3}\cdot\frac{3}{11}+\frac{1}{3}\cdot\frac{3}{11}=\frac{9}{33}=\mathbf{\frac{3}{11}}$.

(c) $P(\textit{two gray}) = \frac{1}{3}\cdot\frac{3}{11}=\frac{3}{33}=\mathbf{\frac{1}{11}}$.

(d) $P(\textit{two same color}) = \mathbf{1}$. There are only three colors among the four socks.

16. (a) "At least as many heads as tails" means:
0 tails and 4 heads, which can occur in 1 way;
1 tail and 3 heads, which can occur in 4 ways;
2 tails and 2 heads, which can occur in 6 ways.
The total number of possible outcomes is $2^4 = 16$.

$P(\textit{as many heads as tails}) = \frac{1+4+6}{16} = \mathbf{\frac{11}{16}}$.

(b) If the quarter is fair (i.e., heads and tails occur the same number of times) then this probability is the same as in (a), $\mathbf{\frac{11}{16}}$.

17. The probability that any smoke detector taken at random will work is $\frac{450}{500} = 0.9$. The probability in any given sample size that at least 1 will work is the complement of the probability that none will work.

Sample size 1: $P(\textit{work}) = 0.9$

Sample size 2: $P(\textit{at least one work}) =$
$1 - P(0\ \textit{work}) = 1-(0.1)^2 = 0.99$.

Sample size 3: $P(\textit{at least one work}) =$
$1 - P(0\ \textit{work}) = 1-(0.1)^3 = 0.999$.

Sample size 4: $P(\textit{at least one work}) =$
$1 - P(0\ \textit{work}) = 1-(0.1)^4 = 0.9999$.

Sample size 5: $P(\textit{at least one work}) =$
$1 - P(0\ \textit{work}) = 1-(0.1)^5 = 0.99999$.

At least four should be installed to assure a probability of at least 99.9% that one will work.

18.

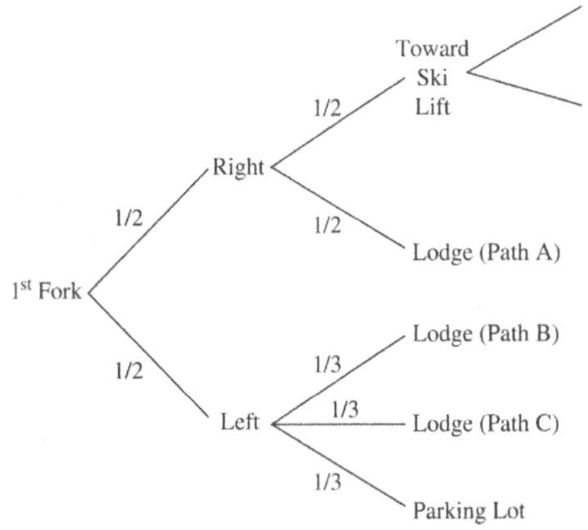

There are three paths to the lodge. These paths represent mutually exclusive events since skiers cannot be on more than one path at a time.
$P(A \cup B \cup C) = P(A) + P(B) + P(C) =$
$\frac{1}{2}\cdot\frac{1}{2}+\frac{1}{2}\cdot\frac{1}{3}+\frac{1}{2}\cdot\frac{1}{3}=\frac{1}{4}+\frac{1}{6}+\frac{1}{6}=$
$\mathbf{\frac{7}{12}} \approx \mathbf{0.58}$.

19. Assuming an analog and not a digital display, the second hand will cover the distance between 3 and 4 in 5 seconds. $P(\textit{between } 3 \textit{ and } 4)$
$= \frac{5\text{ sec}}{60\text{ sec}} = \mathbf{\frac{1}{12}}$.

20. Answers vary. There are $3+4+3=10$ marbles in the bag. Three out of ten are blue so the probability of drawing a blue marble is
$\frac{3}{10}=\frac{30}{100}=30\%$. To change that probability to
$75\%=\frac{75}{100}=\frac{3}{4}$ we could add 18 blue marbles.

There would now be $21+4+3=28$ marbles in the bag. Twenty-one out of twenty eight would be blue, changing the probability of drawing a blue marble to $P(\text{blue})=\frac{21}{28}=\frac{3}{4}=75\%$.

21. **(a)** If the sum was 6 and the first number was 3, the second number has to be 3.

$P(\text{second number is 3})=\frac{1}{5}$.

(b) If the first number is 3 then $P(\text{sum }6)=\frac{1}{6}$

22. P(3 different numbers)

$=\frac{6}{6}\cdot\frac{5}{6}\cdot\frac{4}{6}=1\cdot\frac{5}{6}\cdot\frac{2}{3}=\mathbf{\frac{5}{9}}$

23. **(a)** $P(\text{snowboarding})=\frac{69}{126}=\mathbf{\frac{23}{42}}$.

(b) $P(\text{skiing, given 6th grader})=\frac{18}{45}=\mathbf{\frac{2}{5}}$.

(c) $P(\text{7th grade \& snowboarding})=\mathbf{\frac{22}{38}}$

$P(\text{8th grade \& snowboarding})=\mathbf{\frac{20}{43}}$

$P(\text{both snowboarding})=\frac{22}{38}\cdot\frac{20}{43}=\frac{440}{1634}$
$=\mathbf{\frac{220}{817}}$

(d) P(8th grade & snowboarding, given that 7th grader snowboards) $=\mathbf{\frac{20}{43}}$

24. **(a)** $P(\text{both start})=0.8\cdot0.6=0.48$, or **48%**

(b) P(exactly one will start)=

P(Bel Air starts and the Impala does not start OR Impala starts and Bel Air does not start) =

$0.6\cdot0.2+0.8\cdot0.4=0.12+0.32=0.44=$ **44%**

(c) $P(\text{neither vehicle starts})=0.2\cdot0.4=0.08$ = **8%**

25. **(a)** The tree diagram below illustrates the event paths leading to a white ball:

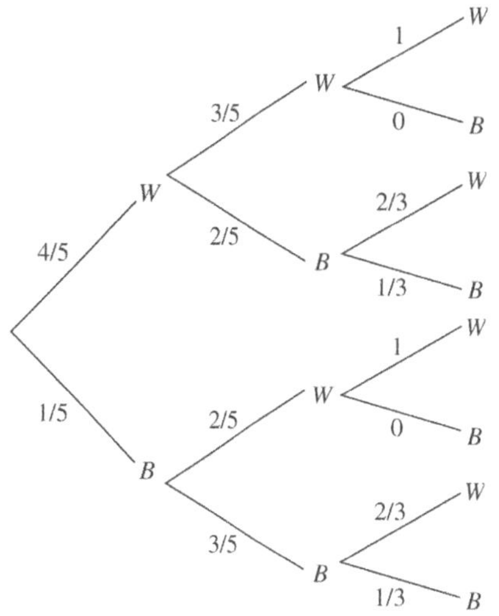

$P(white)=\frac{4}{5}\cdot\frac{3}{5}\cdot1+\frac{4}{5}\cdot\frac{2}{5}\cdot\frac{2}{3}+\frac{1}{5}\cdot\frac{2}{5}\cdot1+\frac{1}{5}\cdot\frac{3}{5}\cdot\frac{2}{3}=\frac{12}{25}+\frac{16}{75}+\frac{2}{25}+\frac{6}{75}=\frac{36+16+6+6}{75}=\mathbf{\frac{64}{75}}$.

(b) $P(black)=1-P(white)=1-\frac{64}{75}=\mathbf{\frac{11}{75}}$.

Mathematical Connections 9-2: Review Problems

19. **(a)** $P(April\,7)=\mathbf{\frac{1}{30}}$.

(b) $P(April\,31)=\mathbf{0}$.

(c) $P(before\ April\,20)=\mathbf{\frac{19}{30}}$.

21. **(a)** $P(\text{black})=\frac{18}{38}=\mathbf{\frac{9}{19}}$.

(b) $P(\text{0 or 00})=P(0)+P(00)$
$=\frac{1}{38}+\frac{1}{38}=\frac{2}{38}=\mathbf{\frac{1}{19}}$

(c) $P(\text{not in 1-12})=1-P(\text{1-12})$
$=1-\frac{12}{38}=\frac{26}{38}=\mathbf{\frac{13}{19}}$

(d) $P(\text{odd or green})=P(\text{odd})+P(\text{green})$

$=\frac{18}{38}+\frac{2}{38}=\frac{20}{38}=\mathbf{\frac{10}{19}}$

Assessment 9-3A: Simulations and Applications in Probability

In discussing methods of using simulations, all the following answers could vary.

1. If the use of the thumbtack was to involve dropping it and then observing whether it landed point up or point down, it would not be possible. The probabilities for birth of boys versus girls are approximately equal, but the thumbtack probabilities are not.
2. One method might be to let the digits 1 and 2 represent Diamonds, the digits 3 and 4 represent Clubs, the digits 5 and 6 represent Hearts, and the digits 7 and 8 represent Spades. Disregard digits 9 and 0, and read one digit at a time.
3. **(a)** Let the numbers 1, 2, 3, 4, 5, and 6 represent numbers on the die; ignore the numbers 0, 7, 8, and 9

   (b) ***(i)*** Number the persons $01, 02, 03, \ldots, 20$

   (ii) Mark off groups of two in a random digit table

   (iii) The three persons chosen are the first three whose numbers appear (ignore two-digit numbers > 20)

   (c) Represent Red by the numbers 0 through 4; Green by the numbers 5, 6, and 7; Yellow by the number 8; and White by the number 9; then pick a number from a random digit table
4. Number the students from 001 to 500; then, in the random digit table, mark off blocks large enough so that 30 three digits from 001 to 500 are in each block (disregard 000 or any numbers >500). These are the numbers of the 30 students who will be chosen.
5. To simulate Monday, let the digits 1 through 8 represent rain and 0 and 9 represent no rain. If rain occurred on Monday, repeat the same process for Tuesday. If it did not rain on Monday, let the digits 1 through 7 represent rain and 0, 8, and 9 represent dry. Repeat a similar process for the rest of the week.

   For example, 60304 13976 was chosen from the random digits table

Day of the week	digit	weather
Mon	6	rain
Tues	0	no rain
Wed	3	rain
Thurs	0	no rain
Fri	4	rain
Sat	1	rain

6. Mark off blocks of two digits and let the digits 00, 01, 02,..., 13, 14 represent contracting the disease; let the digits 15, 16,..., 98, 99 represent not contracting the disease. Mark off three blocks of two digits (for a total of six digits) to represent the three children. If at least one of the two-digit numbers is in the range 00 through 14 it represents a child in the three-child family having contracted strep.
7. $P(< 30) = \frac{n(numbers<30)}{n(possible\ two-digit\ numbers)} = \frac{30}{100} = \mathbf{\frac{3}{10}}.$
8. Assuming an unbiased random sample of fish from the pond is caught, then $\frac{50}{300} = \frac{1}{6}$ of the total population is marked. Let n be the fish population;

   $\frac{1}{6}n = 200 \Rightarrow n = 6 \cdot 200 = \mathbf{1200\ fish}.$
9. **(a)** **7 games.** It is possible for the losing team to win three games while the winning team wins four.

   (b) Answers vary. For example, because the teams are evenly matched use a table of random digits and let a number between 0 and 4 represent a win by Team *A*; let a number between 5 and 9 represent a win by Team *B*.

   Pick a starting spot and count the number of digits it takes before a Team *A* or Team *B* series win is recorded. Repeat the experiment many times **and then base your answers on:**

   (i) $P(four\text{-}game\ series) = \frac{n(four\text{-}game\ series)}{n(total\ series)}.$

   (ii) $P(seven\text{-}game\ series) = \frac{n(seven\text{-}game\ series)}{n(total\ series)}.$

   Given evenly matched teams, the probability of a four-game series would be expected to be low.
10. Answers vary. For example:

    (i) From a random-digit table use numbers 0 through 7 to simulate making the basket and 8 or 9 to simulate missing the basket; or

    (ii) Use a spinner constructed with 80% of its swept angle representing the making of a basket and 20% representing the missing of a basket.

By using the tree diagram shown below, the theoretical estimates for the number of points scored in 25 attempts may be computed and compared with the experimental probability obtained by simulation. Those are shown in the table following the tree diagram:

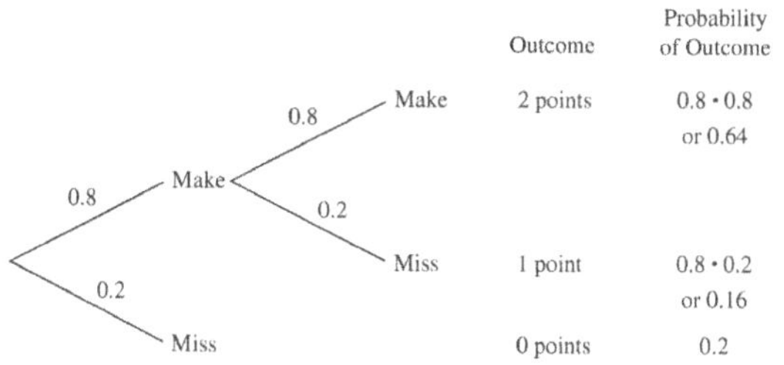

Number of Points	Expected Number of Times Points are Scored in 25 Attempts
0	5
1	4
2	16

11. Answers vary. Let the two-digit numbers 00, 01, 02, 03, …, 57 represent the 58 consonants. Choose a random place to start the random digit table. Read the numbers in pairs. If one of the pairs listed above appears in reading the table, a consonant is found. Here y is counted as a consonant.

12. **(a)** The odds in favor of rolling a sum of 7 are the ways of rolling a sum of 7 to not rolling a sum of 7, which are 6:30 or 1:5.

(b) The odds in favor of rolling a sum greater than 4 are the ways of rolling a sum greater than 4 (5,6,7,8,9,10,11,12) to not rolling a sum greater than 4 (2,3,4), which are 30:6 or 5:1.

(c) The odd in favor of rolling a sum less than or equal to 6 are the ways of rolling a sum less than or equal to 6 (2,3,4,5,6) to not rolling a sum less than or equal to 6 (7,8,9,10,11,12), which are 15:21 or 5:7.

(d) The odds in favor of rolling a sum that is an even number are the ways of rolling an even sum (2,4,6,8,10,12) to the ways of not rolling an even sum (3,5,7,9,11), which are 18:18 or 1:1.

13. If $P(boy) = \frac{1}{2}$ then $P(four\ boys)$ $= \left(\frac{1}{2}\right)^4 = \frac{1}{16}$. Odds against

$= \frac{1-P(four\ boys)}{P(four\ boys)} = \frac{\left(\frac{15}{16}\right)}{\left(\frac{1}{16}\right)} = \frac{15}{1}$

or **15:1**.

14. $\frac{1-P(win)}{P(win)} = \frac{3}{5}$ (given that odds against $= 3:5$).

$5[1 - P(win)] = 3[P(win)] \Rightarrow 5 - 5[P(win)]$ $= 3[P(win)] \Rightarrow 5 = 8[P(win)]$. $P(win)$ $= \mathbf{\frac{5}{8}}$.

15. Given $P(cat) = 0.27$, then the odds against $= \frac{1-P(cat)}{P(cat)} = \frac{0.73}{0.27} = \mathbf{73:27}$.

16. **(a)** $P(red) = \frac{5}{8} \Rightarrow$ odds in

favor$= \frac{P(red)}{1-P(red)} = \frac{\left(\frac{5}{8}\right)}{\left(\frac{3}{8}\right)} = \frac{5}{3}$ or **5:3**.

(b) Odds against

$= \frac{1-P(red)}{P(red)} = \frac{\left(\frac{3}{8}\right)}{\left(\frac{5}{8}\right)} = \frac{3}{5}$ or **3:5**.

17. The odds in favor of the person passing the test are the ways to pass the test to the ways to not pass the test, which are 60:40 or 3:2.

18. By Theorem 9-8, if the odds in favor of an event are $m:n$, then $P(E) = \frac{m}{m+n}$. Thus if the odds in favor of a grand slam are 2:9, then $P(grand\ slam) = \frac{2}{2+9} = \mathbf{\frac{2}{11}}$.

19. The prime numbers of the six on a die are 2, 3, and 5, so $P(prime) = \frac{3}{6} = \frac{1}{2}$. Then odds of a prime are $\frac{P(prime)}{1-P(prime)} = \frac{\left(\frac{1}{2}\right)}{\left(\frac{1}{2}\right)} = \frac{1}{1}$, or **1:1**.

20. Odds against

$= \frac{1-P(event)}{P(event)} = \frac{\left(\frac{5}{93}\right)}{\left(\frac{88}{93}\right)} = \frac{5}{88}$ or **5:88**.

21. If the probability of rain is 90% or 0.9, then the probability of no rain is 10% or 0.1. The odds are 0.9 : 0.1, or equivalently, **9:1**.

22. **(a)** The odds in favor of drawing a red card out a standard deck of 52 cards are the way to draw a red card to the ways to not draw a red card (black card), which are 26:26, or 1:1.

(b) The odds against drawing a spade out of a standards deck of 52 cards is the ways to not draw a spade (diamonds, hearts, clubs) to the ways to draw a spade, which are 39:13 or 3:1.

(c) If the probability of winning the prize is $\frac{1}{100}$, then the way of winning is 1 and the ways of not winning is 99, so the odds in favor of winning are 1:99.

23. $E = \$0.25 \cdot \frac{5}{25} + \$0.10 \cdot \frac{5}{25} + \$0.05 \cdot \frac{5}{25} + \$0.01 \cdot \frac{10}{25} = \$0.084$, or **about 8¢**.

24. Odds in favor of winning $= 5{:}2 \Rightarrow \frac{P(win)}{1-P(win)} = \frac{5}{2} \Rightarrow P(win) = \frac{5}{7}$. $E = \$14{,}000 \cdot \frac{5}{7} =$ **$10,000**

25. The probabilities of rolling all possible values with two dice are tabularized below:

Value	Number of Ways	Probability
2	1	$\frac{1}{36}$
3	2	$\frac{2}{36} = \frac{1}{18}$
4	3	$\frac{3}{36} = \frac{1}{12}$
5	4	$\frac{4}{36} = \frac{1}{9}$
6	5	$\frac{5}{36}$
7	6	$\frac{6}{36} = \frac{1}{6}$
8	5	$\frac{5}{36}$
9	4	$\frac{1}{9}$
10	3	$\frac{1}{12}$
11	2	$\frac{1}{18}$
12	1	$\frac{1}{36}$

$E = \$2 \cdot \frac{1}{36} + \$3 \cdot \frac{1}{18} + \$4 \cdot \frac{1}{12} + \$5 \cdot \frac{1}{9} + \$6 \cdot \frac{5}{36} + \$7 \cdot \frac{1}{6} + \$8 \cdot \frac{5}{36} + \$9 \cdot \frac{1}{9} + \$10 \cdot \frac{1}{12} + \$11 \cdot \frac{1}{18} + \$12 \cdot \frac{1}{36} = \$7.$

No. **Expected gain is less than cost. In the long run you would expect to lose $1 each time you roll the dice.**

26.

Outcome	Value of outcome	Probability of outcome	Product
HHH or TTT	$25	$\frac{2}{8}$	$\frac{50}{8}$
HH	$10	$\frac{3}{8}$	$\frac{30}{8}$
H	$0	$\frac{3}{8}$	0
	Subtotal		$\frac{80}{8} = 10$
	Expected Value (subtotal – cost of playing the game)		$10 - $10 = $0

The expected value of the game is 0. This means that it is a **fair game**.

27. If the ratio of men to women is 5:4 then the odds of a man being chosen is 5:4.

28. If the probability of spilling soup on your tie is $\frac{1}{3}$ then the probability of not spilling soup on your tie is $\frac{2}{3}$ and the odds of not spilling on your tie is 2:1.

29. If 98% of the seeds will germinate 2% will not germinate. The odds of having a seed germinate is 98:2 or 49:1.

30.

Outcome	Value of outcome	Probability of outcome	Product
2 or 12	$20	$\frac{2}{36}$	$\frac{40}{36}$
3 or 11	$10	$\frac{4}{36}$	$\frac{40}{36}$
4 or 10	$7.50	$\frac{6}{36}$	$\frac{45}{36}$
5 or 9	$5	$\frac{8}{36}$	$\frac{40}{36}$
6, 7, or 8	0	$\frac{16}{36}$	
	Subtotal		$\frac{165}{36} \approx 4.58$
	Expected Value (subtotal – cost of playing the game)		$4.58 - $5.00 = $^{-}$$0.42

(a) **The expected value is $^{-}$$0.42.** That means on average, a person playing this game will lose 42 cents each time it is played.

(b) **The game favors the person who is running the game.**

Mathematical Connections 9-3: Review Problems

21. **(a)** $P(club) = \frac{13}{52} = \mathbf{\frac{1}{4}}$. There are 13 cards of each suit in a deck of 52 cards.

(b) $P(queen\ and\ spade) = \mathbf{\frac{1}{52}}$. There is only 1 Queen of Spades in a deck.

(c) $P(not\ a\ queen)$
$= 1 - P(queen) = 1 - \frac{4}{52} =$
$1 - \frac{1}{13} = \mathbf{\frac{12}{13}}$. There are four queens in a deck of 52 cards.

(d) $P(not\ a\ heart)$
$= 1 - P(heart) = 1 - \frac{13}{52} = 1 - \frac{1}{4} = \mathbf{\frac{3}{4}}$.

(e) $P(spade\ or\ heart)$
$= P(spade) + P(heart) =$
$\frac{13}{52} + \frac{13}{52} = \frac{26}{52} = \mathbf{\frac{1}{2}}$. These are mutually exclusive events.

(f) $P(6\ of\ diamonds) = \mathbf{\frac{1}{52}}$. There is only one 6 of Diamonds.

(g) $P(queen\ or\ spade) = P(queen) + P(spade) - P(queen\ and\ spade) = \frac{4}{52} + \frac{13}{52} - \frac{4}{52} \cdot \frac{13}{52} =$
$\frac{1}{13} + \frac{1}{4} - \frac{1}{13} \cdot \frac{1}{4} = \mathbf{\frac{4}{13}}$. These are non-mutually exclusive events.

(h) $P(either\ red\ or\ black) = \frac{1}{2} + \frac{1}{2} = \mathbf{1}$. This is a certain event.

23. If the probability of making a free throw is $\frac{1}{3}$, the probability of missing is $\frac{2}{3}$. The probability of missing three in a row, assuming the player's skill level is unchanged from shot to shot, would be $\frac{2}{3} \cdot \frac{2}{3} \cdot \frac{2}{3} = \mathbf{\frac{8}{27}}$.

25. $P(blue) \cdot P(blue) = \frac{25}{36} = \frac{5}{6} \cdot \frac{5}{6}$ Thus the **blue** section must have $\frac{5}{6} \cdot 360° = \mathbf{300°}$ and the **red** section must have $360° - 300° = \mathbf{60°}$.

Assessment 9-4A: Permutations and Combinations in Probability

1. Each coin has two possible outcomes for each toss: heads or tails. The total number of outcomes is: $2^4 = 16$.

2. **(a)** $2 \cdot 6 = 12$

(b) $2 \cdot 2 \cdot 6 \cdot 6 = 144$

(c) $2 \cdot 2 \cdot 2 \cdot 6 = 48$

(d) $2 \cdot 2 \cdot 52 = 208$

(e) $6 \cdot 6 \cdot 52 = 1872$

(f) $2 \cdot 6 \cdot 52 = 624$

(g) $2 \cdot 2 \cdot 6 \cdot 6 \cdot 52 \cdot 52 = 389{,}376$

(h) $2 \cdot 2 \cdot 6 \cdot 6 \cdot 52 \cdot 51 = 381{,}888$

(i) $52 \cdot 52 \cdot 52 = 140{,}608$

(j) $52 \cdot 51 \cdot 50 = 132{,}600$

3. The event "girls" can occur in 16 ways; the event "boys" can occur in 14 ways. Thus the event "girls *and* boys" can occur as $16 \cdot 14 =$ **224 unique pairings**.

4. The number of ways the four digits after the prefix may be arranged is $10 \cdot 10 \cdot 10 \cdot 10 = 10^4 = 10{,}000$. Thus **10,000 numbers** can be associated with each prefix (assuming all can be used).

5. There are $3 \cdot 15 \cdot 4 =$ **180** possible different three-course meals.

6. **(a)** **True**. $6! = 6 \cdot (5 \cdot 4 \cdot 3 \cdot 2 \cdot 1) = 6 \cdot 5!$.

(b) **False.** $3! + 3! = 3 \cdot 2 \cdot 1 + 3 \cdot 2 \cdot 1 = 12 \neq 6!$.

(c) **False.** $\frac{6!}{3!} = \frac{6 \cdot 5 \cdot 4 \cdot 3 \cdot 2 \cdot 1}{3 \cdot 2 \cdot 1} = 6 \cdot 5 \cdot 4 = 120 \neq 2!$.

7. **(a)** There are 8 unlike letters in SCRAMBLE. $\mathbf{8! = 40{,}320}$.

(b) There are 9 unlike letters in PERMUTATION and 2 like letters. $\mathbf{\frac{11!}{2!} = 19{,}958{,}400}$.

8. Since order is not distinct the number of two-person committees is a combination of six persons taken two at a time.
${}_6C_2 = \frac{6!}{2!(6-2)!} = \frac{6 \cdot 5}{2 \cdot 1} =$ **15 committees**.

9. **(a)** Order is distinct, implying a permutation of 30 persons taken three at a time. ${}_{30}P_3 = \frac{30!}{(30-3)!} = 30 \cdot 29 \cdot 28 = \mathbf{24{,}360}$ ways in which to select the three officers.

(b) Order is not distinct, implying a combination of 30 persons taken three at a time. ${}_{30}C_3 = \frac{30!}{(30-3)!3!} = 29 \cdot 14 \cdot 10 = \mathbf{4060}$ **ways** in which to choose three-person committees.

10. There are nine flags with four (red), three (green), and two (white) repeated. $\frac{9!}{4!3!2!} = \frac{362{,}880}{24 \cdot 6 \cdot 2} =$ **1260 possible** signals.

11. If there were n people at the party there were n combinations of people two at a time shaking hands. ${}_nC_2 = \frac{n!}{(n-2)!2!} = \frac{n(n-1)}{2!} = 28 \Rightarrow$ $n(n-1) = 56$. This is the product of two consecutive whole numbers; look for factors of 56 which yield $8 \cdot 7$. Thus $n(n-1) = 8 \cdot 7$, and $n = \mathbf{8}$ **people** at the party.

12. **(a)** $\text{P(2 spades)} = \frac{13}{52} \cdot \frac{13}{52} = \frac{169}{2704} = \frac{1}{16}$

(b) $\text{P(2 spades)} = \frac{13}{52} \cdot \frac{12}{51} = \frac{156}{2652} = \frac{1}{17}$

(c) $\text{P(club then diamond)} = \frac{13}{52} \cdot \frac{13}{51}$

$= \frac{169}{2652} = \frac{13}{204}$

(d) $\text{P(2 kings)} = \frac{4}{52} \cdot \frac{3}{51} = \frac{12}{2652} = \frac{1}{221}$

13. **(a)** $26 \cdot 10 \cdot 26 \cdot 10 \cdot 26 \cdot 10 = 26^3 \cdot 10^3$

$= 17{,}576{,}000$

(b) $26 \cdot 9 \cdot 26 \cdot 10 \cdot 26 \cdot 10 = 26^3 \cdot 10^2 \cdot 9$

$= 15{,}818{,}400$

14. $31 \cdot 30 = 930$ possible double-decker cones.

15. Order is not important, so the number of ways is ${}_6C_4 = \frac{6!}{4!\cdot 2!} = \frac{720}{24 \cdot 2} =$ **15 ways** of presenting the awards.

16. **(a)** **Permutation** because the order matters. The first prize is different from the second prize and is different from the third prize.

(b) $10 \cdot 9 \cdot 8 = \mathbf{720}$

17. P(no digits or letters repeating)

$= \frac{26 \cdot 25 \cdot 24 \cdot 10 \cdot 9 \cdot 8}{26 \cdot 26 \cdot 26 \cdot 10 \cdot 10 \cdot 10} = \frac{25 \cdot 24 \cdot 9 \cdot 8}{26 \cdot 26 \cdot 10 \cdot 10}$

$= \frac{43{,}200}{67{,}600} = \frac{432}{676} = \frac{108}{169} \approx \mathbf{64\%}$

18. Five volumes may be placed in $5! = 120$ different ways; only one will be in correct order. $P(\textit{correct order}) = \mathbf{\frac{1}{120}}$.

19. **(a)** ${}_{69}C_5 = \frac{69!}{5!(69-5)!}$

$= \frac{69 \cdot 68 \cdot 67 \cdot 66 \cdot 65 \cdot 64!}{5 \cdot 4 \cdot 3 \cdot 2 \cdot 1 \cdot 64!} = 11{,}238{,}513$

(b) ${}_{69}C_5 \cdot {}_{26}C_1 = \frac{69!}{5!(69-5)!} \cdot \frac{26!}{1!(26-1)!}$

$= \frac{69 \cdot 68 \cdot 67 \cdot 66 \cdot 65 \cdot 64!}{5 \cdot 4 \cdot 3 \cdot 2 \cdot 1 \cdot 64!} \cdot \frac{26 \cdot 25!}{1 \cdot 25!}$

$= 11{,}238{,}513 \cdot 26 = 292{,}201{,}338$

(c) $P(\text{winning}) = \frac{1}{292{,}201{,}338}$

20. **(a)** ${}_{25}C_6 = \frac{25!}{6!(25-6)!}$

$= \frac{25 \cdot 24 \cdot 23 \cdot 22 \cdot 21 \cdot 20 \cdot 19!}{6 \cdot 5 \cdot 4 \cdot 3 \cdot 2 \cdot 1 \cdot 19!}$

$= \frac{127{,}512{,}000}{720} = 177{,}100$

(b) ${}_{10}C_2 \cdot {}_8C_2 \cdot {}_7C_2$

$= \frac{10!}{2!(10-2)!} \cdot \frac{8!}{2!(8-2)!} \cdot \frac{7!}{2!(7-2)!}$

$= \frac{10 \cdot 9 \cdot 8!}{2 \cdot 1 \cdot 8!} \cdot \frac{8 \cdot 7 \cdot 6!}{2 \cdot 1 \cdot 6!} \cdot \frac{7 \cdot 6 \cdot 5!}{2 \cdot 1 \cdot 5!}$

$= \frac{90}{2} \cdot \frac{56}{2} \cdot \frac{42}{2} = 45 \cdot 28 \cdot 21 = 26{,}460$

(c) $\frac{26{,}460}{177{,}100} = \frac{189}{1265} \approx 15\%$

21. **(a)** $P(\textit{all girls}) = \frac{{}_{12}C_4}{{}_{22}C_4} = \frac{495}{7315} = \mathbf{\frac{9}{133}}$.

(b) $P(\textit{all boys}) = \frac{{}_{10}C_4}{{}_{22}C_4} = \mathbf{\frac{6}{209}}$.

(c) $P(\textit{at least one girl}) = 1 - P(\textit{all boys}) =$
$1 - \frac{_{10}C_4}{_{22}C_4} = 1 - \frac{210}{7315} = 1 - \frac{6}{209} = \mathbf{\frac{203}{209}}.$

22. **(a)** $P(\textit{two Britons, four Italians, two Danes}) =$
$\frac{_{20}C_2 \cdot {_{21}C_4} \cdot {_4C_2}}{_{45}C_8} = \frac{190 \cdot 5985 \cdot 6}{215{,}553{,}195}$, or about **0.032**.

(b) $P(\textit{no Britons}) = \frac{_{20}C_0 \cdot {_{25}C_8}}{_{45}C_8}$, or about **0.005**.

(c) $P(\textit{at least one Briton}) = 1 - P(\textit{no Britons}) = 1 - \frac{_{20}C_0 \cdot {_{25}C_8}}{_{45}C_8} \approx \mathbf{0.995}.$

(d) $P(\text{all } \textit{Britons}) \; \frac{_{20}C_8 \cdot {_{25}C_0}}{_{45}C_8}$, or about $\mathbf{5.84 \cdot 10^{-4}}$.

23. Seven people can stand in a line $7 \cdot 6 \cdot 5 \cdot 4 \cdot 3 \cdot 2 \cdot 1 = 5040$ ways.

24. **(a)** Since numbers can be repeated, there would be $10^4 = \mathbf{10{,}000}$ possibilities.

(b) If numbers cannot be repeated, there would be $10 \cdot 9 \cdot 8 \cdot 7 = \mathbf{5040}$ possibilities.

(c) There are five choices for each digit. So there are $5^4 = 625$ ways to create four digit numbers where the digits are all even.
$P(\textit{all even}) = \frac{625}{10{,}000} = \mathbf{\frac{1}{16}}.$

25. Order is not important, so the number of choices is $_8C_3 = \frac{8!}{3!5!} = \frac{40{,}320}{6 \cdot 120} =$ **56 choices** for the exercise.

26. **(a)** The word "quick" has 5 letters. If all letters are used $5 \cdot 4 \cdot 3 \cdot 2 \cdot 1 = 120$ arrangements can be made.

(b) If three letters are used, $5 \cdot 4 \cdot 3 = 60$ arrangements can be made.

(c) If four letters are used, $5 \cdot 4 \cdot 3 \cdot 2 = 120$ arrangements can be made.

Mathematical Connections 9-4: Review Problems

13. **(a)** $P(\textit{at least one ace}) = 1 - P(\textit{zero aces}) =$
$1 - \frac{48}{52} \cdot \frac{47}{51} = 1 - \frac{2256}{2652} = 1 - \frac{188}{221}$
$= \mathbf{\frac{33}{221}}.$

(b) $P(\textit{one card red}) = P(\textit{red and black}) + P(\textit{black and red}) = \frac{26}{52} \cdot \frac{26}{51} + \frac{26}{52} \cdot \frac{26}{51} =$
$\frac{13}{51} + \frac{13}{51} = \mathbf{\frac{26}{51}}.$

15. **(*i*)** Possible outcomes:
$S = \{(hh), (ht), (th), (tt)\}.$
$P(\textit{both heads}) = \frac{1}{4}$; $P(\textit{both tails}) = \frac{1}{4}$;
$P(\textit{no match}) = \frac{1}{2}.$

$E = \$5 \cdot \frac{1}{4} + \$3 \cdot \frac{1}{4} + {}^{-}\$4 \cdot \frac{1}{2} =$
$\$\frac{5}{4} + \$\frac{3}{4} + {}^{-}\$\frac{8}{4} = \mathbf{\$0}.$

(*ii*) **Yes**. The expected outcome is 0.

17. We first find the probability that all 40 children have different birthdays and then subtract the result from 1. We get

$$1 - \left(\frac{365-1}{365}\right)\left(\frac{365-2}{365}\right)\cdots\left(\frac{365-39}{365}\right) \approx 0.89.$$

Thus the probability that the friend wins the bet is approximately 0.89 or 89%. For 50 people, the probability rises to approximately 97%.

Chapter 9 Review

1. **(a)** $S =$ **{(*hhh*), (*hht*), (*hth*), (*htt*), (*ttt*), (*tth*), (*thh*), (*tht*)}**. There are $2^3 =$ eight elements
in the sample space.

(b) "At least two" means two or three heads, so $S =$ **{(*hhh*), (*hht*), (*hth*), (*thh*)}**.

(c) $P(\textit{at least two heads})$
$= \frac{n(\textit{at least two heads})}{n(S)} = \frac{4}{8} = \mathbf{\frac{1}{2}}.$

2. **(a)** $S = \{(\textit{Sunday}), (\textit{Monday}), (\textit{Tuesday}), (\textit{Wednesday}), (\textit{Thursday}), (\textit{Friday}), (\textit{Saturday})\}.$

(b) $S = \{(\textit{Tuesday}), (\textit{Thursday})\}.$

(c) $P(\textit{day starting with } T) = \mathbf{\frac{2}{7}}.$

3. Answers vary. For example:

(*i*) $\frac{4}{5} \cdot 1000 = 800$, so there are 800 blue beans;

(*ii*) $\frac{1}{8} \cdot 1000 = 125$, so there are 125 red beans;

(*iii*) $\frac{4}{5} + \frac{1}{8} < 1$, so there are jelly beans in the jar that are neither red nor blue; or (***iv***) There are $1000 - 800 - 125 = 75$ jelly beans that are neither red nor blue.

4. **(a)** Shirts and pants combinations: {(Y,Y), (Y,P), (Y,O), (Y,G), (G,Y), (G,P), (G,O), (G,G), (R,Y), (R, P), (R,O), (R,G)}

(b) $P(Y,O) = \frac{1}{12}$

(c) P(at least 1 item yellow)=

P(1 yellow item) + P(2 yellow items)=

$\frac{5}{12} + \frac{1}{12} = \frac{6}{12} = \frac{1}{2}$

(d) P(purple pants)=$\frac{3}{12} = \frac{1}{4}$

(e) P(shirt and pants are different colors)

$= \frac{10}{12} = \frac{5}{6}$

5. **(a)** $P(black) = \frac{n(black)}{n(S)} = \mathbf{\frac{5}{12}}$.

(b) $P(black\ or\ white) = P(black) + P(white)$ $= \frac{5}{12} + \frac{4}{12} = \frac{9}{12} = \mathbf{\frac{3}{4}}$. These are mutually exclusive events.

(c) *P(neither red nor white)* $= P(black) = \mathbf{\frac{5}{12}}$.

(d) $P(red\ not\ drawn) = P(black\ or\ white)$ $= \mathbf{\frac{3}{4}}$.

(e) $P(black\ and\ white) =$ **0**. An impossible event.

(f) $P(black\ or\ white\ or\ red) =$ **1**. A certain event.

6. **(a)** $P(club) = \frac{n(club)}{n(S)} = \frac{13}{52} = \mathbf{\frac{1}{4}}$.

(b) $P(spade\ and\ 5) = \mathbf{\frac{1}{52}}$. There is only one 5 of spades.

(c) $P(heart\ or\ face\ card) = P(heart) + P(face\ card) - P(heart\ and\ face\ card) =$ $\frac{13}{52} + \frac{12}{52} - \frac{13}{52} \cdot \frac{12}{52} = \frac{1}{4} + \frac{3}{13} - \frac{3}{52} = \mathbf{\frac{11}{26}}$.

(d) $P(jack\ not\ drawn) = 1 - P(jack) =$ $1 - \frac{4}{52} = \frac{48}{52} = \mathbf{\frac{12}{13}}$.

7. **(a)** $P(all\ white) = \frac{4}{9} \cdot \frac{4}{9} \cdot \frac{4}{9} = \mathbf{\frac{64}{729}}$. With replacement the probabilities of each draw are the same.

(b) $P(all\ white) = \frac{4}{9} \cdot \frac{3}{8} \cdot \frac{2}{7} = \frac{24}{504} = \mathbf{\frac{1}{21}}$. Without replacement the probabilities are changed for each draw.

8. Either an L can be drawn from box **1** and an L from box 2 or not an L from box 1 and an L from box 2.
$P(L) = \frac{1}{5} \cdot \frac{2}{5} + \frac{4}{5} \cdot \frac{1}{5} = \frac{2}{25} + \frac{4}{25} = \mathbf{\frac{6}{25}}$.

9. $P(A\ from\ any\ box) = P(choosing\ that\ box) \cdot P(A\ from\ chosen\ box)$. The probabilities are added because the probabilities of A from each of the boxes are mutually exclusive events.
$P(A) = \frac{1}{4} \cdot \frac{0}{2} + \frac{1}{4} \cdot \frac{1}{4} + \frac{1}{4} \cdot \frac{1}{4} + \frac{1}{4} \cdot \frac{1}{5} =$
$0 + \frac{1}{16} + \frac{1}{16} + \frac{1}{20} = \mathbf{\frac{7}{40}}$.

10. **(*i*)**

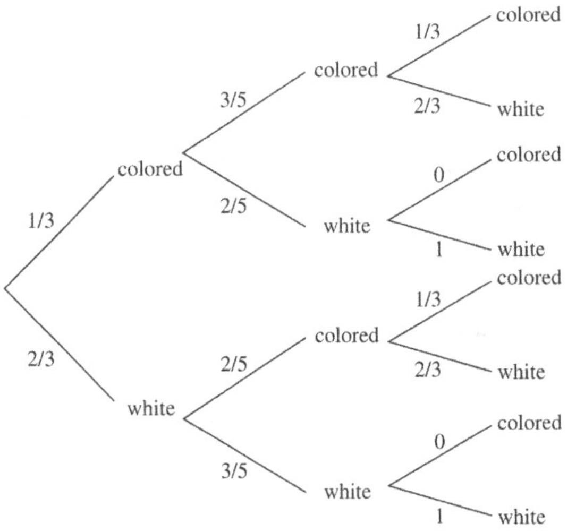

(*ii*) ***P(colored)***
$= \frac{1}{3} \cdot \frac{3}{5} \cdot \frac{1}{3} + \frac{1}{3} \cdot \frac{2}{5} \cdot 0 + \frac{2}{3} \cdot \frac{2}{5} \cdot$
$\frac{1}{3} + \frac{2}{3} \cdot \frac{3}{5} \cdot 0 = \frac{3}{45} + 0 + \frac{4}{45} + 0 = \mathbf{\frac{7}{45}}$.

11. $P(jack) = \frac{4}{52} = \frac{1}{13}$. Odds in favor $=$
$\frac{P(jack)}{1-P(jack)} = \frac{\left(\frac{1}{13}\right)}{\left(\frac{12}{13}\right)} = \frac{1}{12}$ or **1:12.**

12. $P(prime\ number) = \frac{3}{6} = \frac{1}{2}$. Odds against a prime number $= \frac{1-P(prime\ number)}{P(prime\ number)} = \frac{\left(\frac{1}{2}\right)}{\left(\frac{1}{2}\right)} = \frac{1}{1}$ or **1:1**.

13. $\frac{P(event)}{1-P(event)} = \frac{3}{5} \Rightarrow 5.P(event) = 3 \cdot$
$[1 - P(event)] \Rightarrow 5 \cdot P(event) = 3 - 3 \cdot$
$P(event) \Rightarrow 8 \cdot P(event) = 3 \Rightarrow P(event) = \mathbf{\frac{3}{8}}$.

Note that by Theorem 9-8 if the odds in favor of an event are m to n, the probability of that event is $\frac{m}{m+n}$.

14. $P(double\ 1\text{'s}) = \frac{1}{36}$; $P(double\ 6\text{'s}) = \frac{1}{36}$.
$E = \$7.20 \cdot \frac{1}{36} + \$3.60 \cdot \frac{1}{36} + \$0 \cdot \frac{34}{36} =$
\$0.30 or **30¢**.

15. $P(win) = \frac{1}{3000}$. $E = \$1000 \cdot \frac{1}{3000} = \$\frac{1}{3} = 33\frac{1}{3}$¢.
There is **no actual value,** to the nearest cent, that would produce a fair game.

16. $S = 10^4 \Rightarrow P(win) = \frac{1}{10{,}000}$. $E = \$15{,}000 \cdot \frac{1}{10{,}000} =$ **\$1.50**. If the ticket costs \$2, expected earnings are $\$1.50 - \$2.00 = {}^{-}50$¢.

17. There are $9 \cdot 10 \cdot 10 \cdot 1 =$ **900** possible different numbers.

18. Order is not distinct, implying a combination of 10 people taken 3 at a time.
${}_{10}C_3 = \frac{10!}{(10-3)!3!} =$ **120** possible ways.

19. Order is distinct, implying a permutation of 10 flags taken 4 at a time. ${}_{10}P_4 = \frac{10!}{(10-4)!} =$
$10 \cdot 9 \cdot 8 \cdot 7 =$ **5040** possible different ways.

20. $P(both\ blue) = \frac{2}{5} \cdot \frac{1}{4} = \frac{2}{20} = \mathbf{\frac{1}{10}}$.

21. **(a)** Order is distinct, implying a permutation of 5 finishes taken 3 at a time. ${}_5P_3 = \frac{5!}{(5-3)!} =$
$5 \cdot 4 \cdot 3 =$ **60** possible different finishes.

(b) There are ${}_5P_2 = 20$ possible ways for a first/second place finish; there is only one way for a Deadbeat/Bandy finish. *P*(*Deadbeat*/ *Bandy finish*) $= \mathbf{\frac{1}{20}}$.

(c) There are ${}_5P_3 = 60$ possible first-, second-, and third-place finishes; there is only one way for a Deadbeat/Egglegs/Cash finish.
P(*Deadbeat*/*Egglegs*/*Cash finish*) $= \mathbf{\frac{1}{60}}$.

22. If they roll the same number no one wins so half of all possible outcomes minus the ties is how many ways one can win. The possible outcomes of Al's roll and Ruby's roll are: {(1, 2), (1, 3), (1, 4), (1, 5), (1, 6), (2, 3), (2, 4), (2, 5), (2, 6), (3, 4), (3, 5), (3, 6), (4, 5), (4, 6), (5, 6)}, so there are 15 ways for Ruby to roll a higher number than Al. $P(Ruby > Al) =$
$\frac{15}{36} = \mathbf{\frac{5}{12}}$.

23. Order is not distinct, implying a combination of 5 questions taken 3 at a time. There are ${}_5C_3$ ways of selecting three questions out of five; since the number 1 is not to be selected, there are four options left and three of them will be selected. Thus *P*(1 *not chosen*)
$= \frac{{}_4C_3}{{}_5C_3} = \frac{4}{10} = \mathbf{\frac{2}{5}}$.

24. Since fourth and last bats are fixed, there are 7 remaining positions to consider. $7! =$ **5040** ways.

25. $P(all\ green) = 0.3 \cdot 0.3 \cdot 0.3 =$ **0.027**.

26. $P(success) = 1 - P(failure)$. *P*(*second stage success*) $= 1 - \frac{1}{8} = \frac{7}{8}$; *P*(*third stage success*) $= 1 - \frac{1}{10} = \frac{9}{10}$. *P*(*success given stage one*) $= \frac{7}{8} \cdot \frac{9}{10} = \mathbf{\frac{63}{80}}$.

27. Answers vary. For example:

(a) Randomly select digits 1 through 6 from a random digit table; discard digits 0 and 7 through 9.

(b) Block off twelve two-digit blocks in a random digit table (discarding any other than 01 through 12); randomly select three.

(c) Let random digits 0 through 2 represent red; digits 3 through 5 represent white; digits 6 through 8 represent blue; discard any nines.

28. The events are not equally likely. The sample space consists of eight possibilities, so computing the probabilities yields: *P*(3 *heads*) $= \frac{1}{8}$;
P(2 *heads*) $= \frac{3}{8}$; *P*(1 *head*) $= \frac{3}{8}$; and
P(0 *heads*) $= \frac{1}{8}$.

29. My expected value is $\left(\frac{1}{6}\right) \cdot 2 = \frac{1}{3} = 0.\overline{3}$. My friend's expected value is $\left(\frac{1}{12}\right) \cdot 3 = \frac{1}{4} = 0.25$.
The game is not fair.

30. $d(\overline{MQ}) = 20$; $d(\overline{NO}) = 8$.
P(between *N* and *O*) $= \frac{8}{20} = \mathbf{\frac{2}{5}}$.

31. *P*(at least one face card)=1 – *P*(no face card) $= 1 - \frac{{}_{40}C_3}{{}_{52}C_3} = 1 - \frac{\frac{40!}{3!37!}}{\frac{52!}{3!49!}} = \frac{47}{85} \approx 0.553$

32. There are 5 shades of yellow in a box of 64 Crayola crayons. The probability that a crayon chosen at random from the box of crayons is not a shade of yellow

$$P(\text{not a shade of yellow}) = 1 - P(\text{shade of yellow})$$
$$= 1 - \frac{5}{64} = \frac{59}{64}$$

33. The probability of drawing two crayons without replacement that are the most used colors is P(2 of the most used colors)=

1-P(none of the most used colors)

$$=1-\left(\frac{60}{64}\cdot\frac{59}{63}\right)=1-\frac{3540}{4032}=\frac{492}{4032}\approx 0.122$$

34. The probability of choosing the United States out of the two countries is $\frac{1}{2}$. The probability of choosing a well-known zoo in the United States is $\frac{1}{350}$. The probability of choosing a well-known zoo in the United States if you choose one of the two countries at random and one of the zoos is chosen at random is $\frac{1}{2}\cdot\frac{1}{350}=\frac{1}{700}$.

35. **(a)** P(1st prize) = $\frac{1}{1000}$; P(2nd prize) = $\frac{2}{1000}=\frac{1}{500}$; P(3rd prize) = $\frac{100}{1000}=\frac{1}{10}$;

P(any prize) = $\mathbf{\frac{103}{1000}}$

(b) Expected Value =

$$\$100\left(\frac{1}{1000}\right)+\$50\left(\frac{2}{1000}\right)+\$2\left(\frac{100}{1000}\right)=$$
$$\frac{100}{1000}+\frac{100}{1000}+\frac{200}{1000}=\frac{400}{1000}=\frac{2}{5}=0.40.$$

The expected value is **40 cents**.

(c) \$0.40 - \$0.50 = **⁻\$0.10**.

36. If the probability that an engine fails is 0.01, then the probability of an engine working is 1- 0.01 = 0.99.

(a) Probability of a successful flight with 2 engines is equal to one or two engines working which is the complement of no engines working.

$$1 - P(\text{no engines working}) = 1-(0.01\cdot 0.01)$$
$$= 1-(0.0001) = 0.9999$$

(b) Probability of a successful flight with 4 engines is two or three or four engines working which is the complement of no engines working or one engine working.

P(no engines working) = $0.01\cdot 0.01\cdot 0.01\cdot 0.01 = 0.00000001$

P(one engine working)=

P(1st engine working)=$0.99\cdot 0.01\cdot 0.01\cdot 0.01$=0.00000099

+P(2nd engine working)=$0.01\cdot 0.99\cdot 0.01\cdot 0.01$=0.00000099

+P(3rd engine working)=$0.01\cdot 0.01\cdot 0.99\cdot 0.01$=0.00000099

+P(4th engine working)=$0.01\cdot 0.01\cdot 0.01\cdot 0.99$=0.00000099

=0.00000099+0.00000099+0.00000099+0.00000099

=0.00000396

P(no engines working) or P(one engine working)=

0.00000001+0.00000396=0.00000397

P(successful flight)=1-0.000000397=0.99999603.

37. The probability of Tamara getting the same number of heads as Alan is equal to $\frac{3}{8}$. Tamara has four possible outcomes (HH, HT, TH, TT), Alan has 2 possible outcomes (H or T). Together there are 8 possible outcomes (Alan tosses heads then Tamara tosses HH, HT, TH, or TT; Alan tosses tails and Tamara tosses HH, HT, TH, or TT). Out of these 8 possible outcomes 3 have the same number of heads (One head: Alan tosses Heads, Tamara tosses HT or TH; No head: Alan tosses Tails, Tamara tosses TT).

38. $5\cdot 5\cdot 5\cdot 5\cdot 5\cdot 5\cdot 5\cdot 5\cdot 5\cdot 5 = 5^{10}$
$= 9{,}765{,}625$

39. The books can be in the following order: ABC, ACB, BAC, BCA, CAB, CBA. Only ABC is in alphabetical order. P(alphabetical order) = $\frac{1}{6}$.

CHAPTER 10

DATA ANALYSIS/STATISTICS: AN INTRODUCTION

Assessment 10-1A: Designing Experiments/Collecting Data

This entire assessment is subject to varying answers. Each of the following are representative possibilities.

1. **Either an experimental or observational study.** The best type of study to use would be an experimental study. However, because some parents might object to their children using calculators in the classroom, an experimental study may not be possible. Therefore, an observational study would be best, which would have calculators available for student use, but not require it.
2. Some students may look at the measuring cup as being "exactly one cup" at a different level than other students, due to the curvature of the liquid in the measuring cup.
3. It depends on what is being measured, how accurate the measures must be, and the experience that first-graders may have in using a ruler. In general, it would be best to ask for precise measures, but if a student does not know how to accurately read a ruler, having them round to the nearest inch would be fine. Once they gain experience rounding to the nearest inch, they could then measure and round to the nearest half inch, then the nearest quarter inch, and work their way to more accurate measures in this fashion.
4. The sample in **(i)** is more likely to be random since it assures that students in different grades are asked. The sample in **(ii)** is likely to be biased because friends sitting together in the library are likely to be in the same grade and may have similar tastes.
5. **(a)** These students might arrive early for a specific reason (they all rode the same bus, they need to report to detention or a sport/activity), and might over-represent a certain segment of students.

   (b) Students who line up in a row do so oftentimes for a specific reason; they are from one class, or they are a large group of friends, etc. In any case, the potential to over-represent a certain segment of students is high.

   (c) Not all students from the school will attend the soccer game (for example, students who work Friday nights, or students who do not have transportation, or students who simply do not like soccer). In any case, the sample would not be random.

   (d) Not all students will bother to take the time to fill out the questionnaire and drop it in the drop box. Usually, only students who feel strongly about a topic will take the time to respond to it in this manner. Also, there is no guarantee that each student who puts a questionnaire in the drop box filled out exactly one.
6. **(a)** The population is the soccer fans at the stadium. The sample is the 50 fans surveyed. The sample is representative of the population.

   (b) The population consists of the students who attend that particular school. The sample is the 75 students surveyed. The sample is **not** representative of the population, since only those students who show up early are surveyed (they may all come on one bus, for example, or all live near the school). It also misses any student who may enter the school by other means than the main gate.

   (c) The population under study are the households in the town. The sample consists of the students who attend that school that day. The sample is **not** representative of the population, since only students are surveyed. This eliminates any household that does not have students attending that particular school.

   (d) The population under study is the students at the school. Sample: 5 students from each grade. The sample **may or may not be** representative of the population. For example, if the school had 150 first graders and only 30 sixth graders, taking 5 students from each class, even if done randomly, would not reflect the population at large. However, if the number of students in each grade are approximately the same, then the method would be representative of the population.

7. Among the questions the class must determine are:

(*i*) Do you count the houses all around the block or only on the side of your house?

(*ii*) Are the houses across the street from you on your block, and do you count them?

(*iii*) What happens if you are in a new part of town where the blocks are not developed yet?

(*iv*) Do you count any businesses that might be on your block?

Data to be collected will be determined by the questions asked, but in the second grade will likely be a frequency count.

The frequency count could be shown in a histogram or a bar graph. Any interpretations would be made about the graph.

8. Among the questions the class must determine are:

(*i*) What is the definition of "active?"

(*ii*) What are the types of activities; does one group do them more than the other?

(*iii*) Are these leisure or working activities?

(*iv*) Does one group sleep more than the other?

(*v*) What is an "adult?"

9. Among the questions the class must determine are:

(*i*) What is the definition of "classroom?" Does the library count as a classroom? What about auditoriums, the lunch room, the gymnasium, offices, hallways, etc.?

(*ii*) When will the temperature be measured in each location…same time in every room? Or, will the measure be a maximum temperature for the room during the class day, regardless of time? Will it be measured several times per day in each room?

(*iii*) What will be used to measure the temperature? How accurate does the measure have to be?

(*iv*) Where in each room will the temperature be measured? Near the middle, near a wall, by each teacher's desk, etc.?

10. The question in **(a) is fair** because it does not make any assumptions about favorite subjects. The question in **(b)** is **biased** because it uses words such as "refreshing" and "ice-cold…on a hot day" to suggest a person should like soda. The question in **(c)** is **fair** because it is not making any assumptions about the subject, and is not using any persuasive words.

11. **(a)** Elementary students might simply blindfold the adults and hand them each two different soda cans to see if they really could taste the difference.

(b) Older and more sophisticated students might pour the sodas into unmarked cups and use a series of taste tests to counter random guessing. A double-blind test might eliminate unintended hints.

12. Among the questions the class must determine are:

(*i*) What is the definition of "visiting?" An airline layover? Overnight? An extended vacation?

(*ii*) What is the definition of "a country?" For example, would Puerto Rico qualify? Would the Cayman Islands (a British territory) qualify?

(*iii*) If a country is now separated from another, such as with the former Soviet Union, are visits before the breakup to now-independent states counted?

(*iv*) If a country was a protectorate of another state when a visit occurred, does it count?

13. A strong criticism of the prediction is that unrepresentative voters were contacted. Not many owned automobiles and/or telephones in 1936.

14. How are "positive" and "negative" comments determined? A better and fairer way would be to list a representative sample of comments on the website.

15. It must be determined how accurate the second-grade observations really are. Second graders may not actually see what shoes are worn, but instead use personal knowledge of each other; e.g., "Alphie always wears tennis shoes."

It might be observed that on Tuesday the most popular are tennis shoes and crocs, but an equally valid interpretation might be that the students just prefer soft-soled shoes.

16. The student is mixing up percentage of effectiveness with the percentage of taking the drug. Also, it says "up to 92% effective," which means it might not be effective at all…so taking more won't make any difference.

Assessment 10-2A: Displaying Data: Part I

1. (a) **Tuesday**.

(b) There were approximately 4500 pieces of mail processed on Tuesday and 3000 on Monday. $4500 - 3000 = \mathbf{1500}$.

(c) **Three**.

2. In the following pictograph, each glass represents 10 glasses, each half-glass symbol represents 5 glasses:

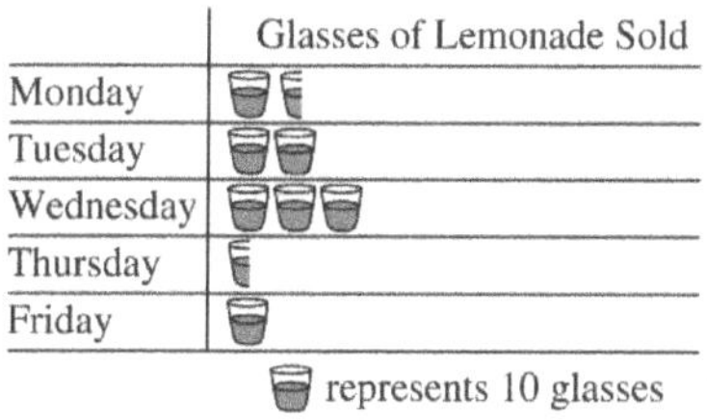

3. **The graph shows that the volume increased by a factor of 8**. If the radius and height for the 2017 cylinder are r and h, respectively, then the radius and height for the 2018 cylinder would be $2r$ and $2h$. The volume of the 2018 cylinder is:

$$V = \pi r^2 h \Rightarrow V = \pi(2r)^2 2h \Rightarrow$$
$$V = \pi(4r^2)(2h) \Rightarrow V = 8\pi r^2 h$$

which is eight times the volume of the 2017 cylinder.

4.

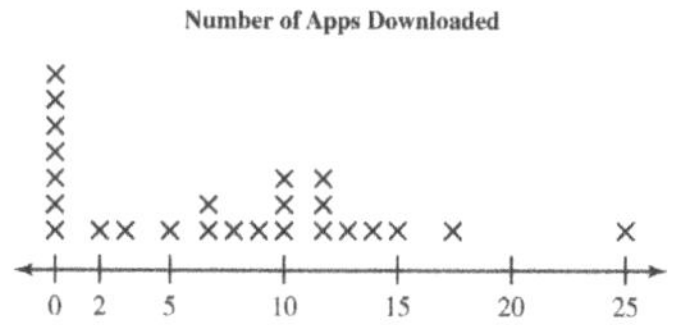

5. Data may vary:

(a) A dot plot (line plot) would have six columns of x's, one for each number on a die. Given a fair die, the six columns would be expected to have approximately the same number of x's, with some variation.

(b) A bar graph would have six bars, one for each number on a die. The scale on the vertical axis should be uniform, and be large enough to show the frequencies of all bars. Given a fair die, the six bars would be expected to be approximately the same height, with some variation.

6. (a) If 7|24 represents weights of 72 and 74 pounds, weights of the fifteen students are: **72, 74, 81, 81, 82, 85, 87, 88, 92, 94, 97, 98, 103, 123, and 125 pounds**.

(b) **72 pounds**.

(c) **125 pounds**.

(d) The histogram below groups weights into 10 pound classes:

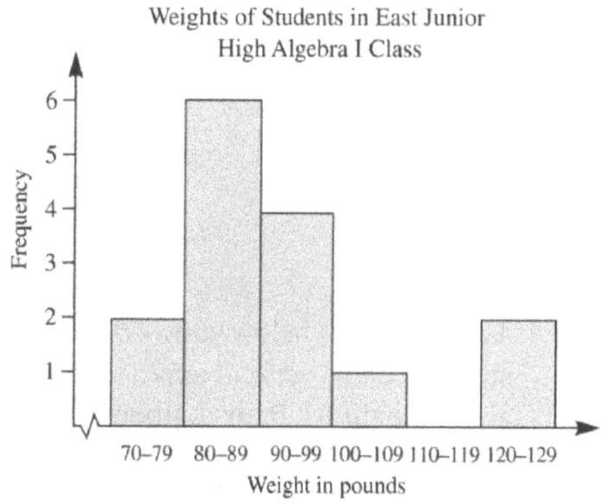

7. (a) Ages of HKM employees:

1	889
2	01113333334566679
3	224447
4	1115568
5	2248
6	233

3 | 4 represents 34 years old

(b) There are more employees in their **40's**. The seven in their 40's are shown by 4|1115568; the four in their 50's are shown by 5|2248.

(c) **20**: three in their teens; seventeen in their 20's.

(d) Seven of the 40 total are 50 or older, $\frac{7}{40} = 0.175 = \mathbf{17.5\%}$.

8. (a) Ordered stem-and-leaf plot showing fall textbook costs. Since each cost is rounded to the nearest $10, the "ones" place is the same for each amount. So, the stem will consist of the "hundreds" digit, and the leaf will consist of the "tens" digit:

1	3 6
2	2 4 4 5 5 6
3	2 4 4 5 5 5 6
4	2 4 4 5 5 6
5	3 4 4 4

1|3 represents \$130

(b) Grouped frequency table showing fall textbook costs:

Classes	Tally	Frequency
\$100 –149	|	1
\$150–199	|	1
\$200–249	|||	3
\$250–299	|||	3
\$300 –349	|||	3
\$350–399	||||	4
\$400 –449	|||	3
\$450–499	|||	3
\$500–549	||||	4

(c) The histogram below shows the number of students on its vertical axis and the dollar amount paid for their textbooks on the horizontal axis:

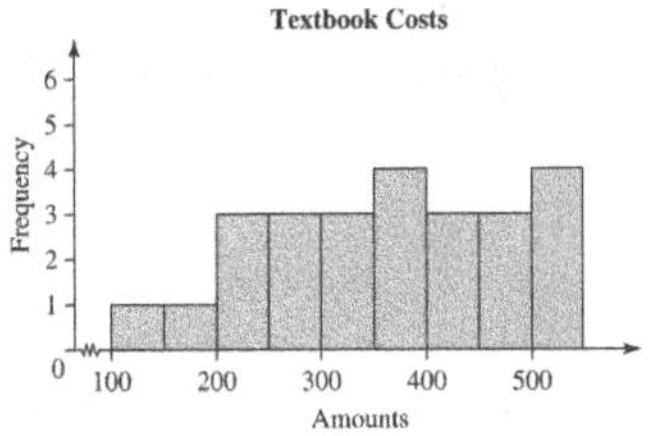

9. (a) **November** (with the highest bar) had the most rain-fall; approximately **30 cm**.

(b) 15 cm (October) + 25 cm (December) + 10 cm (January) = **50 cm**.

10. The bar graph below reflects that two or three heads showed up most frequently. Zero or five heads showed up the least number of times.

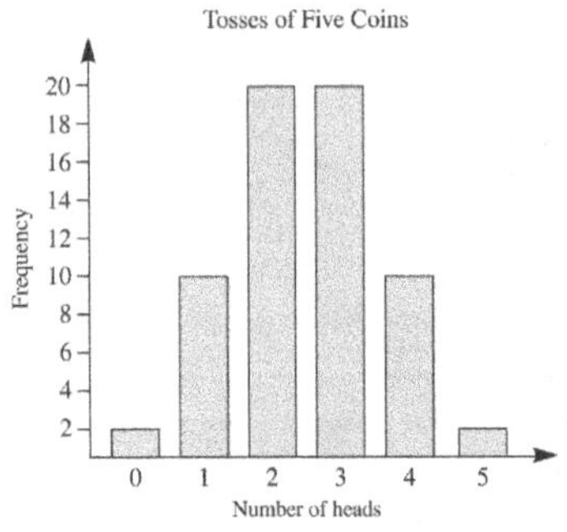

11. (a) Answers may vary. To have a population where, by far, more people in their 80's than in any other age group choose the same type books could mean a more homogeneous group. Perhaps they live together. Additionally, to have more people reading in their 100's than in their 90's is cause for suspicion. In any event, it is questionable data and more information is needed in order to make any reasonable conjecture.

(b) The graph only shows the frequency of choice for a book type—not the number of readers. The mode—i.e., the group having the greatest frequency of choice—is the 80's decade.

12. (a) **Double bar graph**

(b) **Women,** from approximately 55 years in 1925 to approximately 80 years in 2015. Men's life expectancy has changed only from about
53 years in 1925 to about 74 years in 2015.

(c) **One to two years** (as closely as can be approximated from the graph).

(d) **Five to six years** (as closely as can be approximated from the graph).

13. Answers may vary slightly. First, determine the approximate percentage of the bar each item should take up.

\$2050 for rent $\approx$ 38%;

\$1700 for food $\approx$ 31.5%;

\$800 for utilities $\approx$ 14.8%;

\$400 for transportation $\approx$ 7.4%;

\$200 for recreation $\approx$ 3.7%;

and

\$250 for savings $\approx$ 4.6%.

Next, calculate the length of the three-inch rectangle for each percentage:

\$2050 for rent $\approx$ 38%; about 1.14 inches

\$1700 for food $\approx$ 31.5%; about 0.94 in.

\$800 for utilities $\approx$ 14.8%; about 0.44 in.

\$400 for transportation $\approx$ 7.4%; about 0.22 in.

\$200 for recreation $\approx$ 3.7%; about 0.11 in.

and

\$250 for savings $\approx$ 4.6%; about 0.14 in.

Finally, use a ruler to approximate each length on the three inch rectangle and draw the smaller rectangles inside to represent each amount.

Note that percentages below are rounded to the nearest whole percentage.

Family Budget

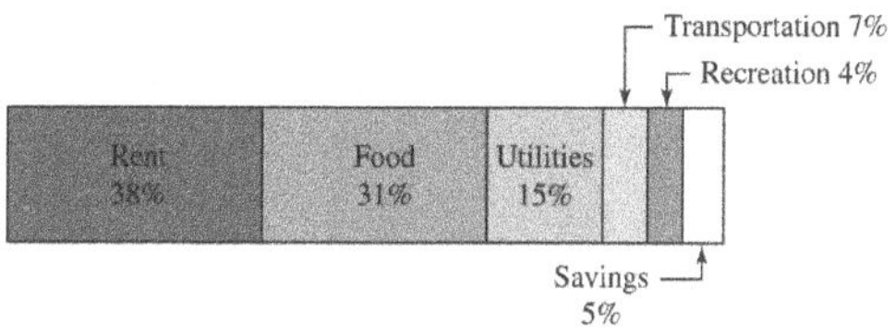

14. **(a)** A $\frac{1}{4}$ – inch rectangle would be one-eighth of the length of two inches. One-eighth of 100% is **12.5%**. Another method of solution is to set up the proportion $\frac{0.25''}{2''} = \frac{x}{100}$ and solve for x.

(b) Using the proportion $\frac{x''}{2''} = \frac{16}{100}$, solving for x yields a solution of **0.32 inch.**

15. The following table illustrates the calculations of the size of each central angle measurement:

Name	Percent	Central Angle
Facebook	64%	$0.64 \times 360 = 230.4^\circ$
Twitter	22%	$0.22 \times 360 = 79.2^\circ$
Myspace	2%	$0.02 \times 360 = 7.2^\circ$
Google +	1%	$0.01 \times 360 = 3.6^\circ$
LinkedIn	1%	$0.01 \times 360 = 3.6^\circ$
none	10%	$0.10 \times 360 = 36^\circ$

From this information, the circle graph is:

Preferred Social Networking Service

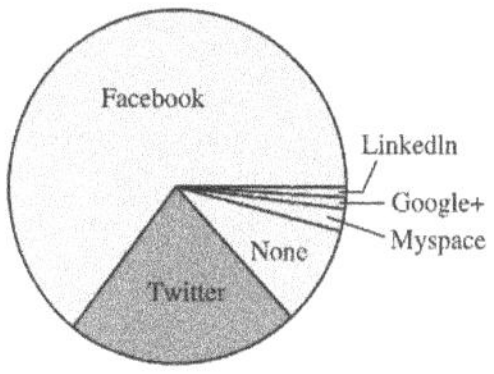

16. **(a)** We need to find out what percentage of the total area (100%) that a sector containing 65° would contain. So, $\frac{65^\circ}{360^\circ} = \frac{x}{100}$; solving for x, you get approximately **18%.**

(b) Using the proportion $\frac{x^\circ}{360^\circ} = \frac{32}{100}$, by solving for x, the solution is **115.2°**

17. Answers are rounded to the nearest whole number.

black: 24% of 241 is $\approx$ **58;**

navy blue: 22% of 241 is $\approx$ **53**;

white: 20% of 241 is $\approx$ **48;**

gray: 17% of 241 is $\approx$ **41;**

maroon: 12% of 241 is $\approx$ **29;**

green: 5% of 241 is $\approx$ **12;**

18. The circle graph for 2018 has four times the area of the circle graph for 2005, making it seem that the number of elementary teaching majors has grown by much more than double in between those years.

Mathematical Connections 10-2: Review Problems

19. Method (i) is biased in that you might only get the opinions of the very satisfied or unsatisfied, as only those with a strong opinion are probably going to mail a postcard back to the restaurant. Method (ii) has similar issues as (i); only those patrons who have a strong opinion will bother visiting the web site to fill out a survey. Method (iii) is probably the least biased of the methods listed here. It is as convenient as possible for patrons to complete and drop off the survey, meaning a better cross-section of patrons will fill it out. Offering the possibility of a prize also encourages all patrons to fill it out. Method (iv) might have some issues, simply because of who is asking the questions. Patrons might be less honest/rate more positively their experience simply because the pollster is working for the restaurant.

Assessment 10-3A: Displaying Data: Part II

1. **(a)** 70% of \$12,000, or **about \$8400.**

(b) (100 − 30)% of \$20,000 = 70% of \$20,000 in depreciation, or **about \$14,000.**

(c) 35% of \$20,000, **about \$7000.**

(d) Right after **two years**. Average trade-in value is about 55% at two years.

2. **(a)** The census that was taken in the year 1800 appears to be just slightly below 6 million people. So, an estimate of when the population was about 6 million people could be **1802**.

(b) 1830 corresponds to about 13, which represents **13,000,000**.

(c) The slope of the line represents change in population over change in years. Since the slope is steeper from **1810 to 1820** the population increased more during this period.

3. **(a)** **January** corresponds to the largest vertical values on both graphs.

(b) **March**.

(c) Although other factors might influence sales, in 4 of the five months illustrated, more snow shovels were sold in **2019**.

4. **Answers may vary**. The graph indicates that the rate of Black Arrests is always higher than the rate of White Arrests. The trend over this time period indicates that the rate of Black Arrests has almost tripled, while the rate of White Arrests, while it has grown, has not grown at the same rate.

5.

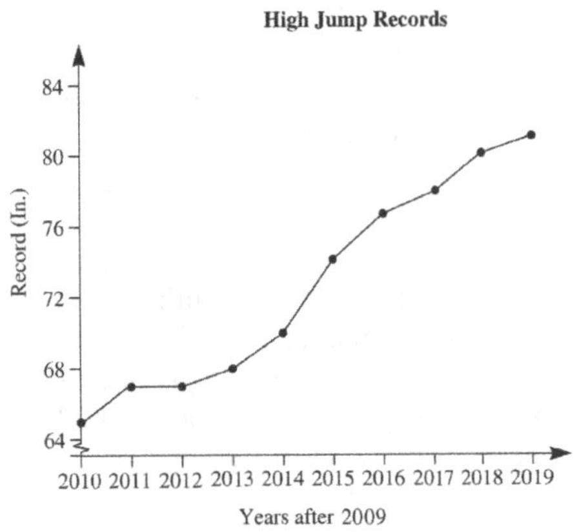

6. The horizontal axis is not labeled; the graph does not have a title; and the intervals on the horizontal axis are not all the same amount.

7. **(a)** **Negative**. The trend line slopes down from left to right.

(b) The number of movies point corresponding to age 25 is **about 10**.

(c) The age point corresponding to 16 movies seen is **about 22**.

8. **(a)** The first six terms and their corresponding values [(1,3), (2,8), (3,13), (4,18), (5,23), and (6,28)] are plotted on the following scatterplot, showing each term on the vertical axis (y) with the value of the term on the horizontal axis (x):

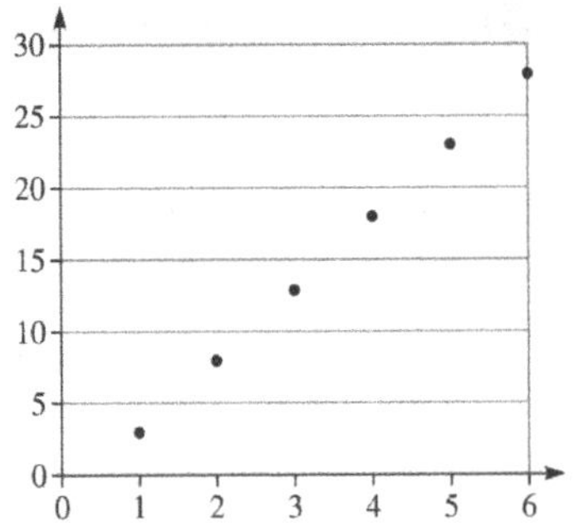

(b) Trend line, depicting the scatterplot points as a set of continuous data:

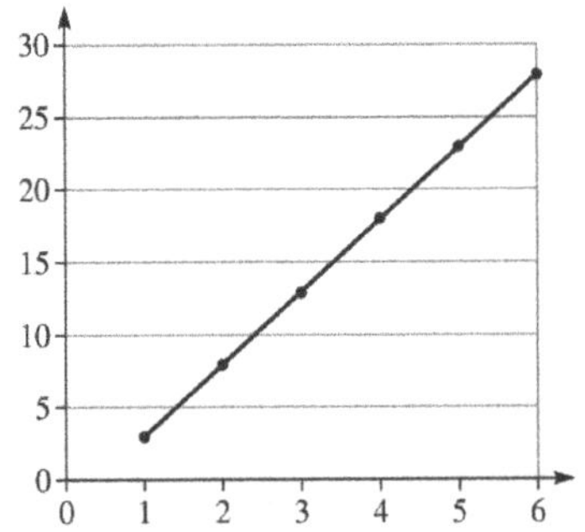

(c) Using the formula for arithmetic sequences,

$$y = 3 + 5(x - 1) \Rightarrow$$
$$y = 3 + 5x - 5 \Rightarrow$$
$$\mathbf{y = 5x - 2.}$$

9. **(a)** **(*i*)** The points generally decrease when moving from left to right, so the association is **negative.**

(*ii*) Answers may vary. A possible trend line is:

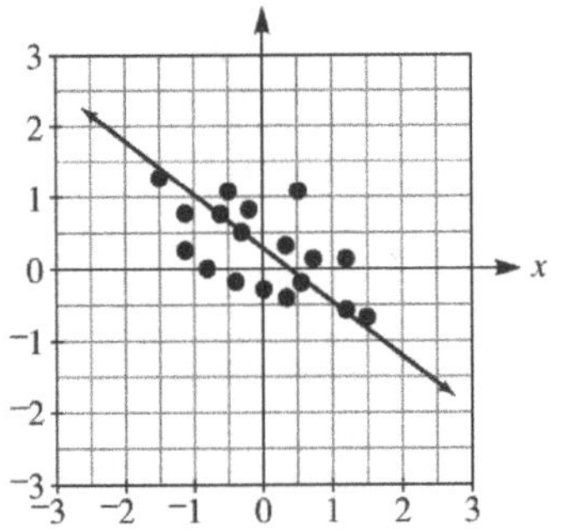

(b) There is no apparent trend, so there will also be no trend line.

10. Answers may vary depending upon the coordinates chosen.

(a) Two representative sets of coordinates appear to be

x	y
$^-2$	$^-1$
3	1

When the x-coordinates of these points go from $^-2$ to 3, an increase of 5, the y-coordinates go from $^-1$ to 1, an increase of 2. So, we have a slope of $\frac{2}{5}$, which means we know that $y = \frac{2}{5}x.$, or $y = 0.4x$. The trend line must be lowered by about 0.2, as nearly as can be determined by looking at the graph; so our equation could be $y = \mathbf{0.4}x - \mathbf{0.2}$.

(b) Two representative sets of coordinates appear to be

x	y
$^-1$	1
1	2

When the x-coordinates of these points go from $^-1$ to 1, an increase of 2, the y-coordinates go from 1 to 2, in increase of 1. So, the slope is $\frac{1}{2} = 0.5$. The trend line must be raised by about 1.5, as nearly as can be determined from the graph, so the equation would be $y = \mathbf{0.5}x + \mathbf{1.5}$.

(c) Two representative sets of coordinates appear to be

x	y
$^-3$	$^-2$
1	1

When the x-coordinates of these points go from $^-3$ to 1, an increase of 4, the y-coordinates go from $^-2$ to 1, an increase of 3. So the slope is $\frac{3}{4} = 0.75$. The trend line must be raised by about 0.25, as nearly as can be determined from the graph, so the equation would be, $y = \mathbf{0.75x + 0.25}$.

11. Answers may vary.

(i)

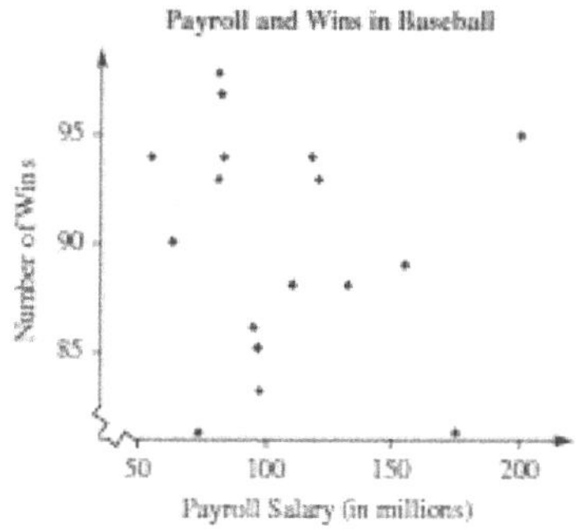

(ii) There doesn't appear to be any trend line. This would indicate that baseball payroll salaries do not necessarily influence number of wins.

12. If the equation is $y = 3.2x - 0.11$:

(a) $y = 3.2(1) - 0.11 = \mathbf{3.09}$

(b) $y = 3.2(0) - 0.11 = \mathbf{^-0.11}$

(c) $y = 3.2(10) - 0.11 = \mathbf{31.89}$

(d) $y = 3.2(20.3) - 0.11 = \mathbf{64.85}$

13. **Negative**. The slope is negative; the values of y will decrease as values of x increase.

14. Answers will vary.

(a) In general there is **no association** between head size and height.

(b) **Negative association**. As the temperatures go down, the amount of electricity used would go up.

(c) **Positive association**. The larger the population of a city, the more automobiles it would have on the streets.

Mathematical Connections 10-3: Review Problems

9. Reducing the miscellaneous category to 10% produces the stacked bar graph below:

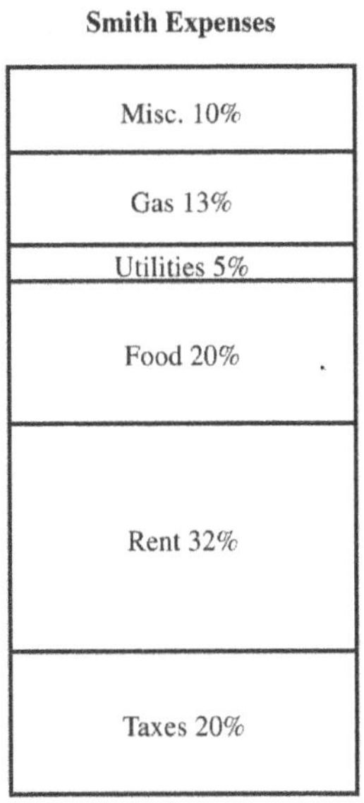

11. (a)

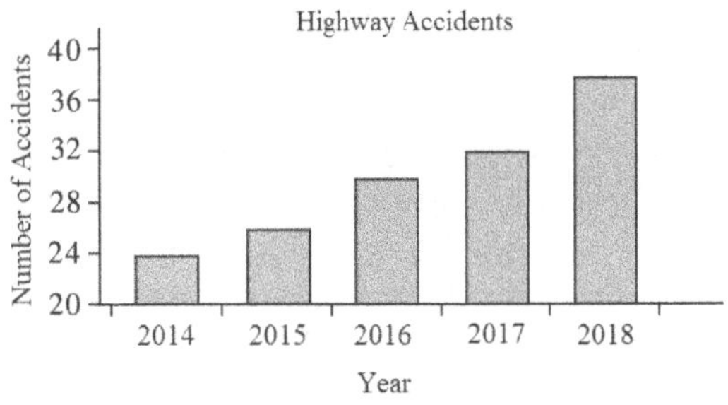

(b)

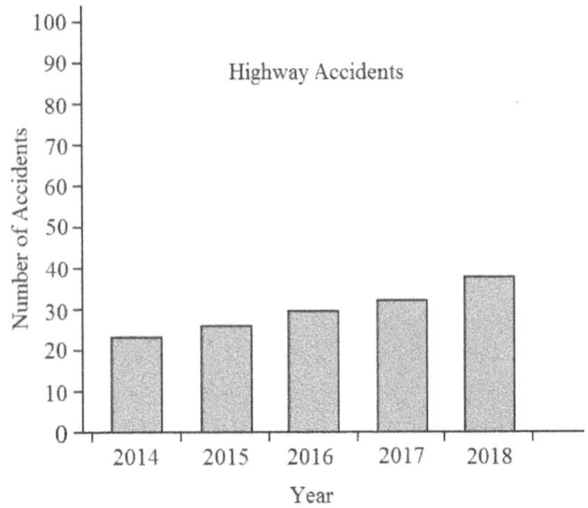

Assessment 10-4A: Measures of Central Tendency and Variation

1. (a) Ordering the data: 3, 4, 6, 7, 7, 7, 8, 11.

(*i*) Mean

$= \frac{3+4+5+7+7+7+8+11}{8} =$

$\frac{53}{8} = \mathbf{6.625}$.

(*ii*) Median is in the 4.5th position, or between the first 7 and the second 7.

So, the median $=$ **7**; or $\frac{7+7}{2}$.

(*iii*) Mode $=$ **7**, the value which occurs most frequently.

(b) Ordering the data: 10.5, 11.5, 12.5, 12.5, 12.5, 14.5, 14.5, 16.5, 20.5.

(*i*)

$\text{Mean} = \frac{10.5+11.5+12.5+12.5+12.5+14.5+14.5+16.5+20.5}{9}$

$= \frac{125.5}{9} = \mathbf{13.9\bar{4}}$.

(*ii*) Median is in the 5th position, which makes it the last **12.5**

(*iii*) Mode $=$ **12.5** the value which occurs most frequently.

(c) Ordering the data: -7.4, -6.2, -2, -1.1, 1.2, 3, 4.2, 8.3.

(*i*) Mean $=$

$\frac{-7.4+(-6.2)+(-2)+(-1.1)+1.2+3+4.2+8.3}{8}$

$= \frac{0}{8} = \mathbf{0}$.

(*ii*) Median is in the 4.5th position, or between -1.1 and 1.2; so

$\frac{-1.1+1.2}{2} = \mathbf{0.05}$.

(*iii*) Mode; **there is no mode**, since each value in the data set occurs exactly once.

(d) Ordering the data: 0, 79, 80, 82, 85, 85, 93

(i) Mean $= \frac{0+79+80+82+85+85+90}{7}$

$= \mathbf{72}$

(ii) Median is in the 4th position, which makes the median $=$ **82**

(iii) Mode $=$ **85**, since this is the value that occurs most frequently.

2. (a) Answers may vary; one example might be 1, 2, 3, 4, 5, 60, 60.

(b) Answers may vary; one example might be 1, 1, 1, 50, 900, 900, 900

3. (a) (*i*) $\frac{3(75)+3(88)}{6} = \mathbf{81.5}$

(*ii*) The median will be found in the 3.5th position, or between the last 75 and the first 88. Average those two middle values: $\frac{75+88}{2} = \mathbf{81.5}$

(*iii*) Bi-modal: **75 and 88.**

(b) Mean $= \frac{\text{sum of test scores}}{\text{number of test scores}}$. So,

$75 = \frac{\text{sum of test scores}}{20}$. So,

sum $= 75 \cdot 20 = \mathbf{1500}$.

4. Mean of 28 scores being 80 implies that the sum of scores is $28 \cdot 80 = 2240$. Adding scores or 60 and 50 yields a new sum of $2240 + 110 = 2350$. New mean $= \frac{2350}{28+2} = \frac{2350}{30} = \mathbf{78.\overline{3}}$.

5. Mean $= \frac{4(98)+11(60)}{15} = \mathbf{70.1\overline{3}}$

6. Mean $= \frac{(m\times 100)+(n\times 50)}{m+n}$; i.e., the sum of the data divided by the number of data points.

7. The number of points for each course is the product of the number of credits and the point value of each. 15 points for math, 12 for English, 10 for physics, 3 for German, and 4 for handball gives a total of 44 points. GPA $= \frac{44 \text{ points}}{17 \text{ credits}}$, or **about 2.59.**

8. Total tackle weight is $7 \cdot 230 = 1610$ pounds. Total backfield weight is $4 \cdot 190 = 760$ pounds. Total player weight is $1610 + 760 = 2370$ pounds. Mean $= \frac{2370 \text{ pounds}}{11 \text{ players}} = \mathbf{215.\overline{45}}$ **pounds.**

9. (a) The total salary for the 50 faculty is shown below:

Salary ($)	No. Faculty	Total ($)
18,000	2	36,000
22,000	6	132,000
32,000	24	768,000
48,000	15	720,000
80,000	2	160,000
150,000	1	150,000
Total	50	1,966,000

Mean annual salary $= \frac{\$1{,}966{,}000}{50} = \mathbf{\$39{,}320}$

(b) Ordering the salaries: \$18,000, 18,000, 22,000, ..., 150,000; the median is between the 20th and 21st and all salaries between the 13th and 25th are \$32,000. Median $= \mathbf{\$32{,}000}$.

(c) **\$32,000**, or the most frequent salary.

10. (a) Salaries are between \$18,000 and \$150,000. The range is thus $\$150{,}000 - \$18{,}000 = \mathbf{\$132{,}000}$.

(b) The mean \$39,320.

Number of Faculty	Salary ($)	Absolute Deviation	No. Faculty × Absolute Deviation
2	18,000	21,320	42,640
6	22,000	17,320	103,920
24	32,000	7,320	175,680
15	48,000	8680	130,200
2	80,000	40,680	81,360
1	150,000	110,680	110,680
50			Sum = 644,480

$MAD = \frac{644{,}480}{50} = \mathbf{\$12{,}889.60}$.

(c) $\overline{x} = \$39{,}320$; n = number of faculty.

x Salary ($)	$x - \overline{x}$	$(x - \overline{x})^2$	$n(x - \overline{x})^2$
18,000	-21,320	454542400	909,084,800
22,000	-17,320	299982400	1,799,894,400
32,000	-7,320	53582400	1,285,977,600
48,000	8,680	75342400	1,130,136,000
80,000	40,680	1,654,862,400	3,309,724,800
150,000	110680	12250,062,400	12,250,062,400
			20,684,880,000

$s = \sqrt{\frac{20684880000}{50}} \approx \mathbf{\$20{,}339.56}$.

(d) Upper extreme = \$150,000;

Upper quartile = \$48,000;

Median = \$32,000;

Lower quartile = \$32,000; and

Lower extreme = \$18,000.

IQR = \$48,000 − \$32,000 = **\$16,000.**

11. (a) Answers may vary. One possible viewpoint might be that the median/mode of \$32,000, coupled with the IQR of \$16,000 would give a realistic picture.

(b) Answers may vary. Based on the above answer for (a), we know that all but 3 salaries are within \$16,000 of \$32,000.

12. Answers may vary. One possible misconception is that people could assume the temperature is always around 82 degrees. Another possible misconception is assuming that this is mean daily **high** temperature, when it is stated as a mean daily temperature.

13. Total trip miles = 43,390 − 42,800 = 590.

Total gasoline used = 12 + 18 = 30 gallons.

Average fuel mileage = $\frac{590}{300}$ = $\mathbf{19\frac{2}{3}}$ **mpg.**

14. 24 + 34 = **58 years old.**

15. Let S be the score needed on the fifth exam. Then

$\frac{84+95+86+94+S}{5} = 90 \Rightarrow \frac{359+S}{5} = 90 \Rightarrow$

$359 + S = 450 \Rightarrow S = \mathbf{91}.$

16. (a) **Theater A: \$25**; Theater **B: \$50**. The median in a box plot is the middle line through the box; i.e., 25 in the Theater A box and 50 in the Theater B box.

(b) **Theater B**. Range is the difference between upper and lower extremes. Theater A range = 40 − 15 = 25; Theater B range = 80 − 15 = 65.

(c) **\$80 at Theater B**, its upper extreme. Theater A's upper extreme is \$40.

(d) Answers may vary. There is significantly more variation and generally higher prices at Theater B.

17. Ordering the data: 20, 69, 70, 72, 75, 80, 83, 88, 90, 92.

Lower extreme = 20; upper extreme = 92;

lower quartile = 70; upper quartile = 88; and

median = $\frac{75+80}{2}$ = 77.5

IQR = 88 − 70 = 18;

1.5 · IQR = 27

Lower quartile − (1.5 · IQR) = 43, so there is a lower outlier of 20.

Upper quartile + (1.5 · IQR) = 115, so there is no upper outlier.

Yields the following box plot:

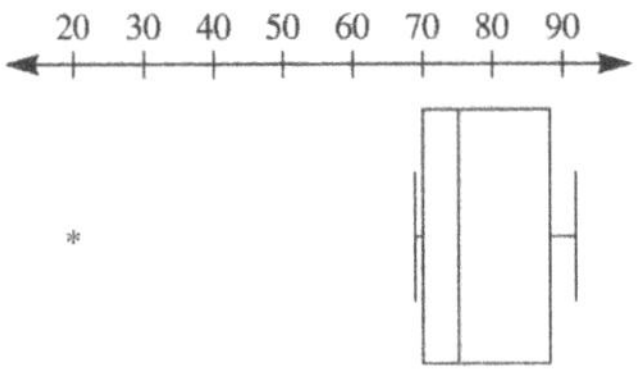

18. (a) *L. A.*: $Q_1 = 729;\ Q_2 = 749;\ Q_3 = 938.$

St. L.: $Q_1 = 398.5;\ Q_2 = 446.5;\ Q_3 = 572.$

Heights of 8 Tallest Buildings in St. Louis and Los Angeles

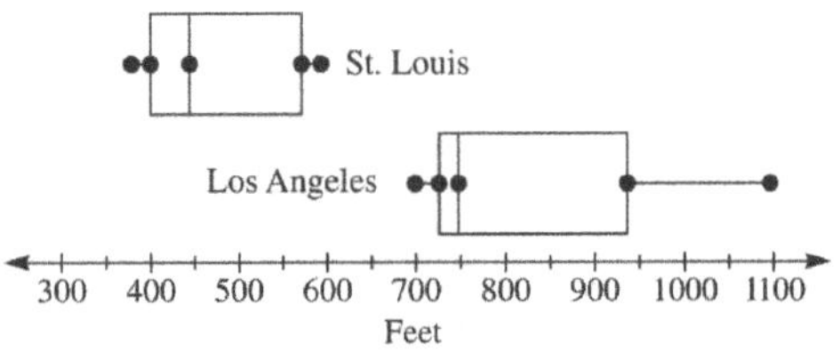

(b) Answers will vary. Of the 8 tallest buildings in each city, all of the tallest 8 buildings in Los Angeles are taller than all of the tallest 8 buildings in St. Louis.

19. $\bar{x} = \frac{175+182+190+180+192+172+190}{7} = \frac{1281}{7} =$ 183 cm. The summation of $(x - \bar{x})^2$ is tabularized below:

x	$x - \bar{x}$	$(x - \bar{x})^2$
175	$^-8$	64
182	$^-1$	1
190	7	49
180	$^-3$	9
192	9	81
172	$^-11$	121
190	7	49
		Total: 374

$s = \sqrt{\frac{374}{7}} \approx$ **7.3 cm.**

20. **(a)** Sum of grades = 96 + 71 + 43 + 77 + 75 + 76 + 61 + 83 + 71 + 58 + 97 + 76 + 74 + 91 + 74 + 71 + 77 + 83 + 87 + 93 + 79 = 1613. $\overline{x} = \frac{1613}{21} \approx$ **76.8.**

(b) Ordering the scores: 43, 58, 61, 71, 71, 71, 74, 74, 75, 76, 76, 77, 77, 79, 83, 83, 87, 91, 93, 96, 97. Median = **76** (the 11th score).

(c) The most frequent score is 71. Mode = **71.**

(d) Upper quartile $= \frac{83+87}{2} = 85$; lower quartile $= \frac{71+71}{2} = 71$.
IQR = 85 − 71 = **14.**

(e) $(x - \overline{x})^2$ is tabularized below:

x	$x - \overline{x}$	$(x - \overline{x})^2$
96	19.2	368.64
71	$^-5.8$	33.64
43	$^-33.8$	1142.44
77	0.2	0.04
75	$^-1.8$	3.24
76	$^-0.8$	0.64
61	$^-15.8$	249.64
83	6.2	38.44
71	$^-5.8$	33.64
58	$^-18.8$	353.44
97	20.2	408.04
76	$^-0.8$	0.64
74	$^-2.8$	7.84
91	14.2	201.64
74	$^-2.8$	7.84
71	$^-5.8$	33.64
77	0.2	0.04
83	6.2	38.44
87	10.2	104.04
93	16.2	262.44
79	2.2	4.84
		Total: 3293.24

$v = \frac{3293.24}{21} \approx$ **156.8.**

(f) $s = \sqrt{v} = \sqrt{156.8} \approx$ **12.5.**

(g) The mean, $\overline{x}$, is 76.8.

x	$\lvert x - \overline{x} \rvert$
96	19.2
71	5.8
43	33.8
77	0.2
75	1.8
76	0.8
61	15.8
83	6.2
71	5.8
58	18.8
97	20.2
76	0.8
74	2.8
91	14.2
74	2.8
71	5.8
77	0.2
83	6.2
87	10.2
93	16.2
79	2.2
	Total: 189.8

$\text{MAD} = \frac{189.8}{21} \approx$ **9.04.**

21. 95% of the area under the normal curve lies with ± 2 standard deviations from the mean. The range corresponding to ± 2 standard deviations is between $65.5 - 2 \cdot 2.5 =$ **60.5** and $65.5 + 2 \cdot 2.5 =$ **70.5** in.

22. 1 minute is two standard deviations below the mean. Approximately 95% of the area under the normal curve lies within 2 standard deviations of the mean. So, **100% − 95% =5%** of the data lies outside 2 standard deviations of the mean; since a normal distribution is symmetrical, $\frac{\mathbf{5\%}}{\mathbf{2}}$ **= 2.5%** of the calls should last less than 1 minute.

23. **(a)** Q_2 represents the number of scores falling below the median, which in the normal distribution equals the mean. Thus $Q_2 =$ **65.**

(b) P_{16} represents the bottom 16% of all scores, which in the normal distribution falls one standard deviation below the mean. Thus $P_{16} = 65 - 12 =$ **53.**

(c) P_{84} represents the value below which are 84% of the scores, which in the normal distribution falls one standard deviation above the mean. Thus $P_{84} = 65 + 12 = 77.$

(d) D_5 is the decile ranking, below which is 50% of the scores; i.e., the same as the mean in the normal distribution, or **65.**

24. \$10.75 is one standard deviation below the mean; \$12.25 is one standard deviation above the mean. About 68% of the area under the normal curve lies within ± 1 standard deviation from the mean. Therefore the probability of having a wage between \$10.75 and \$12.25 is **68%** or **0.68**

25. Two standard deviations above the mean is $79 + (2 \cdot 5.5) = 79 + 11 = 90$. Thus a **90** is the lowest score a person could receive and still earn an A.

26. About 68% of the Values 1 is within are standard deviation of the mean an a normal distribution plot. Al's score is exactly equal to the mean plus one standard deviation. Thus, $68\% + \frac{32\%}{2} = 68\% + 16\% = 84\%$ of the students scored below Al, or $.84 \cdot 10{,}000 =$ **8400 students.**

27. If Al is ranked 14th, 31 students rank below him $\frac{31}{45} = 0.6888...$ Thus, Al is in the **69th percentile.**

28. Jack's percentile ranking is $\frac{200-70}{200} = \frac{130}{200} = 0.65$, meaning the 65th percentile. Thus, **Jill's** standing is higher than Jack's.

Mathematical Connections 10-4: Review Problems

25. Small: 28% of $360° \approx$ **101°.**

Midsize: 48% of $360° \approx$ **173°.**

Large: 7% of $360° \approx$ **25°.**

Luxury: 17% of $360° \approx$ **61°.**

27. More information is probably needed. For example, how do the number of miles driven for males and females compare? Or the number of miles driven in stressful conditions (bad weather, night driving, heavy traffic, etc.)...how do they compare?

Chapter 10 Review

1. *(i)* If the average is reported as 2.41 children, the mean is probably being used. The mode would not have a decimal number, and the median would be either a whole number or one ending in .5.

(ii) If the average is 2.5, then either the mean or the median might have been used.

2. Ten part-timers with average salary of \$100 = a total part-time payroll of \$1000. \$1000 leaves \$3000 for full timers.

$\frac{\$3000 \text{ payroll}}{\$250 \text{ per worker}} =$ **12 full-time workers.**

(Assuming that the meaning of the term "average" as used in this question for part time employees is the same as in the term "mean".)

3. (a) *(i)* Mean $= \frac{10+50+30+40+10+60+10}{7} =$ **30.**

(ii) Ordering the data: 10, 10, 10, 30, 40, 50, 60. Median = **30**, the middle data point.

(iii) Mode = **10**, the most frequent data point.

(b) *(i)* Mean $= \frac{5+8+6+3+5+4+3+6+1+9}{10} =$ **5.**

(ii) Ordering the data: 1, 3, 3, 4, 5, 5, 6, 6, 8, 9. Median $= \frac{5+5}{2} =$ **5.**

(iii) Modes = **3, 5,** and **6.**

(c) *(i)* Mean = **150**. This can be found using the formula for the mean $\frac{4(100)+4(200)}{8} = 150$, but it is also easy to calculate by realizing that each 100 pairs with exactly one 200, and the mean of those two values is 150.

(ii) Median = **150**. The median is between the fourth and fifth values, or the mean of the last 100 and the first 200.

(iii) Mode: two modes, **100 and 200** (bimodal).

(d) *(i)* Mean $= \frac{208.2}{16} = \mathbf{13.0125}$.

(ii) Median: Ordering the data: 10; 10.1; 10.7; 10.9; 11.1; 11.5; 11.9; 12.1; 12.2; 12.3; 12.9; 13.2; 13.5; 13.6; 13.7; 28.5. Median is between the 8th and 9th values, so median = **12.15**.

(iii) Mode: **there is no mode**, since each value in the data set occurs exactly once.

4. **(a)** *(i)* Range $= 60 - 10 = \mathbf{50}$.

(ii) Sum of $|x_n - \overline{x}| = 120$. MAD $= \frac{120}{7} \approx \mathbf{17.1}$.

(iii) Upper quartile $= 50$; lower quartile $= 10$. IQR $= 50 - 10 = \mathbf{40}$.

(iv) Sum of $(x - \overline{x})^2 = 2600$. $v = \frac{2600}{7} \approx$ **371.4**.

(v) $s = \sqrt{v} = \sqrt{\frac{2600}{7}} \approx \mathbf{19.3}$.

(b) *(i)* Range $= 9 - 1 = \mathbf{8}$.

(ii) Sum of $|x_n - \overline{x}| = 18$. MAD $= \frac{18}{10} = \mathbf{1.8}$.

(iii) Upper quartile $= 6$; lower quartile $= 3$. IQR $= 6 - 3 = \mathbf{3}$.

(iv) Sum of $(x - \overline{x})^2 = 52$. $v = \frac{52}{10} = \mathbf{5.2}$.

(v) $s = \sqrt{v} = \sqrt{5.2} \approx \mathbf{2.28}$.

(c) *(i)* Range $= 200 - 100 = \mathbf{100}$.

(ii) Sum of $|x_n - \overline{x}| = 400$. MAD $= \frac{400}{8} = \mathbf{50}$.

(iii) Upper Quartile = 200;
Lower Quartile = 100.
IQR $= 200 - 100 = \mathbf{100}$.

(iv) $(x - \overline{x})^2 = 20{,}000$
$v = \frac{20000}{8} = \mathbf{2500}$.

(v) $s = \sqrt{v} = \sqrt{2500} = \mathbf{50}$.

(d) *(i)* Range $= 28.5 - 10 = \mathbf{18.5}$.

(ii) Using a mean of 13.0125, Sum of $|x_n - \overline{x}| = 34.875$. MAD $= \frac{34.875}{16} \approx \mathbf{2.18}$.

(iii) Upper quartile = 13.35; lower quartile = 11;
IQR $= 13.35 - 11 = \mathbf{2.35}$.

(iv) Using a mean of 13.0125, $(x - \overline{x})^2 = 277.5175$. So, $v = \frac{277.5175}{16} \approx \mathbf{17.34}$.

(v) $s = \sqrt{v} = \sqrt{17.34} \approx \mathbf{4.16}$.

5. **(a)** Dot plot of Miss Rider's class (masses in kilograms):

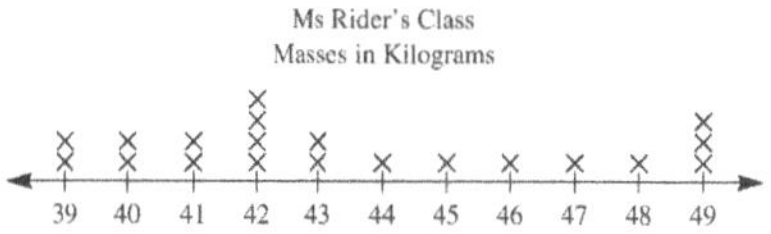

(b) Ordered stem-and-leaf plot of Miss Rider's class (masses in kilograms):

3 | 99
4 | 0011222223345678999   4 | 0 represents 40 kg

(c) Frequency table of Miss Rider's class (masses in kilograms):

Mass	Tally	Frequency
39	||	2
40	||	2
41	||	2
42	||||	4
43	||	2
44	|	1
45	|	1
46	|	1
47	|	1
48	|	1
49	|||	3
		20

(d) Bar graph of Miss Rider's class (masses in kilograms):

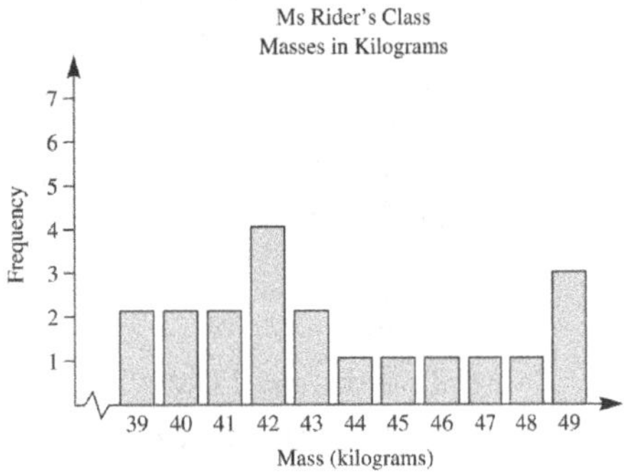

6. (a) Test grade grouped frequency table:

Classes	Talley	Frequency
61–70	𝍸 |	6
71–80	𝍸 𝍸 |	11
81–90	𝍸 ||	7
91–100	𝍸 |	6
		30

(b) Test grade histogram:

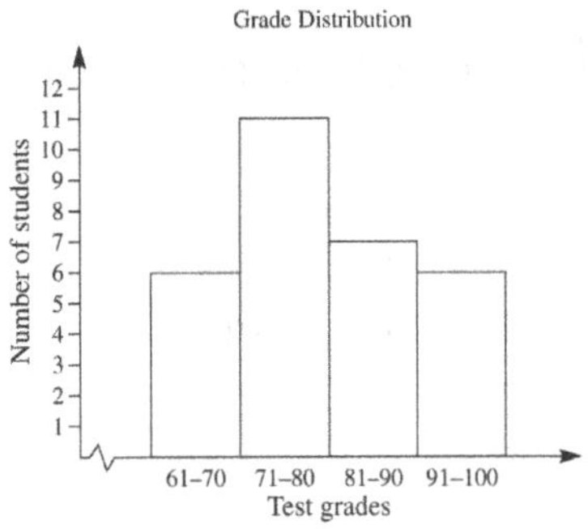

7. Wegetem expenditures, where:

Bribes $= \dfrac{\$600{,}000}{\$2{,}000{,}000} = 0.30$; **30%** of $360° = 108°$.

Legal fees $= \dfrac{\$400.000}{\$2{,}000{,}000} = 0.20$; **20%** of $360° = 72°$.

Public relations $= \dfrac{\$400{,}000}{\$2{,}000{,}000} = 0.20$; **20%** of $360° = 72°$.

Bail money $= \dfrac{\$300{,}000}{\$2{,}000{,}000} = 0.15$; **15%** of $360° = 54°$.

Contracts $= \dfrac{\$300{,}000}{\$2{,}000{,}000} = 0.15$; **15%** of $360° = 54°$.

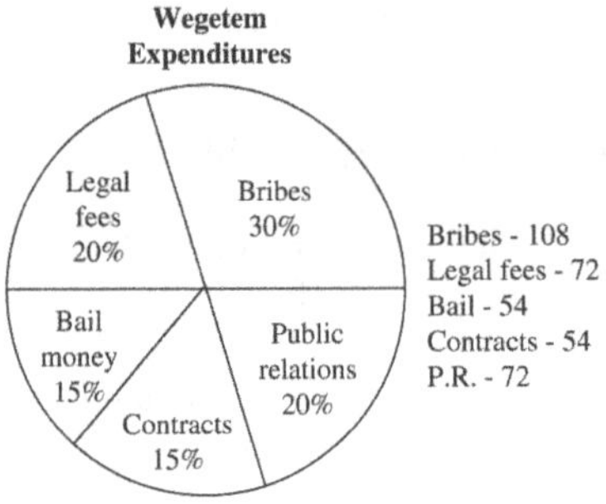

8. The widths of the bars are not uniform. The graph is also missing labels on both axes.

9. Total salary $= 24 \cdot 9000 = \$216{,}000$. New mean $= \dfrac{\$216{,}000 + \$80{,}000}{24+1} = \dfrac{296{,}000}{25} = \$11{,}840$. The mean was increased by $\$11{,}840 - \$9000 =$ **$2840**.

10. Total Gold medals = 83

Norway and Germany: $\frac{14}{83} \approx 0.1687$;
16.87% of $360° \approx 60.7°$.

Canada: $\frac{11}{83} \approx 0.1325$;
13.25% of $360° \approx 47.7°$.

United States: $\frac{9}{83} \approx 0.1084$;
10.84% of $360° \approx 39°$.

Netherlands: $\frac{8}{83} \approx 0.0964$;
9.64% of $360° \approx 34.7°$.

Swenden: $\frac{7}{83} \approx 0.0843$;
8.43% of $360° \approx 30.4°$. Switzerland, Korea, Austria, and France $= \frac{5}{83} \approx 0.0602$;
6.0% of $360° \approx 21.7°$.

GOLD METAL COUNTS, 2018 WINTER OLYMPICS

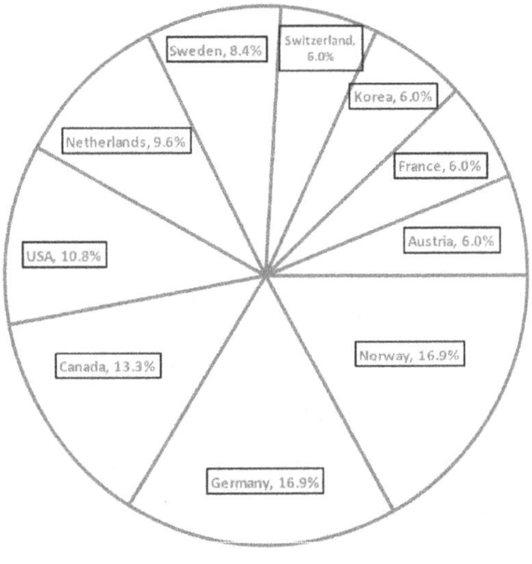

11. **(a)** Life expectancies at birth of males and females (from 1970 through 2011)

Females		Males
	67	1446
	68	28
	69	156
	70	0049
	71	02235578
	72	01145
	73	1689
7	74	123599
9310	75	146
86	76	0233445
88532	77	
9999854332211	78	
99655443210	79	
9642	80	
322210	81	
7 | 74 represents 74.7 years old		|67| 1 represents 67.1 years old

(b) Males: Lower extreme: 67.1
Lower quartile: 70.4
Median: 72.1
Upper quartile: 74.9
Upper extreme: 76.5

Females: Lower extreme: 74.7
Lower quartile: 77.8
Median: 78.9
Upper quartile: 79.9
Upper extreme: 81.3

Life expectancies at birth of males and females (from 1970 through 2015):

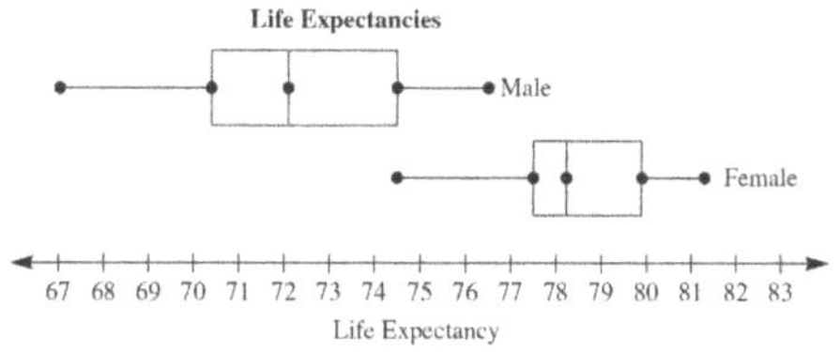

12. **Larry**. Larry's GPA $= \frac{4 \cdot 4 + 4 \cdot 4 + 3 \cdot 3 + 3 \cdot 2 + 1 \cdot 2}{4+4+3+3+1} =$ $3.2\overline{6}$;

Marc's GPA $= \frac{4 \cdot 2 + 4 \cdot 2 + 3 \cdot 3 + 3 \cdot 4 + 1 \cdot 4}{4+4+3+3+1} = 2.7\overline{3}$.

So Larry does have a higher GPA; and if "better grades" is defined as having a higher GPA, then Larry won the bet.

13. **(a)** Ordering the data: 160, 180, 330, 350, 360, 380, 450, 460, 480. Median = **360** yards, the middle data point.

(b) There is **no mode**, since no two or more lengths are the same.

(c) $\bar{x} = \frac{160+180+330+350+360+380+450+460+480}{9}$
= **350 yards.**

(d) The sum of the $(x - \bar{x})^2$ terms = 105,400.

$s = \sqrt{\frac{105{,}400}{9}} \approx$ **108.2** yards.

(e) Range = 480 − 160 = **320** yards.

(f) Lower quartile $= \frac{180+330}{2} = 255$;
upper quartile $= \frac{450+460}{2} = 455$.
IQR = 455 − 255 = **200** yards.

(g) Variance = the sum of the $(x - \bar{x})^2$ terms divided by 9
$= \frac{105{,}400}{9} = \mathbf{11711.\overline{1}\ yd^2}$.

(h) The sum of the $|x_n - \bar{x}|$ terms = 760.
MAD $= \frac{760}{9} =$ $\mathbf{84.\overline{4}}$ yards.

14. **(a)** Ordering the data: 45, 54, 56, 58, 58, 60, 62, 64, 64, 64, 65, 65, 65, 66, 67, 67, 67, 67, 68, 68, 69, 72, 74, 74, 75, 75, 82, 86, 88, 90.

Median $= \frac{67+67}{2} =$ **67** mph.

(b) Lower quartile = **64,** the 8th data point.
Upper quartile = **74,** the 23rd data point.

(c) Lower extreme = 45; upper extreme = 90.
Inter-quartile range (IQR) = 74 − 64 = 10. Lower quartile $-1.5 \cdot$ IQR $= 49 \Rightarrow$ outlier at 45.
Upper quartile $+1.5 \cdot$ IQR $= 89 \Rightarrow$ outlier at 90.
Box plot of car speeds:

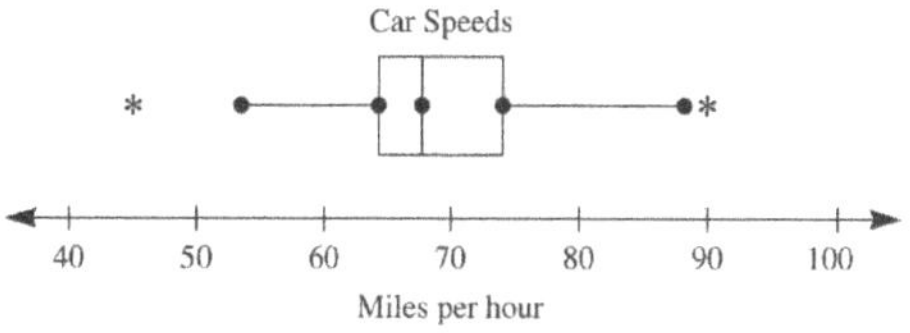

(d) Nine drivers exceeded 70 mph.
$\frac{9}{30} = 0.30 =$ **30%** received tickets.

(e) There are fewer speeds close to 67 mph (the median) in the upper quartile than in the lower. About 50% of the people drove between 64 and 74 mph. At least 25% speed.

(f) $\mathbf{P_{25} = 64; P_{75} = 74; D_5 = 67.}$

15. **Answers may vary**. The complaints are only in a single month; the data could change drastically from month to month. In general American, Delta, and United have many more flights than Express Jet, Virgin America, and Alaska Airlines, so they would probably get more complaints.

16. **Answers may vary.** The erroneous conclusion could be that the more money a state spends on education, the worse their state performs on the SAT. However, this is only looking at two states; to get a more accurate picture, all states should be included. Other factors may contribute to low test scores too, such as school size, student to teacher ratio, and so on.

17. **Only part (b)** might not be helpful in determining the type of automobile that is the most economical to drive.

18. **(a)** To find the percent females in Mr. Ramirez's class, let f represent the percentage of females. Then the percentage of males is $100\% - f$. Since the overall mean is in Mr. Ramirez's class is 218:

$$230(f) + 205(100\% - f) = 218 \Rightarrow$$

$$230(f) + 205(1 - f) = 218 \text{ (since } 100\% = 1). \Rightarrow$$

$$230f + 205 - 205f = 218 \Rightarrow$$

$$25f = 13 \Rightarrow f = \frac{13}{25} \text{ or } f = 0.52.$$

So **52%** of the students are females in Mr. Ramirez's class.

In Ms. Jonsey's class, note that if 88% of the class is female, then 12% of the class is male. Let m be the mean score of males. So

$$224(0.88) + m(0.12) = 221 \Rightarrow$$

$$197.12 + 0.12m = 221 \Rightarrow$$

$$0.12m = 23.88 \Rightarrow m = \mathbf{199.}$$

(b) **Yes**, as found in part (a), The mean for males is higher in Mr. Ramirez's class (205 to 199) and the mean for females is also higher (230 to 224) but the overall mean is lower (218 to 221).

19. **(a)** **Positive** association. Weights tend to increase as heights increase.

(b) **170 pounds**; i.e., the data point corresponding to 72 inches on the horizontal axis.

(c) **67 inches**; i.e., the data point corresponding to 145 pounds on the vertical axis.

(d) **64 inches**. There are four girls 64 inches tall; more than for any other height.

(e) 170 pounds − 120 pounds = **50 pounds**.

20. **(a)** **Collette**. There is less deviation from the mean in her scoring.

(b) (*i*) Collette: From $24 - 6 = \mathbf{18}$ to $24 + 6 = \mathbf{30}$.

(*ii*) Rudy: From $24 - 14 = \mathbf{10}$ to $24 + 14 = \mathbf{38}$.

(c) **Collette** scored more than her average upper range value of 30; Rudy scored less than his average upper range value of 38.

21. **2.5%** of the area under the standard normal curve lies more than two standard deviations above the mean. Thus the probability that a student would score more than 2 standard deviations above the mean = **0.025**.

22. **(a)** The mean would be the common score.

(b) The median would be the common score.

(c) The mode would be the common score.

(d) The standard deviation would be **0**.

(e) The mean absolute deviation would be **0**.

23. The length of the columns in the bar graph should be approximately the same for heads and tails.

24. **(a)** $\frac{647}{649} \approx 0.9969$, so about **99.7%**.

(b) $\frac{622}{649} \approx 0.9584$, so about **95.8%**.

(c) (*i*) $\frac{647}{1269} \approx 0.5099$, so about **51%**.

(*ii*) $\frac{2}{29} \approx 0.069$, so about **6.9%**.

(d) The evidence is not conclusive. Almost half of the smokers do not have lung cancer, while it is possible for non-smokers to get lung cancer.

25. For 100 meter freestyle, low extreme = 52.70, Lower quartile = 53.83, median = 54.79, Upper quartile = 58.59, and high extreme = 61.20.

For the 100 meter butterfly, low extreme = 55.48, Lower quartile = 56.73, median = 59.13, Upper quartile = 63.34 and high extreme = 69.50.

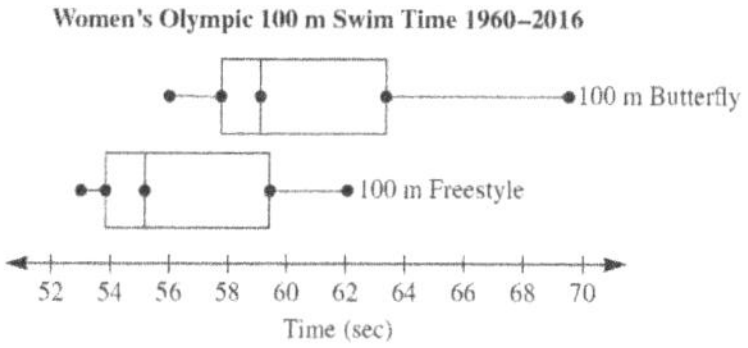

26. (a) $175 - 26 \cdot 6 = $ **19 pounds**.

(b) **No**. This is an example of using a mathematical model where it is no longer valid.

(c) The original statement reads "up to." Any average loss between 0 and 6 pounds would fit that description.

27. The **bar graph** is more appropriate. Line graphs are used to show change over time. Moreover, the points on the line graph should not be connected since there are no values "between" colors.

28. **Probably not**, unless the diving pool is about the same depth as the swimming pool.

29. Answers may vary. One method might be to pull boxes at random, weigh them, and then determine whether some predetermined percentage of the boxes fall within a predetermined weight range.

30. Answers may vary. A high school graduate might want to be with other students of comparable age, and would thus choose the first. If the student would want to go to school with those more mature, though, the student would choose the second.

31. Answers may vary. The area of both sections is about the same size, so you might conclude you need to eat as much of both; also, with the "sweets" on top, you might conclude it is more important than dairy products.

32. Answers may vary; e.g.,

(a) One way would be to leave the television set on even if no one is watching.

(b) The networks air very popular shows during "rating sweeps" periods.

33. Answers may vary. Graphs may show area or volume instead of relative size. Another way is to select a horizontal or vertical baseline that will support the point trying to be made.

34. (a) **About 3 billion**.

(b) **About 6 billion**.

(c) The **population doubled**.

(d) **About 9.3 billion**.

(e) **About 55% percent**. Comparing to (c), the growth factor is about 1.55.

35. (a) 68 inches is two standard deviations above the mean, and 2.5% of the area under the normal curve lies above 2 standard deviations above the mean. 2.5% of 1000 is **25 girls**.

(b) 60 inches is two standard deviations below the mean.
$13.5\% + 34\% = 47.5\%$ of the area under the normal curve lies between the mean and 2 standard deviations below the mean. 47.5% of 1000 is **475 girls**.

(c) 66 inches is one standard deviation above the mean. 50% of the data lies below the mean and 34% lies between the mean and one standard deviation above the mean. This is $50\% + 34\% = 84\%$ of the data, so $100\% - 84\% =$ **16%** is the probability that a girl will be over 66 inches tall.

36. 750 is two standard deviations above the mean. $34\% + 13.5\% = 47.5\%$ of the area under the normal curve lies between the mean and 2 standard deviations above the mean. 47.5% of 1000 is **475 students**.

37. (a) P_{16} represents the bottom 16% of all scores, which in the normal distribution falls one standard deviation below the mean. Thus $P_{16} = 600 - 75 = \mathbf{525}$.

(b) D_5 is the decile ranking below which is 50% of the data; i.e., the mean, or **600**.

(c) P_{84} represents the value below which are 84% of the scores, which in the normal distribution falls one standard deviation above the mean. Thus
$P_{84} = 600 + 75 = \mathbf{675}$.

38. The first person with the calculator types in a large random number (such as a typical annual salary, but not his/hers). The first person then memorizes or records that number, then adds his/her salary to that number on the calculator, passing it to the next person when she/he is done. That person adds her/his salary to the total, and passes it to the next person. They continue in this manner until all 5 people have entered their salaries. When the calculator is returned to the first person, he/she subtracts the random number that was entered, then divides by 5 to obtain the mean salary.

CHAPTER 11

INTRODUCTORY GEOMETRY

Assessment 11-1A: Basic Notions

1. (a) **True;** Two lines in three-dimensional space are coplanar if there is a plane that includes them both. This is true if the lines are parallel or if they intersect each other. Two lines that are not coplanar are called skew lines.
   (b) **True;** Three points are always coplanar, and if the points are distinct and non-collinear, the plane they determine is unique.
   (c) **False;** they intersect in a line, the empty set, or a plane (if two planes are the same plane).
   (d) **True;** for any two points on any line, you can always find a point between them.
   (e) **False;** for example, the union of two rays could form an angle.

2. 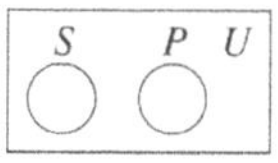

3. (a) Answers vary, $\overrightarrow{BC}, \overrightarrow{AC}, \overrightarrow{CB}, \overrightarrow{CA}$ are examples of rays that contain $\overline{BC}$.
   (b) Answers vary, $\overrightarrow{EG}, \overrightarrow{CG}, \overrightarrow{GE}, \overrightarrow{GC}$ are examples of rays that contain $\overline{EG}$.

4. (a) Given four collinear points A, B, C, and D with B between A and C, and C between B and D, the following rays can be made: $\overrightarrow{AB}, \overrightarrow{BA}, \overrightarrow{CB}, \overrightarrow{BC}, \overrightarrow{DC}$, and $\overrightarrow{CD}$. So **six** rays can be made with four collinear points.
   (b) Given five collinear points A, B, C, D, and E with B between A and C, and C between B and D, and D between C and E, the following rays can be made: $\overrightarrow{AB}, \overrightarrow{BA}, \overrightarrow{CB}, \overrightarrow{BC}, \overrightarrow{DC}, \overrightarrow{CD}, \overrightarrow{DE}$, and $\overrightarrow{ED}$. So **eight** rays can be made with five collinear points.
   (c) Each time a collinear point is added, the number of rays increase by two. Also, the number of rays is two less than twice the number of collinear points. So if there are n collinear points, the number of rays is $\mathbf{2n - 2}$ or $\mathbf{2(n - 1)}$.

5. (a)

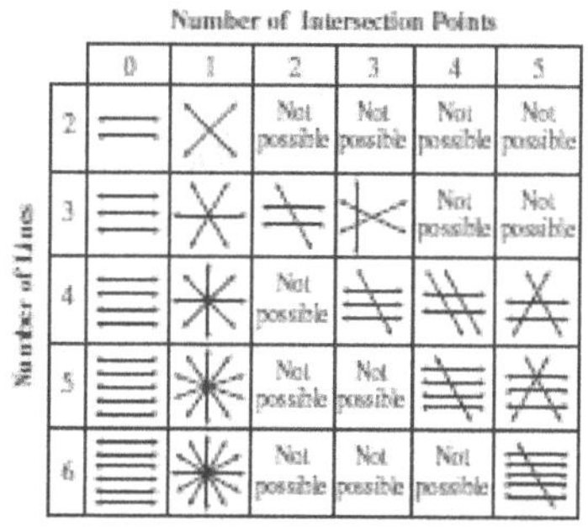

   (b) When there is only one line, there are no intersection points; when there are two lines, a maximum of one intersection point exists; when there are three lines, maximum of three intersection points exist. Although the table from 5(a) doesn't show it, there are a maximum of six intersection points for four lines (shown below):

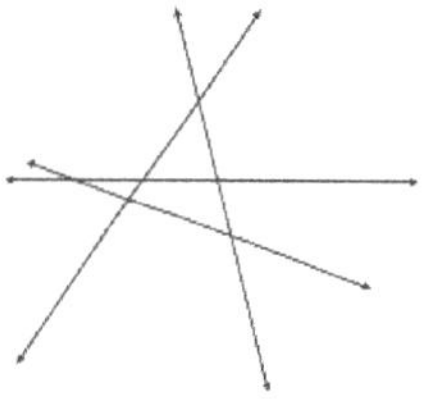

When a fifth line is added, there will be four more possible intersections, bringing the total number of intersections to ten:

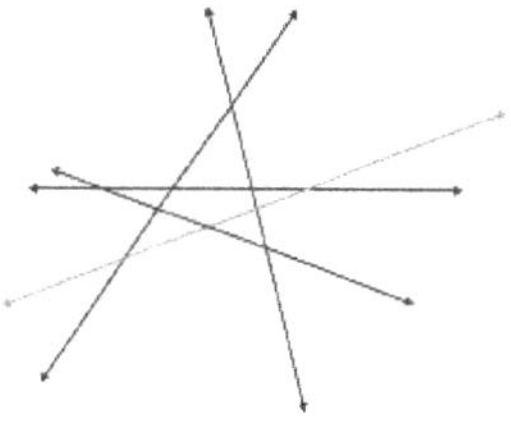

In general, when line n is added, the greatest possible number of intersections will be $1+2+3+...+(n-1)$. This is the sum of the first $(n-1)$ numbers, or

$\frac{n(n-1)}{2}$. Another way of looking at it: the maximum number of intersections is the combination of n lines taken two at a time,

$$\text{or } {}_nC_2 = \frac{n!}{2!(n-2)!} = \frac{n(n-1)(n-2)(n-3)\cdots}{2(n-2)(n-3)\cdots} = \frac{n(n-1)}{2}.$$

6. **(a)** Different order implies different names; i.e., $\overleftrightarrow{AB}, \overleftrightarrow{AC}, \overleftrightarrow{AD}, \overleftrightarrow{BC}, \overleftrightarrow{BD}, \overleftrightarrow{CD}, \overleftrightarrow{BA}, \overleftrightarrow{CA}, \overleftrightarrow{DA}, \overleftrightarrow{CB}, \overleftrightarrow{DB}, \overleftrightarrow{DC}$, which is the permutation of four points taken two at a time.

   ${}_4P_2 = \frac{4!}{(4-2)!} =$ **12 ways.**

   (b) Different order implies different names; i.e., a plane is determined by three non-collinear points. The number of different names $= {}_5P_3 = \frac{5!}{(5-3)!} =$ **60 ways.**

7. **(a)** Answers may vary. Skew lines are two lines that cannot be contained in the same plane; e.g., $\overleftrightarrow{AB}$ and $\overleftrightarrow{FH}$, $\overleftrightarrow{AD}$ and $\overleftrightarrow{GH}$, $\overleftrightarrow{BC}$ and $\overleftrightarrow{DH}$, $\overleftrightarrow{AE}$ and $\overleftrightarrow{BD}$, etc.

   (b) **Parallel**. *BDHF* is a rectangle.

   (c) **No**. They are non-coplanar; i.e., skew.

   (d) **None**. $\overleftrightarrow{BD}$ is parallel to plane *EFG*, so there are no points in common.

   (e) $\overleftrightarrow{FH}$ is the line of intersection of the two planes.

   $\overleftrightarrow{DH}$ is perpendicular to plane *FGH* because it is perpendicular to $\overleftrightarrow{HG}$ and $\overleftrightarrow{EH}$ in that plane. Thus $\overleftrightarrow{DH}$ is perpendicular to any line through *H* in the plane $FGH \Rightarrow \overleftrightarrow{DH} \perp \overleftrightarrow{FH}$.

   If a line ℓ through *H* is constructed in the plane *FGH* and perpendicular to $\overleftrightarrow{FH}$, $\overleftrightarrow{DH}$ will be perpendicular to ℓ Then one of the dihedral angles created by the two planes *BDH* and *FHG* measures 90°.

   Thus the planes are perpendicular because a dihedral angle measures 90°.

8. **(a)** **None**. The lines are parallel.

   (b) **Point *C***. Three distinct planes intersecting at one common point.

   (c) **Point *A***. The intersection of two non-parallel lines.

   (d) Answers may vary. $\overleftrightarrow{AC}$ **and** $\overleftrightarrow{BE}$ or $\overleftrightarrow{AB}$ **and** $\overleftrightarrow{CE}$ are two sets.

   (e) $\overleftrightarrow{AC}$ **and** $\overleftrightarrow{DE}$ or $\overleftrightarrow{AD}$ **and** $\overleftrightarrow{CE}$ are parallel; i.e., opposite sides of a rectangle.

   (f) **Planes *BCD*** or ***BEA***; i.e., planes bisecting the pyramid.

9. Perspectives may vary; e.g.,

   (a)

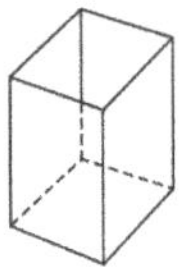

   (b)

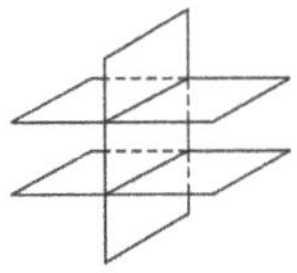

   (c)

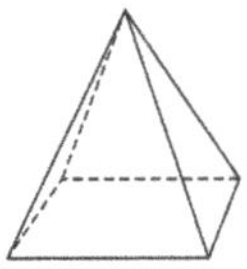

10. Answers may vary; e.g.,

   (a) A real-world example would be a paddle wheel:

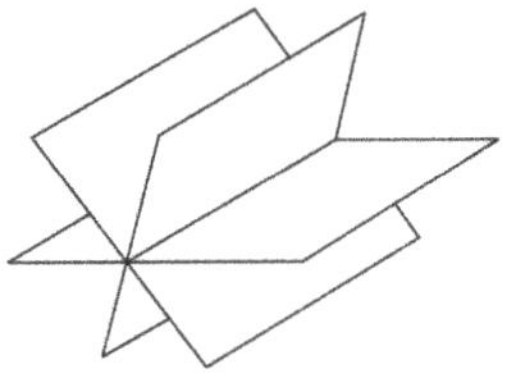

   (b) A real-world example would be the intersection of floor and two adjacent walls of a house:

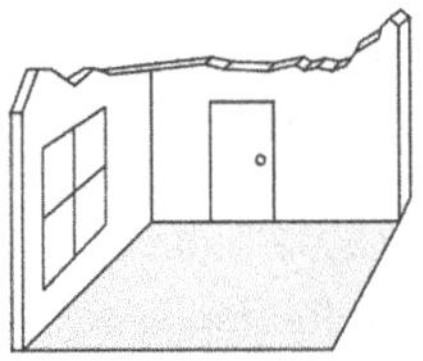

   (c) A real-world example would be a field irrigated with a central pivot:

Field Irrigates with a Central Pivot

11. **(a)** (plane *AFD*) $\cap$ (plane *XYE*) =

   $\overleftrightarrow{AE}, \overleftrightarrow{AD}$, or $\overleftrightarrow{DE}$

(b) (Plane XYE) $\cap \overleftrightarrow{AE} = \overleftrightarrow{\mathbf{AE}}$

(c) $\overline{BE} \cap \overline{CE} = \{\mathbf{E}\}$

(d) $\overline{CE} \cap \angle BEC = \overline{\mathbf{CE}}$

(e) $\overrightarrow{AE} \cap \overset{\circ\quad\circ}{\overline{DE}} = \overset{\circ\quad\circ}{\overline{\mathbf{DE}}}$

(f) $\overrightarrow{EB} \cap \overrightarrow{EC} = \{\mathbf{E}\}$

(g) interior $\angle BEC \cap \overleftrightarrow{BE} = \boldsymbol{\varnothing}$

12. **14 pairs**; angles *AOB*, *AOC*, *AOD*, *AOE*, *BOC*, *BOD*, *BOE*, *BOF*, *COD*, *COE*, *COF*, *DOE*, *DOF*, and *EOF*. This is the combination of six lines taken two at a time (${}_6C_2 = 15$), not including the right angle *AOF*.

13. Answers may vary. Some possibilities:

(a) Edges of a room; vertical and horizontal parts of a window frame; intersecting crossroads.

(b) Branches in a tree; clock hands at 7:30; angle on a yield sign.

(c) The top of a coat hanger; clock hands at 7:15.

14. As measured on the protractor:

(a) **110°**.

(b) $110° - 70° = \mathbf{40°}$.

(c) $180° - 160° = \mathbf{20°}$.

(d) $160° - 30° = \mathbf{130°}$.

15. (a) (*i*) $18°35'29'' + 22°55'41''$ $= 40°90'70'' = 40°91'10'' = \mathbf{41°31'1}$ $\mathbf{0''}$.

(*ii*) $15°29' - 3°45' = 14°89' - 3°45' =$ $\mathbf{11°44'}$

(b) (*i*) $0.9° = (0.9 \cdot 60)' = \mathbf{0°54'00''}$.

(*ii*) $15.13° = 15° + (0.13 \cdot 60)'$ $= 15°7.8' = 15°7' + (0.8 \cdot 60)'' = \mathbf{1}$ $\mathbf{5°7'48''}$.

16. (a) (*i*) The hour hand will be pointed directly at the 3, so it will have moved $\frac{1}{4}$ of $360° = \mathbf{90°}$.

(*ii*) The hand will be $\frac{25}{60}$ of the way from the 12 to the 1, or $\frac{25}{60}$ of the 30° between numbers $= \mathbf{12.5° = 12°30'}$.

(*iii*) The hand will have moved 6 whole spaces at 30° each plus $\frac{50}{60}$ of another 30°, or $180° + 25° = \mathbf{205°}$.

(b) Each minute moves the hour hand $\frac{1}{60}$ of 30° $= 0.5°$. 1:15 P.M. is 75 minutes, so the hour hand will have moved $0.5 \cdot 75 = 37.5°$ past 12.

In 15 minutes the minute hand moves to 90° past 12. $90° - 37.5° = \mathbf{52.5°} =$ $\mathbf{52°30'}$ between the hands.

17. (a) If $m(\angle AOB) = \frac{1}{3}m(\angle COD)$, then $m(\angle COD) = 3m(\angle AOB)$.

Let $x = m(\angle AOB)$. Then $x + 90° + 3x = 180°$ $\Rightarrow 4x = 90° \Rightarrow x = 22.5°$.

$m(\angle AOB) = \mathbf{22.5°}$;

$m(\angle COD) = 3(22.5°) = \mathbf{67.5°}$.

(b) Let $x = m(\angle BOC)$. Then $x + (3x - 35°) = 90°$ $\Rightarrow 4x = 125° \Rightarrow x = 31.25°$.

$m(\angle BOC) = \mathbf{31.25°}$.

$m(\angle AOB) = 3(31.25°) - 35° = \mathbf{58.75°}$.

(c) Assume that if the position of $\overline{OE}$ is changed, the other rays will be adjusted so that all x's are congruent and all y's are congruent. $3x + 3y$ will have different values, and thus x and y will have different values. $3x + 3y$, though, will remain 180° $\Rightarrow 3(x + y) = 180°$ $\Rightarrow x + y = 60° \Rightarrow m(\angle BOC) = \mathbf{60°}$.

18. (a) $\overparen{BC}$ is the arc associated with the central angle $\angle BOC$, whose measure is 70°. Thus, $m(\overparen{BC}) = \mathbf{70°}$.

(b) $m(\overparen{CD}) = m(\angle DOC) = \mathbf{110°}$.

(c) $m(\overparen{AD}) = m(\angle AOD) = \mathbf{70°}$.

(d) $m(\overparen{AB}) = \mathbf{110°}$.

(e) $m(\overparen{ADC}) = \mathbf{180°}$.

19. $\angle A = 30° + 4 \cdot \angle B$ and $\angle A + \angle B = 170°$.

Substitute for $\angle A$: $30° + 4 \cdot \angle B + \angle B = 170°$

Solve for $\angle B$: $5\angle B = 140° \Rightarrow \angle B = \mathbf{28}°$.

The solve for ∠A:

$\angle A + 28° = 170° \Rightarrow \angle A = \mathbf{142}°$.

20. Answers may vary. The best way to measure the angles of the bevel is to use a protractor. Note that angles x and y should sum to 270 degrees (the face of the trim has 5 sides, so the sum of the interior angles is 540 degrees; and the three angles not marked x or y are all right angles).

21. Answers will vary. The drawing should depict a person looking towards the top of a tree with the horizontal line (from the person's eyes) drawn straight across to the tree, and the line of sight drawn from the person's eyes to where the person is looking at the top of the tree.

22. **(a)** Each angle would be $360^\circ \div 2 = \mathbf{180^\circ}$

(b) Each angle would be $360^\circ \div 4 = \mathbf{90^\circ}$

(c) Each angle would be $360^\circ \div 10 = \mathbf{36^\circ}$

(d) Each angle would be $360^\circ \div n = \dfrac{\mathbf{360^\circ}}{\mathbf{n}}$

Assessment 11-2A: Curves, Polygons, and Symmetry

1. **(a)** **Convex**. A segment connecting any two points would lie fully inside the figure.

(b) **Concave**. It is possible to connect two points of the figure with a segment that lies partially or fully outside the figure.

(c) **Convex**.

(d) **Concave**.

2. **No.** By definition, a regular polygon is a convex polygon that is equiangular (all interior angles have the same measure) and equilateral (all sides have the same measure).

3. **(a)** **1, 4, 6, 7, 8**. A simple closed curve does not cross itself and begins and ends at the same point.

(b) **1, 6, 7, 8**. Polygons are polygonal curves (i.e., made up entirely of line segments) which are both simple and closed.

(c) **6, 7**. All segments connecting any two points of the polygon are inside the polygon; i.e., the region is nowhere dented inwards.

(d) **1, 8**. It is possible to draw a segment between two points of the polygon such that part of the segment lies outside the polygon.

4. A segment can pass through at most two sides of a triangle. If each side of the quadrilateral passes through two sides of the triangle there will then be **eight intersections**. One such figure might be:

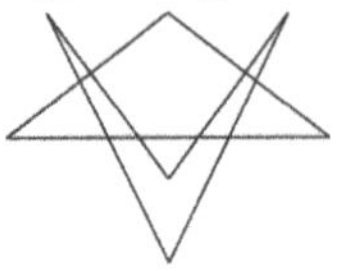

5. A **concave polygon**. In a concave polygon it is possible for a line segment connecting two interior points to contain a point or points outside the interior of the polygon..

6. **Squares**. For a shape to be in the shaded region it must be both a rectangle and a kite. Squares are rectangles with two adjacent sides congruent, which is a property of a kite.

7. **(a)** **Possible**; three sides of different lengths with one obtuse (i.e., greater than 90°) angle.

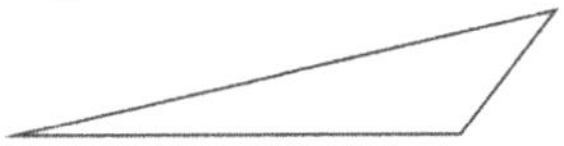

(b) **Possible**; three sides of different lengths with a right angle.

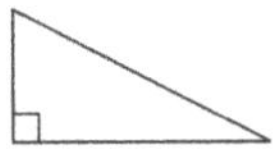

(c) **Impossible**. An equilateral triangle has three 60° angles.

(d) **Impossible**. An equilateral triangle has three 60° angles.

(e) **Possible**; two sides of equal length forming an obtuse angle.

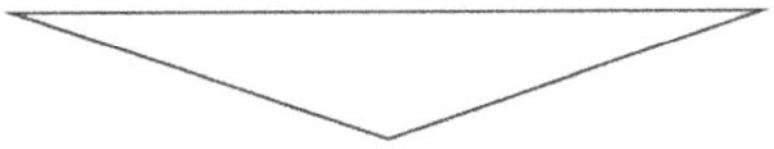

8. **(a)** **Equilateral** and **isosceles**; three congruent sides.

(b) **Isosceles**; exactly two congruent sides.

(c) **Scalene**; no sides congruent.

9. **(a)** The number of ways that all the vertices in a pentagon can be connected two at a time is the number of combinations of five vertices chosen two at a time, or ${}_5C_2 = 10$. This number of segments includes both diagonals and sides. Subtracting the number of sides, 5, from 10 yields **5 diagonals** in a pentagon.

(b) The number of ways that all the vertices in a decagon can be connected two at a time is ${}_{10}C_2 = 45$. Subtracting the number of sides, 10, from 45 yields **35 diagonals** in a decagon.

(c) The number of ways that the vertices in a 20-gon can be connected two at a time is ${}_{20}C_2 = 190.$ Subtracting the number of sides yields **170 diagonals** in a 20-gon.

(d) The number of ways that all the vertices in an n-gon can be connected two at a time is the number of combinations of n vertices chosen two at a time, or

${}_nC_2 = \frac{n!}{(n-2)!2!} = \frac{n(n-1)}{2}.$

This number of segments includes both the number of diagonals and the number of sides. Subtracting the number of sides n from $\frac{n(n-1)}{2}$ yields $\frac{n(n-1)}{2} - n =$

$\frac{n^2-n}{2} - \frac{2n}{2} = \frac{n^2-3n}{2} = \boldsymbol{\frac{n(n-3)}{2}}.$

10. (a)

(b)

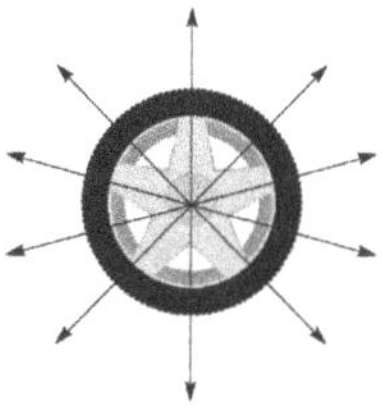

11. (*i*) Line symmetry: (*a*) and (*b*) both have line symmetry since they each can be reflected about a vertical line without changing their appearance; *(a)* has other line symmetries as well.

(*ii*) Turn symmetry: (*a*) has turn symmetry since
it can be "turned" 90° (or 180° or 270°) without changing its appearance.

(*iii*) Point symmetry: (*a*) has point symmetry since it can be "turned" or rotated 180° without changing its appearance.

12.

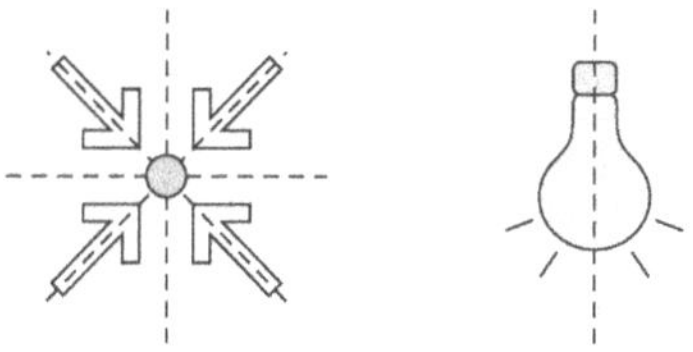

(a) (b)

13. Answers will vary. Some examples:

(a)

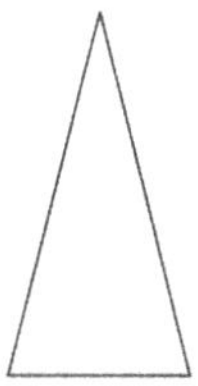

(b)

(c) The capital letter **N** has 180° rotational symmetry and no line of symmetry.

14. (a)

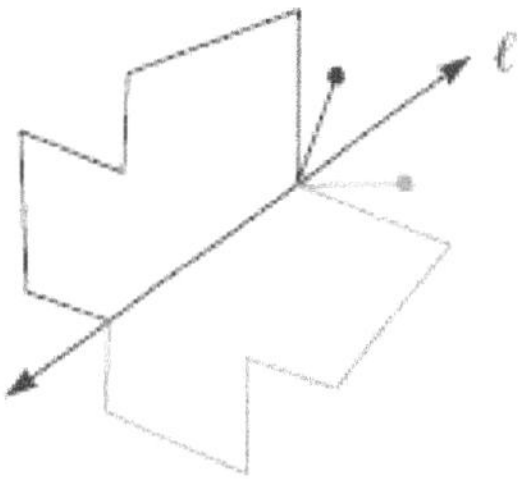

(b)

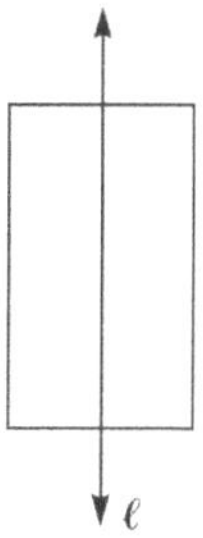

15. (a) 0 has vertical and horizontal line symmetry; if drawn like a circle, 0 could have infinitely many line symmetries. 3 has horizontal line symmetry. 1 could have line symmetry depending on how the number is written.

(b) 0 has point symmetry; 1 could, depending on how the number is written.

16. (a) There is one angle for turn symmetry: **180°**.

(b) There are nine sides to this regular polygon. In a full rotation of 360°, there would be turn symmetry at **40°, 80°, 120°, 160°, 200°, 240°, 280°, 320°**.

17. (a)

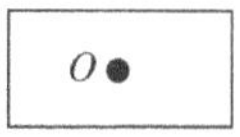

(b)

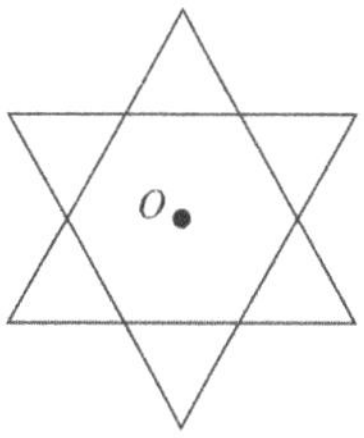

18. **(a)** A **square** has exactly 4 lines of symmetry.

(b) A **kite** has a line of symmetry through a pair of opposite vertices.

(c) A **parallelogram** has 180 degree rotational symmetry.

19. **Answers will vary**. There are examples of line symmetry, turn symmetry, and point symmetry throughout the tiling.

20. **(a)** One **line symmetry**.

(b) Ignoring color, there is rotational symmetry (**180**$^\circ$) and point symmetry. If color is taken into account, there are no symmetries.

(c) Ignoring color, there is rotational symmetry (**180**$^\circ$) and point symmetry. If color is taken into account, there are no symmetries.

Review Problems

17. **(a)** **False**. A ray has one endpoint, continuing forever from that point.

(b) **True**. Both lines continue forever in each direction through both points.

(c) **False**. Skew lines cannot be contained in a single plane.

(d) **True** if lines or line segments; **False** if rays. $\overrightarrow{MN}$ has endpoint M and extends in the direction of point N; $\overrightarrow{NM}$ has endpoint N and extends in the direction of point M.

(e) **True**.

(f) **False**. Their intersection is a line.

19. The bottoms of the three legs of a three-legged stool determine a plane. Since they are all on the same plane, they are fixed and are 'unwiggly'.

Assessment 11-3A: More About Angles

1. The illustration below shows three lines intersecting at one point, forming **six** non-overlapping angles:

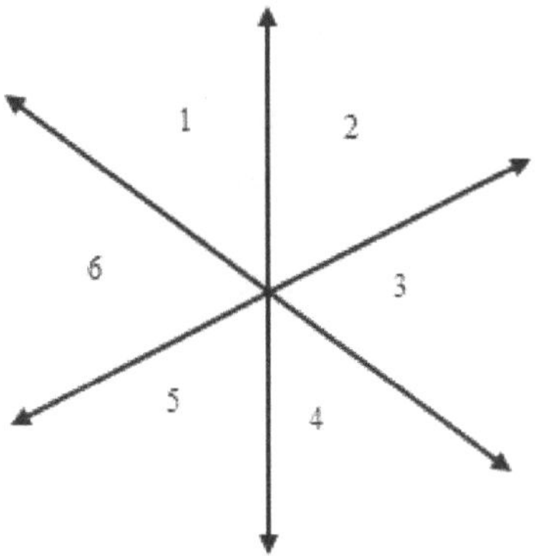

This shows **three pairs** of vertical angles: $\angle 1$ and $\angle 4$; $\angle 2$ and $\angle 5$; and $\angle 3$ and $\angle 6$.

2. The measures of the interior angles of every triangle add to 180°. Subtract the sum of the given angles from 180° to find the third angle.

(a) $180^\circ - (70^\circ + 50^\circ) = \mathbf{60^\circ}$.

(b) $180^\circ - (90^\circ + 45^\circ) = \mathbf{45^\circ}$.

(c) $180^\circ - (90^\circ + 30^\circ) = \mathbf{60^\circ}$.

(d) $180^\circ - (60^\circ + 60^\circ) = \mathbf{60^\circ}$.

3. **(a)** **(i)** **Complementary** angle:
$90^\circ - m(\angle A) = 90^\circ - 30^\circ = \mathbf{60^\circ}$

(ii) Supplementary angle:
$180^\circ - m(\angle A) = 180^\circ - 30^\circ = \mathbf{150^\circ}$

(iii) Vertical angle: 30°.

(b) **(i)** **Complementary** angle:
$90^\circ - m(\angle B) = \mathbf{90^\circ - x^\circ}$

(ii) Supplementary angle:
$180^\circ - m(\angle B) = \mathbf{180^\circ - x^\circ}$

(iii) Vertical angle: x°.

4. $m(\angle B) = 3 \cdot m(\angle A)$ and $m(\angle C) = 2 \cdot m(\angle B)$. So $m(\angle A) = \dfrac{m(\angle B)}{3}$ and in a triangle the sum of the interior angles is 180 degrees so using substitution:

$$m(\angle A) + m(\angle B) + m(\angle C) = 180^\circ$$
$$\frac{m(\angle B)}{3} + m(\angle B) + 2 \cdot m(\angle B) = 180^\circ$$
$$m(\angle B) + 3 \cdot m(\angle B) + 6 \cdot m(\angle B) = 540^\circ$$
$$10m(\angle B) = 540^\circ$$
$$m(\angle B) = \mathbf{54^\circ}$$

Then $54^\circ = 3 \cdot m(\angle A)$ and $m(\angle C) = 2 \cdot 54^\circ$, so $m(\angle A) = \mathbf{18^\circ}$ and $m(\angle C) = \mathbf{108^\circ}$

5. The supplement of a 150° is $180^\circ - 150^\circ = 30^\circ$, the complement of 30° is $90^\circ - 30^\circ = \mathbf{60^\circ}$.

6. The supplement is

$$\begin{aligned} 180^\circ - 23^\circ 17'18'' &= 179^\circ 59'60'' \\ &\underline{-23^\circ 17'18''} \\ &\mathbf{156^\circ 42'42''} \end{aligned}$$

7. $9x^\circ + (5x+62)^\circ = 90^\circ$

$14x^\circ = 28^\circ \Rightarrow x = 2^\circ$

$9(2^\circ) = \mathbf{18^\circ}$ and $5(2^\circ) + 62^\circ = \mathbf{72^\circ}$

8. **(a)** **Yes**. A pair of corresponding angles are 50° each (note that the supplementary angles formed by lines n and ℓ are 130° and 50°).

(b) **Yes**. A pair of corresponding angles are 70° each (note that the supplementary angles formed by lines n and ℓ are 110° and 70°).

(c) **Yes**. A pair of alternate interior angles are 40° each.

(d) **Yes**. A pair of corresponding angles are 90° each.

(e) **Yes**. The marked angles are alternate exterior angles. $k \parallel \ell$ by the alternate exterior angle theorem.

(f) **Yes**. Label as follows:

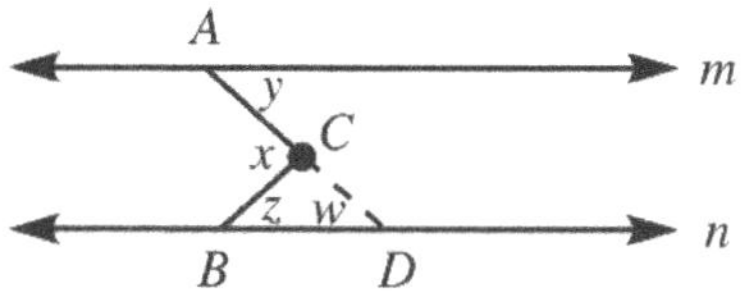

$x = 180^\circ - [180^\circ - (z + w] = z + w.$

It is given that $x = y + z$, so

$z + w = y + z \Rightarrow y = w.$

Because w and y are measures of alternate interior angles, $k \parallel \ell$.

9. The given ratio could be written $7x : 2x$. The angles must add to 90°, thus $7x + 2x = 90$ $\Rightarrow x = 10$. The angles are $7(10) = \mathbf{70^\circ}$ and $2(10) = \mathbf{20^\circ}$.

10. Each exterior angle $= 180^\circ - 162^\circ = 18^\circ$. Since the sum of the measurements of the exterior angles $= 360^\circ$, then $360^\circ \div 18^\circ = 20$ angles, thus **20 sides**.

11. Since the sum of the angles in a triangles is 180°,

$m(\angle 1) + m(\angle 3) + m(\angle 5) + m(\angle 2) + m(\angle 4) + m(\angle 6) = 180^\circ + 180^\circ = \mathbf{360^\circ}$.

12. The general form of an arithmetic sequence is $a_n = (n - 1)\,d + a_1$. Since $a_6 = 130^\circ$, we have that $130^\circ = 5d + a_1$. Since the sum of the interior angles of a hexagon is 720°, we have that $a_1 + (d + a_1) + (2d + a_1) + (3d + a_1) + (4d + a_1) + 130^\circ = 720^\circ$. Simplifying, $5a_1 + 10d = 590^\circ$. Combining this information:

$$\begin{aligned} 5a_1 + 10d &= 590^\circ \\ \underline{-2a_1 - 10d} &\underline{= 260^\circ} \\ 3a_1 \quad\quad &= 330^\circ \Rightarrow a_1 = 110^\circ. \end{aligned}$$

Thus,
$130^\circ = 5d + 110^\circ \Rightarrow 5d = 20^\circ \Rightarrow d = 4.$
Therefore, the angles are as follows:
110°; 114°; 118°; 122°; 126°; 130°.

13. **(a)** $m(\angle 3) = 180^\circ - [m(\angle 1) + m(\angle 2)] = 180^\circ - (45^\circ + 65^\circ) = \mathbf{70^\circ}$.

(b) $ABCD$ is parallelogram, thus opposite angles are equal. $m(\angle D) = m(\angle 3) = \mathbf{70^\circ}$.

(c) $AECB$ is a parallelogram, thus $m(\angle E) = m(\angle 2) = \mathbf{65^\circ}$.

(d) $ACFB$ is a parallelogram, thus $m(\angle F) = m(\angle 1) = \mathbf{45^\circ}$.

14. **(a)** In the triangle formed by the intersecting lines, the left base angle is 40° as it is a vertical angle the angle marked 40°; the top angle of the triangle is 100°, since it is supplementary to the 80° angle. So $x = \mathbf{40^\circ}$.

(b) $x + 4x = 90^\circ$ (complementary angles). $5x = 90 \Rightarrow x = \mathbf{18^\circ}$.

(c) Since $\angle CAD$ and $\angle BAC$ are complementary angles, their sum $= 90^\circ$. Since we are given $m(\angle CAD) = 2m(\angle BAC)$, $m(\angle BAC) = 30^\circ$. Similarly, $m(\angle ABC) = 30^\circ$. Since these two angles and angle x are the interior angles of a triangle and must sum to 180°, $x = \mathbf{120^\circ}$.

15. **(a)** Let the unknown angle $= x$. Then

$m(\angle x) + \frac{1}{2}m(\angle x) = 90° \Rightarrow$

$\frac{3}{2}m(\angle x) = 90°$

$\Rightarrow m(\angle x) = \frac{2}{3}(90°)$. $m(\angle x) = \mathbf{60°}$.

(b) m(third angle) $= 180° - 90°$ **(the sum of the complementary angles)** $=$ **90°**.

16. The six angles surrounding the center point sum to 360°. They are three pairs of vertical angles with the angles contained by triangles equal in measure to those not contained in the triangles. That means the contained angles must sum to half of $360° = 180°$. The sum of the angles in the three triangles is $3 \cdot 180° = 540°$. Thus the numbered angles sum to $540° - 180° = \mathbf{360°}$.

17. The sum of the measures of the interior angles of a hexagon $= (6 - 2) \cdot 180° = 720°$. $m(\angle x) = 720° - (110° + 105° + 142° + 122° + 130°) =$ **111°**.

18. ***(i)*** $m(\angle APT) = 90°$. $m(\angle 1) = 90° - 30° = \mathbf{60°}$.

(ii) $m(\angle 2) = \mathbf{30°}$ (alternate interior angles).

(iii) In ΔAPR, $m(\angle 3) = 180° - (30° + 40°) = \mathbf{110°}$.

19. $m(\angle ACB) = 180° - (70° + 30°) = \mathbf{80°}$.

$m(\angle 1) = 180° - 80° = \mathbf{100°}$ (supplementary angles).

20. If $y = 2x$, then $x = \frac{1}{2}y$. If $y = \frac{1}{2}z$, then $z = 2y$. If $y = \frac{1}{3}w$, then $w = 3y$.

Thus

$\frac{1}{2}y + y + 2y + 3y = 180° \Rightarrow \frac{13}{2}y = 180° \Rightarrow$

$y = \frac{360°}{13} = 27\frac{9°}{13} \approx \mathbf{27.69°}$;

$x = \frac{1}{2}\left(\frac{360}{13}\right)° = \frac{360°}{26} = 13\frac{11°}{13} \approx \mathbf{13.85°}$;

$z = 2\left(\frac{360}{13}\right)° = \frac{720°}{13} = 55\frac{5°}{13} \approx \mathbf{55.38°}$; and

$w = 3\left(\frac{360}{13}\right)° = \frac{1080°}{13} = 83\frac{1°}{13} \approx \mathbf{83.08°}$.

21. $\boldsymbol{m(\angle 1) = 120°}$, (supplementary with the 60° angle)

$\boldsymbol{m(\angle 4)} = \mathbf{120°}$ (vertical to $\angle 1$)

$\boldsymbol{m(\angle 5)} = \mathbf{60°}$ (vertical to the 60° angle)

$\boldsymbol{m(\angle 8)} = \mathbf{60°}$ (alternate interior angle to $\angle 5$)

$\boldsymbol{m(\angle 9)} = \mathbf{70°}$ (vertical to the 70° angle)

$\boldsymbol{m(\angle 10)} = \mathbf{50°}$ (supplementary $\angle$ to $\angle 8 + \angle 9$)

$\boldsymbol{m(\angle 11)} = \mathbf{50°}$ (vertical to $\angle 10$)

$\boldsymbol{m(\angle 12)} = \mathbf{60°}$ (vertical to $\angle 8$)

$\boldsymbol{m(\angle 6)} = \mathbf{50°}$ (alternate interior angle to $\angle 10$)

$\boldsymbol{m(\angle 7)} = \mathbf{130°}$ (supplementary $\angle$ with $\angle 6$)

$\boldsymbol{m(\angle 3)} = \mathbf{50°}$ (vertical to $\angle 6$)

$\boldsymbol{m(\angle 2)} = \mathbf{130°}$ (vertical to $\angle 7$)

22. $m(\angle 1) + m(\angle 2) = 180°$ (straight angle)

$m(\angle 3) + m(\angle 4) = 180°$ (straight angle)

Then $m(\angle 1) + m(\angle 2) = m(\angle 3) + m(\angle 4)$ and since $m(\angle 2) = m(\angle 3)$, we have $m(\angle 1) + m(\angle 3) = m(\angle 3) + m(\angle 4)$. Subtracting $m(\angle 3)$ from both sides, we have $m(\angle 1) = m(\angle 4)$.

23. **(a)** A STOP sign is a regular octagon. The sum of the measures of the interior angles of any octagon is $(8-2)180° = (6)180° = 1080°$ So, each angle measures $\frac{1080°}{8} = \mathbf{135°}$.

(b) As shown in part (a) above, the sum of the measures of all the interior angles of a STOP sign is **1080°**.

24. The measure of each interior angle of a regular polygon is found by $\frac{(n-2)180°}{n}$. Therefore,

$\frac{(n-2)180°}{n} = 156 \Rightarrow$

$180n - 360° = 156n \Rightarrow 24n = 360 \Rightarrow n = 15$.

There are **15 sides** of the polygon.

25. To answer this question would be the same as finding the measure of the central angle of the regular octagon. So $\frac{360°}{8} = \mathbf{45°}$.

26. An isosceles trapezoid is a quadrilateral; all quadrilaterals have total interior angle measure of 360°. Therefore, $360° - 108° = \mathbf{252°}$.

Review Problems

21. **(a)** (*i*) Two sets of parallel sides: *A*, *B*, *C*, *D*, *E*, *F*, *G*.

(*ii*) One set of parallel sides: *I*, *J*.

(*iii*) No parallel lines: *H*.

(b) (*i*) Four right angles: *D*, *F*, *G*.

(*ii*) Two right angles: *I*.

(*iii*) No right angles: *A*, *B*, *C*, *E*, *H*, *J*.

(c) (*i*) Four congruent sides: *B*, *C*, *F*,*G*.

(*ii*) Two pairs of congruent sides: *A*, *D*, *E*, *H*.

(*iii*) One pair of congruent sides: *J*.

(*iv*) No congruent sides: *I*.

23. Answers vary.

(a) The figure has 180° turn symmetry but no line symmetry.

(b) The figure has point symmetry, 90° turn symmetry, and vertical and horizontal line symmetry.

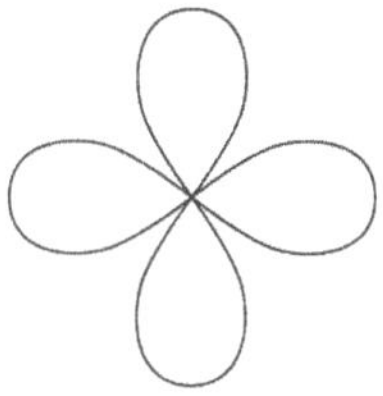

25. **(a)** Turn symmetries of 90°, 180° (also called point symmetry), and 270° about the center of the square.

(b) Point symmetry about the center of the square.

Assessment 11-4A: Geometry in Three Dimensions

1. **(a)** **Quadrilateral pyramid;** possibly rectangular or square pyramid. This is a polyhedron determined by a simple closed polygonal region, a point not in the plane of the region, and triangular regions determined by the point and each pair of consecutive vertices of the polygonal region.

(b) **Quadrilateral prism;** possibly trapezoidal or right trapezoidal prism. This is a polyhedron in which two congruent polygonal faces lie in parallel planes and the other faces are bounded by parallelograms.

(c) **Pentagonal pyramid**. The closed polygonal region is a pentagon.

2. **(a)** (*i*) There are $4 + 2 =$ **6 cubes** in the stack.

(*ii*) Each cube has six faces, so there are 36 total faces. From the left and right only three faces are seen; from front, back, top or bottom only four are seen. Thus there are $4 \cdot 4 + 2 \cdot 3 = 22$ visible faces. Of the 36 total faces, therefore, there must be **14** glued together.

(b) (*i*) There are $2 \cdot 3 \cdot 4 =$ **24** cubes in the stack.

(*ii*) Each cube has six faces, so there are 144 total faces; $12 + 24 + 16 = 52$ are visible. $144 - 52 =$ **92 faces** glued together.

3. **(a)** Vertices: ***A***, ***D***, ***R***, ***W***.

(b) Edges: $\overline{\boldsymbol{AR}}, \overline{\boldsymbol{RD}}, \overline{\boldsymbol{AD}}, \overline{\boldsymbol{AW}}, \overline{\boldsymbol{WR}}, \overline{\boldsymbol{WD}}$.

(c) Faces: $\triangle \boldsymbol{ARD}, \triangle \boldsymbol{DAW}, \triangle \boldsymbol{AWR}, \triangle \boldsymbol{DRW}$.

(d) Intersection of $\triangle DRW$ and $\overline{RA}$: point ***R***.

(e) Intersection of $\triangle DRW$ and $\triangle DAW$: $\overline{\boldsymbol{DW}}$.

4. There is **one** possible pair of bases that lie in parallel planes.

5. **(a)** If it is a prism with *n* sides on the base, then it has ***n* lateral faces**. The top and bottom base are not counted in lateral faces. So the number of lateral faces is equal to the number of sides of the base.

(b) The total number of faces is then ***n* + 2** because the total number of faces includes the top and bottom base.

6. **(a)** **Five;** triangular prism.

(b) **Four;** triangular pyramid.

(c) **Four;** tetrahedron.

7. **(a)** **True**. This is the definition of a right prism.

(b) **False**. No pyramid is a prism; i.e., a pyramid has one base and a prism two bases.

(c) **True**. The definition of a pyramid starts with the fact that it is a polyhedron.

(d) **False**. They lie in parallel planes.

(e) **False.** A prism must have two bases that lie in parallel planes.

(f) **True.** A cylinder has congruent circles for bases that lie in the parallel plane.

8. Answers may vary, but all are possible; e.g.,

(a) An oblique square prism. Some faces are not bounded by squares.

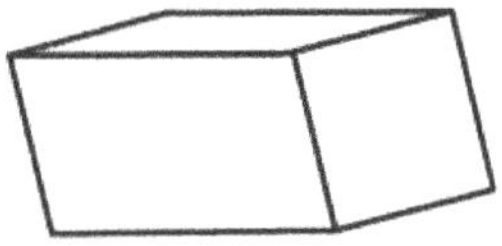

(b) An oblique square pyramid. The base is a square but a vertical line segment from the vertex does not intersect the base at its center.

9. (a) Top: ○; Side △; Front △.

(b) Top: ○; Side: ○; Front: ○.

10.

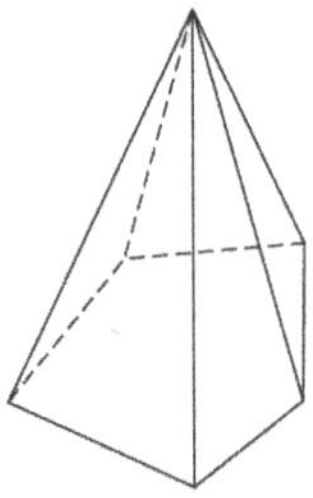

11.

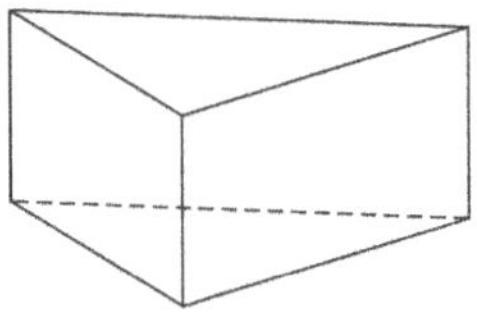

12. (a) *iv*. The cross-section of a cylinder is a rectangle.

(b) *ii*. The cross-section of a cone is a triangle.

13. (a) *i* (end view), *ii* (top view), *iii* (side view).

(b) *i*, *ii*, *iii*, *iv* (top views).

14. (a) The intersection is a **triangle:**

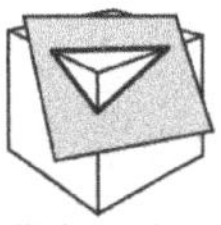

Triangle

(b) The intersection is a **rectangle:**

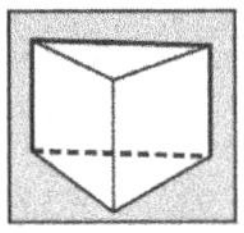

Rectangle

(c) The intersection is a **circle:**

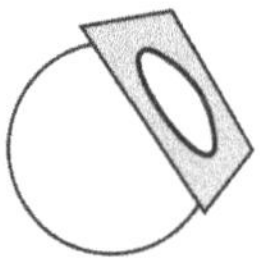

Circle

15. (a) There will be three pairs of parallel faces, determined by any three pairs of opposite sides of the hexagonal base, such as

$AA'B'B$ and $EE'D'D$.

(b) **120°**, e.g., $\angle FAB$ is a dihedral angle between two adjacent faces (each of the interior angles of a regular hexagon measure 120°).

16. (a) **Right regular hexagonal pyramid**. Assume the base is a regular hexagon and the triangles are congruent. Points, if folded up, would meet at a vertex.

(b) **Right square pyramid**. Assume the base is a square and triangles are congruent. Points would meet at a vertex.

(c) **Cube**. All six faces are squares; assume they are congruent.

(d) **Right square prism**. Assume the lateral bases are congruent and the ends congruent. Lateral faces of the prism would all be bounded by rectangles.

(e) **Right regular hexagonal prism**. Assume both bases are congruent regular hexagons and the lateral faces are congruent.

17. (a) **No, one face is missing**. An extra triangular face would be the net of a triangular pyramid.

(b) **Yes**, it is the net of a rectangular prism.

18. (a) (i) The bottom has **4 dots**.

(ii) The left side of the die has **5 dots**.

(b)

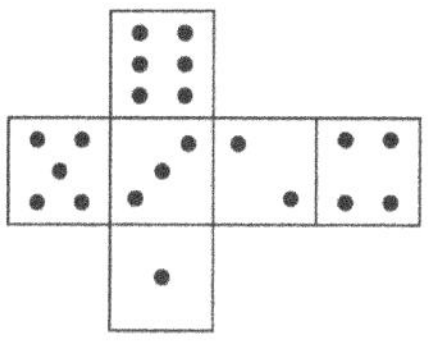

19. The relationship $V + F - E$ is known as Euler's formula, and always equals 2:

(a) $V = 10, F = 7,$ and $E = 15.$
$V + F - E = 10 + 7 - 15 = \mathbf{2}.$

(b) $V = 9, F = 9,$ and $E = 16.$
$V + F - E = 9 + 9 - 16 = \mathbf{2}.$

(c) Each triangle has 3 vertices: $8 \cdot 3 = 24$. Each octagon has 8 vertices: $6 \cdot 8 = 48$. Each vertex of a triangle is shared with a vertex of an octagon. The total number of vertices is then $48 - 24 = \mathbf{24}$.

20. (a) **Object iv.** It cannot be i or ii because the top point of the triangle is pointing to the circle with the line segments in it on the net. It cannot be iii because the black circle and the triangle must be on opposite sides.

(b) **Object ii.** The orientation of the numbers is not correct on i or iv, and on iii, the 6 is not below the 1 with that orientation.

Review Problems

19. The supplement is

$$180^\circ - 18^\circ 13' 42'' = 179^\circ 59' 60'' \\ \underline{-18^\circ 13' 42''} \\ \mathbf{161^\circ 46' 18''}$$

21. **No**, two adjacent angles cannot be vertical angles, because adjacent angles share a common side, and vertical angles do not.

23. A decagon has **35 diagonals**. In a decagon (10 sides), start with a vertex and draw all possible diagonals (8) then move to the adjacent vertex and draw all possible diagonals (7), continue this until all diagonals have been drawn.
$8 + 7 + 6 + 5 + 4 + 3 + 2 = 35$.

25. Two angles of the triangle are given so the third angle is $180^\circ - 30^\circ - 80^\circ = 70^\circ$. The angle adjacent to 70° is $180^\circ - 70^\circ = 110^\circ$. Angle $m(\angle 1) = \mathbf{110^\circ}$ (corresponding angles).

Chapter 11 Review

1. (a) $\overleftrightarrow{AB}, \overleftrightarrow{BC}, \overleftrightarrow{AC}$. One line can be drawn through any two given points.

(b) $\overrightarrow{BA}$ and $\overrightarrow{BC}$. A ray contains one endpoint and all points on the line on one side of the endpoint.

(c) $\overline{AB}$, the line segment between A and B.

(d) $\overline{AB}$, the only line segment containing all points common to both rays.

2. (a) Answers may vary; e.g., $\overleftrightarrow{PQ}$ **and** $\overleftrightarrow{AB}$ are skew. They do not intersect and are non-coplanar.

(b) **Any plane containing** $\overleftrightarrow{PQ}$ is perpendicular to α. Planes APQ and BPQ are two such.

(c) $\overleftrightarrow{AQ}$ is common to each.

(d) **No.** $\overleftrightarrow{AB}$ and $\overleftrightarrow{PQ}$ are skew lines, so no single plane contains them.

3. (a) $113^\circ 57' + 18^\circ 14' = 131^\circ 71' = \mathbf{132^\circ 11'}$.

(b) $84^\circ 13' - 27^\circ 45' = 83^\circ 73' - 27^\circ 45'$
$= \mathbf{56^\circ 28'}$.

(c) $113^\circ 57' + 18.4^\circ = 113^\circ 57' + 18^\circ 24'$
$= 131^\circ 81' = \mathbf{132^\circ 21'}$.

(d) $0.75^\circ = 0^\circ + 0.75(60)' = \mathbf{0^\circ 45' 0''}$.

(e) $\mathbf{35^\circ 8' 35''}$.

4. (a) The measure of one of the dihedral angles formed by planes α and γ is $m(\angle APS) = 90^\circ$, as given. The measure of one of the dihedral angles formed by planes β and γ is $m(\angle BPQ) = 90^\circ$, since it is given that $\overline{PQ} \perp \overline{AB}$.

(b) $\overline{AB}$ is perpendicular to $\overline{PQ}$ and $\overline{PS}$ in plane γ.

5. Examples may vary.

(a) A simple closed curve:

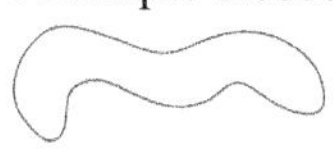

(b) A closed curve that is not simple:

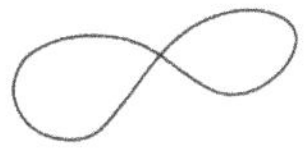

(c) A concave hexagon:

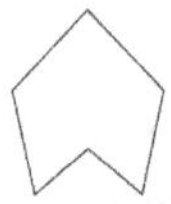

(d) A convex decagon:

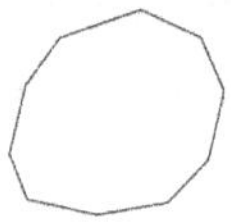

(e) Draw a regular pentagon $ABCDE$. Draw a line CD and reflect CD and DE about the line CD. The will create an equilateral pentagon that is not equiangular.

(f) Draw a rectangle that is not a square.

6. **(a)** **No.** The sum of the measures of two obtuse angles is greater than 180°, which is the sum of the measures of all the angles of any triangle.

(b) **No.** The sum of the measures of the four angles in a parallelogram must be 360°. If all the angles are acute the sum would be less than 360°.

7. Let α be the measure of the smallest angle. Then $\alpha + 2\alpha + 7\alpha = 180° \Rightarrow 10\alpha = 180°$. $\alpha = \mathbf{18°}$, $2\alpha = \mathbf{36°}$, $7\alpha = \mathbf{126°}$.

8. $180° - (90° + 42°) = \mathbf{48°}$.

9. $\frac{(n-2)\cdot 180}{n} = 176 \Rightarrow 180n - 360 = 176n \Rightarrow$ $4n = 360$. $n =$ **90 sides**.

10. $m(\angle 2) = m(\angle 3) = 45°$ because of alternate interior angles. Likewise, $m(\angle 1) = m(\angle 4) = 45°$.
Thus $\mathbf{m(\angle 3) = m(\angle 4) = 45°}$.

11. **(a)** $m(\angle 3) = m(\angle 1) = \mathbf{60°}$ (vertical angles).

(b) $m(\angle 5) = m(\angle 3) = 60°$ (alternate interior angles) and $m(\angle 6) = 180° - m(\angle 5)$ (supplementary angles).
$m(\angle 6) = 180° - 60° = \mathbf{120°}$.

(c) $m(\angle 8) = m(\angle 6) = \mathbf{120°}$ (vertical angles).

12. $m(\angle ABC) = 180° - (90° + 55°) = 35°$.
$m(\angle x) = 180° - (90° + 35°) = \mathbf{55°}$.

13. **(a)** $x + 30° + (180° - 70°) = 180°$
$\Rightarrow x + 30° + 110° = 180°$
$\Rightarrow \mathbf{x = 40°}$.

(b) Extend a line segment as follows.

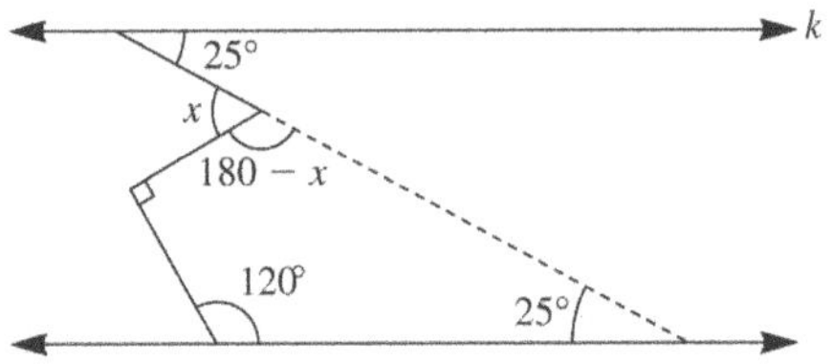

$90° + 120° + 180° - x + 25° = 360°$
$\Rightarrow -x + 415° = 360°$
$\Rightarrow -x = 360° - 415°$
$\Rightarrow -x = -55° \Rightarrow \mathbf{x = 55°}$.

14. **(a)** Consider ΔABC. This tells us that $x + y + 70° = 180°$. Also, $\mathbf{x = 50°}$ because $\overline{AB} \parallel \overline{CD}$. Thus, $\mathbf{y = 60°}$.

(b) $x + (180° - 125°) + 42° = 180°$
$\Rightarrow x + 97° = 180° \Rightarrow \mathbf{x = 83°}$.
$180° - 83° = 97°$; so $\mathbf{y = 97°}$.

15. **(*i*)** $m(\angle 1) = 180° - (70° + 45°) = \mathbf{65°}$.

(*ii*) $m(\angle 2) = m(\angle 1) = \mathbf{65°}$
(alt interior and corresponding angles).

(*iii*) $m(\angle 3) + m(\angle 4) = 360° - (2 \cdot 65°) = 230°$.
$m(\angle 3) = m(\angle 4)$
(opposite $\angle$s of a parallelogram)
$m(\angle 3) = \frac{230°}{2} = \mathbf{115°}$.

(*iv*) $m(\angle 4) = m(\angle 3) = \mathbf{115°}$.

(*v*) The third angle of $\Delta BDF = 65°$ (alternate interior angle with $\angle 2$).
$m(\angle 5) = 180° - (65° + 45°) = \mathbf{70°}$.

16. **(a)** $(5 \cdot 180°) - 360° = 3 \cdot 180° = \mathbf{540°}$.

(b) The measure of all the angles with vertices at $P = 360°$. If there are n triangles then there is a total of $n \cdot 180°$ minus the 360° surrounding P, or
$n \cdot 180° - 360° = \mathbf{(n - 2) \times 180°}$.

(c) Answers may vary. One way would be to connect B with E and connect A with F. There will then be two quadrilaterals and one triangle, or $2 \cdot 360° + 180° = \mathbf{900°}$.

17. **(a)** Alternate interior angles are congruent by construction. $\overline{AB} \parallel \overline{BC}$ by the alternate interior angle theorem.

(b) Corresponding angles are congruent.

(c) $m(\angle B) + m(\angle C) + m(\angle BAC)$
$= m(\angle BAD) + m(\angle DAE) + m(\angle BAC)$
$= 180°$.

18. Given three parallel lines with line l intersecting them:

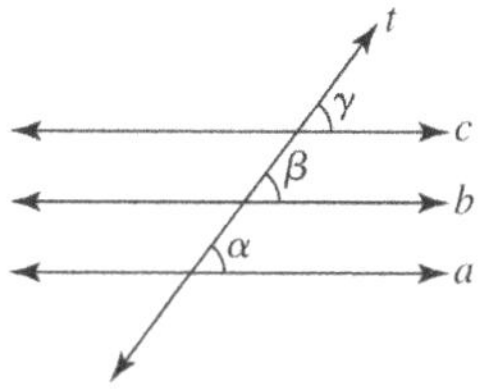

$a \parallel b \Rightarrow \alpha = \beta$. Then $b \parallel c \Rightarrow \beta = \gamma$. Thus $\alpha \parallel \gamma \Rightarrow a \parallel c$ (congruent corresponding angles).

19. **Correct**. If it is assumed that a rectangle has four right angles for a total angle measurement of $360°$ and that a diagonal divides it into two congruent triangles, then the sum of the measures in each right triangle is $180°$.

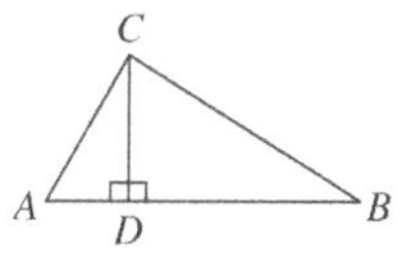

Thus the sum of the measures of the angles in ΔACD and ΔBCD is $2 \cdot 180° = 360°$. In this sum all the angles of the original triangle are included as well as the two right angles at D, so the sum of the measures of the angles in the original triangle is $360° - 2 \cdot 90° = 180°$.

20. One approach would be to draw a line c through A as labeled below,

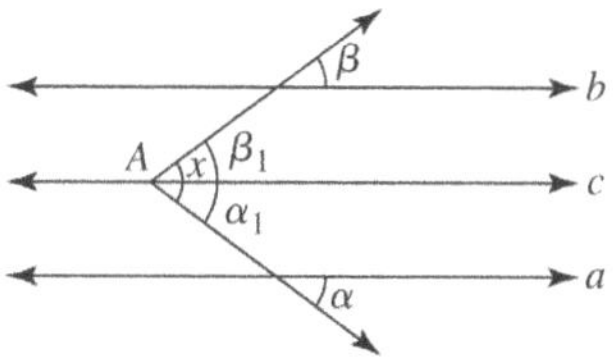

parallel to a and then prove it to be parallel to b. If $c \parallel a$, then $\alpha_1 = \alpha$ (corresponding angles). Since $x = \alpha + \beta$ and
$x = \alpha_1 + \beta_1 = \alpha + \beta_1$ then
$\beta_1 = \beta \Rightarrow c \parallel b$. By transitivity, $a \parallel b$.

21. $m(\angle 3) = 180° - m(\angle 1)$ and
$m(\angle 4) = 180° - m(\angle 2)$. Since
$m(\angle 1) < m(\angle 2)$ then

$^{-}m(\angle 1) > {}^{-}m(\angle 2)$.
$180° - m(\angle 1) > 180° - m(\angle 2)$
$\Rightarrow m(\angle 3) > m(\angle 4)$.

22. **(a)** $m(\angle ABO) + 60° + 90° = 180°$
$\Rightarrow m(\angle ABO) = \mathbf{30°}$.

(b) $m(\angle ABD) = m(\angle ABO) + m(\angle DBO)$
$= \mathbf{60°}$.

23. **(a)** A triangle requires three points. The order of the points is not important, thus the number of possible triangles is the combination of ten points taken three at a time, or ${}_{10}C_3 = \frac{10!}{(10-3)!3!} = \mathbf{120}$.

(b) ${}_nC_3 = \frac{n!}{(n-3)!3!} = \frac{n(n-1)(n-2)}{6}$.

24. **(a)** $\frac{360°}{20} = 18$, so there are **18 sides**.

(b) 25 does not divide $360°$; such a regular polygon **does not exist**.

(c) **Does not exist**; the sum is always $360°$.

(d) **Does not exist**; the equation $\frac{n(n-3)}{2} = 4860$ has no natural number solution.

25. **(a)** **4**

(b) **1**

(c) **1**

(d) **None**

(e) **2**

(f) **2**

26. **(a)** **Line and turn**

(b) **Line, turn, and point**

(c) **Line**.

27. Answers may vary.

(a)

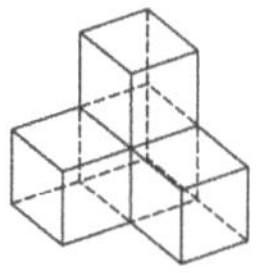

(b)

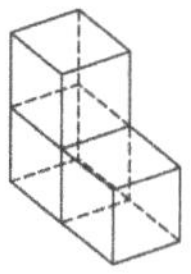

(c)

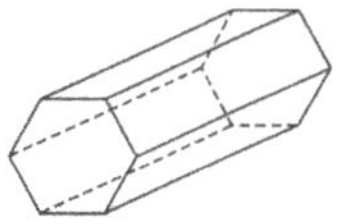

(d)

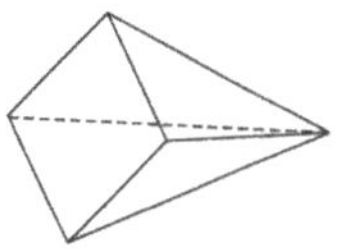

(e)

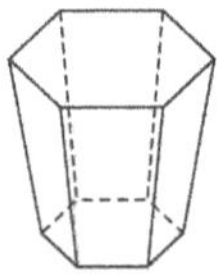

(f)

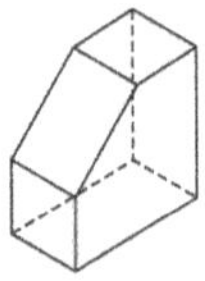

28. **(*i*)** Figure *d* had 5 vertices, 8 edges and 5 faces. $V - E + F = 5 - 8 + 5 = 2$.

(*ii*) Figure *e* has 12 vertices, 18 edges and 8 faces. $V - E + F = 12 - 18 + 8 = 2$.

(*iii*) Figure *f* has 10 vertices, 15 edges and 7 faces. $V - E + F = 10 - 15 + 7 = 2$.

29. **(a)** $180° - 30° - 90° =$ **60°**.

(b) **30°**.

(c) $180° - 120° - 30° =$ **30°**.

(d) $180° - 120° =$ **60°**.

30. There are as many lateral faces as sides. Thus, there are **eight** lateral faces.

31. **(a)**

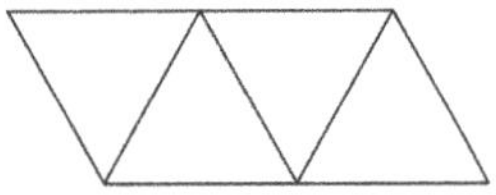

(b)

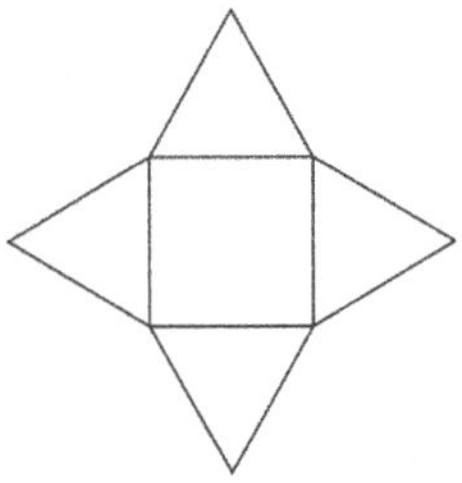

(c)

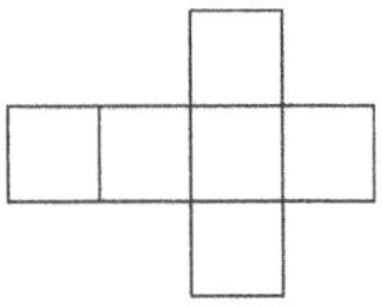

(d)

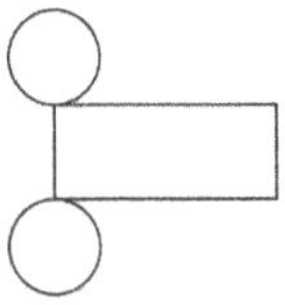

32. **(a)** One possibility is:

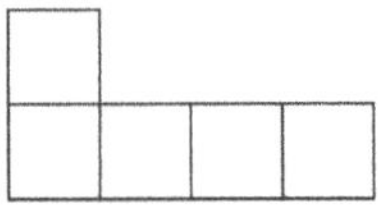

(b) **Eight different nets** are possible, i.e., the bottom may be placed in any of eight positions.

CHAPTER 12

CONGRUENCE AND SIMILARITY WITH CONSTRUCTIONS

Assessment 12-1A: Congruence through Construction

1. (a) **True**; Corresponding Parts of Congruent Triangles are Congruent (CPCTC).
   (b) **False**; the two angles are not corresponding angles, so there is no guarantee that they are congruent.
   (c) **False**; The two line segments are not corresponding line segments, so there is no guarantee that they are congruent.
2. The two triangles are congruent by the Side, Side, Side (***SSS***) congruent condition, so therefore, $\angle R \simeq \angle \mathbf{B}$ by CPCTC.
3. (a) Note that $\overline{BD} \cong \overline{DB}$ since they are the same segment. Therefore $\Delta ABD \cong \Delta CDB$
   (b) $\Delta ABD \cong \Delta CDB$ by the Side, Angle, Side (SAS) property: $\angle D \cong \angle B$, $AD \cong CB$, $BD \cong DB$. Therefore $\angle A \simeq \angle C$, $\angle B \cong \angle D$, and $\angle B \cong \angle D$; also, $\overline{AB} \cong \overline{CD}$ $\overline{AD} \cong \overline{CB}$ and $\overline{BD} \cong \overline{DB}$ by CPCTC.
4. Since both are 20 units long, $\overline{AB} \cong \overline{BC}$; since both angles are 60 degrees, $\angle ABD \simeq \angle CBD$; and $\overline{BD} \cong \overline{BD}$. So by SAS, $\Delta ABD \cong \Delta CBD$.
5. They are not necessarily congruent. Angle, Angle, Angle (AAA) is **not** a congruence property. It is possible that one triangle is larger than the other triangle. Without knowing anything about the side lengths, it cannot be determined if the triangles are congruent.
6. We are given that $\overline{BC} \cong \overline{DC}$ and that $\overline{AC} \cong \overline{EC}$. Also, $\angle ACB \simeq \angle ECD$, because they are vertical angles, and vertical angles are always congruent. So $\Delta ACB \simeq \Delta ECD$ by **SAS**. That means $\angle CAB \simeq \angle CED$ by CPCTC. Note that $\overline{AE}$ is a transversal, with $\angle CAB$ and $\angle CED$ alternate interior angles. Since the alternate interior angles are congruent, that means $\overline{AB} \parallel \overline{DE}$.
7. (a) Since l is the perpendicular bisector of $\overline{AC}$ we know that $\overline{AD} \cong \overline{CD}$ and $\angle ADB \simeq \angle CDB$. Since $\overline{BD} \cong \overline{BD}$, then $\Delta ADB \cong \Delta CDB$ by **SAS**. Therefore, $\overline{AB} \cong \overline{CB}$ by CPCTC. So, set their equations equal to each other and
   $$6x + 15 = 12x - 9$$
   solve for x: $\Rightarrow 24 = 6x$
   $$\Rightarrow \mathbf{4} = x$$
   (b) Since l is the perpendicular bisector of $\overline{AC}$ we know that $\overline{AB} \cong \overline{CB}$ and $\angle ABD \simeq \angle CBD$. Since $\overline{BD} \cong \overline{BD}$, then $\Delta ABD \cong \Delta CBD$ by **SAS**. Therefore, $\overline{AD} \cong \overline{CD}$ by CPCTC. So, set their equations equal to each other and
   $$6x - 3 = 3x + 15$$
   solve for x: $\Rightarrow 3x = 18$
   $$\Rightarrow x = \mathbf{6}$$
8. We are given that $\overline{AB} \cong \overline{BC}$ and $\overline{AD} \cong \overline{CD}$. Also, $\overline{BD} \cong \overline{BD}$. Therefore $\Delta ABD \cong \Delta CBD$ by **SSS**. That means $\angle ABD \simeq \angle CBD$ by CPCTC.
9. If the 3 points are collinear, then it is not possible. If the 3 points are not collinear, construct a triangle whose vertices are the 3 points. Now choose any 2 sides and construct perpendicular bisectors. The intersection of these bisectors will be a point equidistant from the 3 points.
10. **Obtuse triangles**. While it will not produce a general argument, constructing an obtuse triangle and then constructing the perpendicular bisectors to the sides adjacent to the obtuse angle will demonstrate why this is so.
11. (a) The angle opposite $\overline{BC}$ is greater than the angle opposite $\overline{AC}$; i.e., $\boldsymbol{m(\angle A) > m(\angle B)}$.
    (b) In any triangle, the side of greater length is opposite the angle of greater measure.

12. ΔABD and ΔCDB share a common side, $\overline{BD}$. Since both triangles are right triangles and $\overline{AB} \cong \overline{CD}$, $\Delta ABD \cong \Delta CDB$ by the H-L theorem. Thus, $\angle ABD \cong \angle CDB$. Since these are alternate interior angles created by transversal $\overline{BD}$, $\overline{AB} \parallel \overline{CD}$. Similarly, $\overline{BC} \parallel \overline{AD}$ by noting that $\angle ADB \cong \angle CBD$. Since opposite sides are parallel, $ABCD$ is a parallelogram.

13. **(a)** $\overline{AD}$ was created with a compass as shown below:

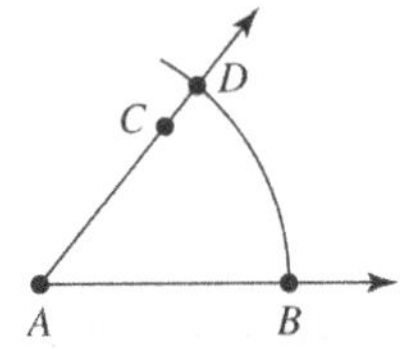

(b) One strategy is to construct a line segment for any of the three lengths, for example 2 cm. Then use a compass to draw circles of radii 3 cm and 4 cm from either endpoints. Where the circles intersect is a third vertex for a triangle with the desired properties. Alternative, cut three pieces of string of the desired lengths and arrange to form a triangle.

(c) Use a strategy similar to that in part (b).

(d) Since $4 + 5 < 10$, it is **not possible.**

(e) Use a strategy similar to that in part (b).

(f) Use protractor to create an angle whose measure is $75°$. Along the either ray of this angle draw line segment from the vertex of length 6 cm and 7 cm. Connect the end points of these line segments.

(g) Use *SSA* construction. This triangle is not unique.

(h) Use *SSA* construction. This triangle is not unique.

(i) Use a strategy similar to that in (f), i.e., *SAS* construction.

(j) (b). **Unique** by *SSS*. If the three sides of one triangle are congruent, respectively, to the three sides of a second triangle, then the triangles are congruent. I.e., all triangles with sides 2 cm, 3 cm, and 4 cm are the same. The figure below is similar to a 2 cm, 3 cm, 4 cm triangle.

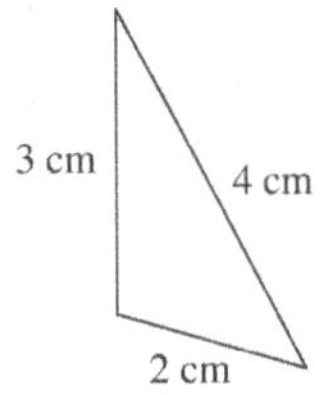

(c). **Unique** by *SSS*. A scalene right triangle.

(d) **No triangle**. The triangle inequality states that the sum of the measurements of any two sides of a triangle must be greater than the measure of the third side, and $10 > 4 + 5$.

(e) **Unique** by *SSS*.

(f) **Unique** by *SAS*. If two sides and the included angle of one triangle are congruent to two sides and the included angle of another triangle, respectively, then the two triangles are congruent. The figure below is similar to a *SAS* 6 cm-75°-7 cm triangle.

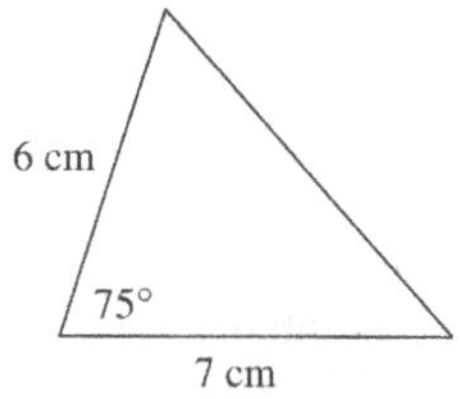

(g) **Not unique**. Three different triangles are possible; those below are representative.

(i)

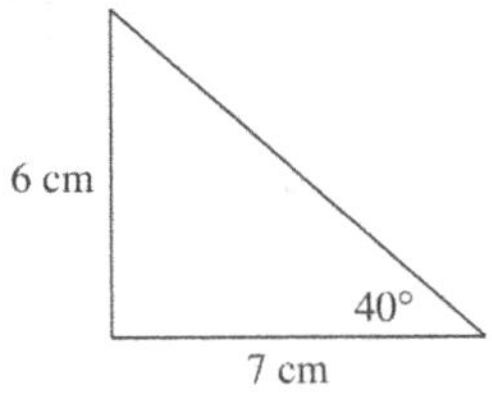

(ii)

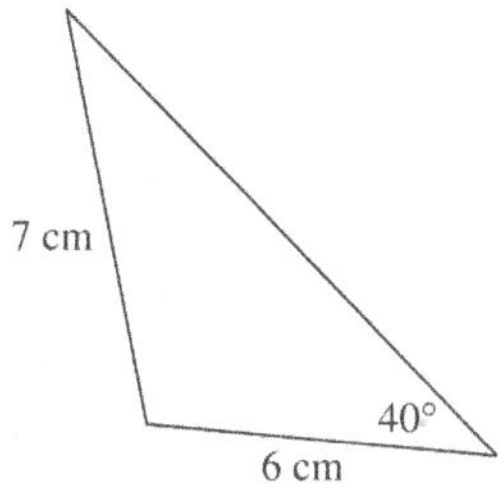

(iii)

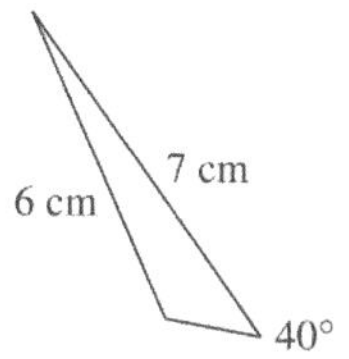

(h) **Unique** by *SAS*. With one angle given in an isosceles triangle the others are determined.

(i) **Unique** by *SAS* with an included right angle.

14. Construct a right isosceles triangle. The angle opposite the base is the right angle, and the perpendicular bisector of the hypotenuse (the base) bisects the right angle, creating two 45 degree angles.

15. **(a)** Use the procedure of Figure 11 in the text, marking the span of $\angle B$ from the point where the span of $\angle A$ fell on the arc.

(b) Use the same procedure as in part (a), but mark the span of $\angle A$ back toward the starting point from where the span of $\angle B$ fell on the arc.

16. **No**. If all sides are congruent then all angles must be congruent; thus no right angle is possible.

17. **(a)** Construction. The triangles *ABO*, *BCO*, *CDO*, and *DAO* are congruent (*SAS*) isosceles right triangles. Therefore the congruent angles in each triangle measure $45°$. Consequently all the angles in *ABCD* measure $90°$ and all the sides are congruent.

(b) Because the arcs $\overset{\frown}{BE}$ and $\overset{\frown}{EC}$ each measure $45°$, the chords *BE* and *EC* are congruent.

(c) Bisecting each of the right central angles we get the vertices of the regular octagon.

18. This point can be found only if *A* and *B* are not points on a line perpendicular to *m*. Assuming *A* and *B* have this property, construct the line segment $\overline{AB}$. Use the process illustrated in Figure 19 to construct the perpendicular bisector to $\overline{AB}$. By Theorem 12-3, the intersection of *m* and this perpendicular bisector is a point on *m* equidistance from *A* and *B*.

19. The center of the circle is the intersection of the $\perp$ bisectors of any two adjacent sides of the square. The radius of the circle is the distance from the center to any of the vertices of the square. Alternatively, the center can be found as the intersection of the diagonals.

20. **(a)** Use the procedure of Figure 10 in the textbook with all sides the length of $\overline{AB}$. Each angle will measure $60°$.

(b) The construction in part (a) produces three $60°$ angles.

(c) Extend one of the sides of the triangle in (a). The exterior angle to the equilateral triangle formed will be supplementary to a $60°$ **angle** and thus $\mathbf{120°}$.

21. **(a)** $\Delta ACB \cong \Delta BCA$

(b) From part (a) $\angle A \cong \angle B$, by corresponding parts of congruent triangles.

Assessment 12-2A: Additional Congruence Theorems

1. Constructions may vary; those below are representative.

(a) *ASA*:

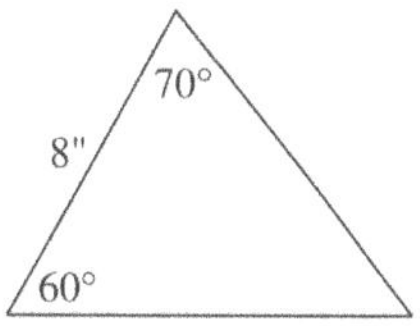

(b) *AAS*:

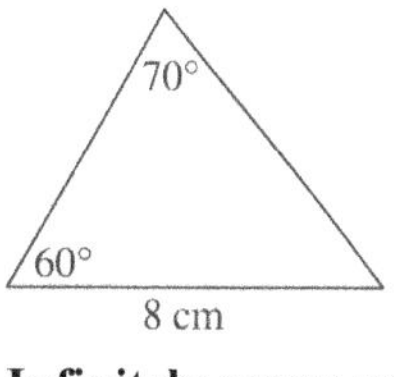

(c) **Infinitely many** are possible. All would be similar; *AAA* determines a unique shape, but not size.

2. **(a)** **No**. The triangle is unique by *ASA*.

(b) **No**. The triangle is unique by *AAS*.

(c) **Yes**. *AAA* determines a unique shape, but not size.

3. Compare the following triangles:

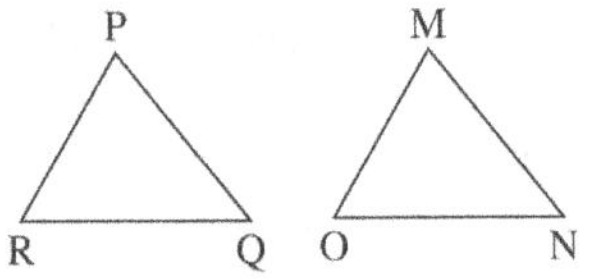

(a) **Yes**. Congruent by *ASA*.

(b) **Yes**. Congruent by *AAS*.

4. At any setting of the ruler, it is given that $\overline{AB} \cong \overline{DC}$ and $\overline{AC} \cong \overline{BD}$. Note that $\overline{BC} \cong \overline{CB}$. Then $\Delta ABC \cong \Delta DCB$ by *SSS*. $\angle ABC \cong \angle DCB$ by CPCTC. Because these are alternate interior angles formed by lines $\overleftrightarrow{AB}$ and $\overleftrightarrow{CD}$ with transversal line $\overleftrightarrow{BC}$, $\overline{AB} \parallel \overline{DC}$.

5. (a) $\angle ABD \cong \angle CBD$ is given. Note that D is a point on $\overline{AC}$ such that $A - D - C$. Since $\angle CDB$ is a right angle, then $\angle ADB$ is also a right angle (the two angles are supplementary); therefore they are congruent. Finally, $\overline{BD} \cong \overline{BD}$. So, by **ASA**, $\Delta ABD \cong \Delta CBD$

(b) There are no congruent triangles in this figure, based on the information that is given. AAA is not a congruence property.

6. Since the student has $\angle B \cong \angle E$ and $\overline{BC} \cong \overline{EF}$, congruence can be established in two different ways. If the student can show that $\angle C \cong \angle F$ then the triangles will be congruent by the **ASA** congruence property. If the student can show that $\overline{AB} \cong \overline{DE}$, then the triangles will be congruent by the **SAS** congruence property.

7. (a) Since $\Delta ABC \cong \Delta DEF$, $\angle E$, corresponds to $\angle B$, meaning they must have the same measure. Therefore $m\angle E = 85°$.

(b) By Theorem 11.3, it is known that the sum of the interior angles of a triangle add to $180°$. So, since $m\angle A = 40°$, and $m\angle B = 85°$, then $m\angle C = 55°$. And, since $\Delta ABC \cong \Delta DEF$, $\angle F$, corresponds to $\angle C$, meaning they must have the same measure. Therefore, $m\angle F = 55°$.

(c) Since $\Delta ABC \cong \Delta DEF$, $\overline{DE}$, corresponds to $\overline{AB}$, meaning they must have the same measure. Therefore $m\overline{DE} = 6$.

(d) Since $\Delta ABC \cong \Delta DEF$, $\overline{EF}$, corresponds to $\overline{BC}$, meaning they must have the same measure. Therefore $m\overline{EF} = 5$.

8. We are given that $\angle ABC \cong \angle DEF$, $\overline{AB} \cong \overline{DE}$ and $\overline{BC} \cong \overline{EF}$. So $\Delta ABC \cong \Delta DEF$ by **SAS**. That means $\angle BAC \simeq \angle EDF$ by CPCTC.

9. We are given $\overline{AD} \parallel \overline{EC}$ and $\overline{BC} \cong \overline{BD}$. Since $\overline{AD} \parallel \overline{EC}$, alternate interior angles $\angle DAB$ and $\angle CEB$ are congruent. Also $\angle ABD \simeq \angle EBC$ since they are vertical angles. Therefore, $\Delta ABD \cong \Delta EBC$ by **AAS**.

10. No. Two non-congruent kites can have correspondingly congruent diagonals. For example, consider the two kites below.

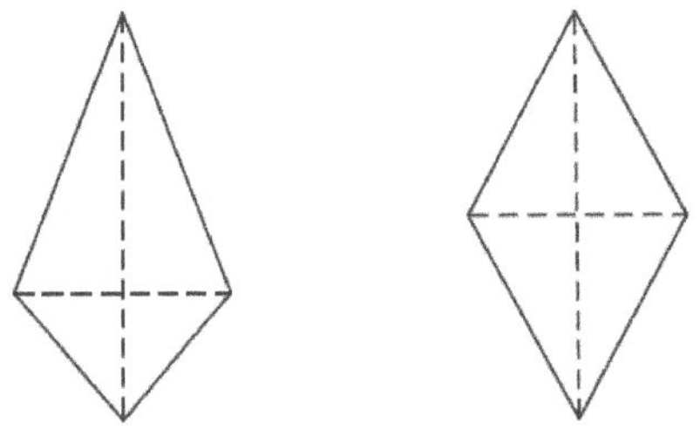

Kite that is not a rhombus — Rhombus

11. See Table 1 for properties.

(a) **Parallelogram**, by properties *d* and *e*.

(b) **None**. It must be known that the quadrilateral is a parallelogram before it can be known that it is a rectangle. Otherwise it could be an isosceles trapezoid.

(c) **None**. The quadrilateral could be a parallelogram, rectangle, or trapezoid even if the diagonals are not perpendicular.

12. An isosceles trapezoid is formed. The angles formed with one base as a side are congruent, as are the angles formed on the other base.

The congruent angles are sometimes referred to as the base angles of the isosceles trapezoid.

13. See Table 1.

(a) **True**, by property.

(b) **True**, by definition.

(c) **True**, by definition.

(d) **False**. A trapezoid may have only one pair of parallel sides, while in a parallelogram each pair of opposite sides must be parallel. A trapezoid may also have two consecutive right angles with the other two not right angles.

14. **(a)** Constructions may vary; possibilities include:

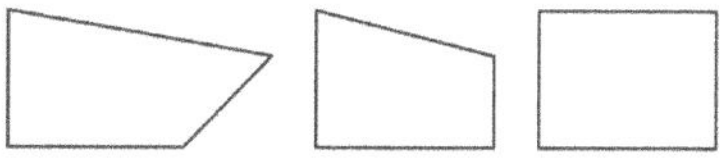

(b) **No**. The sum of the measures of the angles in a quadrilateral must be 360°. If the quadrilateral has three right angles, then the fourth must also be a right angle.

(c) **No**. Any parallelogram with a pair of right angles must have right angles as its other pair. Thus it must also be a rectangle (a rectangle
is a parallelogram with right angles).

15. *ABCD* is a kite.

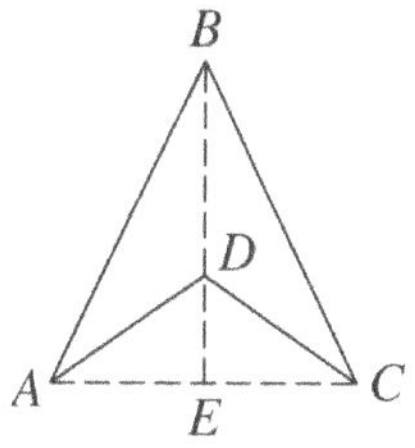

Because $AB = BC$, B is equidistant from A and C. By Theorem 12-3, point B is on the perpendicular bisector of $\overline{AC}$. Likewise, D is on the perpendicular bisector of $\overline{AC}$. Because two points determine a unique line $\overleftrightarrow{BD}$ is the perpendicular bisector of $\overline{AC} \Rightarrow \overleftrightarrow{BD} \perp \overleftrightarrow{AC}$.

16. **(a)** **Rectangle.**
$\overline{AB} \parallel \overline{ED}$ and $\overline{AE} \parallel \overline{BD}$. Because $\Delta AFE \cong BCD$ adjacent angles are congruent.

(b) **Isosceles trapezoid**. $\overline{BE} \parallel \overline{AF}$ but $\overline{AB} \nparallel \overline{EF}$; $EF = AB$.

17. A **rhombus**. All sides are congruent.

18. Either the arcs or the central angles must have the same measure. Radii are already the same, since the sectors are part of the same circle.

19. **(a)** A **kite**.
$\overline{DX} \cong \overline{AX}$ (given). $\angle D$ and $\angle A$ are right angles of the rectangle. $\overline{DP} \cong \overline{AQ}$ (given). Thus $\Delta PDX \cong \Delta QAX$ by *SAS* and $\overline{PX} \cong \overline{QX}$ by CPCTC.

$\overline{DC} \cong \overline{BA}$ as opposite sides of the rectangle. Since $\overline{DP} \cong \overline{AQ}$ then $\overline{CP} \cong \overline{BQ}$.

$\overline{CY} \cong \overline{BY}$ (given). $\angle C$ and $\angle B$ are right angles of the rectangle. Therefore $\Delta CYP \cong \Delta BYQ$ by *SAS* so that $\overline{QY} \cong \overline{PY}$.

By definition, then, *PXQY* is a kite.

(b) The quadrilateral is **still a kite**. When *P* and *Q*, however, are midpoints of $\overline{DC}$ and $\overline{AB}$ respectively, *PXQY* is a rhombus.

20. Construct the first kite by constructing a segment to become a diagonal and then construct two isosceles triangles with the segment as a common base (see below).

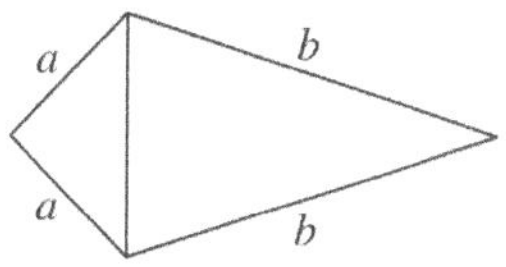

Construct the second kite starting with a segment not congruent to the first. Construct two isosceles triangles with that segment as their common base but the sides congruent to the corresponding sides of the isosceles triangles in the first construction (see below).

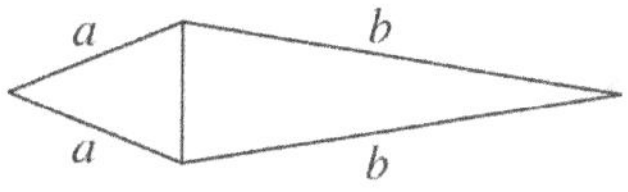

21. **Rhombus**. Use *SAS* to prove that $\Delta ECF \cong \Delta GBF \cong \Delta EDH \cong \Delta GAH$. Then $\overline{EF} \cong \overline{GF} \cong \overline{EH} \cong \overline{GH}$.

22. **(a)** The lengths of one side of each square must be congruent.

(b) Two adjacent sides of one must be congruent to the corresponding sides of the other.

(c) Answers may vary. E.g., one solution is that two adjacent sides and the included angle of one must be congruent to the other.

23. **(a)** In the figure, $\overline{BC} \parallel \overline{AD}$ and $\overline{BC} \cong \overline{AD}$.

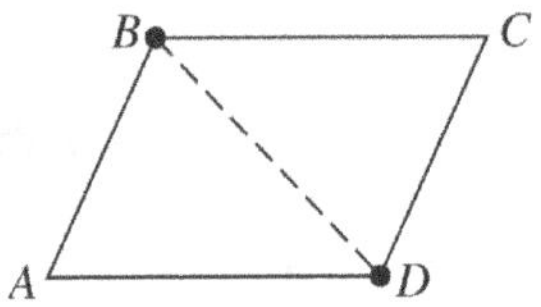

Since $\overline{BC} \parallel \overline{AD}$, $\angle ADB \cong \angle CBD$. Also, $\overline{BD} = \overline{DB}$. Thus, by SAS, $\Delta ADB \cong \Delta CBD$. Thus, corresponding parts are congruent, and $\overline{AB} \cong \overline{CD}$ and $\angle ABD \cong \angle CDB$. Since these last angles are alternate interior angles for transversal $\overline{BD}$, $\overline{AB} \parallel \overline{CD}$. Since opposite sides are parallel and congruent, the quadrilateral is a parallelogram.

(b) If the diagonals bisect each other, then $AO = OC$ and $BO = OD$ in the figure below.

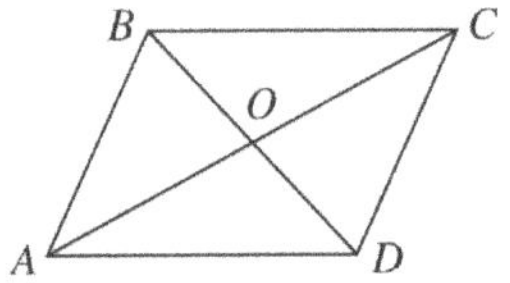

In addition, vertical angles $\angle BOC$ and $\angle DOA$ are congruent. Thus, by *SAS*, $\Delta BOC \cong \Delta DOA$. $\angle OBC \cong \angle ODA$ because they are corresponding parts of $\cong$ Δ's, and $\overline{BC} \parallel \overline{AD}$, because the angles are alternate interior angles formed by transversal $\overline{BD}$. Since quadrilateral *ABCD* has a pair of opposite sides parallel and congruent, it must be a parallelogram by part (a).

24. Answers vary. For example: the polygons must have the same number of sides with one pair congruent. All regular polygons with the same number of sides are similar, so if they have the same number of sides with one pair congruent they are congruent.

Review Problems

19. Triangles ***BCD, CDE, DEA***, and ***EAB***. By definition, all five sides and angles are congruent. Therefore, the triangles are congruent by *SAS*.

21. Follow the procedure of Figure 10, using the given segment for all three sides.

23. First construct an equilateral triangle. Use your straightedge to draw line segment *AB*. Place your compass point on A and measure the distance to point B. Using your compass, draw a circle with radius of this size and center *A*. Construct another circle with center *B*. and the same radius. Create a point at the intersection of the two circles and label it *C*. Connect *A* to *C* and *B* to *C*.

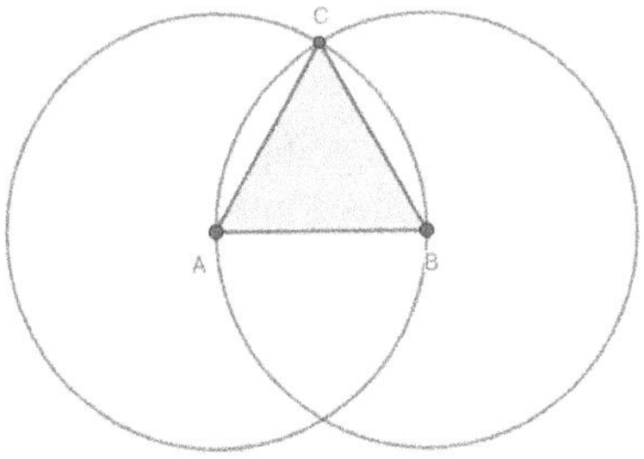

Each angle inside an equilateral triangle is 60°. Bisecting one angle will construct a 30° angle. Place the point of the compass on *A*, and swing an arc *ED*. Then, with *D* as center and *DE* as radius, draw an arc. Keeping the same radius and with E as center, draw an arc that will intersect the first; call that point of intersection *F*, and draw *AF*. Then *AF* bisects angle *BAC* and angle *CAF* is 30°.

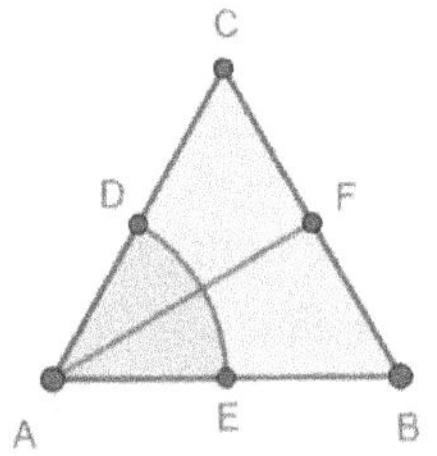

Assessment 12-3A: Additional Constructions

1. **(a)** Refer to Figure 36. Draw a line through *P* that intersects ℓ, but copy α as an alternate interior angle (i.e., vertical to α as shown in the Figure).

(b) Refer to Figure 35. Draw the line through *P* that intersects ℓ, and use the construction shown to create a rhombus.

2. Draw a line l and a point P not on l. Let B be any point on line l and construct $\overline{PB}$. Let C be any other point on line l. Copy $\angle PBC$

to create new angle $\angle BPD$ (see page 744, Figure 11 on how to copy an angle) so that the vertex of the new angle is point P and the side $\overline{PD}$ of the new angle $\angle BPD$ is such that points D and C are on opposite sides of $\overline{PB}$.

Now extend $\overline{PD}$ to create line m. If this is done correctly, you should have $l \parallel m$, since $\angle PBC$ and $\angle BPD$ are congruent alternate interior angles.

3. Copy $\angle A$ using the directions for copying from Figure 11. Now, place the compass point on vertex A and draw an arc that intersects both sides of the angle. Label the points of intersection B and C. Keeping the compass opening the same as before, place the point of the compass on point B. Construct an arc in the interior of $\angle BAC$. Place the compass point on point C, and construct another arc in the interior of $\angle BAC$, so that the two arcs intersect. Label the point of intersection D.

Construct ray $\overrightarrow{AD}$ using a straightedge. This ray should be the bisector of $\angle BAC$, and $\angle BAD \cong \angle CAD$.

4. **(a)** A **right triangle**. Assuming the ground is level, the cable will hang perpendicularly to the ground.

(b) The altitude is the extension of the cable from the ground to vertex *A*. It will lie outside the triangle.

5. **(a)** The perpendicular bisectors of the sides of an acute triangle meet inside the triangle.

(b) The perpendicular bisectors of the sides of a right triangle meet at the midpoint of the hypotenuse.

(c) The perpendicular bisectors of the sides of an obtuse triangle meet outside the triangle.

(d) Use the point of intersection of the perpendicular bisectors as the center of the circle, then use the distance from the intersection to any of the vertices as the radius of the circle.

6. **(a)** Construct the perpendicular bisectors of any two sides of the triangle. The intersection of the two is a point *P*, equidistant to *A*, *B*, and *C*.

Any point on the perpendicular bisector of a segment is equidistant from the endpoints of the segment. Point *P* is equidistant from all vertices because it is on two perpendicular bisectors. Being at the intersection of two of the perpendicular bisectors forces the point to be equidistant from all three vertices.

(b) Same as part (a), except that point *P* will be outside the obtuse triangle. In part (a) it was inside.

7. Extend $\overline{AB}$ and construct a right angle at *A*, using the method of Figure 44.

Measure $\overline{AB}$ with a compass and use it to mark off side $\overline{AC}$ along the perpendicular.

From *C* and *B*, mark off the length of $\overline{AB}$ by drawing compass arcs at the approximate location of the final vertex. Call the intersection of the two arcs point *D*.

Construct $\overline{AD}$ and $\overline{CD}$ to form square *ABCD*.

8. **(a)** Make an arc of radius *BC* with center *A* and one with radius *AB* and center *C* so that the two intersect. This intersection is the location of the fourth vertex.

Another technique would be to construct a parallel to $\overline{BC}$ through *A*, then mark off the distance *BC* from *A*.

(b) Since one use of a compass or straight edge is not sufficient to find the fourth vertex, we must use the tools twice to locate the fourth vertex. Then, we must use the straight edge twice to construct the remaining sides of the parallelogram after the fourth vertex is located. The cheapest way is **40¢**.

9. See Figure 43 on how to construct the perpendicular bisector of a segment. Following the direction on those pages as given, and drawing segments $\overline{AC}$, $\overline{CB}$, $\overline{BD}$ and $\overline{DA}$, the quadrilateral that will be created should be a rhombus, since the four segments just drawn should be congruent. Further, since the diagonals of the rhombus bisect each other, the figure is also a parallelogram.

10. Place the compass point on point B and the pencil part of the compass on point A. Draw an arc that intersects A. Now, draw points C and D on the arc. Draw segments $\overline{BC}$, $\overline{BD}$ and $\overline{CD}$. If done correctly, ΔBCD should be an isosceles triangle (since $\overline{BC}$ and $\overline{BD}$ are radii of the same circle) with altitude $\overline{AB}$ and apex point B.

11. Because *O* is the incenter, the three angle bisectors of triangle *ABC* intersect at *O*. By Theorem 12-10, *OD* is equal to the distance from O to $\overline{AC}$. Label the point of intersect of $\overline{AC}$ and a line perpendicular to $\overline{AC}$ through O as P. $OPCD$ forms a rectangle. Since opposites sides of rectangles are congruent, $\boldsymbol{OD = OP = CD = 2''}$.

12. The side opposite the $7''$ side is also $7''$ by parallelogram properties. Let $x = EF$. Then $5 + 5 - x = 7 \Rightarrow \boldsymbol{x = 3''}$.

13. **(a)** If the parallelogram is not a rectangle, cut along any altitude. If the parallelogram is a rectangle, cut along any line through the point where the diagonals meet. The line must not be a diagonal nor parallel to any side.

(b) Make a copy of the given trapezoid and put it upside down next to $\overline{CD}$; i.e.,

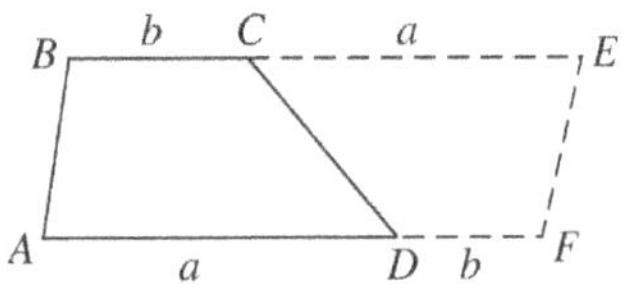

extend $\overline{BC}$ so that $CE = a$ and extend $\overline{AD}$ so that $DF = b$. Since $\overline{BE} \parallel \overline{AF}$ and $\overline{BE} = \overline{AF}$ (the length of each is $a + b$), $ABEF$ is a parallelogram.

14. As **close to 26 inches** as the jack can close; i.e., twice the length of a side.

15. **(a)** **Possible**. Draw line $\overline{AB}$ the length of a side of the square, then construct a right angle at A using the method of Figure 44. Measure $\overline{AB}$ with a compass and use it to mark off side $\overline{AC}$ along the perpendicular.

From C and B, mark off the length of $\overline{AB}$ by drawing compass arcs at the approximate location of the final vertex. Call the intersection of the two arcs point D.

Construct $\overline{AD}$ and $\overline{CD}$ to from square $ABCD$.

(b) **No unique rectangle**. The endpoints of two segments bisecting each other and congruent to the given diagonal determine a rectangle, but since the segments may intersect at any angle there are infinitely many such rectangles.

(c) **Not possible**. The sum of the measurements of the angles would be greater than 180°.

(d) **No unique parallelogram**. Given three right angles, the fourth angle must also be a right angle. The parallelogram would be a rectangle or square, an infinite number of which could be constructed.

16. **(a)** Construct an equilateral triangle and bisect one of its angles.

(b) Add 30° and 15° angles or bisect a right angle.

(c) Add 60° and 15° angles or 45° and 30° angles.

17. Make arcs of the same radius from A and B above $\overline{AB}$ and label their intersection C. Repeat the process with a new radius, labeling this intersection D. $\overleftrightarrow{CD}$ is the perpendicular bisector of $\overline{AB}$.

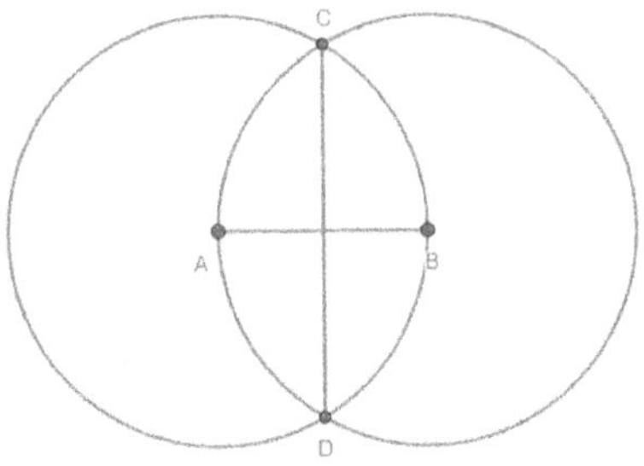

18. **(a)** **Possible**. The point is determined by the intersection of the angle bisector of $\angle A$ and the perpendicular bisector of $\overline{BC}$. Since the point is on the angle bisector of $\angle BAD$, it is equidistant from its sides. It is on the perpendicular bisector of $\overline{BC}$, thus it is equidistant from B and C.

(b) **Possible**. The point is determined by the intersection of the angle bisectors of $\angle A$ and $\angle B$.

19. Answers vary. One method is to construct a circle with center O and radius r. Then draw a diameter and label one of the intersections of the diameter and the circle as D. Construct an arc with radius r and center O. This will determine an equilateral triangle whose vertex O determines a central angle of 60°.

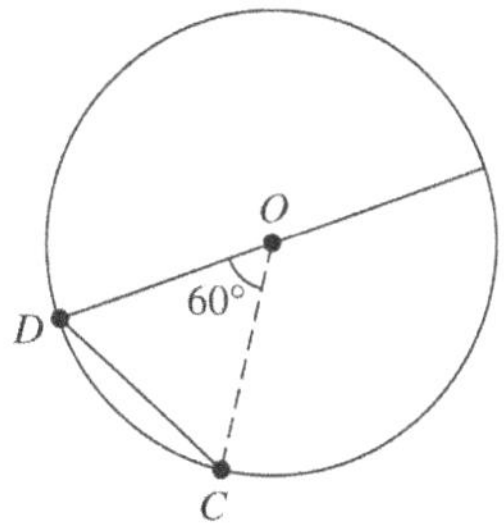

This method can be continued to construct six congruent triangles as shown below. Two adjacent triangles determine a central angle of 120° with vertex O. Because $\Delta AOB, \Delta BOC,$ and ΔAOC are congruent by SAS, $\overline{BC}, \overline{BA},$ and $\overline{AC}$ are congruent and hence we have constructed an equilateral inscribed triangle.

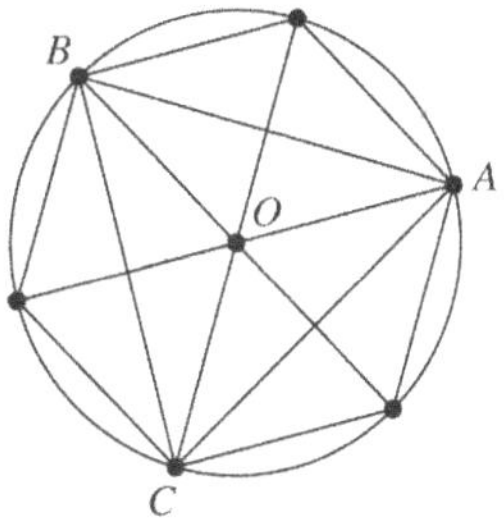

Review Problems

13. $\Delta ABC \cong \Delta DEC$ by ASA ($\overline{BC} \cong \overline{CE}$; $\angle ACB \cong \angle ECD$ as vertical angles, and $\angle B \cong \angle E$ as alternate interior angles formed by the parallels $\overline{AB}$ and $\overline{ED}$ and the transversal $\overleftrightarrow{EB}$). Thus $\overline{AB} \cong \overline{DE}$ by CPCTC.

15. If $\angle A$ is not the right angle and congruent pieces are corresponding, the triangles are congruent. If the right angles are at A and D, $\overline{AC} \cong \overline{DF}$ and $DE \neq AB$, then the triangles would not be congruent.

17. **(a)** Consider $\triangle ABC$ and $\triangle DEF$. If is given that $AB \cong DE$, then the least amount of information that we need to conclude that the triangles are congruent using SSS is $AC \cong DF$, $BC \cong EF$.

(b) The least amount of information that we need to conclude that the triangles are congruent using SAS is $AC \cong DF$, $\angle A \cong \angle D$, or $BC \cong EF$, $\angle B \cong \angle E$.

19. $\angle ADB \cong \angle ADC$, because these angles are the supplements of congruent angles (marked at D). Because $AD \cong AD$, $\Delta ABD \cong \angle ACD$ by ASA. Thus $BD \cong CD$.

Assessment 12-4A: Similar Triangles and Other Similar Figures

1. A scale of 1 cm: 110 km, means that every 1 cm of shoelace represents 110 km of river. Since the shoelace is 61 cm long, to find the length of the Nile River, multiply $61 \times 110 = 6710$ km. Another solution method would be to create a proportion, and solve for the unknown variable x:

$$\frac{1\text{ cm}}{110\text{ km}} = \frac{61\text{ cm}}{x\text{ km}}$$

2. A scale of 1 in: 100 ft means every inch of the drawing should represent 100 feet of the Washington Monument. Since the Washington Monument is 555 feet tall, to find the height of the drawing divide $555 \div 100 = 5.5$ inches. Another solution method would be to create a proportion, and solve for the unknown variable x: $\frac{1''}{100'} = \frac{x''}{555'}$

3. **Yes.** The left/right sides of the small rectangle are of length 4, while the left/right side of the large rectangle is of length $4 + 4 = 8$; so the length of the left/right sides of the large rectangle is twice that of the small rectangle. Similarly, the top/bottom sides of the small rectangle are of length 8.5, while the top/bottom sizes of the large rectangle are of length $17 = 8.5 + 8.5$; so once again those lengths are double that of the small rectangle. Since all sides are in proportion, the two rectangles are similar.

4. In a rhombus, adjacent angles are supplementary. Since $\angle BAD \cong \angle B_1A_1D_1$ we know that $\angle ABC \cong \angle A_1B_1C_1$. Thus, the corresponding angles are congruent. Since, in a rhombus, adjacent sides are congruent, $\frac{AD}{AB} = 1 = \frac{A_1D_1}{A_1B_1}$. This implies that $\frac{AD}{A_1D_1} = \frac{AB}{A_1B_1}$, in other words, corresponding sides are proportional. Therefore, the rhombuses are similar.

5. **(a)** **Always similar**, by AA. All angles measure 60°.

(b) **Always similar**. Sides are proportional and angles congruent.

(c) **Not** always similar. Consider a square and a nonsquare rectangle.

(d) **Always similar**. The ratio of corresponding sides is 2 and the angles are congruent.

6. Make all dimensions three times as long. For example, in part (c) each side would be three diagonal units long. One possible solution set is below.

(a)

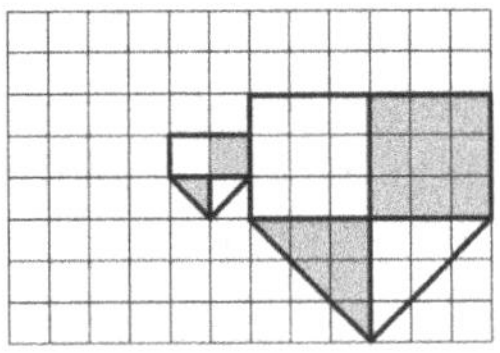

(b)

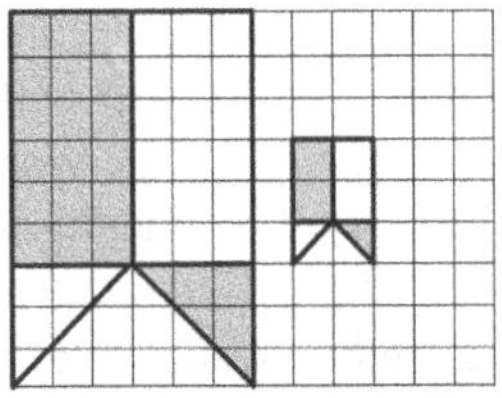

(c)

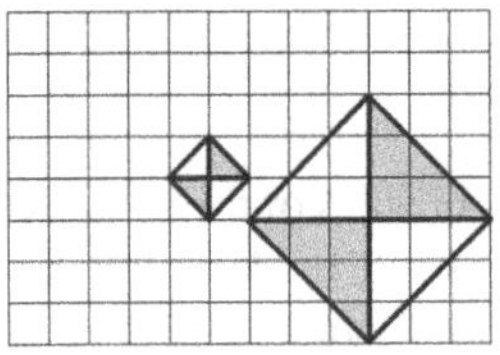

(d)

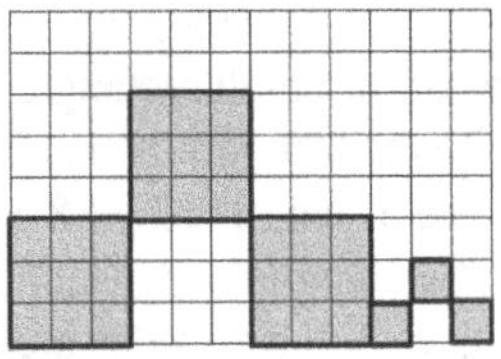

7. **(a)** *(i)* $\Delta ABC \sim \Delta DEF$ by *AA*.

(ii) $\Delta ABC \sim \Delta EDA$ by *AA*.

(iii) $\Delta ACD \sim \Delta ABE$ by *AA*. $\angle A$ is in both triangles.

(iv) $\Delta ABE \sim \Delta DBC$ by *SAS*, since $\frac{2}{3} = \frac{3}{4.5}$ and the vertical angles at *B* are congruent.

(b) *(i)* $\frac{2}{3}$. Sides of length 2 and length 3 are corresponding.

(ii) $\frac{1}{2}$. Corresponding sides of ΔAED are twice the length of ΔACB.

(iii) $\frac{3}{4}$. Corresponding sides are of length 6 and $6 + 2 = 8$; $\frac{6}{8} = \frac{3}{4}$.

(iv) $\frac{2}{3}$. Corresponding sides are *AB* and *BD*. The lengths of their sides are in the ratio 2 : 3.

8. **(a)** $\frac{\text{Short side}}{\text{Long side}} = \frac{5}{10} = \frac{x}{x+7} \Rightarrow 5(x + 7) = 10x \Rightarrow 5x + 35 = 10x$. $\boldsymbol{x = 7}$.

(b) $\frac{3}{x} = \frac{7}{8} \Rightarrow 7x = 3 \cdot 8$. $\boldsymbol{x = \frac{24}{7}}$.

9. The two triangles are similar by **AA** similarity; it is given that one pair of angles is congruent, and the two vertical angles are congruent. Therefore, the corresponding sides will be in proportion. Note that the side of length 10 on the left triangle corresponds with the side of length 15 of the large triangle. Using that information, these proportions can be set up to solve for sides x and y.

(Note that it is possible to set up other correct proportions):

$$\frac{10}{15} = \frac{x}{21} \qquad \frac{10}{15} = \frac{12}{y}$$

$$210 = 15x \text{ and } 10y = 180$$

$$14 = x \qquad y = 18$$

10. We know ΔAEB is similar to ΔCED by **AA** similarity; the two triangles share $\angle E$, and corresponding angles (such as $\angle B$ and $\angle EDC$) will be congruent because $\overline{AB} \parallel \overline{CD}$. So the corresponding sides of the two triangles will be in proportion.

(a) $\frac{y}{x} = \frac{9}{6}$, simplifying, $\frac{y}{x} = \boldsymbol{\frac{3}{2}}$.

(b) $\frac{x}{x+y} = \frac{6}{6+9}$; so $\frac{x}{x+y} = \frac{6}{15} = \boldsymbol{\frac{2}{5}}$.

(c) $\frac{6}{x} = \frac{9}{y}$

11. Follow the procedure illustrated by Figures 57 and 58 in textbook.

12. If $\frac{a}{b} = \frac{c}{x}$, then $x = \frac{bc}{a}$. Construct as below, using the technique of Figure 12-56.

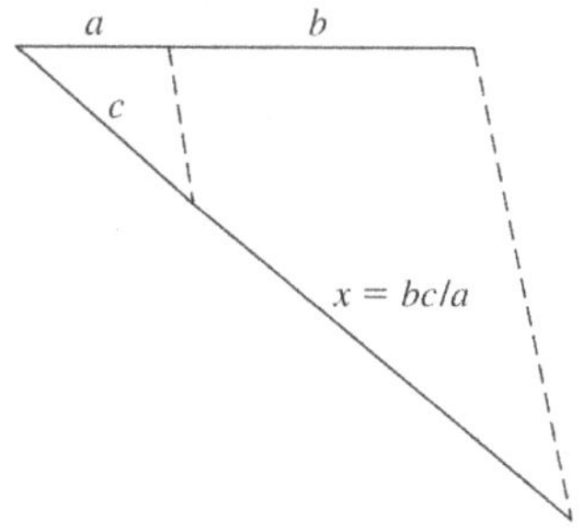

Note that the two dashed lines are parallel.

13. **(a)** Yes, because the angles stay the same and the ratio of corresponding sides are proportional.

(b) Let S_1 and S_2 represent the lengths of a side of the original polygon and the reduced polygon respectively. Then,

$$S_2 = 0.8\,(0.8\,S_1) \Rightarrow \frac{S_2}{S_1} = \mathbf{0.64 : 1}$$

14. Sketches may vary. Make the sides proportional but interior angles not congruent; e. g.:

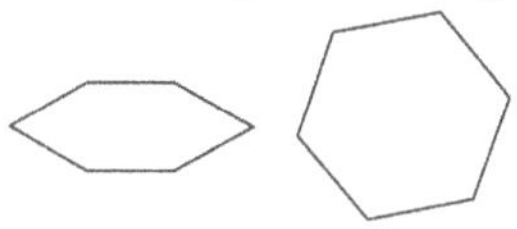

15. The new page will be $\frac{75}{100}$ the size of the original. The ratio to return to the original size would thus be $\frac{100}{75} = \mathbf{133\frac{1}{3}\%}$. (Most copy machines do not include this setting.)

16. Let the height above ground be h and convert all measurements to inches (3 feet $=$ 36 inches: 7 feet$=$ 84 inches) and use similar triangles. $\frac{36}{13} = \frac{h}{84} \Rightarrow 13h = 36 \cdot 84.\ h \approx 232.6$ inches or **about 19.38 feet**.

17. $\Delta DCP \sim \Delta BAP$. Thus $\frac{6}{AB} = \frac{4}{10} \Rightarrow 4AB = 6 \cdot 10.\ AB = \mathbf{15\ m}$.

18. We know ΔACB ΔDCG and ΔFCG are all similar by **AA** similarity; the three triangles share $\angle C$, and corresponding angles (such as $\angle B$, $\angle FGC$, and $\angle DEC$) will be congruent because $\overline{AB} \parallel \overline{DE} \parallel \overline{FG}$. So the corresponding sides of the three triangles will be in proportion.

(a) It is given that F is one-fourth of the distance from C to A. From that, it can be deduced that $\overline{CF}$ is one-fourth the length of $\overline{AC}$. It is given that $AC = 4$; therefore, $\mathbf{CF = 1}$.

(b) It is given that G is one-fourth of the distance from C to B. From that, it can be deduced that $\overline{CG}$ is one-fourth the length of $\overline{BC}$. It is given that $BC = 4.5$; therefore, $\mathbf{CG = 1.125}$.

(c) From parts (a) and (b), we know that the sides of ΔFCG are one-fourth the length of the sides of ΔACB; so $\overline{FG}$ will be one-fourth the length of $\overline{AB}$. It is given that $AB = 3$; **therefore $\mathbf{FG = 0.75}$.**

19. It is given that $\overline{BC} \parallel \overline{ED}$ and $\overline{AB} \parallel \overline{DC}$. Since $\overline{BC} \parallel \overline{ED}$,it is true that $\angle BCA \cong \angle DEC$, since they are alternate interior angles with $\overline{AC}$ as the transversal. Also, since $\overline{AB} \parallel \overline{DC}$, it is true that $\angle BAC \cong \angle DCA$, since they are alternate interior angles, with $\overline{AC}$ as the transversal. So, $\Delta ABC \sim \Delta CDE$ by **AA** similarity.

20. **(a)** A perimeter is the sum of the lengths of the sides, so the ratio of the perimeters is the same as the ratio of the sides.

(b) If $a, b, c,$ and d are the sides of one quadrilateral and $a_1, b_1, c_1,$ and d_1 are the corresponding sides of a similar quadrilateral, then $\frac{a}{a_1} = \frac{b}{b_1} = \frac{c}{c_1} = \frac{d}{d_1} = r$ the scale factor. Then

$$\frac{a+b+c+d}{a_1+b_1+c_1+d_1} = \frac{a_1r+b_1r+c_1r+d_1r}{a_1+b_1+c_1+d_1} = \frac{(a_1+b_1+c_1+d_1)r}{a_1+b_1+c_1+d_1} = r$$

So the ratio of the perimeters is r. An analogous proof works for any two similar n-gons.

21. **(a)** $ABCD$ is a rhombus $\Rightarrow \overline{AB} = \overline{BC} = \overline{CD} = \overline{DA}$, thus all angles of $MNPQ$ are right and it is a **rectangle**.

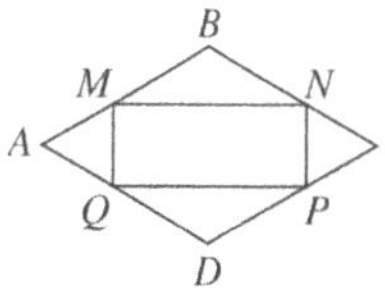

(b) If $ABCD$ is a kite, $MNPQ$ is a **rectangle**.

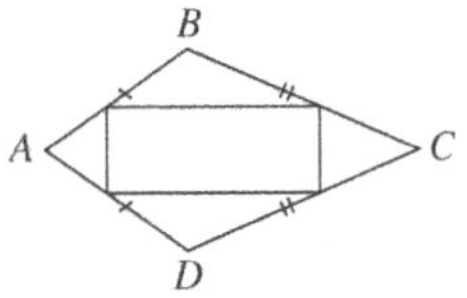

(c) If $ABCD$ is an isosceles trapezoid, $MNPQ$ is a **rhombus**.

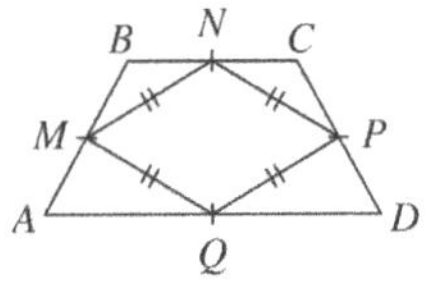

(d) If $ABCD$ is quadrilateral which is not a rhombus nor a kite but whose diagonals are perpendicular to each other, $MNPQ$ is a **rectangle**.

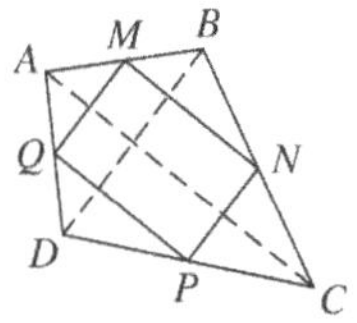

22. **Yes**. All such parallel cross-sections are circular. They have exactly the same shape but are not necessarily the same diameter; i.e., they are similar.

23. 4.2 cm represents the distance from city A to city B on the map: 512 miles represents the distance from city A to city B in reality. This means the distance from city B to city C can be found using the ratio $\frac{4.2\text{ cm}}{512\text{ miles}}$ to solve for the unknown miles:

$$\frac{4.2\text{ cm}}{512\text{ miles}} = \frac{10.1\text{ cm}}{BC\text{ miles}};\quad BC \approx \mathbf{1231\text{ miles}}.$$

Similarly, the distance from city A to city C can be found using the same ratio $\frac{4.2\text{ cm}}{512\text{ miles}}$ as follows:

$$\frac{4.2\text{ cm}}{512\text{ miles}} = \frac{12.2\text{ cm}}{AC\text{ miles}};\quad AC \approx \mathbf{1487\text{ miles}}.$$

Review Problems

23. Start with the given base and construct its perpen-dicular bisector. The vertex of the required triangle must be on this perpendicular bisector. Starting at the point where the perpendicular bisector intersects the base, mark on the perpendicular bisector a segment congruent to the given altitude. The end-point of the segment not on the base is the vertex of the required isosceles triangle.

25. Answers vary. For example: construct a perpendicular line at one of the endpoints of the hypotenuse, giving a $90°$ angle. Bisect this angle to yield a $45°$ angle.

Copy the $45°$ angle at the other endpoint of the hypotenuse, extending until it meets the bisector which formed the first $45°$ angle. These will meet at a $90°$ angle across from the given hypotenuse.

Chapter 12 Review

1. (a) $\Delta ABD \cong \Delta CBD$ by ***SAS***. $\overline{BD}$ is common to both; right angles are congruent; $\overline{AD} \cong \overline{DC}$.

(b) $\Delta GAC \cong \Delta EDB$ by ***SAS***. $\overline{AC} \cong \overline{DB}$; right angles are congruent; $\overline{AG} \cong \overline{DE}$.

(c) $\Delta ABC \cong \Delta EDC$ by ***AAS***. Right angles are congruent; vertical angles; $\angle ACB \cong \angle DCE$; $\overline{AC} \cong \overline{CE}$.

(d) $\Delta BAD \cong \Delta EAC$ by ***ASA***. $\angle A$ is common to both; $\overline{AB} \cong \overline{AE}$; $\angle B \cong \angle E$.

(e) $\Delta ABD \cong \Delta CBD$ by ***AAS***. $\overline{BD}$ is common to both; $\overline{AD} \cong \overline{DC}$; right angles are congruent; $\angle A \cong \angle C$.

(f) $\Delta ABD \cong \Delta CBD$ by ***SAS***. $\overline{AB} \cong \overline{BC}$; $\angle ABD \cong \angle DBC$; $\overline{BD}$ is common to both.

(g) $\Delta ABD \cong \Delta CBE$ by ***SSS***. $\overline{BD} \cong \overline{BE}$; $\overline{AB} \cong \overline{BC}$; $\overline{AD} \cong \overline{EC}$.

(h) $\Delta ABC \cong \Delta ADC$ by ***SSS***; $\Delta ABE \cong \Delta ADE$ by ***SSS*** or ***SAS***; $\Delta EBC \cong \Delta EDC$ by ***SSS*** or ***SAS***. $\overline{BC} \cong \overline{CD}$; $\overline{AB} \cong \overline{AD}$; properties of kites contribute to most of congruency.

2. **Parallelogram**. $\Delta ADE \cong \Delta CBF$ by SAS, so $\angle DEA \cong \angle CFB$; $\angle DEA \cong \angle EAF$ (alternate interior angles between the parallels $\overleftrightarrow{DC}$ and $\overleftrightarrow{AB}$ and the transversal $\overleftrightarrow{AE}$), and $\angle EAF \cong \angle CFB$. Therefore $\overline{AE} \parallel \overline{FC}$.

$\overline{EC} \parallel \overline{AF}$ (parallel sides of the square). Two pairs of parallel opposite sides implies a parallelogram.

3. **(a)** ***(i)*** Use the method illustrated by Figure 40 in the text.
   (ii) Fold the angle down the middle so the sides match and the crease passes through *A*.
   (b) ***(i)*** See text Figure 44.
   (ii) Fold the line on top of itself so that the crease passes through *B*.
   (c) ***(i)*** See text Figure 43.
   (ii) Same as in part (b)(*ii*).
   (d) ***(i)*** See text Figures 36.
   (ii) Make line $k \perp \ell$ through P as in part (b) and (c), then make $m \perp k$ through P as in part (b) $\Rightarrow m \parallel \ell$.

4. **(a)** ***(i)*** $\frac{x}{4} = \frac{6}{3} \Rightarrow 3x = 4 \cdot 6$. $\boldsymbol{x = 8}$ **cm.**
   (ii) $\frac{y}{10} = \frac{3}{6} \Rightarrow 6y = 3 \cdot 10$. $\boldsymbol{y = 5}$ **cm.**
   (b) $\frac{x}{13} = \frac{2.5}{5} \Rightarrow 5x = 2.5 \cdot 13$. $\boldsymbol{x = 6.5}$ **m.**

5. Use the method illustrated by Figure 57 and 58 in the text.

6. $\frac{a}{b} = \frac{c}{d}$. $\frac{a}{b} = \frac{x}{y}$ in ΔABC; $\frac{x}{y} = \frac{c}{d}$ in ΔACD, so $\frac{a}{b} = \frac{c}{d}$ by the transitive property.

7. $\overline{AB}$ must be a chord of the circle, and the perpendicular bisector of a chord passes through the center. Construct this line to locate the center on ℓ, measure the radius to *A* or *B*, and draw the circle with your compass.

8. **(a)** ***(i)*** $\Delta ACB \sim \Delta DEB$ by *AA*.
   (ii) $\frac{x}{3} = \frac{8}{5} \Rightarrow 5x = 8 \cdot 3$.
   $\boldsymbol{x = \frac{24}{5}}$ **inches.**
   (b) ***(i)*** $\Delta AED \sim \Delta ACB$ by *AA*.
   (ii) $\frac{4}{6} = \frac{y+6}{11} \Rightarrow 6(y+6) = 4 \cdot 11$.
   $\boldsymbol{y = \frac{4}{3}}$ **feet.**
   (iii) $\frac{6}{5} = \frac{11}{x} \Rightarrow 6x = 11 \cdot 5$. $\boldsymbol{x = \frac{55}{6}}$ **feet.**

9. We know ΔADE is similar to ΔACB by AA similarity; the two triangles share $\angle A$, and corresponding angles (such as $\angle B$ and $\angle AED$) will be congruent because $\overline{DE} \parallel \overline{CB}$. So the corresponding sides of the two triangles will be in proportion. To find $\overline{AD}$, the proportion $\frac{AE}{EB} = \frac{AD}{DC}$ can be used. Substituting in the proper numbers, the proportion is $\frac{20}{12} = \frac{AD}{9}$; therefore the solution is $AD = 15$.

10. It is given that $l \parallel m$; place a point on l called A. Construct a line (call it n) perpendicular to line l through point A. See Figure 44 on how to do this. Since $l \parallel m$, line n is also perpendicular to line m Now, plot the point of intersection that line n has with line m; label that point B Next, find the perpendicular bisector of $\overline{AB}$. See Figure 43 for an explanation on how this is done. Extend the perpendicular bisector of $\overline{AB}$ to create line p If all is done correctly, line p should be equidistant and parallel to lines l and m.

11. **(a)** The width (left/right sides) of the pool is 16 feet, while the width of the patio is 24 feet. The widths are in the ratio of $\frac{16}{24} = \frac{2}{3}$. The length of the pool is 32 feet, while the length of the patio is 48 feet. They are in the ratio of $\frac{32}{48} = \frac{2}{3}$. Since the ratios are the same, the rectangular pool is similar to the rectangular patio.
   (b) Comparing the patio to the pool is comparing the larger object to the smaller object. Therefore the scale factor of the patio to the pool is $\frac{3}{2}$.
   (c) The total perimeter of the patio is $48 + 24 + 48 + 24 = 144$ feet. The total perimeter of the pool is $32 + 16 + 32 + 16 = 96$ feet. So the ratio of the patio to the po**ol would be** $\frac{144}{96} = \frac{3}{2}$.

12. The two triangles are not congruent. For two triangles to be congruent, their corresponding angles must be congruent (which is true in this situation) and their corresponding sides must be congruent. The corresponding sides cannot be congruent since one triangle is larger than the other. However, we can determine that the two triangles are similar by **AA** similarity.

13. **(a)** Given $\Delta ABC \cong \Delta TRI$, $\angle T$ must correspond with $\angle A$. Since $m\angle A = 70°$, the $m\angle T = 70°$.

(b) Given $\Delta ABC \cong \Delta TRI$, $\angle C$ must correspond with $\angle I$. Since $m\angle I = 62°$, the $m\angle C = 62°$.

(c) Given $\Delta ABC \cong \Delta TRI$, $\angle B$ must correspond with $\angle R$. Since $m\angle R = 48°$, the $m\angle B = 48°$.

14. $\Delta ADO \cong \Delta CDO \cong \Delta OEC \cong \Delta OEB$

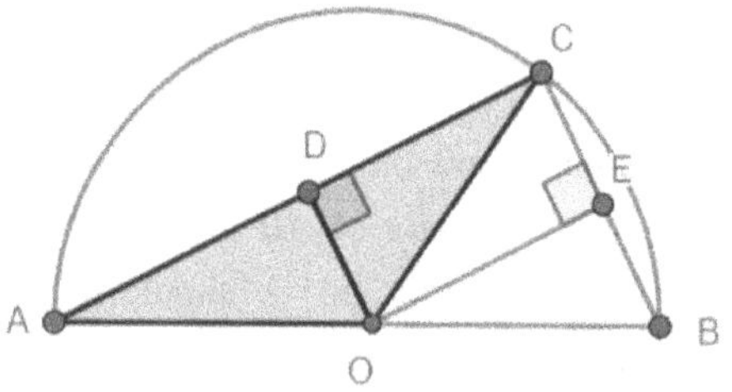

$\boldsymbol{\Delta ADO \cong \Delta CDO}$ are congruent because ΔAOC is isosceles and OD is the perpendicular bisector of AC.

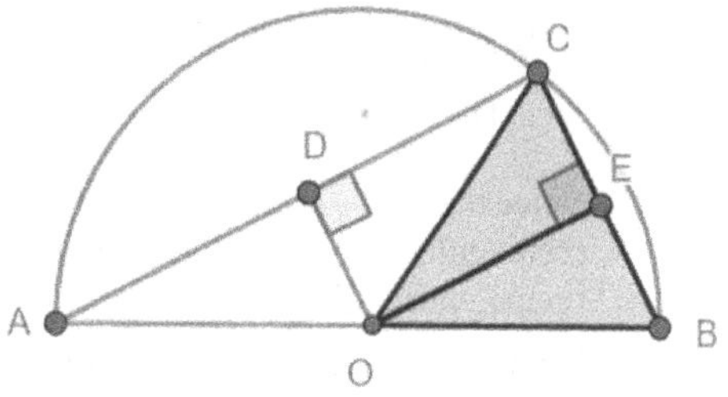

Similarly, $\boldsymbol{\Delta OEC \cong \Delta OEB}$ are congruent because ΔBOC is isosceles and OE is the perpendicular bisector of BC.

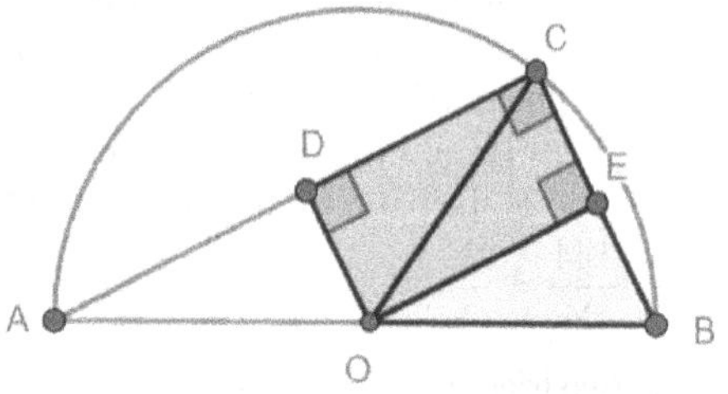

Finally, $\boldsymbol{\Delta CDO \cong \Delta OEC}$ because a diagonal in a rectangle divides the rectangle into two congruent triangles. (Notice that $ODCE$ is a rectangle because $\angle ACB$ is a right angle.).

15. If h is the height of the building: $\dfrac{2 \text{ m tall}}{1 \text{ m shadow}} =$

$\dfrac{h \text{ m high}}{6 \text{ m shadow}} \Rightarrow 1h = 2 \cdot 6.$ $\boldsymbol{h = 12}$ **m.**

16. $\frac{h}{8} = \frac{1.5}{2} \Rightarrow 2h = 1.5 \cdot 8.$ $\boldsymbol{h = 6}$ **m.**

17. $\frac{d}{64} = \frac{16}{20} \Rightarrow 20d = 16 \cdot 64.$ $\boldsymbol{d = \frac{256}{5}}$ **m.**

18. In the figure below:

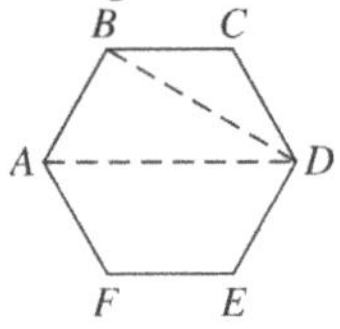

$\angle CBD = 30°$, $\angle C = 120°$, and $\angle CDB = 30°$.

(a) $\overline{AD}$ is a diameter, since it passes through the center of the circumcircle and the midpoint of $\overline{AD}$ is the circumcenter.

(b) $m(\angle CBD) = 30°$, since base angles of an isosceles triangle are congruent. Since $m(\angle ABC) = 120°$, then $m(\angle ABD) = m(\angle ABC) - m(\angle CBD) = 120° - 30° = 90°$. Thus $\angle ABD$ is a right angle.

19. The statement is false since in (i) we have a counterexample.

(*i*) Quadrilateral $ABCD$ below is not a square, even through its diagonals are perpendicular and congruent (i.e., $\overline{AB} \cong \overline{BD}$).

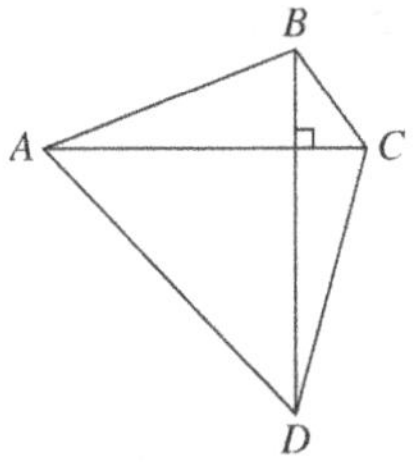

(*ii*) If the diagonals bisect each other, the quadrilateral is a square.

20. Congruent triangles have congruent corresponding angles. Thus they are similar by AA.

21. ΔABC is an isosceles triangle, since $\angle A \cong \angle B$. That means the sides opposite those angles are congruent. Setting them equal to each other,

$$3x - 4 = 2x + 1$$
$$x = 5$$

Now substitute 5 in for x in for all three sides:

$3(5) - 4 = 11$

$2(5) + 1 = 11$

$(5) + 1 = 6$

22. It is given that $\angle 1 \cong \angle 2$ and $\angle 3 \cong \angle 4$. Since $\overline{AC} \cong \overline{AC}$, by **ASA**, $\Delta ACD \cong \Delta ACB$. That means $\overline{BC} \cong \overline{DC}$ by CPCTC. Also, $\overline{CE} \cong \overline{CE}$. Therefore, by **SAS** congruence, $\Delta BCE \cong \Delta DCE$.

23. Diagonals in a rhombus bisect the angles of the rhombus and the diagonals are perpendicular bisectors of each other.

Thus, to construct the rhombus we bisect $\angle C$. Place the point of the compass on C, and swing an arc ED or a circle with center C and a radius smaller than $\overline{AC}$. Then, with D as center and DE as radius, draw an arc or circle. Keeping the same radius and with E as center, draw an arc or circle that will intersect the first; call that point of intersection F, and draw CF. Then CF bisects angle C The bisector meets $\overline{AB}$ at F.

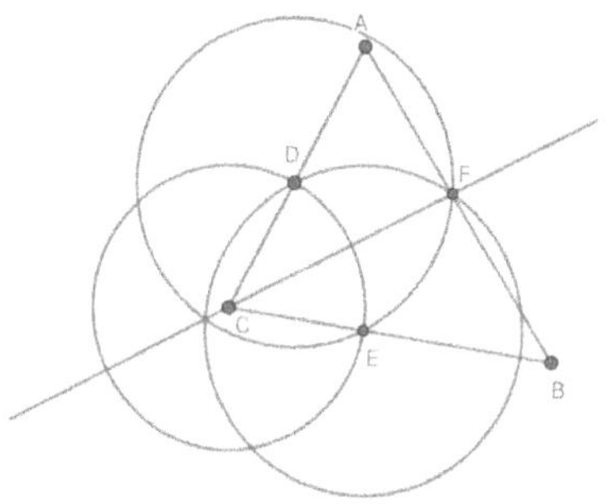

The perpendicular bisector (see Figure 43 for instructions) of CD intersects the sides at E and F.

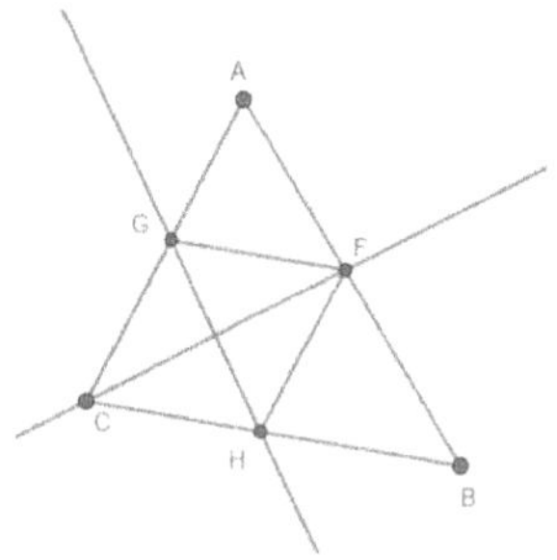

Quadriletral $CGFH$ is a rhombus.

24. The distance along Ocean Drive bordering the lots is 40 m + 34 m + 30 m = 104 m. Let x, y, and z be the ocean frontage for lots A, B, and C, respectively. Using Theorem 12-12, we can set up the following proportions:

Lot A: $\frac{104}{130} = \frac{40}{x}$ and $104x = 5200m$.

$x = \frac{5200}{104}m. \rightarrow x = 50m.$

Lot B: $\frac{104}{130} = \frac{34}{y}$ and $104y = 4420m$.

$y = \frac{4420}{104}m. \rightarrow y = 42.5m.$

Lot C: $\frac{104}{130} = \frac{30}{z}$ and $104z = 3900m$.

$z = \frac{3900}{104}m. \rightarrow z = 37.5m.$

Therefore Lot A has the most ocean frontage and hence the highest selling price.

CHAPTER 13

AREA, PYTHAGOREAN THEOREM, AND VOLUME

Assessment 13-1A: Linear Measure

1. **(a)** $AB = $ **1.0 cm.**

   (b) $DE = 4.5 \text{ cm} - 3.5 \text{ cm} = $ **1.0 cm.**

   (c) $CJ = 10.0 \text{ cm} - 2.0 \text{ cm} = $ **8.0 cm.**

   (d) $EF = 5.0 \text{ cm} - 4.5 \text{ cm} = $ **0.5 cm.**

   (e) $IJ = 10.0 \text{ cm} - 9.3 \text{ cm} = $ **0.7 cm.**

   (f) $AF = 5.0 \text{ cm} - 0 \text{ cm} = $ **5.0 cm.**

   (g) $IC = 9.3 \text{ cm} - 2.0 \text{ cm} = $ **7.3 cm.**

   (h) $GB = 6.2 \text{ cm} - 1.0 \text{ cm} = $ **5.2 cm.**

2. Answers will vary.

   (a) About **3 cubits**, assuming a 60-inch desk and a 20-inch measurement from elbow to fingertips.

   (b) About **2 pencil-lengths,** assuming a mechanical pencil.

   (c) About **25 pencil-widths,** assuming the book is closed.

3. **(a)** 100 inches

   $= \frac{100 \text{ inches}}{1} \times \frac{1 \text{ yard}}{36 \text{ inches}} = \frac{25}{9} = \mathbf{2\frac{7}{9}} \text{ yd.}$

   (b) 400 yards

   $= \frac{400 \text{ yards}}{1} \times \frac{36 \text{ inches}}{1 \text{ yard}} = \mathbf{14{,}400} \text{ in.}$

   (c) $300 \text{ feet} = \frac{300 \text{ feet}}{1} \times \frac{1 \text{ yard}}{3 \text{ feet}} = \mathbf{100} \text{ yd.}$

   (d) 372 inches

   $= \frac{372 \text{ inches}}{1} \times \frac{1 \text{ foot}}{12 \text{ inches}} = \mathbf{31} \text{ ft.}$

4. The following lines are one-half scale (e.g., a 10 cm line will measure 5 cm below):

   (a) 10 mm (or 1 cm):

   (b) 100 mm (or 10 cm):

   (c) 1 cm (or 10 mm):

   (d) 10 cm (or 100 mm):

5. Answers may vary depending on your estimate; e.g.,

   (a) About **81 mm.**

   (b) About **8.1 cm.**

6. Estimates will vary.

   (a) **Centimeters**. A new pencil measures about 19 cm.

   (b) **Millimeters** or **centimeters**. The diameter is about 21 mm or 2 cm.

   (c) **Centimeters** or **meters**. A desk is normally about 120 cm or 1.2 meters wide.

7. Estimates will vary.

   (a) **Inches**. 19 cm is about 7.5 inches.

   (b) **Inches**. 21 mm is about 0.8 or $\frac{13}{16}$ inch.

   (c) **Feet**. 1.2 m is about 3 feet 11 inches.

8. In each case, note that:

   From m to cm move the decimal point two places to the right;
   From cm to mm move the decimal point one place to the right;
   From mm to cm move the decimal point one place to the left; and
   From cm to m move the decimal point two places to the left.

Item	m	cm	mm
(a) Length of a piece of paper	**0.35**	35	**350**
(b) Height of a woman	1.63	**163**	**1630**
(c) Width of a filmstrip	**0.035**	**3.5**	35
(d) Length of a cigarette	**0.1**	**10**	100
(e) Length of two meter sticks laid end to end	2	**200**	**2000**

9. **(a)** Dime = \$0.10 = Decidollar

   (b) Penny = \$0.01 = Centidollar

(c) $10 = Dekadollar
(d) $100 = Hectodollar
(e) $1000 = Kilodollar

10. (a) **13.50 mm**. 10 mm is about 0.4 inch.
(b) **0.770 m**. 0.77 m is about 30 inches.
(c) **10.0 m**. 10 m is about 33 feet.
(d) **15.5 cm**. 15.5 cm is about 6 inches.

11. Convert each measure to cm to list in decreasing order. 5218 mm = 521.8 cm; 91 mm = 9.1 cm; 6 m = 600 cm; 700 mm = 70 cm. So, along with 8 cm and 245 cm, the order is: **6m, 5218 mm, 245 cm, 700 mm, 91 mm, 8 cm.**

12. If the circumference is π, using $C = 2\pi r$ we have: $\pi = 2\pi r \Rightarrow 1 = 2r \Rightarrow r = \frac{1}{2}$ Since the radius = 0.5 inch, then the diameter = **1 inch.**

13. (a) Where $r = 1$ unit and a semicircle is one-half the circumference, $\frac{1}{2}C = \frac{1}{2}(2\pi r) = \boldsymbol{\pi}$ **units.**
(b) Where $r = \frac{1}{2}$ unit, $\frac{1}{2}C = \frac{1}{2}\left(2\pi \cdot \frac{1}{2}\right) =$ $\boldsymbol{\frac{\pi}{2}}$ **units.**

14. Estimates will vary. Listing lengths of sides starting clockwise from the top of each figure (measurements are approximate):
(a) About $1.75 + 1.75 + 1.75 + 1.75 =$ **7 cm.**
(b) About $3.7 + 0.8 + 1.8 + 1 + 1.8 + 1.9 =$ **11 cm.**

15. See Figure 13-2. In each case below, to move from a smaller unit to a larger move the decimal point to the left, and from a larger unit to a smaller move the decimal point to the right.
(a) 10 mm = **1 cm**.
(b) 262 m = **0.262 km**.
(c) 3 km = **3000 m**.
(d) 30 mm = **0.03 m**.

16. Answers for the lengths of sides will vary. Consider the following triangle, in which $AB \approx$ 20 mm. $BC \approx 24$ mm, and $AC \approx 17$ mm. The sum of the lengths of any two sides of the triangle is greater than the length of the third side:

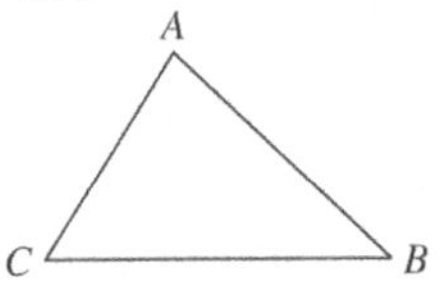

(a) $\boldsymbol{AB + BC} \approx 44$ mm $> \boldsymbol{AC} \approx 17$ mm.
(b) $\boldsymbol{BC + AC} \approx 41$ mm $> \boldsymbol{AB} \approx 20$ mm.
(c) $\boldsymbol{AB + AC} \approx 37$ mm $> \boldsymbol{BC} \approx 24$ mm.

17. Use the triangle inequality: The sum of the lengths of any two sides of a triangle is greater than the length of the third side.
(a) **Yes.** $23 + 50 > 60$.
(b) **No.** $10 + 40 \not> 50$.

18. (a) 31 + 85 = 116, so the third side must be less than or equal to **115 cm.**
(b) 85 – 31 = 54, so the third side must be greater than or equal to **55 cm.**

19. The hypotenuses of the resultant right triangles are $\sqrt{4.25^2 + 11^2} \approx 11.8$ inches.
(a) Form an isosceles triangle with the two congruent sides formed by the diagonals and the base formed by the two short $\left(4\frac{1}{4}\text{ inch}\right)$ sides. The perimeter is approximately **32.1 inches.**

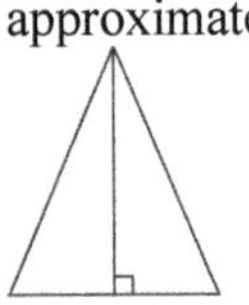

(b) Form the base with the two long (11 inch) sides. The perimeter is about **45.6 inches.**

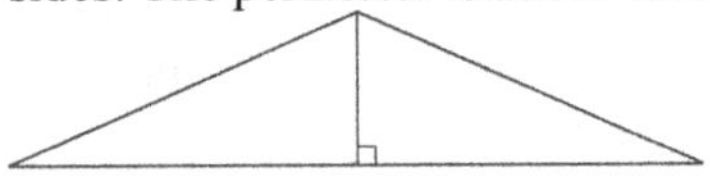

20. Circumference $(C) = 2\pi \cdot$ radius $(r) \Rightarrow r = \frac{C}{2\pi}$.
(a) $r = \frac{12\pi}{2\pi} =$ **6 cm.**
(b) $r = \frac{6}{2\pi} = \boldsymbol{\frac{3}{\pi}}$ **m.**

21. Circumference $(C) = \pi \cdot$ diameter (d) or $C = 2\pi \cdot$ radius (r).
(a) $C = \pi \cdot 6 = \boldsymbol{6\pi}$ **cm.**
(b) $C = 2\pi \cdot \frac{2}{\pi} =$ **4 cm.**

22. The relationship between the two measures is linear; i.e., when the radius increases, the circumference will increase by the same factor. Thus the circumference will **double.**

23. (a) Mach $3 \times \frac{344\text{ m}}{\text{sec}} =$ **1032 m/sec.**
(b) $\text{M} = \frac{\text{speed of aircraft}}{\text{speed of sound}} = \frac{\frac{5000\text{ km}}{\text{hr}}}{\frac{0.344\text{ km}}{\text{sec}} \cdot \frac{3600\text{ sec}}{\text{hr}}}$
$\approx$ **Mach 4.04.**

24. Greatest possible error is one-half the whole-number measurement unit used.
(a) 23 m implies GPE of **0.5 m.**
(b) 3.6 cm = 36 mm implies GPE of 0.5 mm = **0.05 cm.**
(c) 3.12 m = 312 cm implies GPE of 0.5 cm = **0.005 m.**

25. **(a)** Length of an arc $= 2\pi r \cdot \frac{\textit{central angle}}{360^\circ}$, so if the arc is 36° and the radius is 6 cm, then the length of the arc is

$$2\pi 6 \text{ cm} \cdot \frac{36^\circ}{360^\circ} = 12\pi \cdot \frac{1}{10} \text{cm} = \mathbf{\frac{6\pi}{5} cm.}$$

(b) The arc has to be longer because 8 cm is greater than 6 cm, and the radius is in the numerator of the fraction in the computation.

(c) The arc length is

$$2\pi 8 \text{ cm} \cdot \frac{36^\circ}{360^\circ} = 16\pi \cdot \frac{1}{10} \text{cm} = \mathbf{\frac{8\pi}{5} cm.}$$

26. If the arc length is $14\pi\, m$ then the radius of the circle containing an 80° arc is

$$2\pi r \text{ m} \cdot \frac{80^\circ}{360^\circ} = 14\pi \text{ m}$$

$$r = \frac{360^\circ}{80^\circ} \cdot \frac{14\pi \text{ m}}{2\pi \text{ m}} \Rightarrow r = \frac{9}{2} \cdot 7 \text{ m} = \mathbf{\frac{63}{2} m.}$$

27. Use 2, 3, and 4 as the radii and find the arc

$$2\pi 2 \cdot \frac{60^\circ}{360^\circ} : 2\pi 3 \cdot \frac{60^\circ}{360^\circ} : 2\pi 4 \cdot \frac{60^\circ}{360^\circ}$$

lengths: $4\pi \cdot \frac{1}{6} : 6\pi \cdot \frac{1}{6} : 8\pi \cdot \frac{1}{6}$

$$\frac{4}{6}\pi : \frac{6}{6}\pi : \frac{8}{6}\pi \Rightarrow \frac{2}{3}\pi : \frac{3}{3}\pi : \frac{4}{3}\pi$$

The ratio of the arc length is 2:3:4.

Assessment 13-2A: Areas of Polygons and Circles

1. Answers may vary. E.g., for a 30-inch by 60-inch desktop, and if the $8\frac{1}{2}$ by 11-inch notebook papers were to be oriented in portrait fashion, the desktop would be about 7 papers by $2\frac{3}{4}$ papers $= 19\frac{1}{4}$ square notebook papers.

2. **(a)** **(*i*)** $\mathbf{cm^2}$; about 21.6 cm by 28 cm.

(*ii*) $\mathbf{in.^2}$; $8\frac{1}{2}$ in. by 11 in.

(b) **(*i*)** $\mathbf{mm^2}$ **or** $\mathbf{cm^2}$; diameter about 2.3 cm or 23 mm.

(*ii*) $\mathbf{in.^2}$; diameter about $\frac{15}{16}$ in.

(c) **(*i*)** $\mathbf{cm^2}$ **or** $\mathbf{m^2}$; about 61cm $= 0.61$m by 152 cm $= 1.52$ m.

(*ii*) $\mathbf{in.^2}$ **or** $\mathbf{yd^2}$; about 24 in. $= \frac{2}{3}$ yd by 60 in. $= 1\frac{2}{3}$ yd.

(d) **(*i*)** $\mathbf{m^2}$; assume about 4.5 m by 9 m.

(*ii*) $\mathbf{ft.^2}$ **or** $\mathbf{yd^2}$; assume about 5 yd by 10 yd.

3. Answers may vary. Some possible approximate measures are:

(a) A 30 in. by 78 in. door (approximately 0.8 m by 2.0 m), **about 1.6** $\mathbf{m^2}$.

(b) A 30 in. by 60 in. desktop (approximately 0.76 m by 1.5 m), **about 1.1** $\mathbf{m^2}$.

4. In each case, note that:

From m^2 to cm^2, move the decimal point four places to the right;

From cm^2 to mm^2, move the decimal point two places to the right;

From mm^2 to cm^2, move the decimal point two places to the left; and

From cm^2 to m^2, move the decimal point four places to the left;

Item	m^2	cm^2	mm^2
(a) Area of sheet of paper	**0.0588**	588	**58,800**
(b) Area of a cross section of a crayon	**0.000192**	**1.92**	192
(c) Area of a desktop	1.5	**15,000**	**1,500,000**
(d) Area of a dollar bill	**0.01**	100	**10,000**

5. **(a)** $\frac{4000 \text{ ft}^2}{1} \cdot \frac{1 \text{ yd}^2}{9 \text{ ft}^2} = \mathbf{444\frac{4}{9}\ yd^2.}$

(b) $\frac{10^6 \text{ yd}^2}{1} \cdot \frac{1 \text{ mi}^2}{3.0976 \cdot 10^6 \text{ yd}^2} \approx \mathbf{0.32\ mi^2.}$

(c) $\frac{10 \text{ mi}^2}{1} \cdot \frac{640 \text{ acre}}{1 \text{ mi}^2} = \mathbf{6400\ acres.}$

(d) $\frac{3 \text{ acre}}{1} \cdot \frac{4840 \text{ yd}^2}{1 \text{ acre}} \cdot \frac{9 \text{ ft}^2}{1 \text{ yd}^2} = \mathbf{130,680\ ft^2.}$

6. **(a)** Triangle with base 3 and height 2: $A = \frac{1}{2} \cdot 3 \cdot 2 = \mathbf{3\ units^2.}$

(b) Construct a rectangle around the figure and subtract the areas of region A, B, C, and D:

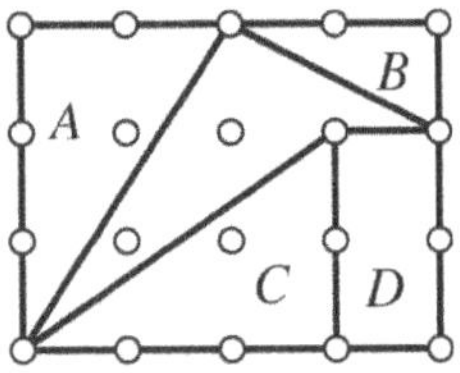

Total area $= 4 \cdot 3 = 12$ units^2.

Area of $A = \frac{1}{2} \cdot 3 \cdot 2 = 3$ units^2;

Area of $B = \frac{1}{2} \cdot 2 \cdot 1 = 1$ unit^2;

Area of $C = \frac{1}{2} \cdot 3 \cdot 2 = 3$ units^2; and

Area of $D = 2 \cdot 1 = 2$ units^2.

Area of figure $= 12 - (3 + 1 + 3 + 2) =$

3 units^2.

(c) Triangle with base 2 and height 2:
$A = \frac{1}{2} \cdot 2 \cdot 2 = \mathbf{2\,units^2}$.

(d) Construct a rectangle around the figure and subtract the areas of regions A, B, C, and D:

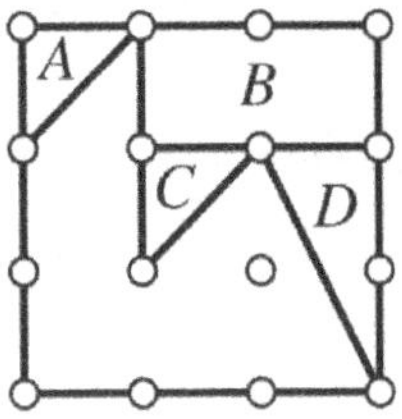

Total area $= 3 \cdot 3 = 9$ units^2.

Area of $A = \frac{1}{2} \cdot 1 \cdot 1 = \frac{1}{2}$ unit^2;

Area of $B = 2 \cdot 1 = 2$ units^2;

Area of $C = \frac{1}{2} \cdot 1 \cdot 1 = \frac{1}{2}$ unit^2; and

Area of $D = \frac{1}{2} \cdot 2 \cdot 1 = 1$ unit^2.

Area of figure $= 9 - \left(\frac{1}{2} + 2 + \frac{1}{2} + 1\right)$

$= \mathbf{5\,units^2}$.

7. (a) $49\text{ m} \cdot 100\text{ m} = \mathbf{4900\text{ m}^2}$.

(b) $\frac{2\text{ fields}}{1} \cdot \frac{4900\text{ m}^2}{1\text{ field}} \cdot \frac{1\text{ a}}{100\text{ m}^2} = \mathbf{98\text{ a}}$.

(c) $\frac{98\text{ a}}{2\text{ fields}} \cdot \frac{1\text{ ha}}{100\text{ a}} = \mathbf{0.98\text{ ha}}$.

8. (a) $A = \frac{1}{2} \cdot 10 \cdot 4 = \mathbf{20\text{ cm}^2}$.

(b) $A = \frac{1}{2} \cdot 5 \cdot 3 = \mathbf{7\frac{1}{2}\text{ m}^2}$.

9. The greatest area of the triangle would occur if drawn as below:

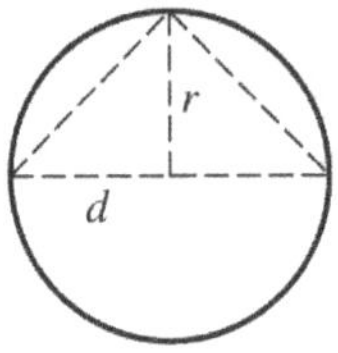

In this case, $A = \frac{1}{2}(2r)(r) = \mathbf{r^2}$.

10. Measurement of area is in square units, so the ratio of the areas is $\left(\frac{a}{b}\right)^2 = \frac{a^2}{b^2}$, or $\mathbf{a^2 : b^2}$.

11. (a) $A = 3 \cdot 3 = \mathbf{9\text{ cm}^2}$.

(b) $A = 8 \cdot 12 = \mathbf{96\text{ cm}^2}$.

(c) $A = 5 \cdot 4 = \mathbf{20\text{ cm}^2}$.

(d) $A = \frac{1}{2} \cdot 7 \cdot (14 + 10) = \mathbf{84\text{ cm}^2}$.

12. (a) *(i)* $A = 1.3\text{ km} \times 1.5\text{ km} = \mathbf{1.95\text{ km}^2}$.

(ii) $A = \frac{1.95\text{ km}^2}{1} \times \frac{10^6\text{m}^2}{1\text{ km}^2} \times \frac{1\text{ ha}}{10^4\text{m}^2} =$

195 ha.

(b) *(i)* $A = 1300\text{ yd} \times 1500\text{ yd}$

$= 1.95 \cdot 10^6\text{ yd}^2$.

$\frac{1.95 \cdot 10^6\text{ yd}^2}{1} \times \frac{1\text{ mi}^2}{3.0976 \cdot 10^6\text{ yd}^2}$

$\approx \mathbf{0.63\text{ mi}^2}$.

(ii) $\frac{0.63\text{ mi}^2}{1} \times \frac{640\text{ acres}}{1\text{ mi}^2} \approx$ **403 acres.**

(c) Answers may vary. The metric system is easier (mathematically) because it is necessary only to move the decimal point to convert units. However, most Americans have very little number sense in the metric system, so the English system might be preferred.

13. (a) **True**. It is not possible to determine a height when only side lengths are known.

(b) The area could be 60 cm^2 if the parallelogram is a rectangle; however the assertion is **false**.

(c) **False**. The area cannot be greater than 60 cm^2 since the maximum is 60 cm^2 if the parallel-gram is a rectangle.

(d) **False**. The area could equal 60 cm^2.

14. A rhombus is a parallelogram with all sides congruent, thus the diagonals are perpendicular. The height of ΔABC in the rhombus below is then $\frac{b}{2}$. Area of $\Delta ABC = \frac{1}{2} \cdot a \cdot \frac{b}{2} = \frac{ab}{4}$; since there are two such triangles the area of the rhombus is

$2 \cdot \frac{ab}{4} = \frac{ab}{2} = \frac{12\,\text{cm} \cdot 5\,\text{cm}}{2} = \mathbf{30\,cm^2}$.

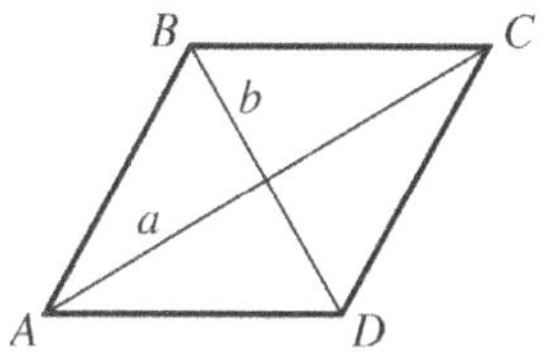

15. **(a)** $6.5 \text{ m} \times 4.5 \text{ m} = 29.25 \text{ m}^2$. $29.25 \text{ m}^2 \times$ \$13.85 per $\text{m}^2 = \mathbf{\$405.11}$.

(b) $15 \text{ ft} \times 11 \text{ ft} = 165 \text{ ft}^2$. $\left(\frac{165 \text{ ft}^2}{1}\right) \cdot \left(\frac{1 \text{ yd}^2}{9 \text{ ft}^2}\right)$

$\frac{165}{9} \text{ yd}^2. = 18\frac{1}{3} \text{ yd}^2 \times \$30 \text{ per yd}^2$

$= \mathbf{\$550}$.

16. **(a)** $A = \pi \cdot 5^2 = \mathbf{25\pi\ cm^2}$.

(b) $A = \frac{60°}{360°} \cdot \pi \cdot 4^2 = \mathbf{\frac{8}{3}\pi\ cm^2}$.

17. Bathroom area

$= 300 \text{ cm} \times 400 \text{ cm} = 120{,}000 \text{ cm}^2$. Each tile is

$10 \text{ cm} \times 10 \text{ cm} = 100 \text{ cm}^2$, thus

$\frac{120{,}000 \text{ cm}^2}{100 \text{ cm}^2} =$ **1200 tiles** (assuming no waste).

18. The area of a regular polygon is $\frac{1}{2}ap$, where a is the apothem (height of one of the triangles of a regular polygon) and p is the perimeter.

(a) $a = 2\sqrt{3}$; $p = 6 \cdot 4 = 24$.

$A = \frac{1}{2} \cdot 2\sqrt{3} \cdot 24 = \mathbf{24\sqrt{3}\ cm^2}$.

(b) The area of a regular triangle is $\frac{1}{2}$ as, where s is the length of a side.

$a = 3\sqrt{3}$; $s = 6$, so

$A = \frac{1}{2} \cdot 3\sqrt{3} \cdot 6 = \mathbf{9\sqrt{3}\ cm^2}$.

19. **(a)** $C = 2\pi r \Rightarrow r = \frac{C}{2\pi} = \frac{8\pi}{2\pi} = 4$.

$A = \pi r^2 = \pi(4)^2 = \mathbf{16\pi\ cm^2}$.

(b) $A_{\text{circle}} = A_{\text{square}} \Rightarrow \pi r^2 = s^2 \Rightarrow$

$r^2 = \frac{s^2}{\pi}$.

$\mathbf{r = \frac{s}{\sqrt{\pi}}}$.

20. **(a)** Radius of large circle

$= 2$ cm; $A_{large\ circle} = \pi(2)^2 = 4\pi \text{ cm}^2$.

$A_{\text{each small circle}} = \pi(1)^2 =$

$\pi \text{ cm}^2$. $A_{\text{shaded}} = 4\pi - 2 \cdot \pi = \mathbf{2\pi\ cm^2}$.

(b) $A_{\text{semicircle}} = \frac{1}{2}(\pi \cdot 1^2) = \frac{1}{2}\pi \text{ cm}^2$.

$A_{\text{triangle}} = \frac{1}{2} \cdot 2 \cdot 1 = 1 \text{ cm}^2$.

$A_{\text{shaded}} = \mathbf{\left(\frac{1}{2}\pi + 1\right) cm^2}$.

(c) If the 1 cm radius were to be extended it would be the diameter of the large circle, cutting off a small shaded semicircle the same size as the small white semicircle. The shaded area is thus equal to half the large circle with radius 2 cm.

$A_{\text{shaded}} = \frac{1}{2}(\pi \cdot 2^2) = \mathbf{2\pi\ cm^2}$.

(d) Consider half the figure, as shown below. Areas $A + B =$ area C:

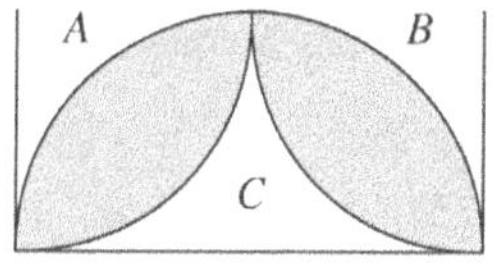

10 cm square

Areas $A + B =$ rectangle $-$ semicircle $=$

$5 \cdot 10 - \frac{1}{2}(\pi \cdot 5^2) = 50 - \frac{25}{2}\pi$.

$A + B + C$ is twice this, or $100 - 25\pi$. Considering both halves of the figure, total white area is

$2(100 - 25\pi) = (200 - 50\pi) \text{ cm}^2$.

The shaded area is that of the square less the white area. $A_{shaded} = 10^2 -$

$(200 - 50\pi) = \mathbf{(50\pi - 100)\,cm^2}$.

(e) ratio of the white portion to shaded portion: (a), ratio is **1:1**, since both areas are $2\pi \text{ cm}^2$; (b), area of white triangle is 1 cm^2, so using the answer from (b) above for the area of the shaded region, the ratio is $\mathbf{1 : \left(\frac{1}{2}\pi + 1\right)}$; (c), ratio is **1:1**, since both areas are $2\pi \text{ cm}^2$; (d), from (d) above, the area of the white area is $200 - 50\pi$, the area of the shaded region is $50\pi - 100$, so the ratio is $(200 - 50\pi) : 50\pi - 100 = 1 : \frac{50\pi - 100}{200 - 50\pi} = \mathbf{1 : \frac{\pi - 2}{4 - \pi}}$.

21. The flower bed with its encircling sidewalk forms a circle with radius $(3 + 1) = 4$ m.

$A_{\text{encircled bed}} = \pi \cdot 4^2 = 16\pi \text{ m}^2$. $A_{\text{flower bed}} = \pi \cdot 3^2 = 9\pi \text{ m}^2$.

$A_{\text{side walk}} = A_{\text{encircled bed}} - A_{\text{flower bed}} = 16\pi - 9\pi = \mathbf{7\pi \text{ m}^2}$.

22. **(a)** $A_{square} = 144 \text{ cm}^2$. Length per side $= \sqrt{144} = 12$ cm. Perimeter $= 4 \cdot 12 =$ **48 cm.**

(b) $P_{square} = 32$ cm. Length per side $= \frac{32}{4} = 8$ cm. Area $= 8^2 = \mathbf{64 \text{ cm}^2}$.

23. **(a)** Area is **quadrupled**. Sides of length s mean area $= s^2$. Sides of length $2s$ mean area $= (2s)^2 = 4s^2$, which is quadruple the original area.

(b) **1 : 25**. Sides of first square $= 1s$, so area $= (1s)^2 = s^2$. Sides of second square $= 5s$, so area $= (5s)^2 = 25s^2$—or a ratio of **1 : 25**.

24. Draw diameters connecting points of tangency. The shaded area is the area of a 20 m by 16 m rectangle minus that of two half-circles, each with radius 8.

$A_{shaded} = 20 \cdot 16 - 2 \cdot \frac{1}{2} \cdot \pi(8)^2 = \mathbf{(320 - 64\pi) \text{ m}^2}$.

25. The radius of the total area is $148\pi = \pi r^2 \Rightarrow r = \sqrt{148} = 2\sqrt{37}$. If the radius of the sidewalk is 0.5 m then the radius of the flower bed is $2\sqrt{37} - 0.5 = \mathbf{\left(2\sqrt{37} - \frac{1}{2}\right) \text{m}}$.

26. Label the triangle as shown:

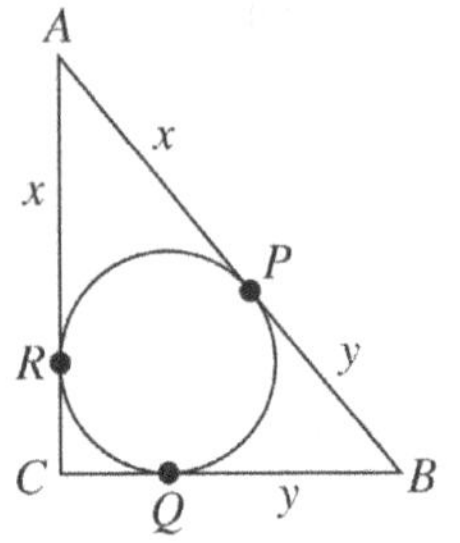

If two tangents are drawn to a circle from the same point on the exterior of the circle, the distances from the common point to the points of tangency are equal. Thus $RC = 4 - x$ and $QC = 3 - y$. Since $QC = RC$, then $4 - x = 3 - y$, or $x - y = 1$.
$AB = 5 \Rightarrow x + y = 5$.
Solving this system of equations:

$$x - y = 1$$
$$\underline{x + y = 5}$$
$$2x \quad = 6$$
$$x = 3$$

Since $x = 3$, then RC, and the radius of the inscribed circle, $= 4 - 3 = 1$. The radius of the circular glass is **1 ft**.

27. **(a)** $s = \frac{3+4+5}{2} = 6$.

$A = \sqrt{6(6-3)(6-4)(6-5)} = \mathbf{6 \text{ cm}^2}$.

(b) $s = \frac{5+12+13}{2} = 15$.

$A = \sqrt{15(15-5)(15-12)(15-13)} = \mathbf{30 \text{ cm}^2}$.

28. The area of the cross section is $\pi r^2 = \pi(6378)^2 = 40{,}678{,}884\pi \, km^2$ or $127{,}796{,}483.1 \text{ km}^2 \approx \mathbf{128{,}000{,}000 \text{ km}^2}$.

29. **(a)** The area of the sector is

$$\pi r^2 \cdot \frac{\text{central angle}}{360^\circ} = \pi 1^2 \cdot \frac{1^\circ}{360^\circ} = \mathbf{\frac{\pi}{360^o} m^2}.$$

(b) The area of the circle is $\pi 1^2 = \mathbf{\pi \text{ m}^2}$.

(c) 360 sectors $= 360 \cdot \frac{\pi}{360^\circ} m^2 = \mathbf{\pi \text{ m}^2}$.

30. The shaded portion is $\frac{1}{4}$ of the difference of the outer circle and the inner circle $=$

$$\frac{1}{4}\left(\pi 4^2 - \pi 2^2\right) = \frac{1}{4}(16\pi - 4\pi) = \frac{1}{4}12\pi = \mathbf{3\pi \text{ cm}^2}.$$

31. **(a)** If each of the nine squares is divided into 4 equal parts then 24 out of 36 parts are shaded. $\frac{24}{36}=\frac{2}{3}$ of the square is shaded. If the area of the square is $6^2 = 36\,\text{cm}^2$ then the area of the shaded quilt pattern is $\frac{2}{3}\cdot 36 = \mathbf{24\,cm^2}$.

(b) If each of the nine squares is divided into 4 equal parts then 28 out of 36 parts are shaded. $\frac{28}{36}=\frac{7}{9}$ of the square is shaded. If the area of the square is $6^2 = 36\,\text{cm}^2$ then the area of the shaded quilt pattern is $\frac{7}{9}\cdot 36 = \mathbf{28\,cm^2}$.

(c) If each of the nine squares is divided into 2 equal parts then 8 out of 18 parts are shaded. $\frac{8}{18}=\frac{4}{9}$ of the square is shaded. If the area of the square is $6^2 = 36\,\text{cm}^2$ then the area of the shaded quilt pattern is $\frac{4}{9}\cdot 36 = \mathbf{16\,cm^2}$.

(d) If each of the nine squares is divided into 2 equal parts then 6 out of 18 parts are shaded. $\frac{6}{18}=\frac{1}{3}$ of the square is shaded. If the area of the square is $6^2 = 36\,\text{cm}^2$ then the area of the shaded quilt pattern is $\frac{1}{3}\cdot 36 = \mathbf{12\,cm^2}$.

32. Half of the parallelogram is shaded. The area of the parallelogram is $b\cdot h$, so the area of the white staircase is $\frac{\mathbf{b\times h}}{\mathbf{2}}$ square units.

33. **(a)** The area of the rectangle is $xy = \mathbf{24\,cm^2}$.

(b) Answers vary. Possible answers are: **1 and 24; 2 and 12; 3 and 8; 4 and 6; 6 and 4; 8 and 3; 12 and 2; and 24 and 1.**

34. The area of the trapezoid is $\frac{(b_1+b_2)h}{2}=\frac{(15+35)20}{2}=(50)(10)=500\,cm^2$. The side of the square with the same area is $s^2 = 500\,cm^2 \Rightarrow s=\sqrt{500\,cm^2}=\mathbf{10\sqrt{5}\,cm}$.

35. If the hexagon is regular then the side of the hexagon is equal to the radius of the circle because the hexagon consist of six equilateral triangles that all have equal sides. The area is then $\pi 4^2 = \mathbf{16\pi\,cm^2}$.

Review Problems

25. **(a)** $C = 2\pi r \Rightarrow r = \frac{C}{2\pi} = \frac{39750}{2\pi}$, about **6330 km**.

(b) This arc length is $\frac{1}{4}$ the circumference, or $\frac{39750}{4} = \mathbf{9937.5\,km}$.

27. **(a)** Using sides 1 and 3 their respective perimeters are (1 + 1 + 1) = 3 and (3 + 3 + 3) = 9. The ratio of their perimeters is 3:9 or **1:3**.

(b) $\frac{1}{3}=\frac{x}{36}\Rightarrow 36=3x\Rightarrow x=\mathbf{12\,cm}$.

29. **(a)** $2\text{ m} = 2\text{ m}\cdot\frac{100\text{ cm}}{1\text{ m}} = \mathbf{200\ cm}$.

(b) $2\text{ in.} = \frac{2}{12}=\mathbf{\frac{1}{6}\ ft}$.

(c) $250\text{ cm}\cdot\frac{1\text{ km}}{100000\text{ cm}} = \mathbf{0.0025\ km}$.

(d) $2500\text{ yd}\cdot\frac{3\text{ ft}}{1\text{ yd}}\cdot\frac{1\text{ mi}}{5280\text{ ft}}\approx \mathbf{1.42\ mi}$.

Assessment 13-3A: The Pythagorean Theorem, Distance Formula, and Equation of a Circle

1. **(a)** $d = \sqrt{4^2+2^2} = \sqrt{20} = 2\sqrt{5} \approx \mathbf{4.5}$.

(b) $d = \sqrt{4^2+1^2} = \sqrt{17} \approx \mathbf{4.1}$.

2. **(a)** $x^2+8^2=10^2 \Rightarrow x^2 = 10^2-8^2 \Rightarrow x=\sqrt{100-64}=\sqrt{36}$. $\mathbf{x = 6}$.

(b) $x^2 = (4a)^2+(3a)^2 = 16a^2+9a^2 \Rightarrow x = \sqrt{a^2(16+9)} = \sqrt{25a^2}$. $\mathbf{x = 5a}$.

(c) $x^2+5^2=13^2 \Rightarrow x^2=13^2-5^2 \Rightarrow x=\sqrt{169-25}=\sqrt{144}$. $\mathbf{x = 12}$.

(d) $x^2 + \left(\frac{s}{2}\right)^2 = s^2 \Rightarrow x^2 = s^2 - \left(\frac{s}{2}\right)^2 =$

$s^2 - \frac{s^2}{4} = \frac{4s^2 - s^2}{4} = \frac{3s^2}{4} \Rightarrow x = \sqrt{\frac{3s^2}{4}}.$

$\boldsymbol{x = \frac{s\sqrt{3}}{2}}.$

(e) $x^2 + 1^2 = y^2$ and $x^2 = 1^2 + 1^2$. Since

$\boldsymbol{x = \sqrt{2}}$, $y^2 = (\sqrt{2})^2 + 1^2$. Thus,

$y^2 = 2 + 1 \Rightarrow \boldsymbol{y = \sqrt{3}}.$

(f) $x^2 = 10^2 + 15^2$ and $y^2 = x^2 + 7^2$.

Since

$x = \sqrt{100 + 225} = \boldsymbol{\sqrt{325}},\ y^2 =$

$\sqrt{325}^2 + 7^2 = 374$. Thus, $\boldsymbol{y = \sqrt{374}}$.

3. (a) Make a right triangle with **sides 2 and 3**.

Hypotenuse $= d = \sqrt{2^2 + 3^2} = \sqrt{13}.$

(b) Make a right triangle with **sides 1 and 2**.

Hypotenuse $= d = \sqrt{1^2 + 2^2} = \sqrt{5}.$

4. Answers may vary. If the distance between dots is one unit, d each side $= \sqrt{2}$ units. $P = 4\sqrt{2}$ units.

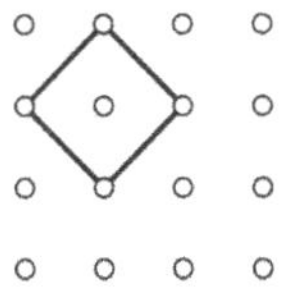

5. The hypotenuse has length $\sqrt{4^2 + 4^2} = 4\sqrt{2}.$

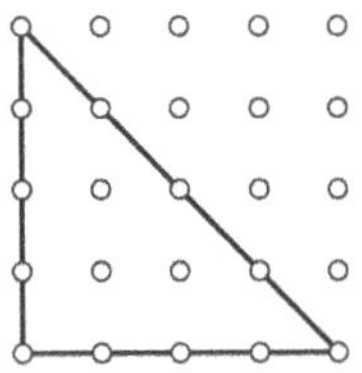

$P = \boldsymbol{8 + 4\sqrt{2} \approx 13.66}$ **units.**

6. Let x and $2x$ be the lengths of the two legs.

$x^2 + (2x)^2 = 30^2 \Rightarrow 5x^2 = 900 \Rightarrow$

$x = \sqrt{180} = \sqrt{36 \cdot 5} = \boldsymbol{6\sqrt{5}}$ and $2x = \boldsymbol{12\sqrt{5}}.$

7. For the answer to be yes, the number must satisfy the Pythagorean theorem.

(a) No. $24^2 \neq 10^2 + 16^2.$

(b) Yes. $34^2 = 16^2 + 30^2.$

(c) Yes. $2^2 = (\sqrt{2})^2 + (\sqrt{2})^2.$

8. Let x be the diagonal of a face. Then

$x^2 = 9^2 + 12^2 \Rightarrow$

$x = \sqrt{81 + 144} = \sqrt{225}.\ x = 15.$

Let d be the diagonal of the prism. Then

$d^2 = d = \sqrt{225 + 225} = \sqrt{225(1 + 1)}.$

$15^2 + 15^2 \Rightarrow \boldsymbol{d = 15\sqrt{2}}$ **cm.**

9. The boat is 10 miles south and 5 miles east of A. The distance from A is

$d = \sqrt{10^2 + 5^2} = \sqrt{125}$, so

$\boldsymbol{\sqrt{125} = 5\sqrt{5}}$ **miles** or **about 11.2 miles.**

10. Let h be the height of the ladder's top.

$h^2 + 3^3 = 15^2 \Rightarrow$

$h = \sqrt{15^2 - 3^2} = \sqrt{216} = 6\sqrt{6}$. So,

$\boldsymbol{6\sqrt{6}}$ **feet**, or about **14.7 feet**.

11. The tall pole stands 10 m above the short pole. Draw a horizontal line from the top of the short pole to form a right triangle, where d is the distance between the poles.

$d^2 + 10^2 = 14^2 \Rightarrow$

$d^2 = 14^2 - 10^2 \Rightarrow d = \sqrt{196 - 100} = \sqrt{96} =$

$\sqrt{16 \cdot 6} = \boldsymbol{4\sqrt{6}}$ **meters**, or **about 9.8 meters.**

12. (a) $h = \sqrt{15^2 - 9^2} = 12.$

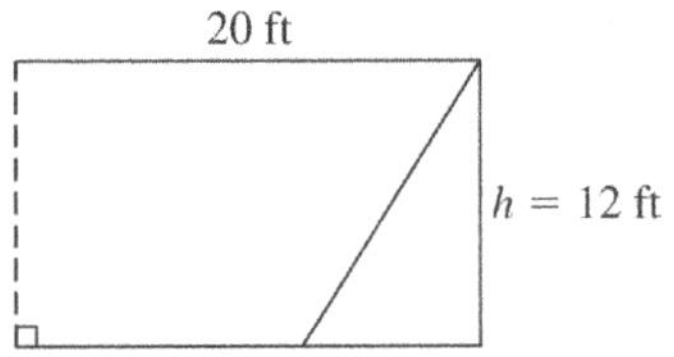

Thus, the area is $20 \cdot 12 = \boldsymbol{240\ \text{ft}^2}$.

(b) $A = \frac{1}{2}bh \Rightarrow$

$A = \frac{1}{2} \cdot 2\sqrt{40^2 - 30^2} \cdot 30 \Rightarrow$

$A = \sqrt{1600 - 900} \cdot 30 \Rightarrow$

$A = 30\sqrt{700} \Rightarrow$

$A = 30\sqrt{100 \cdot 7} \Rightarrow$

$A = \boldsymbol{300\sqrt{7}\ \text{ft}^2 \approx 793.7\ \text{ft}^2}$

13. **(a)** $A = \frac{1}{2} \cdot$(length of diagonal 1)$\cdot$(length of diagonal 2) $= \frac{1}{2}(8)\left(2 \cdot \sqrt{10^2 - 4^2}\right) = 8\sqrt{84} = 8\sqrt{4 \cdot 21} = \mathbf{16\sqrt{21}\ cm^2}$.
Alternatively, observe that the figure is made of 4 triangles, each of which have area $\frac{1}{2}(4) \cdot \sqrt{10^2 - 4^2} = 2\sqrt{84} = 4\sqrt{21}\text{cm}^2$.
So total area $= 4 \cdot 4\sqrt{21} = 16\sqrt{21}\text{cm}^2$.

(b) The figure is made of four triangles, each having area $\frac{1}{2}(2)\sqrt{10^2 - 2^2} = \sqrt{96} = 4\sqrt{6}$.
Thus, the total area is $4 \cdot 4\sqrt{6} = \mathbf{16\sqrt{6}\ cm^2}$.

14. **(a)** **(*i*)** In the large triangle: $x^2 = 4^2 + (4\sqrt{3})^2 = 16 + 16 \cdot 3 \Rightarrow x = \sqrt{16 + 48} = \sqrt{64}$. $\mathbf{x = 8}$.

(*ii*) Area$_{large\ triangle}$ may be expressed in two ways:
$A = \frac{1}{2}(4)(4\sqrt{3}) = 8\sqrt{3}$ and
$A = \frac{1}{2}xy = \frac{1}{2}(8)y = 4y$

Equating: $4y = 8\sqrt{3}$. $\mathbf{y = 2\sqrt{3}}$.

(b) Use the special relationship for $45°$-$45°$-$90°$ right triangles. If the length of each leg is a, the hypotenuse has length $a\sqrt{2}$.

(*i*) Hypotenuse $= 2\sqrt{2}$, so $a = 2$.
$x = 2a \Rightarrow \mathbf{x = 4}$.

(*ii*) $y = \frac{1}{2}x$. $y = \frac{1}{2} \cdot 4 \Rightarrow \mathbf{y = 2}$.

15. Let ℓ be the length of the lake. $\ell^2 + 150^2 = 180 \Rightarrow \ell^2 = 180^2 - 150^2 \Rightarrow \ell = \sqrt{9900} = \sqrt{100 \cdot 99} = 10\sqrt{99} = \mathbf{30\sqrt{11}\ ft}$, or approximately **99.5 ft**.

16. **(a)** Using the figure, label the non-right angles in the top shaded triangle y and z. Since the shaded triangle on top and the shaded triangle on the bottom are congruent (they are congruent by Side-Side-Side), the non-right angles in the bottom triangle can also be labeled y and z. See the illustration below.

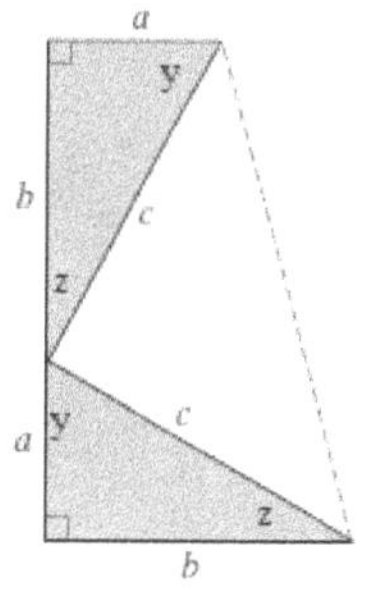

Note in both triangles
$y + z + 90° = 180°$, so
$y + z = 90°$. The angle formed by the two sides of length c is supplementary to the other two angles y and z where they meet; since $y + z = 90°$, the angle $= 90°$, and is a right angle. Therefore, the isosceles triangle with sides c is a right triangle.

(b) The area of the trapezoid is equal to the sum of the areas of the three triangles:

$\frac{1}{2}(a + b)(a + b) = \frac{1}{2}ab + \frac{1}{2}ab + \frac{1}{2}c^2$;
$\frac{1}{2}(a^2 + 2ab + b^2) = ab + \frac{1}{2}c^2$; so
$\frac{a^2}{2} + ab + \frac{b^2}{2} = ab + \frac{c^2}{2} \Rightarrow$
$\frac{a^2}{2} + \frac{b^2}{2} = \frac{c^2}{2}$.
multiplying both sides by 2 yields
$a^2 + b^2 = c^2$.

17. The area of the large square equals the sum of the areas of the small square and the four right triangles:

$(a + b)^2 = c^2 + 4\left(\frac{1}{2}ab\right) \Rightarrow$
$a^2 + 2ab + b^2 = c^2 + 2ab \Rightarrow$
$a^2 + b^2 = c^2$.

The inside quadrilateral is in fact a square, since each side is the hypotenuse of a triangle with the same length sides, and since at each of its vertices there are three angles whose measures sum to 180°— two of which are complementary. Thus the angles of the quadrilateral are right angles.

18. **(a)** The equilateral triangle of base 4 cm has height $= \sqrt{4^2 - 2^2} = \sqrt{12} = 2\sqrt{3}$ cm. Thus, the area is $\frac{1}{2}(4)(2\sqrt{3}) = \mathbf{4\sqrt{3}\ cm^2}$.

Working clockwise, the area of the triangles are

$$\frac{1}{2}\cdot 5\cdot\sqrt{5^2 - \left(\frac{5}{2}\right)^2} = \frac{1}{2}\cdot 5\cdot\sqrt{25 - \frac{25}{4}} =$$

$$\frac{5}{2}\cdot\sqrt{\frac{75}{4}} = \frac{5}{2}\cdot\frac{5\sqrt{3}}{2} = \mathbf{\frac{25\sqrt{3}}{4}\ cm^2} \text{ and}$$

$$\frac{1}{2}\cdot 3\cdot\sqrt{3^2 - \left(\frac{3}{2}\right)^2} = \frac{3}{2}\cdot\sqrt{9 - \frac{9}{4}} =$$

$$\frac{3}{2}\cdot\sqrt{\frac{27}{4}} = \mathbf{\frac{9}{4}\sqrt{3}\ cm^2}.$$

(b) The areas of the smaller triangles (those attached to the legs of the shaded triangle) sum to the area of the largest triangle (attached to the hypotenuse).

$$4\sqrt{3} + \frac{9}{4}\sqrt{3} = \frac{16+9}{4}\sqrt{3} = \frac{25}{4}\sqrt{3}.$$

19. We can position the figure on a coordinate system with the figure positioned so that the origin is at the midpoint of the two right angles, as shown below.

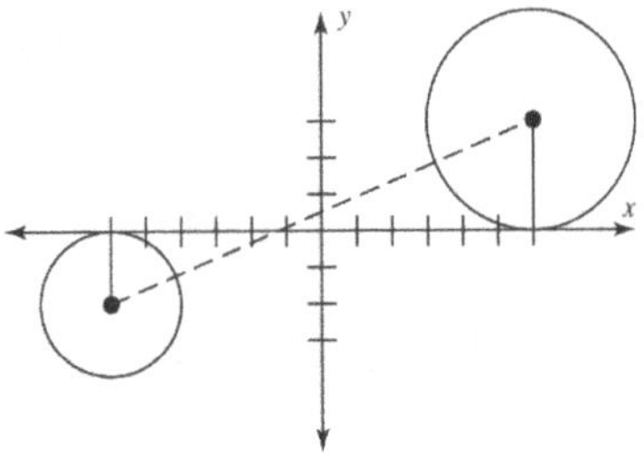

Thus, the distance is

$$\sqrt{(6-(-6))^2 + (3-(-2))^2}$$

$$= \sqrt{12^2 + 5^2} = \sqrt{144 + 25} = \mathbf{13}.$$

20. **(a)** $AB = \sqrt{(0-0)^2 + (7-3)^2} = \sqrt{0+16} = \mathbf{4}$.

(b) $AB = \sqrt{(4-0)^2 + (0-3)^2} = \sqrt{16+9} = \mathbf{5}$.

(c) $AB = \sqrt{(3 - {}^-1)^2 + ({}^-4 - 2)^2} = \sqrt{16+36} = \sqrt{52} = \mathbf{2\sqrt{13}}$, or **about 7.2**.

21. From the special properties of a 30°-60°-90° right triangle, the short leg (opposite the 30° angle) is half the hypotenuse, or $\frac{1}{2}\cdot\frac{c}{2} = \frac{c}{4}$. The longer leg (opposite the 60° angle) is $\sqrt{3}$ times the short leg. Thus the side opposite the 60° angle $= \sqrt{3}\cdot\frac{c}{4} = \mathbf{\frac{c\sqrt{3}}{4}}$.

22. Given the triangle, draw and label it as follows:

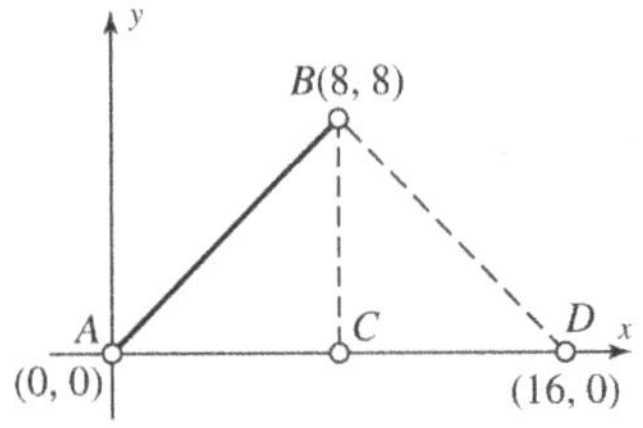

(a) Draw the line segment $\overline{BC}$ so that it is perpendicular to the x-axis. Reflect the triangle ABC about $\overline{BC}$ to form isosceles triangle ABC with third vertex at **(16, 0)**.

(b) $AB^2 = AC^2 + BC^2 \Rightarrow$

$AB = \sqrt{8^2 + 8^2} = 8\sqrt{2}$. Thus the sides are $\mathbf{8\sqrt{2}, 8\sqrt{2},}$ and **16**.

(c) $(8\sqrt{2})^2 + (8\sqrt{2})^2 = 128 + 128$

$= 256 = \mathbf{16^2}$.

23. Given that the equation of a circle with (h, k) and radius r is

$(x-h)^2 + (y-k)^2 = r^2$:

(a) $h = {}^-3$ and $k = 4 \Rightarrow (x - {}^-3)^2 + (y-4)^2 = 4^2$, or $\mathbf{(x+3)^2 + (y-4)^2 = 16}$.

(b) $h = {}^-3$ and $k = {}^-2 \Rightarrow$

$(x - {}^-3)^2 + (y - {}^-2)^2 = (\sqrt{2})^2$, or

$\mathbf{(x+3)^2 + (y+2)^2 = 2}$.

24. **(a)** If $(x-0)^2 + (y-0)^2 = 4^2$, then the center of the circle is **(0,0)** and the radius is **4**.

(b) If $(x-3)^2 + (y-2)^2 = 10^2$, then the center of the circle is **(3, 2)** and the radius is **10**.

(c) If $(x - {}^-2)^2 + (y-3)^2 = (\sqrt{5})^2$, then the center of the circle is **($^-$2, 3)** and the radius is $\mathbf{\sqrt{5}}$.

(d) If $(x-0)^2 + (y - {}^-3)^2 = 3^2$, the center of the circle is **(0, $^-$3)** and the radius is **3**.

25. AB forms the hypotenuse of a $3 - 1 = 2$, $2 + 4 = 6$, h right triangle. Then $h = \sqrt{2^2 + 6^2} = \mathbf{2\sqrt{10}\ km} \approx \mathbf{6.3\ km}$.

26. **(a)** The distance from first base to third base is the diagonal of the square. Using Pythagorean Theorem: $90^2 + 90^2 = d^2 \Rightarrow$ $d = \sqrt{16200} = \mathbf{90\sqrt{2}\ ft} \approx \mathbf{127\ ft}$.

(b) The distance from halfway between first and second base and third base can be found using Pythagorean Theorem:
$45^2 + 90^2 = d^2 \Rightarrow$
$d = \sqrt{10125} = \mathbf{45\sqrt{5}\ ft} \approx \mathbf{101\ ft}$.

(c) The distance from halfway between first and second base and home plate can be found using Pythagorean Theorem:
$45^2 + 90^2 = d^2 \Rightarrow$
$d = \sqrt{10125} = \mathbf{45\sqrt{5}\ ft} \approx \mathbf{101\ ft}$.

27. One side of the rhombus creates a right triangle with half of each of the diagonals. One side is $s^2 = 4^2 + 6^2 \Rightarrow s = \sqrt{52} = \mathbf{2\sqrt{13}\ in} \approx \mathbf{7.2\ in}$.

28. Similar triangles will have sides $3k, 4k,$ and $5k$ and $(3k)^2 + (4k)^2 = (5k)^2$.

Review Problems

25. **(a)** Change all measurements to mm and draw horizontal lines to form three rectangles: 75 mm by 25 mm, 25 mm by 30 mm, and 35 mm by 20 mm.

The respective areas are
$75 \cdot 25 = 1875\ \text{mm}^2$,
$25 \cdot 30 = 750\ \text{mm}^2$, and $35 \cdot 20 = 700\ \text{mm}^2$.
$1875 + 750 + 700 = 3325\ \text{mm}^2 = \mathbf{33.25\ cm^2}$.

(b) $A = \frac{1}{2} \cdot 10 \cdot 6 = \mathbf{30\ cm^2}$.

(c) Change 600 cm to 6 m.
$A = \frac{1}{2} \cdot (6 + 10) \cdot 4 = \mathbf{32\ m^2}$.

27. Given $C = 10, 2\pi r = 10 \Rightarrow r = \frac{5}{\pi}$. $A = \pi \cdot \left(\frac{5}{\pi}\right)^2 = \mathbf{\frac{25}{\pi}\ m^2}$.

29. The U.S. penny has a diameter of 0.75 in. = 19.05 mm. The circumference of the penny is $2\pi r = \pi d = \pi(19.05) \approx 59.85\ \text{mm} \approx 5.985\ \text{cm}$ or **about 6 cm**.

Assessment 13-4A: Surface Areas

1. (b) and (d) can form cubes. The other figures have one of the faces out of order.

2. Where SA represents surface area:

(a) SA of a cube $= 6e^2$ (where e is the length of each edge). $SA = 6(4\ \text{cm})^2 = \mathbf{96\ cm^2}$.

(b) SA of a right cylinder $= 2\pi r^2 + 2\pi rh$ (where r is the radius of the top and bottom circles and h is the height of the cylinder).
$SA = 2\pi(6\ \text{cm})^2 + 2\pi(6\ \text{cm})(12\ \text{cm}) = 72\pi + 144\pi = \mathbf{216\pi\ cm^2}$.

(c) SA of a right rectangular prism is the sum of the lateral surface area and the area of the bases. The lateral surface area is ph (where p is the lateral perimeter), or $2\ell + 2w$, and h is the height. The sum of the bases is $2B$ (where B is length times width. $SA = (2 \cdot 8\ \text{cm} + 2 \cdot 5\ \text{cm}) \cdot 6 + 2(8\ \text{cm} \cdot 5\ \text{cm}) = 26 \cdot 6 + 2 \cdot 40 = 156 + 80 = \mathbf{236\ cm^2}$.

(d) SA of a sphere $= 4\pi r^2$ (where r is the radius of the sphere).
$SA = 4\pi(4\ \text{cm})^2 = \mathbf{64\pi\ cm^2}$.

(e) SA of a right circular cone $= \pi r^2 + \pi r\ell$ (where r is the radius of the base and ℓ is the slant height from any point of the base to the vertex of the cone).
To find ℓ, use the Pythagorean theorem, i.e., $\ell = \sqrt{r^2 + h^2}$ (where h is the height of the vertex above the base).
$SA = \pi(6\ \text{cm})^2 + \pi(6\ \text{cm})\sqrt{(6\ \text{cm})^2 + (8\ \text{cm})^2} = 36\pi + 60\pi = \mathbf{96\pi\ cm^2}$.

3. Area of walls
$= 2(6\ \text{m})(2.5\ \text{m}) + 2(4\ \text{m})(2.5\ \text{m}) = 50\ \text{m}^2$.
Paint needed is $\left(\frac{50\ \text{m}^2}{1}\right) \cdot \left(\frac{1\ \text{L}}{20\ \text{m}^2}\right) = \mathbf{2.5\ liters}$, **so buy 3 liters**.

4. Change all units to mm. The ring then has an inner radius of 20 mm and a height of 30 mm.

The outer ring has a radius of 22 mm and a height of 30mm.

$LSA = 2\pi rh = 2\pi(22)(30) = 1320\pi\ \text{mm}^2$.

The inner ring has a radius of 20 mm and a height of 30 mm.

$LSA = 2\pi(20)(30) = 1200\pi\ \text{mm}^2$.

The area of the top and bottom rings is the area of a circle with radius 22 mm minus the area of a circle radius 20 mm. There are two base rings.

$A = 2\pi(r_{\text{outer}}^2 - r_{\text{inner}}^2) = 2\pi(22^2 - 20^2) = 168\pi\ \text{mm}^2$.

Total $SA = (1320 + 1200 + 168)\pi\ \text{mm}^2 =$ $\mathbf{2688\pi\ mm^2}$.

5. $SA = 4\pi(6370\ \text{km})^2 = \mathbf{162{,}307{,}600\pi\ km^2}$.

6. $SA_{\text{large cube}} = 6e^2 = 6(6\ \text{cm})^2 = 216\ \text{cm}^2$.

$SA_{\text{small cube}} = 6(4\ \text{cm})^2 = 96\ \text{cm}^2$.

Ratio of surface areas $= 96:216 = \mathbf{4:9}$.

7. SA of a right pyramid is $B + \frac{1}{2}p\ell$, where B is the area of the base, p is the perimeter of the base, and ℓ is the slant height from the base to the apex.

$B = \frac{1}{2}ap$, where a is the apothem (the height of the triangles forming the base of a regular polygon). Since a hexagon is composed of equilateral triangles about the centre, the distance from each vertex to the center is the same as the length of each edge (see below).

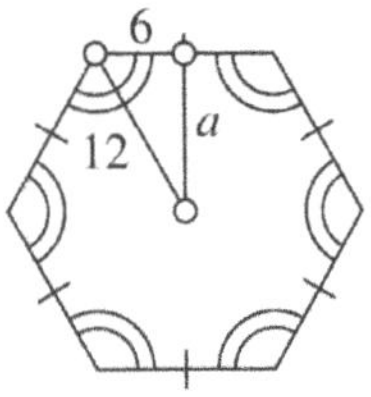

Regular hexagon

$a = \sqrt{12^2 - 6^2} = \sqrt{108} = 6\sqrt{3}$ and $B = \frac{1}{2}(6\sqrt{3})(6 \cdot 12) = 216\sqrt{3}$.

The apothem $= 6\sqrt{3}$ and altitude $= 9$.

$\ell = \sqrt{9^2 + (6\sqrt{3})^2} = \sqrt{189} = 3\sqrt{21}$.

Thus $SA = 216\sqrt{3} + \frac{1}{2}(6 \cdot 12)(3\sqrt{21}) =$ $\mathbf{(216\sqrt{3} + 108\sqrt{21})\,m^2}$.

8. $LSA = 2\pi rh = \pi dh = \pi\left(2\tfrac{5}{8}\ \text{in.}\right)(4\ \text{in.}) =$ $\mathbf{10.5\pi\ in.^2}$, or about $\mathbf{32.99\ in^2}$.

9. (*i*) The top could be $(1 \cdot 88)\ \text{cm}^2$, $(2 \cdot 44)\ \text{cm}^2$, $(4 \cdot 22)\ \text{cm}^2$, or $(8 \cdot 11)\ \text{cm}^2$.

(*ii*) One side could be $(1 \cdot 32)\ \text{cm}^2$, $(2 \cdot 16)\ \text{cm}^2$ or $(4 \cdot 8)\ \text{cm}^2$. The other side could be $(1 \cdot 44)\ \text{cm}^2$, $(2 \cdot 22)\ \text{cm}^2$, or $(4 \cdot 11)\ \text{cm}^2$.

The only dimensions shared by two groups each are $(8 \cdot 11)\ \text{cm}^2$, $(4 \cdot 8)\ \text{cm}^2$, and $(4 \cdot 11)\ \text{cm}^2$. Thus the box is **4 cm by 8 cm by 11 cm**.

10. **(a)** Lateral surface area is proportional to slant height. If slant height is tripled, lateral surface area is **tripled, or multiplied by 3**.

(b) Lateral surface area is proportional to the radius of the base. If radius is tripled, lateral surface area is **tripled, or multiplied by 3**.

(c) Lateral surface area is proportional to the product of slant height and base radius. If both slant height and radius are tripled, lateral surface area is **multiplied by 9**.

11. $SA = B + \frac{1}{2}p\ell$, where

$B = 100$, $p = 4\sqrt{100} = 40$ (i.e., the length of each of the four sides of the square base is the square root of the area), and

$\ell = \sqrt{20^2 + 5^2} = \sqrt{425} = 5\sqrt{17}$. $SA = \left(100 + \frac{1}{2} \cdot 40 \cdot 5\sqrt{17}\right) = \mathbf{(100 + 100\sqrt{17})\,cm^2}$, or about $\mathbf{512.3\ cm^2}$.

12. $\ell = 1.5$ m. Circumference of base $= \frac{240}{360}[2\pi(1.5)] = 2\pi\ \text{m}$. $2\pi = 2\pi r$ (where r is the radius of the base of the cone), thus $r = 1$ m.

(a) Lateral surface area $= \pi r\ell = \pi(1)(1.5) =$ $\mathbf{1.5\pi\ m^2}$.

(b) $SA = \pi r^2 + \pi r\ell = \pi(r^2 + r\ell) = \pi(1^2 + 1 \cdot 1.5) = \mathbf{2.5\pi\ m^2}$.

13. **(a)** $2\pi r = 6\pi \Rightarrow \frac{6\pi}{2\pi} = \mathbf{3\ units}$.

(b) $\ell =$ sector radius $=$ **5 units**.

(c) $h = \sqrt{\ell^2 - r^2} = \sqrt{25 - 9} = \mathbf{4\ units}$.

(d) $C_{full\,circle} = 2\pi(5) = 10\pi$ units. Then $m(\angle_{sector}) = \frac{6\pi}{10\pi}(360°) = \mathbf{216°}$.

14. $C = \pi d = 2.5\pi$ in. Then $SA = 4(2.5\pi) = \mathbf{10\pi\ in^2}$.

15. Surface area is proportional to the square of the radius, thus their ratio would be $\left(\frac{1}{2}\right)^2 = \frac{1}{4}$ or **1 : 4**.

16. The earth's surface area is approximately $4\pi\left(\frac{13000}{2}\right)^2 = 169{,}000{,}000\pi$ km^2. 70% of this area is $.7(169{,}000{,}000\pi) = \mathbf{118{,}300{,}000\pi\ km^2} \approx \mathbf{371{,}650{,}410.9\ km^2}$.

17. $A_{cube} = 6s^2$, where s is the length of a side. Thus $s = \sqrt{\frac{A}{6}}$. The ratio of the edges of the cubes would be $\sqrt{\frac{\left(\frac{64}{6}\right)}{\left(\frac{36}{6}\right)}} = \sqrt{\frac{16}{9}} = \frac{\mathbf{4}}{\mathbf{3}}$ or **4 : 3**.

18. (a) $SA = 6s^2$ (where s is the length of a side of the cube). Then $s = \sqrt{\frac{SA}{6}} = \sqrt{\frac{\mathbf{10{,}648}}{\mathbf{6}}}$ **cm**, or about **42.13 cm**.

(b) Consider the base square with sides s. Based on the properties of 45°-45°-90° triangles, the length of the diagonal of the base is $s\sqrt{2}$.

Using the Pythagorean theorem, take the length of the diagonal and the length of a vertical side, s, to find $d = \sqrt{s^2 + (s\sqrt{2})^2} = \sqrt{3s^2} = s\sqrt{3} = \sqrt{\frac{10648}{6}} \cdot \sqrt{3} = \mathbf{\sqrt{5324}\ cm} \approx \mathbf{72.97\ cm}$.

19. The surface area of the square pyramid is the area of the base $10^2 = 100\,cm^2$ plus the area of the lateral faces. If the height of the pyramid is 10 cm then the slant height of the lateral face is $10^2 + 5^2 = h^2 \Rightarrow h = \sqrt{125} = 5\sqrt{5}\,cm$ and the area is $\frac{1}{2}10 \cdot 5\sqrt{5} = 25\sqrt{5}\,cm^2$ and the total surface area is $100 + 4 \cdot 25\sqrt{5} = 100 + 100\sqrt{5}\,cm^2$. The surface area of the right circular cone is the area of the base $\pi 5^2 = 25\pi\,cm^2$ and the area of the "lateral face" $\pi(5) \cdot 5\sqrt{5} = 25\pi\sqrt{5}\,cm^2$. The total surface area of the cone is $25\pi + 25\pi\sqrt{5}$. The numerical difference in the surface area of the square pyramid and the cone is $\mathbf{100 + 100\sqrt{5} - (25\pi + 25\pi\sqrt{5}) \approx 69.44\ cm^2}$.

20. The surface area of a hexagon minus the bases is the perimeter times the height. The perimeter of the hexagon is $4 \cdot 6 = 24\,cm$ and the height is 48 cm. The surface area of the inside and outside is $2 \cdot 24 \cdot 48 = \mathbf{2304\,cm^2}$.

21. Each glass plate has a surface area of $2(48 \cdot 96) + 2(48 \cdot \frac{1}{4}) + 2(96 \cdot \frac{1}{4})$ $= 9216 + 24 + 48 = 9288\,in.^2$ and the total surface area is $54 \cdot 9288 = 501{,}552\,in.^2 = \mathbf{3483\,ft^2}$.

Review Problems

17. $d = \sqrt{10^2 + 20^2} = \sqrt{500} = \mathbf{10\sqrt{5}\ cm}$.

19. (a) Change 0.6m to 60 cm. Hypotenuse = $\sqrt{60^2 + 80^2} = 100cm$.

(i) Perimeter = 60 + 80 + 100 = **240 cm.**
(ii) Area = $\frac{1}{2}(80)(6) = \mathbf{2400\ cm^2}$.

(b) Drawing an altitude to the end point of the top base formals a $45° - 45° - 90°$ isosceles triangle. The bottom base is 5 cm longer on each side than the top. Thus one leg of the isosceles triangle = 5cm. The height is also 5 cm. Use the Pythagorean Theorem to find the length of each leg of the trapezoid to be $\sqrt{5^2 + 5^2} = 2\sqrt{2}$cm.

(i) Perimeter = $20 + 10 + 2(5\sqrt{2})$ $= \mathbf{\left(30 + 10\sqrt{2}\right)cm.}$

(ii) Area = $\frac{1}{2}(10 + 20)(5) = \mathbf{75\ cm^2}$.

21. The diagonal of the square at the top of the cube is $\sqrt{1^2 + 1^2} = \sqrt{2}$ units. The diagonal of the cube is the hypotenuse of the right triangle created by the diagonal of any one of the faces and the edge of two adjacent faces. Thus, the diagonal of the cube has length

$\sqrt{1^2 + \sqrt{2}^2} = \sqrt{3}$ **units.**

23. **(a)** $150\text{m} \cdot \frac{1\text{km}}{1000\text{m}} = \mathbf{0.15km}$.

(b) $0.002\text{cm} \cdot \frac{10\text{mm}}{1\text{cm}} = \mathbf{0.02mm}$.

(c) $1.44\text{yd}^2 \cdot \frac{9\text{ft}^2}{1\text{yd}^2} \cdot \frac{144\text{in}^2}{1\text{ft}^2} = \mathbf{1866.24in^2}$.

(d) 1 in = **2.54cm.**

Assessment 13-5A: Volume and Mass

1. **(a)** $\frac{8\text{ m}^3}{1} \cdot \frac{(10\text{ dm})^3}{1\text{ m}^3} = \mathbf{8000\ dm^3}$.

(b) $\frac{675000\text{ m}^3}{1} \cdot \frac{(0.001\text{ km})^3}{1\text{ m}^3} = \mathbf{0.000675\ km^3}$.

(c) $\frac{7000\text{ mm}^3}{1} \cdot \frac{(0.1\text{ cm})^3}{1\text{ mm}^3} = \mathbf{7\ cm^3}$.

(d) $\frac{400\text{ in.}^3}{1} \cdot \frac{1\text{ yd}^3}{(36\text{ in.})^3} =$

$\mathbf{\frac{25}{2916} yd^3} \approx \mathbf{0.00857\ yd^3}$.

(e) $\frac{0.2\text{ ft}^3}{1} \cdot \frac{(12\text{ in.})^3}{1\text{ ft}^3} = \mathbf{345.6\ in.^3}$

2. $\frac{60\text{ min}}{1\text{ hr}} \cdot \frac{24\text{ hr}}{1\text{ day}} \cdot \frac{30\text{ days}}{1\text{ mo}} \cdot \frac{15\text{ drops}}{1\text{ min}} \cdot \frac{1\text{ mL}}{20\text{ drops}} \cdot$

$\frac{1\text{ L}}{1000\text{ mL}} =$ **32.4 L** per month water wasted.

3. $V_{Great\ Pyramid} = \frac{1}{3}(771^2)(486)$

$= 96{,}299{,}442\text{ ft}^3$.

$V_{each\ apartment} = (35)(20)(8) = 5600\text{ ft}^3$.

$\frac{96{,}299{,}442\text{ ft}^3}{5600\text{ ft}^3} \approx 17{,}196.36$, or the equivalent volume of about **17,197 rooms.**

4. **(a)** $V_{sphere} = \frac{4}{3}\pi r^3$ so volume of hemispherical portion $= \frac{1}{2}\left(\frac{4}{3}\right)\pi(4^3) = \left(\frac{128}{3}\right)\pi\text{ cm}^3$. $V_{cone} = \frac{1}{3}\pi r^2 h$ so volume of conical portion $= \frac{1}{3}\pi(4^2)(8) = \left(\frac{128}{3}\right)\pi\text{ cm}^3$.

Total volume $= 2\left(\frac{128}{3}\pi\right) = \mathbf{\left(\frac{256}{3}\right)\pi\ cm^3}$.

(b) The volume of a prism is the area of its base times its height.

$V_{right\ triangular\ prism} = Bh$, where B is the area of the triangular base and h is height.

$B = \frac{1}{2}bh_{triangle} = \frac{1}{2}(6)(6) = 18$; $h_{prism} = 12$.

$V = (18)(12) = \mathbf{216\ cm^3}$.

(c) $V_{right\ circular\ cone} = \frac{1}{3}\pi r^2 h$, where $r = 3$ and $h = 5$. $V = \frac{1}{3}\pi(3^2)(5) = \mathbf{15\pi\ cm^3}$.

(d) $V_{sphere} = \frac{4}{3}\pi r^3$, where $r = 10$. $V = \frac{4}{3}\pi(10^3) = \mathbf{\frac{4000}{3}\pi\ cm^3}$.

(e) Volume of hemispherical portion $= \frac{1}{2}\left(\frac{4}{3}\pi r^3\right)\ \frac{1}{2}\left(\frac{4}{3}\right)\pi(10^3) = \frac{2000}{3}\pi\text{ ft}^3$.

Volume of cylindrical portion $= \pi r^2 h = \pi(10^2)(60) = 6000\pi\text{ ft}^3$. $V_{total} = \frac{2000}{3}\pi + 6000\pi = \mathbf{\frac{20{,}000}{3}\pi\ ft^3}$.

5. $\text{cm}^3 \times 0.001 = \text{dm}^3$; $\text{dm}^3 = \text{L}$;

$\text{L} \times 1000 = \text{mL}$ (i.e., $\text{cm}^3 = \text{mL}$ and $\text{dm}^3 = \text{L}$).

	(a)	(b)	(c)	(d)	(e)	(f)
cm^3	**2000**	500	**1500**	**5000**	750	4800
dm^3	2	**0.5**	**1.5**	5	**0.750**	**4.8**
L	**2**	**0.5**	1.5	**5**	**0.750**	**4.8**
mL	**2000**	**500**	**1500**	5000	**750**	**4800**

6. **(a)** A paper cup holds about **200.0 mL** (i.e., the average paper cup holds about a fifth of a liter).

(b) A regular soft drink bottle holds about **0.320 L** (i.e., the 12-ounce size).

(c) A quart milk container holds about **1.0 L** (a quart and a liter are roughly the same amount).

(d) A teaspoonful of cough syrup is about **5.00 mL** (or 5 cc).

7. $V_t = 4^3 = 64$; $V_2 = 6^3 = 216$. $V_1 : V_2 = 64 : 216 = \mathbf{8 : 27}$. (If the side lengths of the two cubes have the ratio $m : n$, their volumes will have the ratio $m^3 : n^3$.)

8. Volume is **multiplied by a factor of 8**. Volume of a sphere is proportional to the cube of the radius, so if the radius is doubled then volume is multiplied by a factor of $2^3 = 8$.

9. For each right rectangular prism, $V = \ell wh \Rightarrow h = \frac{V}{\ell w}$. Note that in (b) and (d), units must be matched:

	(a)	(b)	(c)	(d)
Length	20 cm	10 cm	2 dm	15 cm
Width	10 cm	2 dm	1 dm	2 dm
Height	10 cm	3 dm	**2 dm**	**2.5 dm** or **25 cm**
Volume (cm^3)	**2000**	**6000**	**4000**	**7500**
Volume (dm^3)	**2**	**6**	**4**	7.5
Volume (L)	**2**	**6**	4	**7.5**

10. Volume of a sphere is proportional to the cube of the radius. A sphere with 4 times the radius of another has $4^3 = 64$ times its volume, or a ratio of **64 : 1**.

11. $V = \ell wh = (50)(25)(2) = 2500 \text{ m}^3 =$ **2,500,000 L.** $(1 \text{ m}^3 = 1000 \text{ L.})$

12. The radius of the straw is 2 mm = 0.2 cm. $V = \pi r^2 h = \pi(0.2^2)(25) = \pi \text{ cm}^3 = \boldsymbol{\pi} \textbf{ mL}$.

13. (a) Volume is **multiplied by $2^3 = 8$.**

(b) Volume would be **multiplied by $3^3 = 27$.**

(c) Volume will be **multiplied by n^3.**

14. Volume of any pyramid $= \frac{1}{3}$ (area of base) (height).

$V_{\text{Great Pyramid}} = \frac{1}{3}\left[\left(\frac{940}{4}\right)^2\right](148) =$

$2{,}724{,}433.\overline{3}$ m^3;

$V_{\text{Transamerica Bldg}} = \frac{1}{3}\left[\left(\frac{140}{4}\right)^2\right](260) =$

$106{,}167 \text{ m}^3$.

$\frac{2{,}724{,}433.\overline{3} \text{ m}^3}{106{,}167 \text{ m}^3} \approx$ **25.66 times** as much.

15. A cross-section of the cone filled to half its height shows two similar 30° - 60° - 90° triangles. The radius at a height of 4 cm is 2 cm. Each dimension of the cup is halved, thus

$\frac{1}{2} \cdot \frac{1}{2} \cdot \frac{1}{2} = \frac{1}{8}$ **the volume** of the full cup when it is filled to half its height.

16. $100\% + 30\% = 130\% = 1.3. (1.3^3) = 2.197$, the multiple of volume increase, or **119.7%** increase.

17. $V_{\text{can}} = \pi r^2 h = \pi(3.5^2)(2 \cdot 3.5 \cdot 3) = 257.25\pi \text{ cm}^3$.

$V_{\text{tennis balls}} = 3\left(\frac{4}{3}\pi r^3\right) = 3\left(\frac{4}{3}\right)\pi(3.5^3) = 171.5\pi \text{ cm}^3$.

$V_{\text{air}} = (257.25 - 171.5)\pi = 85.75\pi \text{ cm}^3$.

$\frac{85.75\pi \text{ cm}^3}{257.25 \text{ cm}^3} = \frac{1}{3} =$ **$33\frac{1}{3}$% air.**

18. Let r be the radius of each of the cans and h be the height of the box and the cans. The dimensions of the base of the box are $6r$ by $4r$, thus $V_{box} = (6r)(4r)(h) = 24r^2h$;

$V_{6 \text{ cans}} = 6\pi r^2 h$.

$V_{\text{wasted}} = 24r^2h - 6\pi r^2 h = 6r^2h(4 - \pi)$.

$\frac{6r^2h(4-\pi)}{24r^2h} = \frac{4-\pi}{4} \approx 0.215$, or **about 21.5% wasted**.

19. $V_{prism} = AB \cdot BC \cdot AP$. $V_{pyramid} = \frac{1}{3}(AB \cdot BC \cdot AX) = \frac{1}{3}(AB \cdot BC \cdot 3AP) = AB \cdot BC \cdot AP$, or **equal volume**.

20. (a) Answers may vary. One design would be a square base with sides 5 m and height 12 m. Then $V = \frac{1}{3}Bh = \frac{1}{3}(5^2)(12) = 100 \text{ m}^3$.

(b) **Infinitely many**. $V_{pyramid} = \frac{1}{3}a^2h$, where a is the length of a side of the square base. $300 = a^2h$ is an equation having an infinite number of solutions.

21. (a) **Kilograms or metric tons**. A car weighs in the thousands of pounds or tons.

(b) **Kilograms**. An adult human weighs from about 100 to 250 pounds.

(c) **Grams**. Orange juice concentrate is normally weighed in ounces.

(d) **Metric tons**. An adult African elephant can weigh as much as $7\frac{1}{2}$ tons.

22. **(a)** Staple: A fraction of an ounce, or about 340 **milligrams**.

(b) Professional football player: About 250 pounds on the average, or about 110 **kilograms**.

(c) Vitamin tablet: A fraction of an ounce, or about 1100 **milligrams**.

23. In each metric case below, when converting from smaller to larger units move the decimal point to the left. When converting from larger to smaller units move the decimal point to the right.

(a) $15{,}000\,g = \mathbf{15}$ **kg**.

(b) 0.036 kg $=$ **36 g.**

(c) 4320 mg $=$ **4.320** g.

(d) 0.03 t $=$ **30** kg.

(e) $\frac{25\text{ oz}}{1}\cdot\frac{1\text{ lb}}{16\text{ oz}} = \mathbf{1\frac{9}{16}}\text{ lb} = \mathbf{1.5625}\text{ lb}.$

24. **(a)** **No,** assuming lifting by hand. $1{,}000{,}000$ g $=$ 1000 kg.

(b) **Possibly.** $\frac{1{,}000{,}000}{10}\text{ g} = 100\text{ kg}.$

(c) **Yes.** $\frac{1{,}000{,}000}{100}\text{ g} = 10\text{ kg}.$

(d) **Yes.** $\frac{1{,}000{,}000}{1000}\text{ g} = 1\text{ kg}.$

(e) **Yes.** $\frac{1{,}000{,}000}{10{,}000}\text{ g} = 0.1\text{ kg}.$

25. $V = \ell wh = (40)(20)(20) = 16{,}000\text{ cm}^3$. 1 cm^3 of water weights 1 g. $16{,}000\text{ cm}^3 = 16{,}000\text{ g} = \mathbf{16\text{ kg}}$.

26. **(a)** Answers vary. Draw any segment through the intersection point of the two diagonals that has endpoints on the parallel sides. There are infinitely many ways to do this.

(b) Answers vary. Consider any plane through the intersection line that contains the point of intersection of the two diagonal planes.

27. 1 lb = 453.592 grams so there would be about **454** $100 bills in 1 pound.

28. 1 English ton = 2000 lbs. For the alligator that is $\frac{1\text{ Ton}}{2000\text{ lbs}} = \frac{x}{279\text{ lbs}} \Rightarrow x = \frac{279}{2000} = \mathbf{0.1395}.$

1 English ton = 2000 lbs. For the freshwater fish that is $\frac{1\text{ Ton}}{2000\text{ lbs}} = \frac{x}{646\text{ lbs}} \Rightarrow x = \frac{646}{2000} = \mathbf{0.323}.$

29. 500 sheets weigh 300 lbs and each sheet is 22” by 30”. Each sheet is 660 $\text{in}^2 = 4.583\text{ ft}^2$ that is a total of 2291.67 ft^2 in 300 lbs or 7.6389 ft^2 per pound. $\frac{1\text{ lb}}{7.6389\text{ ft}^2}\cdot\frac{1\text{ ft}^2}{0.092903\text{ m}^2}\cdot\frac{453.592\text{ g}}{1\text{ lb}}$ $= \mathbf{639.15\text{ g/m}^2}$.

30. A tile plus grout on one side is 12.125 inches. On the 8 foot (96 inch) side there is grout, then 7 whole tiles plus grout on one side: $0.125 + 12.125\cdot 7 = 85$. This means the last tile will be cut to 10.875 inches, plus it will need grout to equal the 96 inches on that side. A total of 8 tiles across the 8 foot side.

On the 12 foot (144 inch) side there is grout, then 11 whole tiles plus grout on one side: $0.125 + 12.125\cdot 11 = 133.5$. The last tile will be 10.375 inches, plus it will need grout; the total will be 144 inches on that side. A total of 12 tiles are across the 12 foot side.

This means $8\times 12 = \mathbf{96}$ **tiles** will be used, although some will not be whole tiles.

31. $\frac{1}{2}$ lbs per person = 2.5 lbs for 5 people.

$2.5\text{ lbs}\cdot\frac{0.453592\text{ kg}}{1\text{ lbs}} = \mathbf{1.13398\text{ kg}}.$

Review Problems

21. **(a)** 350 mm $=$ **35** cm (1 cm $= 10$ mm).

(b) $1600\text{ cm}^2 = \mathbf{0.16}\text{ m}^2$ $(1\text{ m}^2 = 10{,}000\text{ cm}^2)$.

(c) $0.4\text{ m}^2 = \mathbf{400{,}000}\text{ mm}^2$ $(1\text{ m}^2 = 1{,}000{,}000\text{ mm}^2)$.

(d) $5.2\text{ cm}^2 = \mathbf{0.00052\text{ m}^2}$ $(1\text{ cm}^2 = 0.0001\text{ m}^2)$.

23. **(a)** Radius of base $= \sqrt{50^2 - 40^2} = 30$. $SA = \pi(30^2) + \pi(30)(50) = \mathbf{2400\pi\text{ cm}^2}$.

(b) Hypotenuse of triangle $= \sqrt{33^2 + 65^2} = \sqrt{5314}$. $SA = 2\left[\frac{1}{2}(65)(33)\right] + (33 + 65 + \sqrt{5314})(40) = \mathbf{(6065 + 40\sqrt{5314})\text{ cm}^2}$.

25. From exercise 24: $6\sqrt{3} > 6\sqrt{2}$. The non-face diagonal is the hypotenuse of a right triangle containing the diagonal of the face as a leg. Since the hypotenuse is the longest side in a right triangle, the non-face diagonal has to be larger than the face diagonal.

Chapter 13 Review

1. **(a)** $50 \text{ ft} \cdot \frac{1 \text{ yd}}{3 \text{ ft}} = \mathbf{16\frac{2}{3}\text{ yd}}$.

(b) $947 \text{ yd} \cdot \frac{\text{mi}}{1760 \text{ yd}} = \mathbf{\frac{947}{1760}\text{mi}} \approx \mathbf{0.538\text{ mi}}$.

(c) $0.75 \text{ mi} \cdot \frac{5280 \text{ ft}}{1 \text{ mi}} = \mathbf{3960 \text{ ft}}$.

(d) $349 \text{ in} \cdot \frac{1 \text{ yd}}{36 \text{ in}} = \mathbf{9\frac{25}{36}\text{ yd}} \approx \mathbf{9.694 \text{ yd}}$.

(e) $5 \text{ km} \cdot \frac{1000 \text{ m}}{1 \text{ km}} = \mathbf{5000 \text{ m}}$.

(f) $165 \text{ cm} \cdot \frac{\text{m}}{100 \text{ cm}} = \mathbf{1.65 \text{ m}}$.

(g) $52 \text{ cm} \cdot \frac{10 \text{ mm}}{\text{cm}} = \mathbf{520 \text{ mm}}$.

(h) $125 \text{ m} \cdot \frac{1 \text{ km}}{1000 \text{ m}} = \mathbf{0.125 \text{ km}}$.

2. **(a)** The triangle inequality tells us that $r + q > p$. If $p - q > r$, then $p - q + q > r + q > p$. This would say that $p > p$, which is false.

(b) If $r = p - q$, then $r + q = p$, which would form a line segment not a triangle.

3. By the Pythagorean theorem, the other side of the rectangle is given by $\sqrt{130^2 - 120^2} = 50$ cm. Thus, the perimeter is $2(50 \text{ cm}) + 2(120 \text{ cm}) = \mathbf{340 \text{ cm}}$.

4. $C = 2\pi r \Rightarrow 3m = 2\pi r \Rightarrow r = \mathbf{\frac{3}{2\pi}\text{m}}$.

5. If a circle has radius 6 cm, then its circumference is $2\pi r = 2\pi(6 \text{ cm}) = 12\pi$ cm. Martine is not correct.

6. From Pick's theorem, $A = I + \frac{1}{2}B - 1$, where I = the number of dots inside the polygon and B = the number of dots on the polygon boundary. Alternatively, area could be determined through the rectangle method.

(a) $A = 7 + \frac{1}{2}(5) - 1 = \mathbf{8\frac{1}{2}\text{ cm}^2}$.

(b) $A = 5 + \frac{1}{2}(5) - 1 = \mathbf{6\frac{1}{2}\text{ cm}^2}$.

(c) $A = 2 + \frac{1}{2}(12) - 1 = \mathbf{7\text{cm}^2}$.

7. Rearranging as shown yields a rectangle with width $\frac{h}{2}$ and length $A'B' = b_1 + b_2$. Then $A = \ell w = \frac{1}{2}(b_1 + b_2)$, which must be the area of the initial trapezoid.

8. Area of $\Delta ABC < \Delta ABD = \Delta ABE < \Delta ABF$. All triangles have the same base, so area is proportional only to height. Ordering by height only, ΔABD and ΔABE are equal in area.

9. **(a)** $A = \frac{1}{2}ap$, where $a = \sqrt{6^2 - 3^2} = 3\sqrt{3}$ and $p = 6 \cdot 6 = 36$. $A = \frac{1}{2}(3\sqrt{3})(36) = \mathbf{54\sqrt{3}\text{ cm}^2}$.

(b) $A = \pi r^2 = \pi(6^2) = \mathbf{36\pi\text{ cm}^2}$.

10. **(a)** $A = \pi(4^2) - \pi(2^2) = \mathbf{12\pi\text{ cm}^2}$.

(b) $A_{\text{semicircle}} = \frac{1}{2}\pi(3^2) = 4.5\pi \text{ cm}^2$.

$A_{\text{triangle}} = \frac{1}{2}(6)(4) = 12 \text{ cm}^2$.

$A_{\text{shaded region}} = \mathbf{(4.5\pi + 12)\text{ cm}^2}$.

(c) $A = (6)(4) = \mathbf{24\text{ cm}^2}$.

(d) $A = \frac{40°}{360°}\pi(6^2) = \mathbf{4\pi\text{ cm}^2}$.

(e) $A = \frac{1}{2}(2)(3) + \frac{1}{2}(3)(3) + (3)(15) + (3)(4)$
$= 3 + 4.5 + 45 + 12 = \mathbf{64.5\text{ cm}^2}$.

(f) $A = (8)(18) + \frac{1}{2}(18 + 5)(3) = \mathbf{178.5\text{ m}^2}$.

11. Extend $\overline{AB}$ to make right triangle ΔADC as shown below:

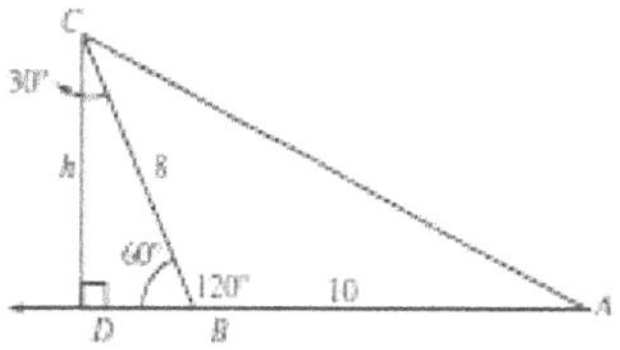

$\overline{BD}$ will have length 4, since it should be half of $\overline{BC}$; use the Pythagorean Theorem to find h, the height of ΔADC and ΔABC; $h^2 + 4^2 = 8^2$; $h^2 + 16 = 64 \Rightarrow h^2 = 48 \Rightarrow h = \sqrt{48}$.

Area $\Delta ABC = \frac{1}{2}(10)\sqrt{48} = \mathbf{5\sqrt{48}} = \mathbf{20\sqrt{3}}$
$\approx$ **34.64 square units**.

12. $AC = \sqrt{1^2 + 2^2} = \sqrt{5}$. $AD = \sqrt{1^2 + (\sqrt{5})^2}$ $= \sqrt{6}$. Each succeeding segment thus adds 1 to the number under the radical; $AG = \sqrt{9} = $ **3.**

13. (a) Yes. $5^2 + 12^2 = 13^2$.

(b) No. $40 + 60 < 104$, thus the measures cannot represent any triangle.

14. $A_{\text{top/bottom}} = 2\left[40^2 - 4(5^2)\right] = 3000 \text{ cm}^2$.

$A_{\text{sides}} = 4(40)(15) = 2400 \text{ cm}^2$.

$A_{\text{total}} = 3000 + 2400 = \mathbf{5400 \text{ cm}^2}$.

15. In each of the following, part (*i*) is surface area and part (*ii*) is volume.

(a) (*i*) $SA = B + \frac{1}{2}p\ell$, where $\ell = \sqrt{6^2 + 4^2} = 2\sqrt{13}$. $SA = (8^2) + \frac{1}{2}(32)(2\sqrt{13}) = \mathbf{32(2+\sqrt{13}) \text{ cm}^2}$.

(*ii*) $V = \frac{1}{3}Bh = \frac{1}{3}(8^2)(6) = \mathbf{128 \text{ cm}^3}$.

(b) (*i*) $SA = \pi r^2 + \pi r\ell$, where $\ell = \sqrt{8^2 + 6^2} = 10$. $SA = \pi(6^2) + \pi(6)(10) = \mathbf{96\pi \text{ cm}^2}$.

(*ii*) $V = \frac{1}{3}\pi r^2 h = \frac{1}{3}\pi(6^2)(8) = \mathbf{96\pi \text{ cm}^3}$.

(c) (*i*) $SA = 4\pi r^2 = 4\pi(5^2) = \mathbf{100\pi \text{ m}^2}$.

(*ii*) $V = \frac{4}{3}\pi r^3 = \frac{4}{3}\pi(5^3) = \mathbf{\frac{500}{3}\pi \text{ m}^3}$.

(d) (*i*) $SA = 2\pi r^2 + 2\pi rh = 2\pi(3^2) + 2\pi(3)(6) = \mathbf{54\pi \text{ cm}^2}$.

(*ii*) $V = \pi r^2 h = \pi(3^2)(6) = \mathbf{54\pi \text{ cm}^3}$.

(e) (*i*) $SA = sB + ph = (2)(4)(10) + (28)(8) = \mathbf{304 \text{ m}^2}$.

(*ii*) $V = \ell wh = (10)(4)(8) = \mathbf{320 \text{ m}^3}$.

16. Lateral surface area $= \pi r\ell$, where $\ell = \sqrt{12^2 + 5^2} = 13$.

Area $= \pi(5)(13) = \mathbf{65\pi \text{ m}^2}$.

17. $V = \pi r^2 h$. If both r and h are doubled, volume is $\pi(2r)^2(2h) = 8\pi r^2 h$, or eight times the original volume. The graph really shows an eight-fold growth, rather than the doubled sales.

18. Sum the areas of the four triangles.

$A = 2\left(\frac{1}{2} \cdot 5 \cdot 12\right) + 2\left(\frac{1}{2} \cdot 12 \cdot \sqrt{20^2 - 12^2}\right) =$

$60 + 192 = \mathbf{252 \text{ cm}^2}$.

19. Change the diagonal measurement to 130 cm. The length of the other side of the rectangle h is $h^2 + 120^2 = 130^2 \Rightarrow h = \sqrt{130^2 - 120^2} \Rightarrow h = \sqrt{2500} \Rightarrow h = 50$.

$A = \ell w = (120)(50) = \mathbf{6000 \text{ cm}^2}$.

20. $\mathbf{2\sqrt{2} \text{ m}^2}$ $\left[h = \sqrt{3^2 - 1^2} = 2\sqrt{2} \Rightarrow A = \frac{1}{2}bh = \frac{1}{2}(2)(2\sqrt{2}) = 2\sqrt{2} \text{ m}^2\right]$.

21. Change the area of printed matter to 2500 cm^2.
$h_{\text{printed matter}} = 74 - 24 = 50$ cm;

$w_{\text{printed matter}} = \frac{2500 \text{ cm}^2}{50 \text{ cm}} = 50$ cm.

$w_{\text{poster}} = 50 + 12 = \mathbf{62 \text{ cm}}$.

22. $V_{\text{cylinder}} = \pi r^2 h = 10\pi r^2$ and $V_{\text{cone}} = \frac{1}{3}\pi r^2 h$.
$10\pi r^2 = \frac{1}{3}\pi r^2 h \Rightarrow 10 = \frac{1}{3}h$. $h = \mathbf{30 \text{ cm}}$.

23. Answers may vary.

(a)

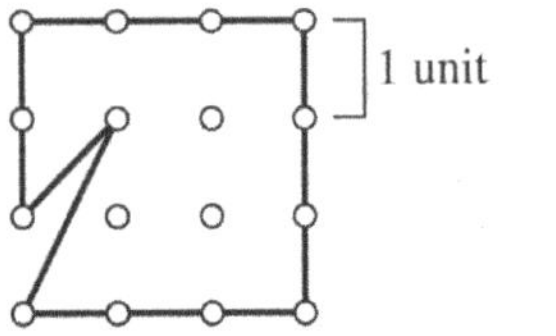

Perimeter $= 15 + \sqrt{2} + \sqrt{3}$ units. Any polygon having sides interior to the edges would have a perimeter greater than 16.

(b)

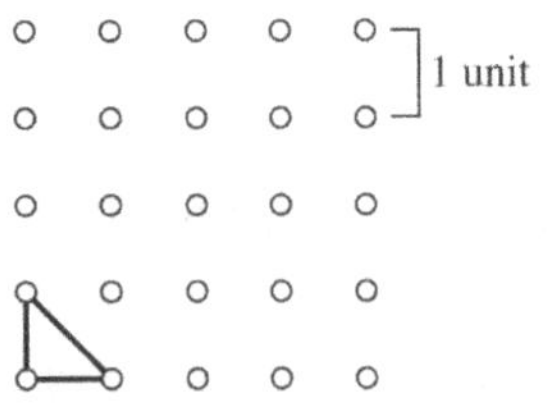

Perimeter $= 1 + 1 + \sqrt{2} = 2 + \sqrt{2}$ units. No other polygon would have a lesser perimeter.

(c)

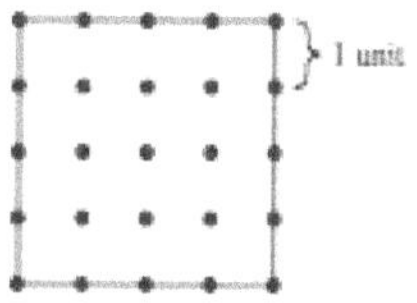

$A = 5 \cdot 5 = 25$ units^2. Any other polygon would have space outside the measured area, and would thus have less.

24. Diagonal $= \sqrt{\left(8\frac{1}{2}\right)^2 + 11^2} = \frac{\sqrt{773}}{2}$

$\approx$ **13.9 inches.**

25. (a) Given that the equation of a circle with center (h, k) and radius r is $(x - h)^2 + (y - k)^2 = r^2$: $h = 3$ and $k = {}^-4 \Rightarrow (x - 3)^2 + (y - {}^-4)^2 = 5^2$, or $\mathbf{(x - 3)^2 + (y + 4)^2 = 25}$.

(b) If $(x - 5)^2 + (y - {}^-3)^2 = 6^2$, then the center of the circle is $\mathbf{(5, {}^-3)}$ and the radius is **6**.

26. (a) **Metric tons**, which corresponds to thousands of pounds, or tons.

(b) $1\text{ cm} \cdot 1\text{ cm} \cdot 1\text{ cm} = \mathbf{1\text{ cm}^3}$.

(c) 1 cm^3 of water weighs **1 gram**.

(d) $\mathbf{1\text{ L} = 1\text{ dm}^3}$, so the two have the same volume.

(e) $\frac{1\text{ L gas}}{12\text{ km}} = \frac{x\text{ L gas}}{300\text{ km}} \Rightarrow 12x = 300.\ x = \mathbf{25\text{ L}}$.

(f) **2000** a (1 ha = 100 a).

(g) **51,800** cm^3 ($1\text{ L} = 1000\text{ cm}^3$)

(h) **10,000,000** m^2 ($1\text{ km}^2 = 1{,}000{,}000\text{ m}^2$).

(i) **50,000** mL (1 L = 1000 mL).

(j) **5.83** L (1000 mL = 1 L).

(k) **25,000** dm^3 ($1\text{ m}^3 = 1000\text{ dm}^3$).

(l) **75,000** mL ($1\text{ dm}^3 = 1000$ mL).

(m) **52.813** kg (1000 g = 1 kg).

(n) **4.8** t (1000 kg = 1 t).

27. (a) $V_{\text{tank}} = \ell wh = (2)(1)(3) = 6\text{ m}^3 = 6{,}000{,}000\text{ cm}^3$. $6{,}000{,}000\text{ cm}^3$ of water = 6,000,000 g = **6000 kg** of water.

(b) $V_{\text{sphere}} = \frac{4}{3}\pi r^3 = \frac{4}{3}\pi(30)^3 = 36{,}000\pi\text{ cm}^3$, or about $113{,}097\text{ cm}^3$. To find the water's height increase from the sphere (converting length and width of the tank to cm): $V = 113{,}097 = (200)(100)h.\ h = \frac{113.097}{20{,}000}$, or about 5.65 cm rise from the sphere's volume. The tank was half full (i.e., $h = 1.5$ m), so with a rise of 5.65 cm = 0.0565 m the new water height is **1.5565 m**.

28. (a) **80 L**, or about 21 gallons.

(b) **82 kg**, or about 181 pounds.

(c) **978 g** = 0.978 kg, or about 2 pounds.

(d) **5 g**, or about 0.2 ounce.

(e) **4 kg**, or about 8.8 pounds.

(f) **1.5 metric tons** = 1500 kg, or about 3300 pounds.

(g) **180 mL** = 0.180 L.

29. (a) The centers of the circles are (2,11) and (11,2). Using the distance formula,

$d = \sqrt{(2-11)^2 + (11-2)^2} = \sqrt{(-9)^2 + 9^2} = \sqrt{162} = \mathbf{9\sqrt{2} \approx 12.73}$.

(b) The length of the sum of the radii of the two circles is $5 + 5 = 10$ which is shorter than $9\sqrt{2}$. The two circles do not intersect.

(c)

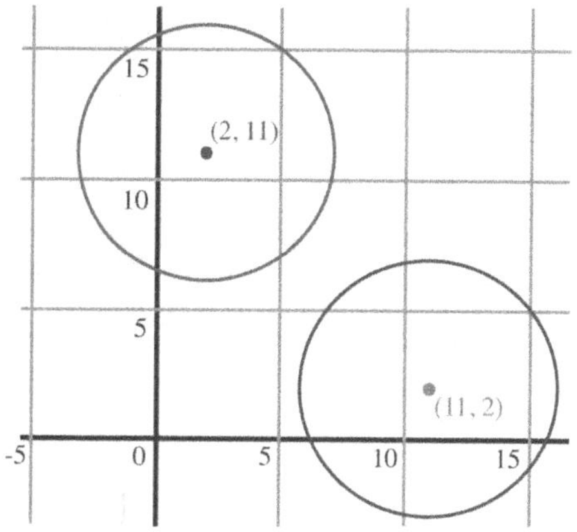

30. (a) **2000 g** ($1\text{ dm}^3 = 1000\text{ cm}^3$, or 1000 g of water).

(b) **1000 g**

($1\text{ L} = 1000\text{ cm}^3$, or 1000 g of water).

(c) **3 g** ($1\text{ cm}^3 = 1$ g).

(d) 0.0042 kg
(1 mL of water = 1 g = 0.001 kg).

(e) $\mathbf{0.0002\ m^3}$ (1 L = 1000 cm^3 = 0.001 m^3).

31. The volume of a sphere is $\frac{4}{3}\pi r^3$. The ratio of the capacities is
$\frac{4}{3}\pi 4^3 : \frac{4}{3}\pi 6^3 \Rightarrow \frac{4}{3}\pi 64 : \frac{4}{3}\pi 216$
$\Rightarrow 64:216 \Rightarrow \mathbf{8:27}$.

32. If the scale factor of two similar figures is k, then the ratio of their volumes is k^3 (Theorem 14-12). If the volumes of two cylinders have a ratio 2:3 then the scale factor is $\sqrt[3]{k}$ and their radii have a ratio of $\mathbf{\sqrt[3]{2} : \sqrt[3]{3}}$.

33. The area of the square is $s^2 = (2r)^2 = 4r^2$. The area of the circle is πr^2. The percent that will be watered is the percent the circle area is of the square area. $\frac{\pi r^2}{4r^2} \approx 0.7854 \approx 78.54\%$ will be watered.

The area of the square is $s^2 = (6r)^2 = 36r^2$.

The area of the nine circles is $9\pi r^2$. The percent that will be watered is the percent the nine circles' area is of the square area.
$\frac{9\pi r^2}{36r^2} \approx 0.7854 \approx 78.54\%$ will be watered.

Both sprinkler systems cover the same percentage of the square field and it does not matter which system is used if the only selection criterion is the amount of land covered by the system.

34. Draw altitudes $\overline{BE}$ and $\overline{DF}$ of triangles BCP and DCP respectively. $\triangle ABE \cong \triangle CDF$ by AAS
$A: \angle E \cong \angle F \cong 90^\circ; A: \angle BAE \cong \angle DCF; \text{S}: \overline{AB} \cong \overline{CD}$
Thus altitudes $\overline{BE} \cong \overline{DF}$. Because $\overline{CP}$ is a base of ΔBCP and ΔDCP, and because the altitudes are the same, the areas must be equal.

CHAPTER 14

TRANSFORMATIONS

Assessment 14-1A: Translations, Rotations, and Tessellations

1. (a) Each corner of the trapezoid moves two dots to the right.

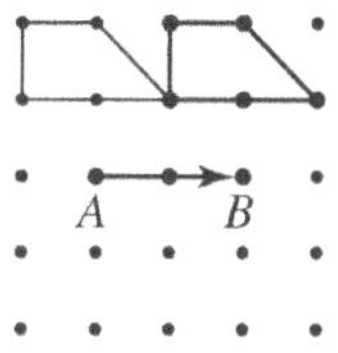

(b) Each corner of the trapezoid moves one dot down and one dot to the right.

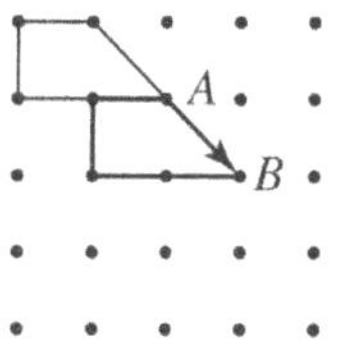

2. (a) Trace $\overline{BC}$ and the line containing the slide arrow on the tracing paper and label the trace of B as B' and the trace of C as C'. Mark on the original paper and on the tracing paper the initial point of the arrow by P and the head of the arrow by Q. Slide the tracing paper along the line $\overleftrightarrow{PQ}$ so that P will fall on Q; trace $\overline{BC}$. The segment $\overline{B'C'}$ is the image of $\overline{BC}$ under the translation.

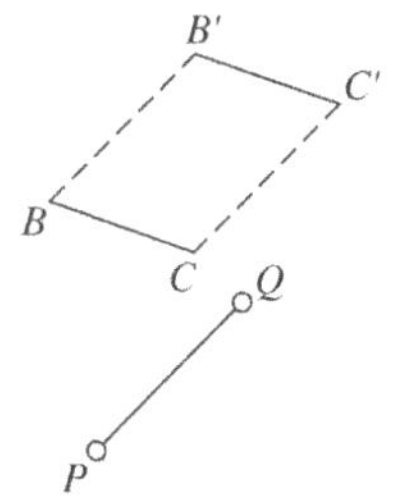

(b) Construct a parallelogram with $\overline{BC}$ and $\overline{B'C'}$ as opposite sides.

3. (a) $(0, 0) \to (0 + 3, 0 - 4) = \mathbf{(3, {}^-4)}$.

(b) $({}^-3, 4) \to ({}^-3 + 3, 4 - 4) = \mathbf{(0, 0)}$.

(c) $({}^-6, {}^-9) \to ({}^-6 + 3, {}^-9 - 4) = \mathbf{({}^-3, {}^-13)}$.

4. To go from an image to the coordinates of the original points, reverse the signs of the translations; i.e.,
$(x', y') \to (x' + 3, y' - 4) = (x, y)$.

(a) $(0, 0) \to (0 + 3, 0 - 4) = \mathbf{(3, {}^-4)}$.

(b) $({}^-3, 4) \to ({}^-3 + 3, 4 - 4) = \mathbf{(0, 0)}$.

(c) $({}^-6, {}^-9) \to ({}^-6 + 3, {}^-9 - 4) = \mathbf{({}^-3, {}^-13)}$.

5. (a) $A({}^-4, 2) \to ({}^-4 + 3, 2 - 4) = A'({}^-1, {}^-2)$;
$B({}^-2, 2) \to ({}^-2 + 3, 2 - 4) = B'(1, {}^-2)$;
$C(0, 0) \to (0 + 3, 0 - 4) = C'(3, {}^-4)$;
$D({}^-2, 0) \to ({}^-2 + 3, 0 - 4) = D'(1, {}^-4)$.

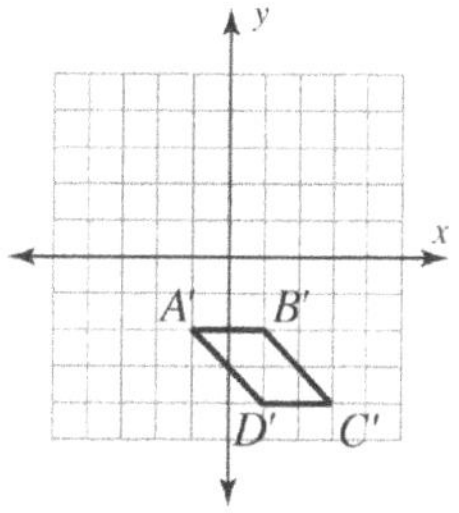

(b) $A({}^-1, 1) \to ({}^-1 + 3, 1 - 4) = A'(2, {}^-3)$;
$B(1, 4) \to (1 + 3, 4 - 4) = B'(4, 0)$;
$C(3, {}^-1) \to (3 + 3, {}^-1 - 4) = C'(6, {}^-5)$.

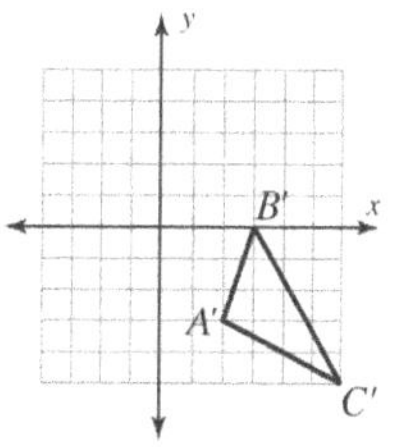

6. $(x', y') \to (x' - 3, y' + 4)$:

Label points $A'({}^-1, 0)$, $B'(2, 0)$, $C'(3, {}^-3)$, and

$D'(0, {}^{-}3)$.

$A'({}^{-}1, 0) \rightarrow ({}^{-}1 - 3, 0 + 4) = A({}^{-}4, 4)$;

$B'(2, 0) \rightarrow (2 - 3, 0 + 4) = B({}^{-}1, 4)$;

$C'(3, {}^{-}3) \rightarrow (3 - 3, {}^{-}3 + 4) = C(0, 1)$;

$D'(0, {}^{-}3) \rightarrow (0 - 3, {}^{-}3 + 4) = D({}^{-}3, 1)$.

The figure whose image was shown is:

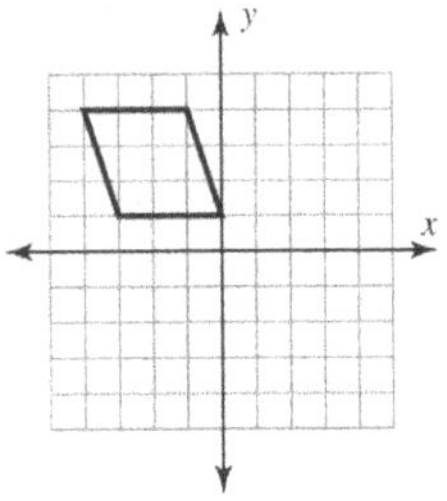

7. Rotate each corner 90° counterclockwise around O to obtain:

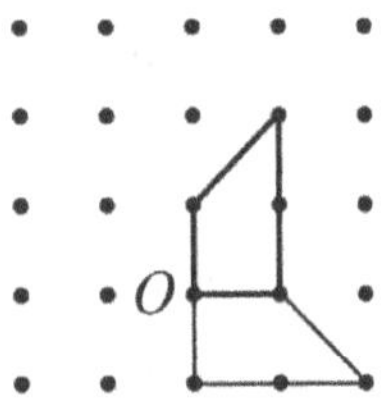

8. Construct a line perpendicular to ℓ passing through O. Denote the intersection of this perpendicular line with ℓ as Q. Draw a circle with radius $\overline{OQ}$ and center O; label point P where the perpendicular line intersects the circle on the opposite side of P. Then construct a perpendicular to $\overleftrightarrow{OQ}$ at $P \Rightarrow \ell'$.

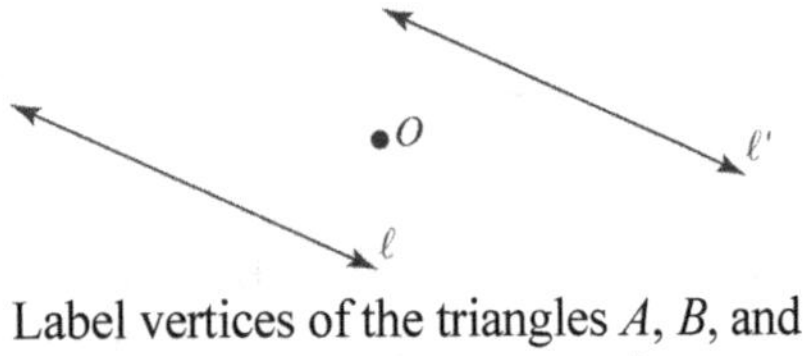

9. Label vertices of the triangles A, B, and C. Find the image of A by drawing $\overrightarrow{AO}$ and marking off $OA' = OA$ so that O is the midpoint of $\overline{AA'}$. Similarly find B' and C'. A', B', and C' are the vertices of the figures under a half-turn about O.

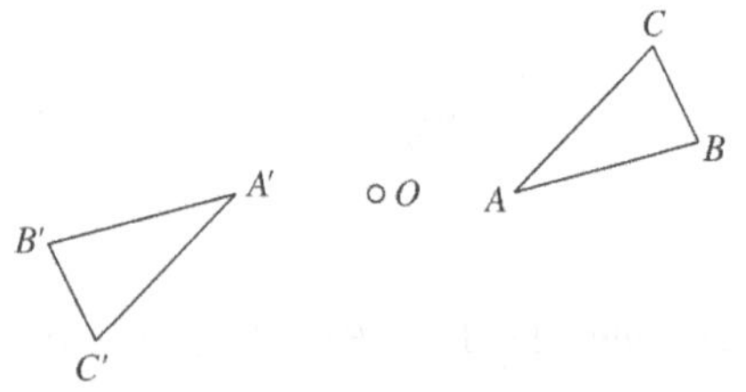

10. (a) $\ell' = \ell$.
    (b) $\ell' \perp \ell$.

11. Reverse the rotation [i.e., counterclockwise] to locate $\overline{AB}$, which is the pre-image.

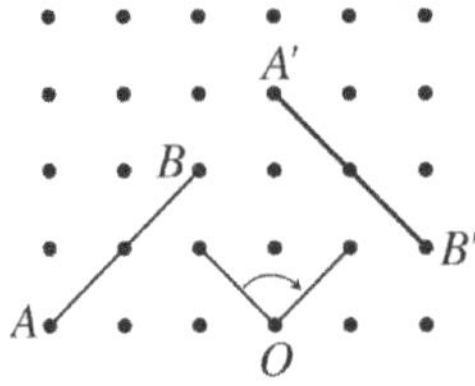

12. Answers may vary; e.g., SOS. H, I, N, O, S, X, or Z could appear in such rotational words; variations in drawing could use M and W in rotational images such as MOW.

13. (a) Connect O to P with a line using a straight edge. Construct a circle with Center O through point P. Construct a circle with Center P through point O. Label the circle intersections Q (left intersection) and R (right intersection). Connect Q to Q with a line using a straight edge. By construction $\overleftrightarrow{OP} \perp \overleftrightarrow{QR}$. Bisect $\angle POQ$ using the steps from "Constructing an Angle Bisector" Figure 40 in Chapter 12. The bisected angle measures 45° because it is half of the right angle created by the perpendicular lines. Put the legs of your compass on O and on P; using the same compass opening put the legs on O and on the line segment that bisected $\angle POQ$ and mark P' to rotate P 45° counterclockwise.

    (b) Construct a circle with Center O through point P. Construct a circle with Center P through point O. Label the circle intersections Q (left intersection) and R (right intersection). Connect O to P and Q to O and Q to P with a segment using a straight edge. The triangle formed is an equilateral triangle and the interior angles measure 60°. We can replace R with P' to rotate P 60° clockwise.

    (c) $105^\circ = 45^\circ + 60^\circ$. Construct the 45° angle using the steps from part (a). Construct the 60° using the steps from part (b) but start using P' created in part (a) instead of P. Label P' from part (b) with P''. The angle $\angle POP''$ is a 105° angle.

14. Signs of the *x* and *y* coordinates will be reversed under a half-turn about the origin.

(a) $(4, 0) \rightarrow$ $\mathbf{(^-4, 0)}$.

(b) $(2, 4) \rightarrow$ $\mathbf{(^-2, ^-4)}$.

(c) $(^-2, ^-4) \rightarrow$ $\mathbf{(2, 4)}$.

(d) $(a, b) \rightarrow$ $\mathbf{(^-a, ^-b)}$.

15. **He ends up at (2,0).**

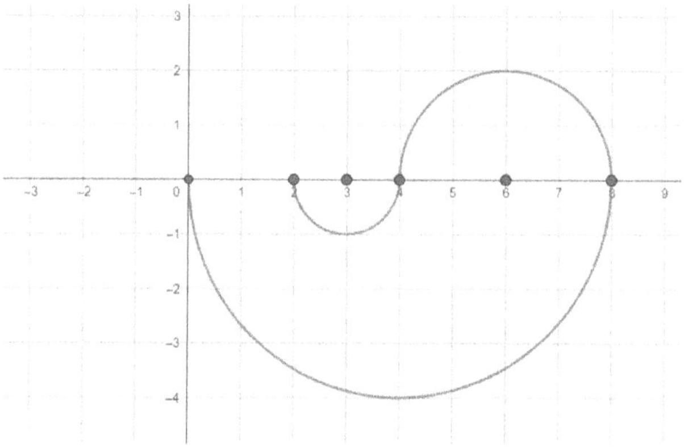

16. The radius of the semicircle with diameter $\overline{BB'}$ is equal to the diagonal $\overline{BD}$ of the piece of chalk. Since the sides of the stamp are all 1 in., we can find the length of the radius using Pythagorean Theorem:

$1^2 + 1^2 = r^2 \Rightarrow 2 = r^2 \Rightarrow r = \sqrt{2}$ in.

The area of the semicircle is half of the area of a full circle of radius

$\sqrt{2}$ is $\frac{1}{2}\pi\left(\sqrt{2}\right)^2 = \frac{1}{2}2\pi = \pi \text{ in}^2$.

The total area is the area of the semicircle plus the areas of the two half-squares. Each square has an area of 1 in^2, so the two halves together have an area of 1 in^2. The total area is $(\pi + 1)\text{in}^2$.

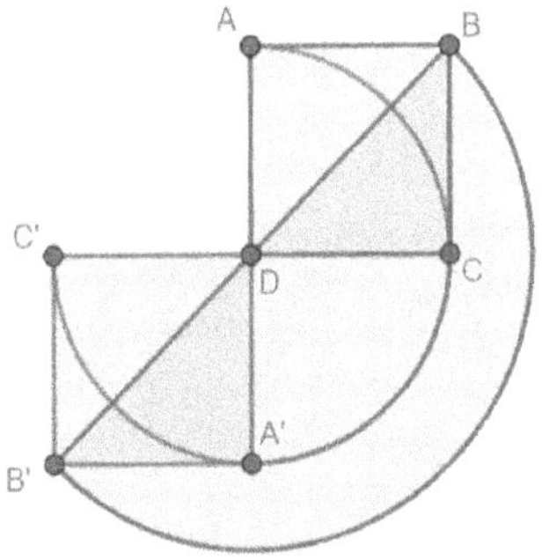

17. **(a)** First rotate ΔABC by angle α to $\Delta A'B'C'$. Then rotate $\Delta A'B'C'$ by angle β to obtain $\Delta A''B''C''$.

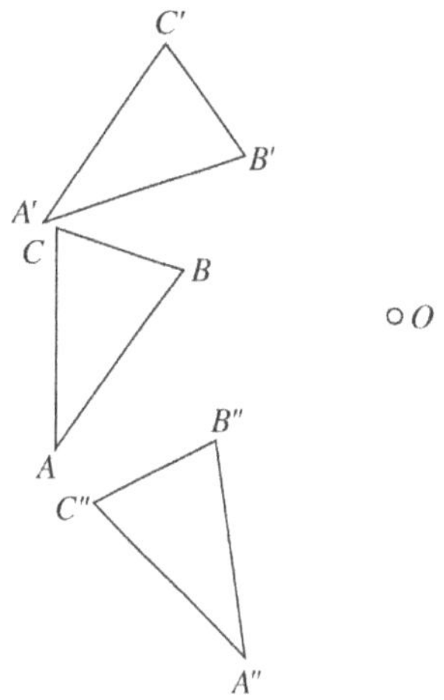

(b) **No**. Either rotation first would produce the same image.

(c) **Yes**. Rotate around *O* by angle $|\alpha - \beta|$ in the direction of the larger of α and β.

18. **(a)** The following figures show 90° counterclockwise rotation about the origin.

(i) $m_\ell = \frac{3}{2} \Rightarrow m_\perp = \frac{2}{^-3}$ because

$(2, 3) \rightarrow$ $\mathbf{(^-3, 2)}$.

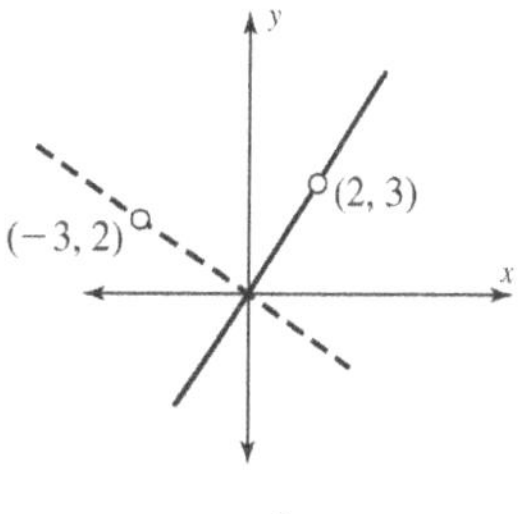

(ii) $m_\ell = \frac{2}{^-1} \Rightarrow m_\perp = \frac{^-1}{^-2}$ because

$(^-1, 2) \rightarrow$ $\mathbf{(^-2, ^-1)}$.

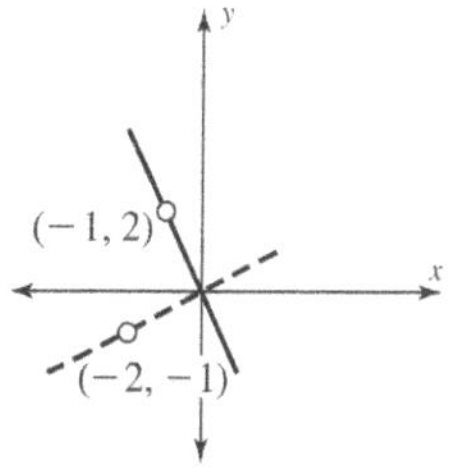

(***iii***) Under 90° counterclockwise rotation about the origin, $(m,n)\rightarrow(^{-}\boldsymbol{n},\boldsymbol{m})$ or $m_{\perp} = \frac{m}{^{-}n}$.

(b) In problem 18(a)(*iii*) the point (a, b) moves to $(^{-}b, a)$. Repeating this 90° counterclockwise rotation would produce a full half-turn and the point would end at $(^{-}a, ^{-}b)$.

(c) $m_{\ell} = \frac{b}{a} \Rightarrow m_{\perp} = -\frac{^{-}a}{b}$ because $(a,b) \rightarrow$ $(\boldsymbol{b}, ^{-}\boldsymbol{a})$.

(***i***) (a, b) rotated 90° counterclockwise:

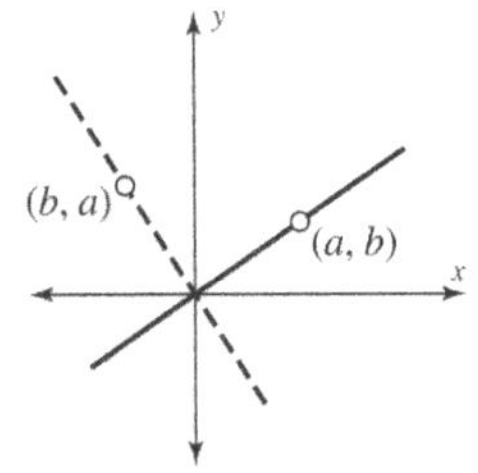

(***ii***) $(^{-}b, a)$ under a half-turn about the origin:

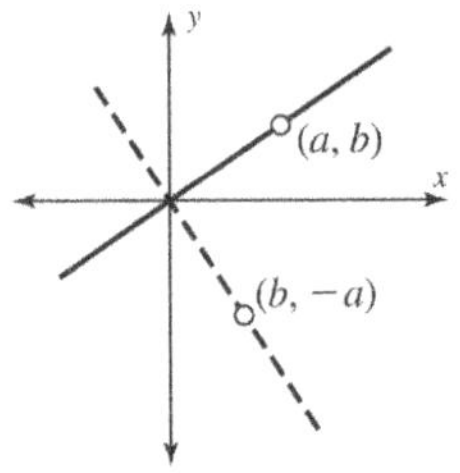

Or $(a,b) \rightarrow (\boldsymbol{b}, ^{-}\boldsymbol{a})$.

19. **(a)** Rotation by 90° or a multiple of 90° about the point of intersection of the diagonals.

(b) Rotation by 72° or a multiple of 72° about O, where O is the point of intersection of the perpendicular bisectors of the sides.

(c) Rotation by $\frac{360^{\circ}}{n}$ or a multiple of $\frac{360^{\circ}}{n}$ about O, where O is the point of intersection about the center by any turn angle.

(d) Infinite turn symmetries.

(e) No rotational symmetries.

20. **(a)** Half-turn symmetry about the point of intersection of the diagonals. A rectangle has half-turn symmetry because the diagonals bisect each other.

(b) Half-turn symmetry about the point of intersection of the diagonals. A parallelogram has half-turn symmetry because the diagonals bisect each other

(c) No half-turn symmetry.

(d) No half-turn symmetry.

21. **(a)** Find the image of the center O by connecting point O and connecting P with a line using a straight edge. Put your compass legs on O and P and using the same compass opening leave one leg on point P and mark point O'. Construct the circle with point O' as the center and the same radius as $\overline{OP}$.

(b) The circles have a single point P in common and they have the same radius $\overline{OP} = \overline{PO'}$.

22. Construct a point O and with any compass opening construct a circle using the point O as its center. Put the legs of the compass on the center O and on any point A on the circle. Using the same compass opening leave one of the legs on A and create another point B counterclockwise from point A on the circle. This is the chord $\overline{AB}$. Then construct A' by using a straight edge and including the center O and point A. The intersection of the straight line and the circle is a half turn of point A with turn center O. Then construct B' by using a straight edge and including the center O and point B. The intersection of the straight line and the circle is a half turn of point B with turn center O. To construct O' put your compass legs on point A and point B. Using the same compass opening leave one leg on point A and mark point O' on the circle clockwise from point A. Using the same compass opening leave one leg on point B and mark point P on the circle counterclockwise from point B. $ABPA'B'O'$ is a regular hexagon.

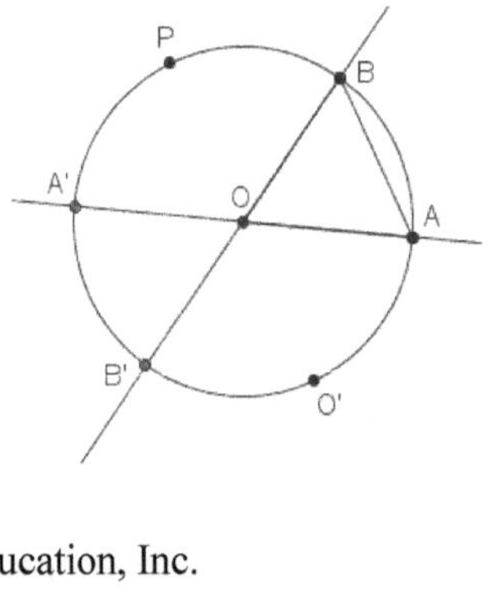

23. Forming rectangles will tessellate the plane.

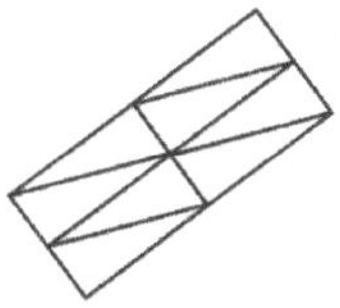

24. **(a)** Perform half-turns about the midpoints of all sides.

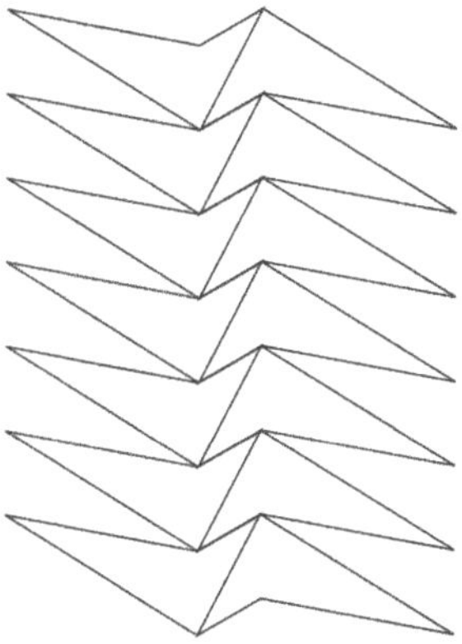

(b) **Yes**. If a polygon tessellates the plane, the sum of the angles around every vertex must be 360°. Successive 180° turns of a quadrilateral about the midpoint of its sides will produce four congruent quadrilaterals around a common vertex, with each of the quadrilaterals angles being represented at each vertex.

These angles must add to 360°, as do angles of any quadrilateral.

25. Experimentation by cutting shapes out and moving them about is one way to learn about these tessellations.

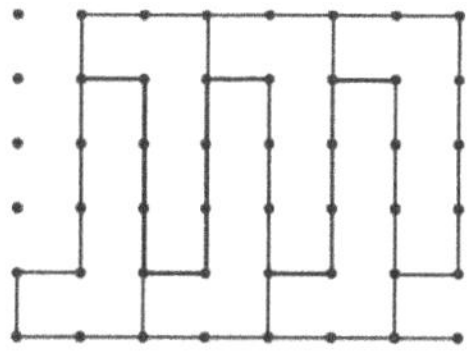

26. The shape will tessellate a plane as shown below, provided that it has a symmetry line as shown below:

27. Pictures may vary; e.g.:

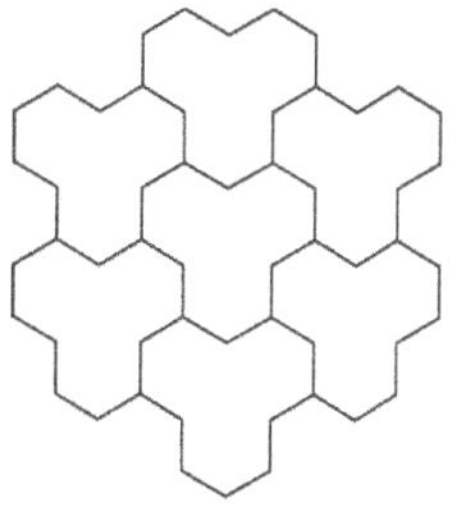

28. The hexagon will tessellate the plane as follows:

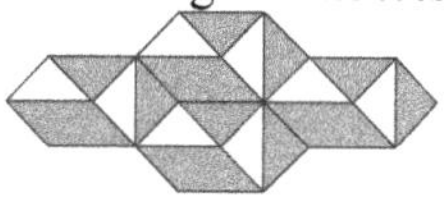

Assessment 14-2A: Reflections and Glide Reflections

1. Locate the image of vertices directly across (perpendicular to) ℓ on the geoboard.

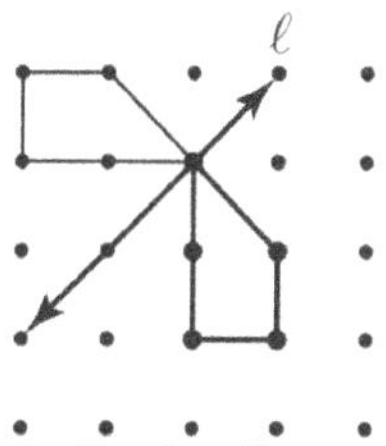

2. Reflecting lines are described for each.

(a) **All diameters**. There are infinitely many.

(b) **Perpendicular bisector** of the segment. The line the segment is contained is a trivial line of reflection

(c) Any **line perpendicular** to the given line or the line itself.

(d) **Perpendicular bisectors** of the sides and lines containing the diagonals.

(e) **None**.

(f) **Perpendicular bisectors** of each of the sides.

(g) **None**.

(h) The **line containing the diagonal** determined by vertices of the noncongruent angles.

(i) **Perpendicular bisectors** of parallel sides and three diagonals determined by vertices on the circumscribed circles; or the **angle bisectors** of the vertices.

3. The original figure is reflected back upon itself. I.e., if the first reflection yields $\Delta A'B'C'$, the second reflection brings $\Delta A'B'C'$ back to ΔABC.

4. Find the image of the center of the circle and one point on the circumference of the circle to determine the image of the circle. One possibility is:

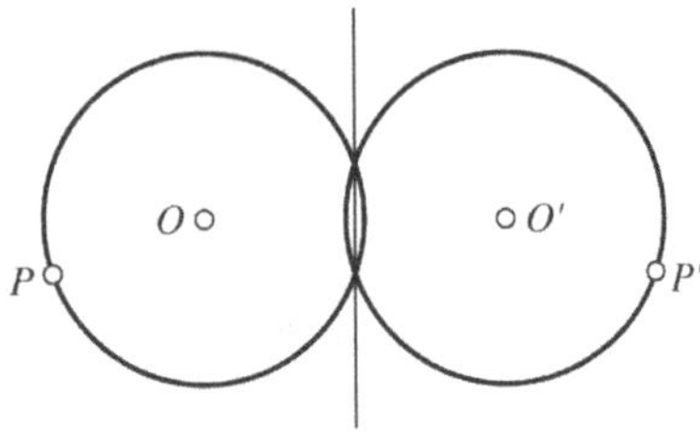

5. (a) **No**. The final images are congruent but in different locations, thus they are not the same.

   (b) A translation determined by a slide arrow from P to R. Let P be any point on ℓ and Q on m such that $\overrightarrow{PQ} \perp \ell$. Point R is on $\overrightarrow{PQ}$ such that $PQ = QR$.

6. (a) Translation double the distance from line j to line k.

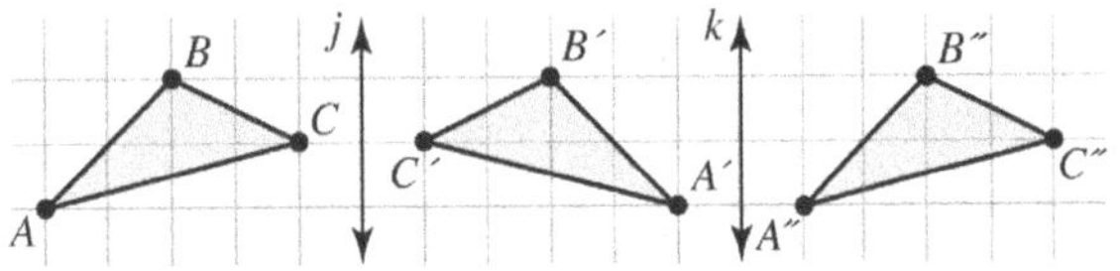

   (b) **Reflection across line *k*.**

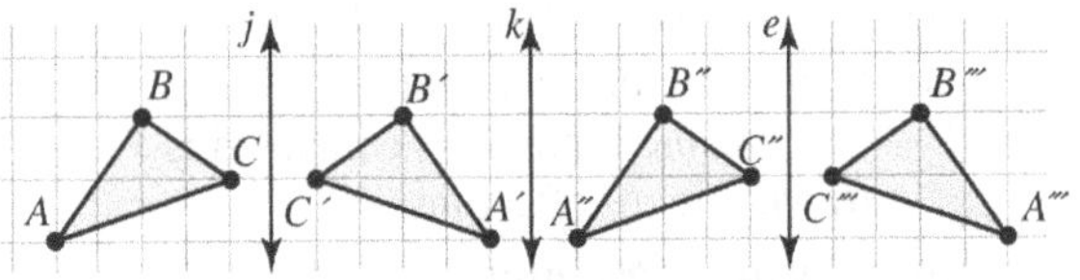

7. (a) Single 120 degree rotation with center O.

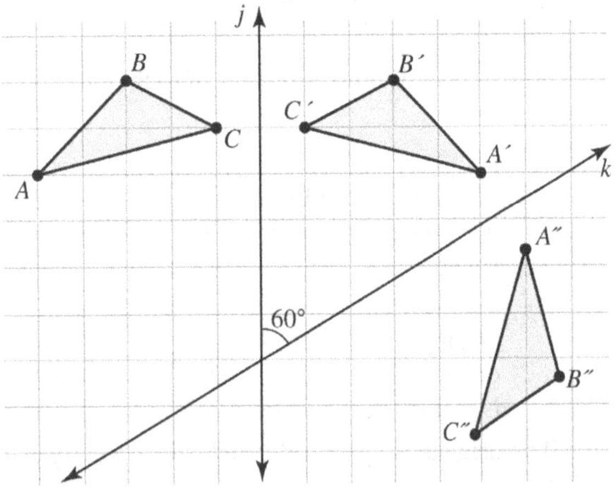

   (b) Single 90 degree rotation with center O.

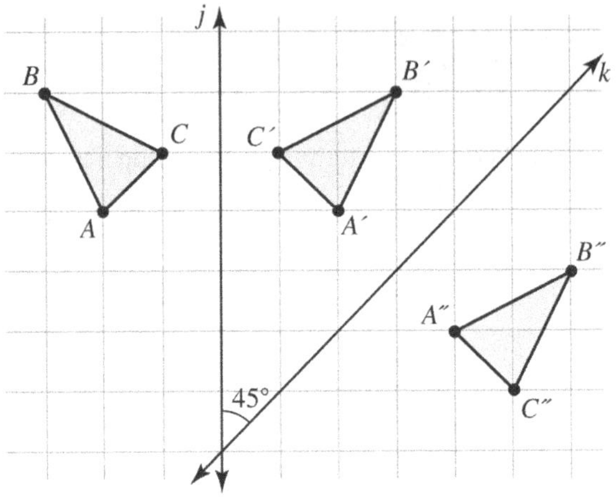

8. The line of reflection is the **perpendicular bisector** of $\overline{AA'}$, $\overline{BB'}$, or $\overline{CC'}$.

9. (a) Examples may vary, but include MOM, WOW, TOOT, HAH, etc..

   (b) (*i*) Examples include BOX, HIKE, CODE, OBOE, etc..

   (*ii*) The letters B, C, D, E, H, I, K (depending on construction), O, and X may be used.

   (c) Others of these numbers include combinations of 1, 8, and 0. Examples are 1, 8, 11, 88, 101, 111, 181, 808, 818, 888, 1001, or 1111.

10. (a) For glide reflections with the translation parallel to the reflection line, the **images are the same** regardless of order of operation.

    (b) **Commutative**, for glide reflections with the translation parallel to the reflection line.

11. None of the images has a reverse orientation, so there are no reflections or glide reflections involved. Thus

    1 to 2 can be viewed as a counterclockwise rotation;

    1 to 3 can be viewed as a clockwise rotation;

    1 to 4 is a translation down;

    1 to 5 is a clockwise rotation followed by a translation down or a rotation about an exterior point;

    1 to 6 is a translation or a clockwise rotation; and

    1 to 7 is a translation or a clockwise rotation.

12. **(a)** When reflecting in the x-axis, $P(x, y) \to P'(x, {}^-y)$. Thus $A(3, 4) \to \mathbf{A'(3, {}^-4)}$; $B(2, {}^-6) \to \mathbf{B'(2, 6)}$; and $C({}^-2, 5), \to \mathbf{C'({}^-2, {}^-5)}$.

(b) When reflecting in the line $y = x$, $P(x, y) \to P'(y, x)$. Thus $A(3, 4) \to \mathbf{A'(4, 3)}$; $B(2, {}^-6) \to \mathbf{B'({}^-6, 2)}$; and $C({}^-2, 5), \to \mathbf{C'(5, {}^-2)}$.

13. **(a)** $P(x, y) \to \mathbf{P'(x, {}^-y)}$ under reflection in the x-axis; $P(x, y) \to \mathbf{P'(y, x)}$ under reflection about $y = x$;

(b) $\mathbf{({}^-x, {}^-y)}$. From part (a) the reason can be seen as below:

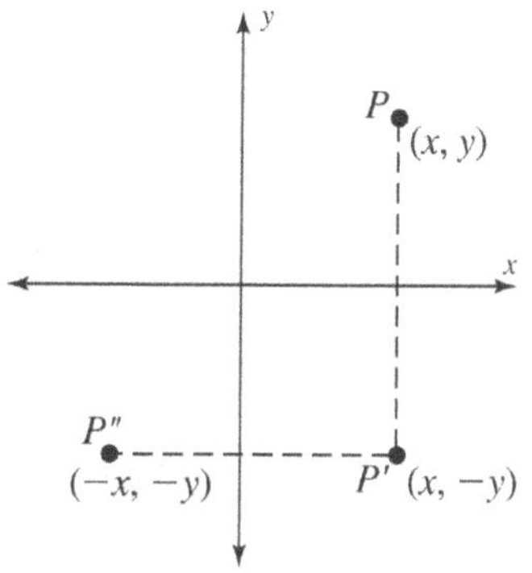

14. $P(x, y) \to P(x, {}^-y)$ when reflected in the x-axis. Thus:

(a) $y = {}^-x + 3 \to {}^-y = {}^-x + 3 \Rightarrow$ $\mathbf{y = x - 3}$.

(b) $y = 0 \to {}^-y = 0 \to \mathbf{y = 0}$.

15. $P(x, y) \to P({}^-x, y)$ when reflected in the y-axis. Thus:

(a) $y = {}^-x + 3 \to y = {}^-({}^-x) + 3 \Rightarrow$ $\mathbf{y = x + 3}$.

(b) $y = 0 \to \mathbf{y = 0}$.

16. **(a)** Given the circles below:

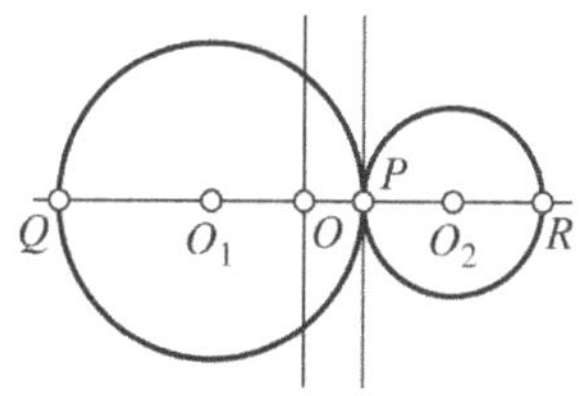

(*i*) The line through P perpendicular to $\overleftrightarrow{O_1O_2}$ will cause the image to be an interior circle. The smaller circle will be tangent to point P.

(*ii*) The line through O, the perpendicular bisector of $\overline{QR}$, will cause the image to be an interior circle. The smaller circle will be tangent to point Q.

(b) $\overleftrightarrow{\mathbf{O_1O_2}}$

17. Given the circles below:

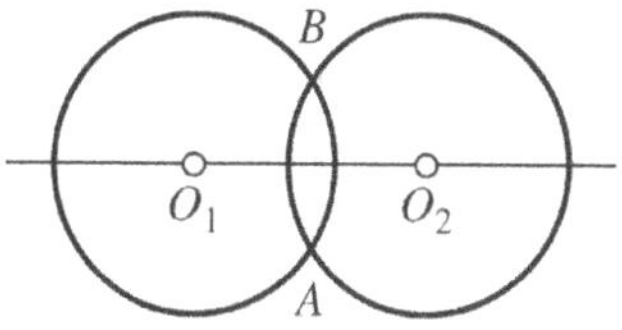

(a) The line $\overleftrightarrow{\mathbf{AB}}$, since the circles are congruent.

(b) **Yes**, a translation taking O_1 to O_2.

18. **(a)**

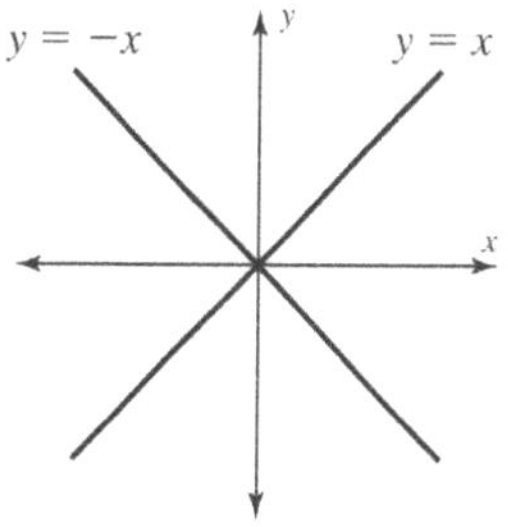

Any point (a, b) on the line $y = x$ will reflect across the vertical axis to the point $({}^-a, b)$ on the line $y = {}^-x$. Any point (a, b) on the line $y = x$ will reflect across the horizontal axis to the point $(a, {}^-b)$ on the line $y = {}^-x$. Thus the lines of reflection are the axes, $\mathbf{x = 0}$ and $\mathbf{y = 0}$.

(b)

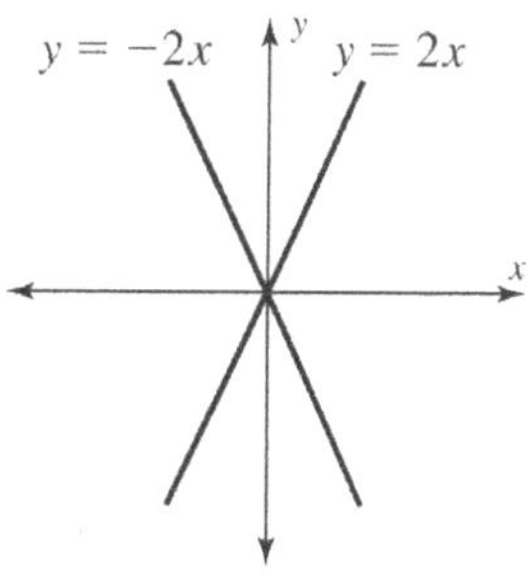

Any point (a, b) on the line $y = 2x$ will reflect across the vertical axis to the point

$(^-a, b)$ on the line $y = {}^-2x$. Any point (a, b) on the line $y = 2x$ will reflect across the horizontal axis to the point $(a, {}^-b)$ on the line $y = {}^-2x$. Thus the lines of reflection are the axes, $x = 0$ and $y = 0$.

19. **(a)** If point P was determined by a glide reflection by the slide arrow $\overrightarrow{AB}$ and line l parallel to $\overleftrightarrow{AB}$, then we can start with P' and work backwards: reflect P' across line l and translate by the slide arrow $\overrightarrow{BA}$ to P.

(b) Construct $\triangle DEF$ Connect A to D, E and F. Connect B to D, E and F. Construct a line parallel to AB through point D. Find point D' such that $AB=DD'$ and $\overrightarrow{DD'}$ is in the same direction as $\overrightarrow{AB}$. Repeat for points E' and F'. Connect D', E', and F' to create $\triangle D'E'F'$ Repeat this procedure to construct $\triangle D''E''F''$ starting from $\triangle D'E'F'$.

A single translation by slide arrow $\overrightarrow{AC}$, where A, B, and C are collinear and B is the midpoint of $\overline{AC}$ takes $\triangle PQR$ to $\triangle P''Q''R''$.

20. **(a)** Any point moves right 2 and reflects across the x-axis. Point (x, y) moves to

$(x+2, {}^-y)$.

(b) If $(x+2, {}^-y) = (3,5)$ then

$x+2=3 \Rightarrow x=1$ and ${}^-y=5 \Rightarrow y={}^-5$ so the coordinates of P are .

Review Problems

17. **(a)** **Yes**

(b) **Yes**

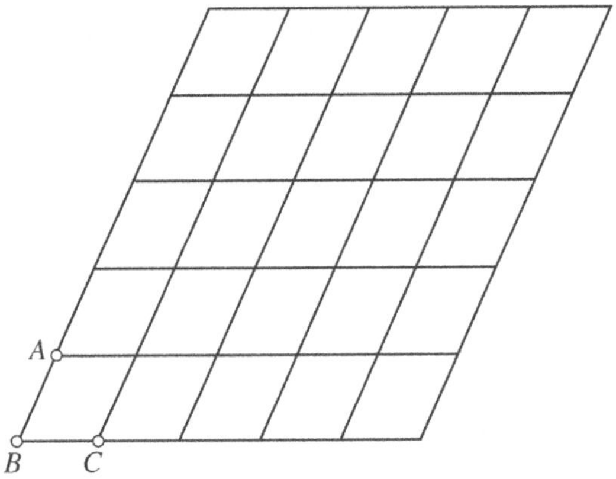

18. **Answers vary. For example:**

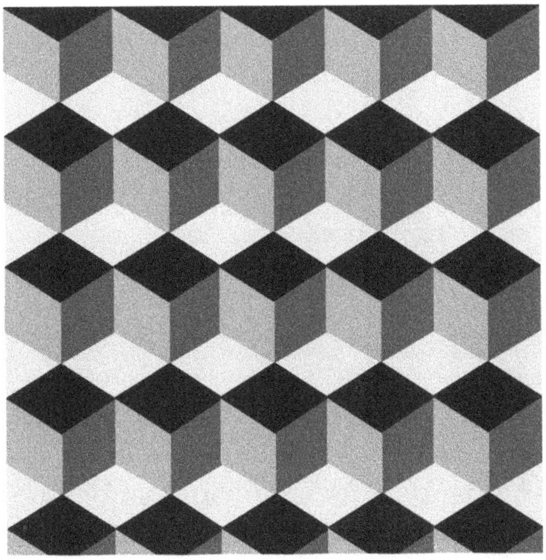

19. **0, 1, and 8,** depending upon how they are drawn. **5 and 2,** if drawn as a digital clock would show them.

20. $(^-a, {}^-b)$. See below, where, for example, (a, b) is in the first quadrant. The relationship is the same regardless of the quadrant in which (a, b) is located.

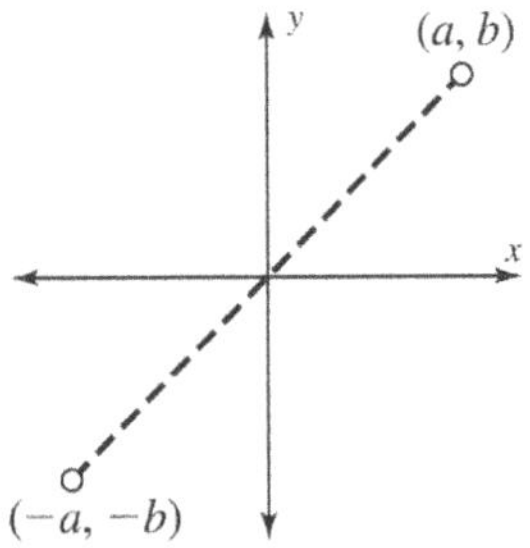

21. A **rotation** of any angle about the center of the circle will transform the circle into itself.

22. Construct $\overline{BE}$ perpendicular to $\overline{AD}$ as shown below:

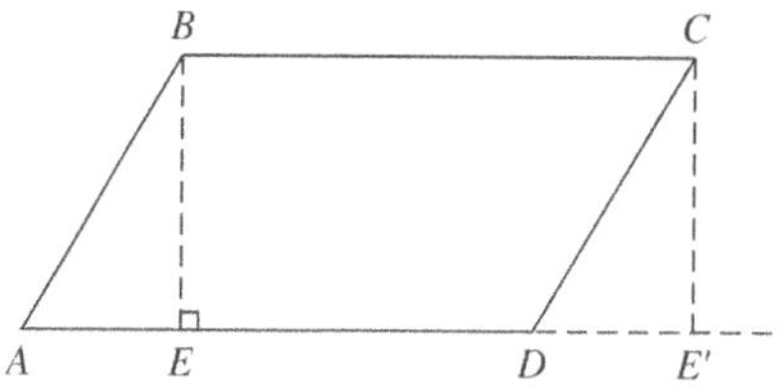

Translate ΔABE by the slide arrow from B to C. The image of ΔABE is $\Delta DCE'$, thus the rectangle $BCE'E$ is the required rectangle.

Assessment 14-3A: Dilations

1. **(a)** Slide the small triangle down three units (translation). Then complete the dilation with scale factor 2 (i.e., the larger triangle has sides twice as long) using the top right vertex as center.

(b) Slide the smaller triangle right 5 and up 1. Then complete the size transformation with scale factor 2 using the top right vertex as center.

2. Vertices A', B', and C' will be half the distance from O as vertices A, B, and C, respectively.

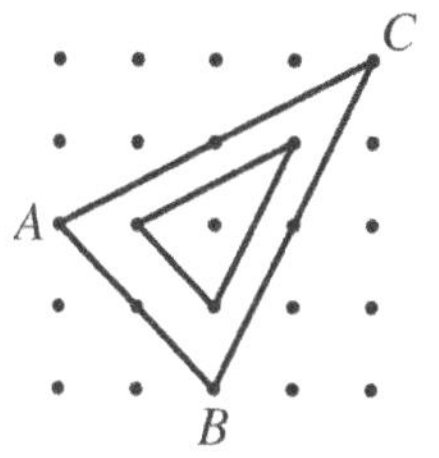

3. Answers vary.

(a) Translate B to B' followed by a size transformation with center B' and scale factor $\frac{A'B'}{AB}$.

(b) Rotate a half-turn with the midpoint of $\overline{AA'}$ as the center followed by a size transformation with center A'.

4. **(*i*)** Scale factor of $\frac{4}{10} = \mathbf{\frac{2}{5}}$.

(*ii*) $x = \frac{2}{5}(15) = \mathbf{6}$. $y = \frac{2}{5}(13) = \frac{26}{5} = \mathbf{5.2}$.

5. $\frac{3 \text{ cm}}{10 \text{ cm}} = \frac{BA \text{ cm}}{40 \text{ cm}} \Rightarrow 10 \cdot BA = 120 \Rightarrow BA =$ the height of the candle $=$ **12 cm**.

6. Given $A(2, 3)$ and $B(^{-}2, 3)$:

(a) $A' = (2 \cdot 3 \cdot 2, 3 \cdot 3 \cdot 2) = \mathbf{(12, 18)}$.

$B' = (^{-}2 \cdot 3 \cdot 2, 3 \cdot 3 \cdot 2) = \mathbf{(^{-}12, 18)}$.

(b) **Same as in (a)**. Order of size transformation is irrelevant.

7. The dilation with center O and scale factor $\frac{1}{r}$ (equivalent to using division to reverse multiplication).

8. **(a)** Answers vary. One explanation is shown by the figure below:

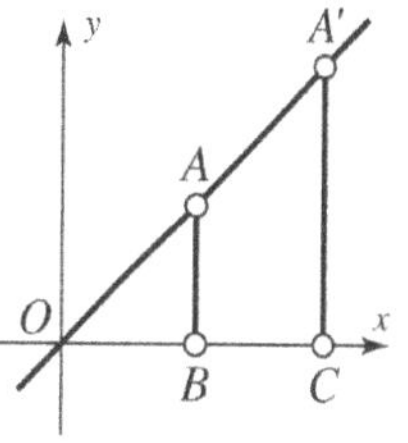

Let $A'(x', y')$ be the image of A under the size transformation. The definition of a size transformation yields $\frac{OA'}{OA} = r$. Since $\Delta OA'C \sim \Delta OAB$, $\frac{OC}{OB} = r$, which implies that $\frac{x'}{x} = r$ and thus $x' = rx$. Similarly, $y' = ry$.

(b) The dilation with center at the origin and scale factor 2, followed by a half turn on the origin.

9. A translation taking O_1 **to** O_2 followed by a size transformation with **center at** O_2 and **scale factor** $\mathbf{\frac{3}{2}}$.

10. The set of images of the set of integers on a number line would be the set of points with coordinates multiples of 3; i.e., **{..., ⁻6, ⁻3, 0, 3, 6,...}**.

11. **(*i*)** Scale factor of $\mathbf{\frac{7}{15}}$.

(*ii*) $x = \frac{7}{15}(14) = \mathbf{\frac{98}{15}}$. $y = 6 \div \frac{7}{15} = \mathbf{\frac{90}{7}}$.

12. **Yes**. The enlargement can be achieved by a size transformation, but there are many potential centers; e.g., one might be the lower left corner of the 2 inch $\times$ 3 inch rectangle.

13. Answers vary. For example:

(a)

Length	Width
1	20
2	10
4	5
5	4
10	2

(b)

Length	Width
2	40
4	20
8	10
10	8
20	4

(c)

Area (a)	Area (b)
20 in^2	80 in^2
20 in^2	80 in^2
20 in^2	80 in^2
20 in^2	80 in^2
20 in^2	80 in^2

(c) The enlargement factor is 4.

(d) $2a \times 2b = 4ab$.

Review Problems

15. (a) The translation given by slide arrow from *N* to *M.*

(b) A counterclockwise rotation of 75° about *O*.

(c) A clockwise rotation of 45° about *A*.

(d) A reflection in *m* and translation from *B* to *A*.

(e) A second reflection in line *n*.

16. The *x*-coordinate is unchanged when reflecting about *m*; the *y*-coordinate is unchanged when reflecting about *n*.

(a) (4, 3) reflects about *m* to (4, 1), then (4, 1) reflects about *n* to (2, 1).

(b) $(0, 1) \to (0, 3) \to (6, 3)$.

(c) $(^-1, 0) \to (^-1, 4) \to (7, 4)$.

(d) $(0, 0) \to (0, 4) \to (6, 4)$.

17. Reflect each part of the figure across the reflecting line:

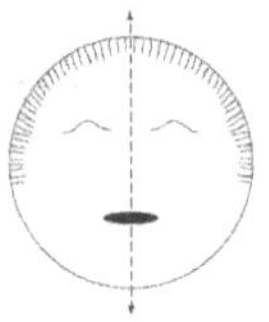

Chapter 14 Review

1. (a) Translate three units down and three units left:

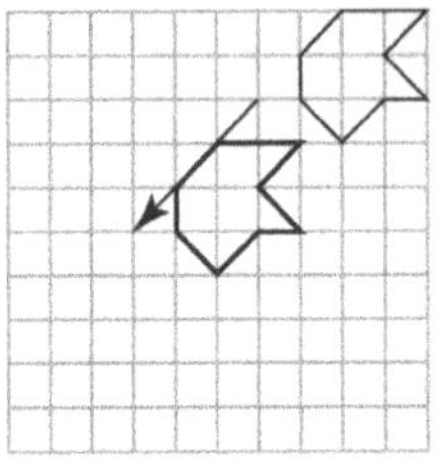

(b) Rotate 90° clockwise with *O* as a center:

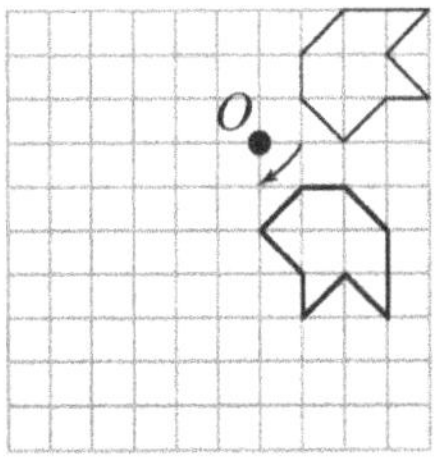

(c) Reflect across ℓ:

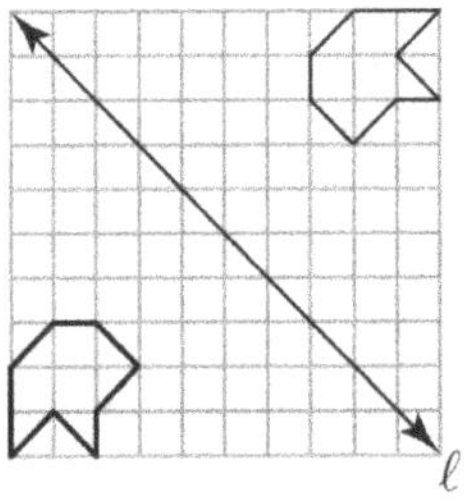

2. In each part find the images of the vertices:

(a) Reflect about ℓ:

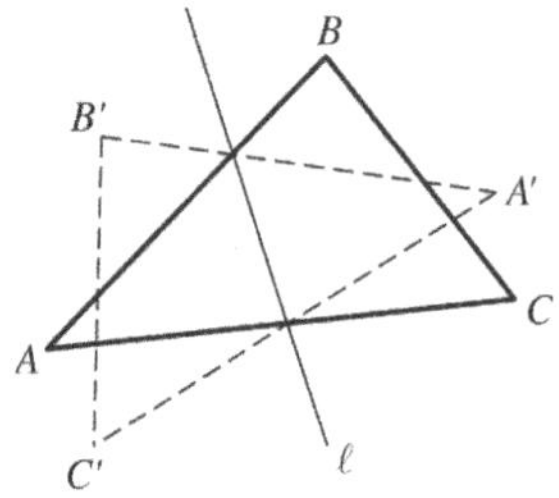

(b) Rotate counterclockwise in *O*:

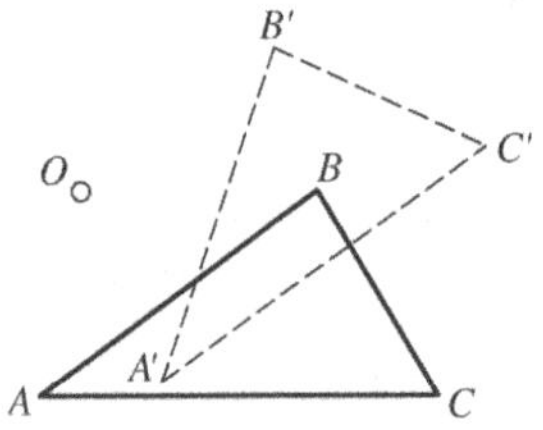

(c) Translate through the arrow pictured:

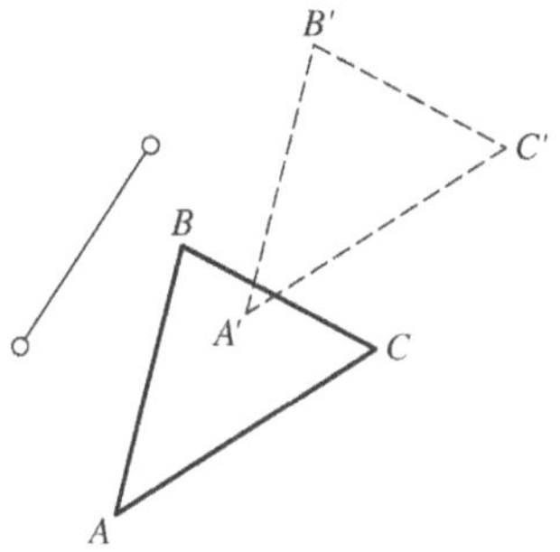

3. **(a)** **Four reflections**, two diagonals, one horizontal, and one vertical, as shown below. Rotations of $90°$, $180°$, and $270°$ also work.

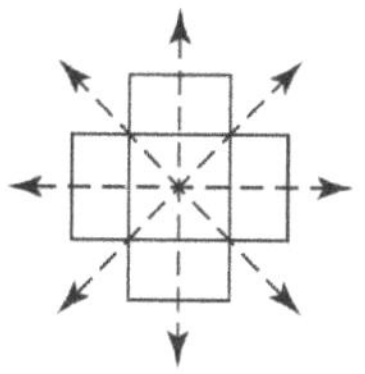

(b) **One**, the diameter bisecting the central angle.

(c) **One**, the bisector of the point angle.

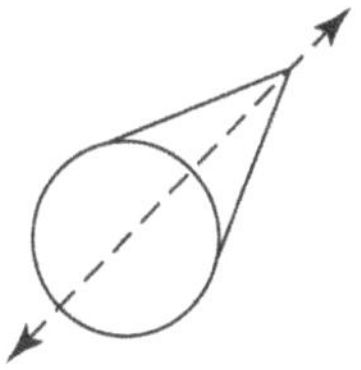

(d) **One:** a half turn through center 0.

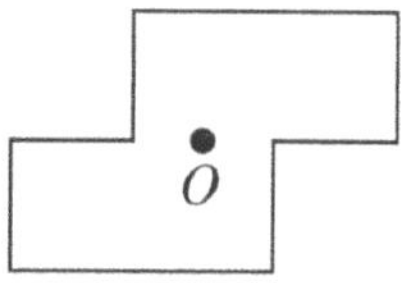

(e) **Three**: A horizontal reflection through the points of intersection, a vertical reflection through the centers, and a half turn through center 0.

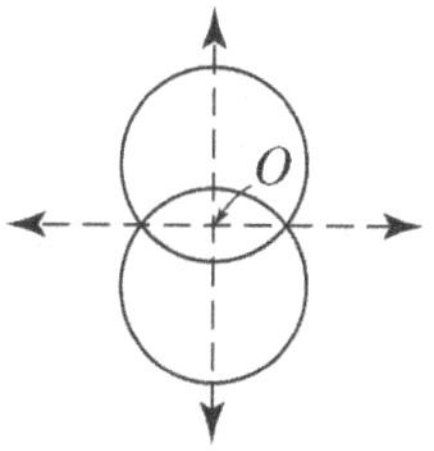

(f) **Three**: Vertically and horizontally reflections through the center and a half turn through the center.

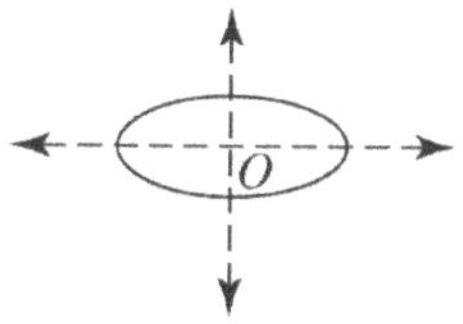

4. $A = A'$, B is the midpoint of $\overline{A'B'}$, and C is the midpoint of $\overline{A'C'}$.

5. **(a)** A **half-turn** about x.

(b) A **half-turn** about x.

6. **(a)** **Rotation by 120°** clockwise about the center of the hexagon.

(b) A **reflection** in the perpendicular bisector of $\overline{BY}$ or a **rotation by 60°** counterclockwise around the center and then a reflection about the perpendicular bisector of $\overline{B'Y'}$.

7. A **reflection** in $\overleftrightarrow{SO}$.

8. Let $\Delta H'O'R'$ be the image of ΔHOR under a half-turn about R. Then ΔSER is the image of $\Delta H'O'R'$ under a size transformation with center R and scale factor $\frac{2}{3}$. Thus ΔSER is the image of ΔHOR under a half-turn about R followed by a $\frac{2}{3}$ size transformation.

9. Rotate ΔPIG $180°$ (i.e., a half-turn) about the mid-point of $\overline{PT}$, then perform a size transformation with scale factor 2 and center $P'(= T)$.

10. **(a)** In each case, $(x', y') \rightarrow (x' - 3, y' + 5)$.

$A'(0, 7.91) \rightarrow (0 - 3, 7.91 + 5) =$ $\mathbf{A(^{-}3, 12.91)}$.

$B'(^{-}5, {}^{-}4.93) \rightarrow (^{-}5 - 3, {}^{-}4.93 + 5) =$ $\mathbf{B(^{-}8, 0.07)}$.

$C'(4.83, 0) \rightarrow (4.83 - 3, 0 + 5) =$ $\mathbf{C(1.83, 5)}$.

(b) Under the translation $\mathbf{(x', y') \rightarrow (x' - 3, y' + 5)}$.

11. $(x, y) \rightarrow (x, y)$. This is an "identity" translation; i.e., $(x - 3, y + 2)$ reverses $(x + 3, y - 2)$ so that every point is its own image.

12. **(a)** The translation from $\Delta A''B''C''$ to ΔABC would be under the translation $(x, y) \rightarrow (x - 1, y - 3)$ followed by the translation $(x, y) \rightarrow (x - 2, y + 1)$.

$A = (0 - 1 - 2, 0 - 3 + 1) = (^-3, {}^-2)$.

$B = (1 - 1 - 2, 5 - 3 + 1) = (^-2, 3)$.

$C = (^-2 - 1 - 2, 7 - 3 + 1) = (^-5, 5)$.

(b) $(x, y) \rightarrow (x + 2 + 1, y - 1 + 3) =$ $(x + 3, y + 2)$.

13. **(a)** *(i)* A translation from A **to** C.

(ii) **Same** as *(i)*. Translations with the same center are commutative.

(b) A rotation about O by $90° - 30° = 60°$ **counterclockwise**.

(c) *(i)* A size transformations with center O and scale factor $3 \cdot 2 = 6$.

(ii) **Same** as *(i)*. Size transformations are commutative as long as the center does not change.

14. As the path is traced, the lengths of $XY + YZ$ is the same as the length AC. We know this because of the $60°$ angles formed as the ball bounces at Y making $\angle BYZ$ have a measure of $60°$. Thus $AXYZ$ is a parallelogram making $\overline{AX} \cong \overline{ZY}$ and $\overline{XY} \cong \overline{AZ}$. Thus $XY + YZ = AX + AZ$. Also $\triangle BYZ$ is equilateral so $\overline{BZ} \cong \overline{ZY}$. Thus $XY + YZ = AZ + ZB$. Hence $XY + YZ = 1m$. Similarly $Zw + WP = BC = 1m$, and $PQ + QX = AC = 1m$. Thus the length of the path is 3(1 m) = 3 m. The perimeter of ΔABC is also 3 m, so the ball travels the length equal to the perimeter.

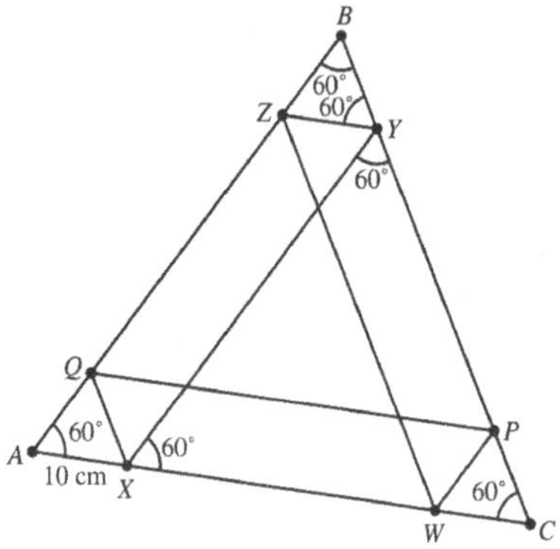

15 **(a)** If $(x, y) \rightarrow (x + 2, y - 3)$ then each point on the line is translated. Thus $(0, 3) \rightarrow (2, 0)$ and $(3, 0) \rightarrow (5, {}^-3)$.

$m = \frac{^-3-0}{5-2} = {}^-1$ and $0 = {}^-1(2) + b \Rightarrow$ $b = 2$. The equation of the image is $y = {}^-x + 2$.

(b) $P(x, y) \rightarrow P(x, {}^-y)$ under reflection in the x-axis. Thus $P(0, 3) \rightarrow P'(0, {}^-3)$ and $P(3, 0) \rightarrow$ $P'(3, 0)$.

The equation of the reflected line is $y = x - 3$.

(c) $P(x, y) \rightarrow P'(^-x, y)$ under reflection in the y-axis. Thus $P(0, 3) \rightarrow P'(0, 3)$ and $P(3, 0) \rightarrow P'(^-3, 0)$.

The equation of the reflected line is $y = x + 3$.

(d) $P(x, y) \rightarrow P'(y, x)$ under reflection in the line $y = x$. Thus $P(0, 3) \rightarrow P'(3, 0)$ and $P(3, 0) \rightarrow P'(0, 3)$.

The equation of the reflected line is $y = {}^-x + 3$.

(e) $P(x, y) \rightarrow P'(^-x, {}^-y)$ under a half-turn about the origin. Thus $P(0, 3) \rightarrow P'(0, {}^-3)$ and $P(3, 0) \rightarrow P'(^-3, 0)$.

The equation of the reflected line is $y = {}^-x - 3$.

(f) $P(x, y) \rightarrow P(2x, 2y)$ under a scale factor of 2. Thus $P(0, 3) \rightarrow P'(0, 6)$ and $P(3, 0) \rightarrow P'(6, 0)$.

The equation of the scaled line is $\mathbf{y = {}^{-}x + 6}$.

16. **(a)** The translation must be reversed; $(x, y) \rightarrow \mathbf{(x - 3, y + 5)}$.

(b) This becomes the identity translation: $\mathbf{(x, y) \rightarrow (x, y)}$.

17. The measure of each exterior angle of a regular octagon is $\frac{360^\circ}{8} = 45^\circ$. Thus the measure of each interior angle is $180^\circ - 45^\circ = 135^\circ$, and $135^\circ \nmid 360^\circ$. Therefore a regular octagon does not tessellate the plane.

18. **(a)**

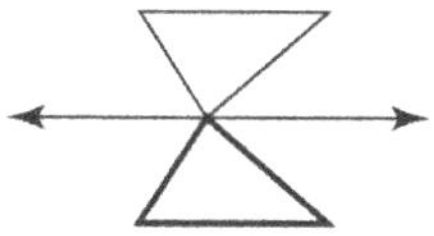

(b)

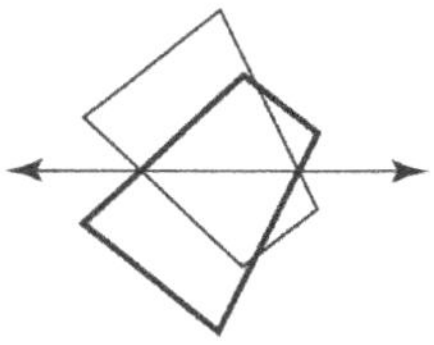

(c)

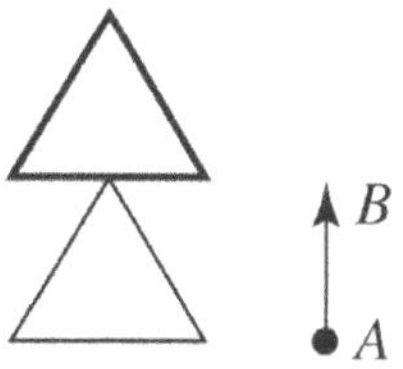

19. **(a)** **Yes**, it will tessellate. Below is a start.

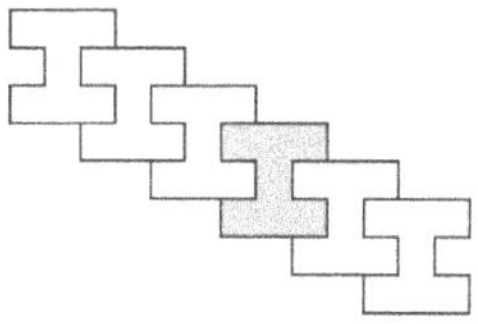

(b) **Yes**, it will tessellate. Below is a start.

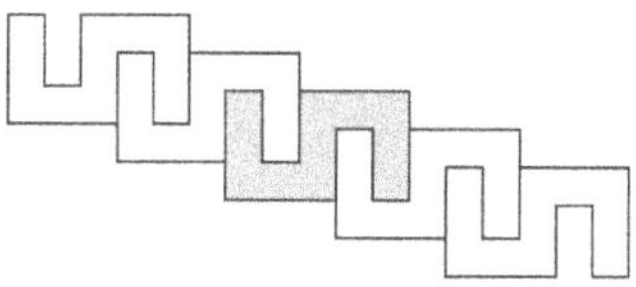

(c) **No**, it will not tessellate.

20. For parts (i) and (ii) assume OA forms one side of possible rhombi—not a diagonal For part (iii) assume OA forms the diagonal.

(a) **(*i*)** $OA = \sqrt{(3 - 0)^2 + (4 - 0)^2} = 5.$ Thus, possible coordinates for point C would be $(0 - 5, 0) = \mathbf{({}^{-}5, 0)}$.

(*ii*) OA, or the length of each side, is 5. Thus $5 = \sqrt{(a - 3)^2 + (4 - 0)^2} = \sqrt{a^2 - 6a + 9 + 16} = \sqrt{a^2 - 6a + 25}.$

Squaring each side of the equation (i.e., if $a = b$, then $a^2 = b^2$): $25 = a^2 - 6a + 25$

$\Rightarrow a^2 - 6a = 0 \Rightarrow a(a - 6) = 0$. So $a = 0$ or a=6. If $a = 0$, the C=O, a contradiction. So another possible coordinates for point C is $\mathbf{(6, 0)}$.

(iii) If OA is the diagonal, then the sides form by OC and CA must be of equal length.

$\sqrt{(a-0)^2 + (0-0)^2} = \sqrt{(3-a)^2 + (4-0)^2} \Rightarrow$

$a^2 = (3-a)^2 + 4 \Rightarrow a = \frac{25}{6}$. Thus, a possible value for C would be $\mathbf{(\frac{25}{6}, 0)}$.

(b) **(*i*)** The possible values for point B would be $(3 - 5, 4) = \mathbf{({}^{-}2, 4)}$. See the following figure:

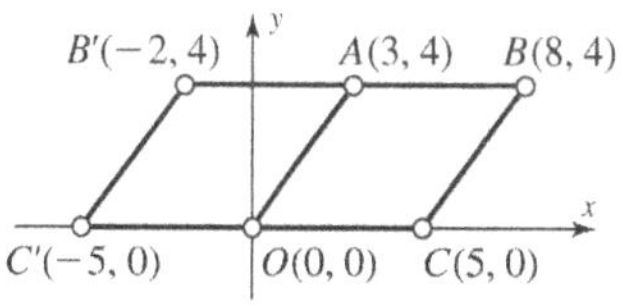

(*ii*) If $O = (0, 0)$, $A = (3, 4)$, and $C = (6, 0)$ then another possible value for point B is $\mathbf{(3, {}^{-}4)}$. See the following figure:

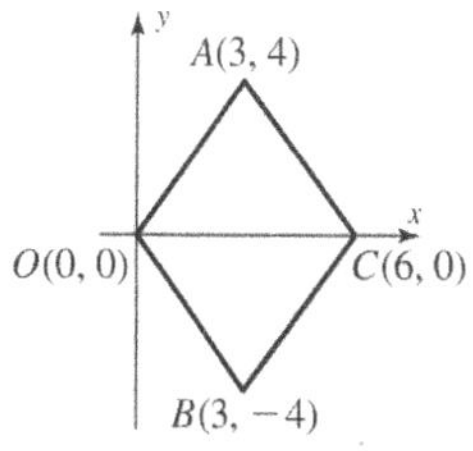

(iii) Reflecting $(\frac{25}{6}, 0)$ across the diagonal OA will help us estimate the coordinates. Then note that the other vertex (x, y) is equal distance from O and A. Thus,

$$\sqrt{(x-0)^2+(y-0)^2}=\sqrt{(x-3)^2+(y-4)^2} \Rightarrow$$

$$x^2+y^2=(x-3)^2+(y-4)^2 \Rightarrow 6x+8y=25.$$

Using the visual estimation and guess and check in the last equation, the solution $(\frac{^-7}{6}, 4)$ can be found.

(c) **(*i*)** $m_{\overline{OB}} = \frac{4-0}{8-0} = \frac{1}{2}$.

$m_{\overline{AC}} = \frac{4-0}{3-5} = {}^-2.$

$m_{\overline{OB}} \cdot m_{\overline{AC}} = {}^-1$, so the diagonals are perpendicular. Similarly, the slopes of diagonals $B'O$ and $C'A$ are also perpendicular to each other.

(*ii*) One diagonal is the *x*-axis and the other is parallel to the *y*-axis $\Rightarrow$ the diagonals are perpendicular to each other.

21. Find the image O' of the center of the circle O under the dilation with center A and a scale factor $\frac{1}{2}$. Then construct the circle with center O' and radius $\frac{1}{2}$ of the radius of the circle.

22. **(a)** *M* slides to *N*:

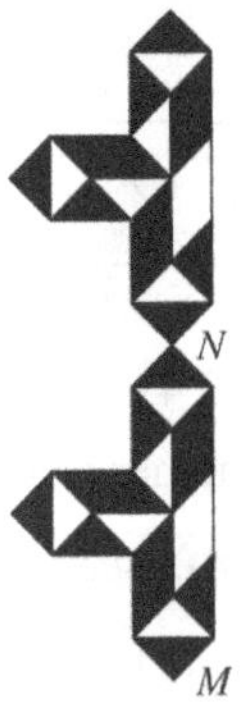

(b) 90° counterclockwise about *M*:

23. **(a)** **90°, 180°** (point), and **270°** symmetries.

(b) **(*i*)** Two **line** symmetries; one through each
of the congruent parallel sides.

(*ii*) **Point** symmetry about the center of the square.

24. **Point** symmetry about the center of the middle dark square.

25.

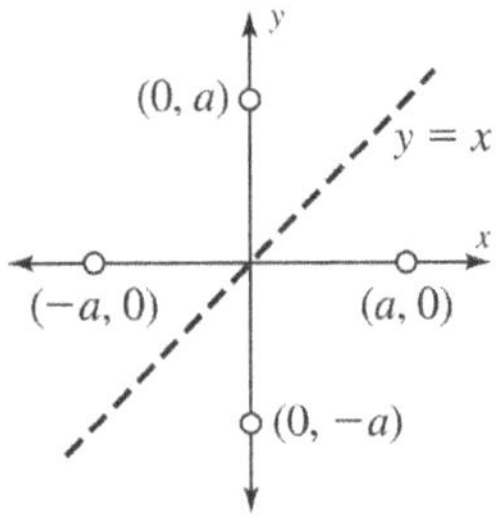

(a) *d*(*a*) from the origin, or intersection of the diagonals, is the same to all vertices. Thus the figure must be a **square**. You could also say that each vertex is a 90° rotation of a previous vertex.

(b) $P(x, y) \rightarrow P(y, x)$ under reflection in the line $y = x$. Thus $(a, 0) \rightarrow$ **$(0, a)$**, $(0, a) \rightarrow$ **$(a, 0)$**, $({}^-a, 0) \rightarrow$ **$(0, {}^-a)$**, and $(0, {}^-a) \rightarrow$ **$({}^-a, 0)$**.

26. If O is on $\overline{AB}$, then the points of $\overline{A'B'}$ lie on $\overleftrightarrow{AB}$ and thus the segments are parallel. If O is not on $\overline{AB}$, then $OA' = r(OA)$; $OB' = r(OB)$; and $\angle O \cong \angle O$. Thus, $\Delta A'OB' \sim \Delta AOB$, which implies $\angle OA'B' \cong \angle OAB$. These are corresponding angles for lines $\overleftrightarrow{A'B'}$ and $\overleftrightarrow{AB}$ and transversal $\overline{OA}$. Thus, $\overleftrightarrow{AB} \parallel \overleftrightarrow{A'B'}$ implying $\overline{AB} \parallel \overline{A'B'}$.

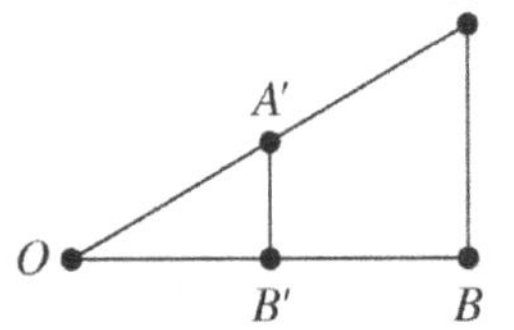

27. Answers vary. If $k \neq 1$, then no dilation exists since the center, a point on $y = x$ other than the center, and the image of that point will not be collinear. One example of a dilation is a dilation of $k=1$ with a center O on the line $y = x$. Any such choice of O will suffice if $k=1$.

www.ingramcontent.com/pod-product-compliance
Lightning Source LLC
Chambersburg PA
CBHW061409210326
41598CB00035B/6156
9780135184202